Telecommunications Engineering: Networks and Architecture

Telecommunications Engineering: Networks and Architecture

Edited by Bernhard Ekman

CLANRYE
INTERNATIONAL
www.clanryeinternational.com

Clanrye International,
750 Third Avenue, 9th Floor,
New York, NY 10017, USA

ISBN: 978-1-63240-597-5

Cataloging-in-Publication Data

Telecommunications engineering : networks and architecture / edited by Bernhard Ekman.
    p. cm.
Includes bibliographical references and index.
ISBN 978-1-63240-597-5
1. Telecommunication. 2. Computer networks. 3. Computer network architectures. 4. Mobile communication systems.
5. Communication and technology. I. Ekman, Bernhard.
TK5101 .T45 2017
621.382--dc23

For information on all Clanrye International publications
visit our website at www.clanryeinternational.com

CLANRYE
INTERNATIONAL

Printed in the United States of America.

Contents

Permissions

List of Contributors

Index

Preface

This book was inspired by the evolution of our times; to answer the curiosity of inquisitive minds. Many developments have occurred across the globe in the recent past which has transformed the progress in the field.

Telecommunications engineering studies the development of networks that are used for communication between separate groups. Technological innovations in this field, both in terms of devices as well as transmission, have led to designing of advanced telecommunications networks. This book is compiled in such a manner, that it will provide in-depth knowledge about the theory and practice of telecommunications engineering. It aims to present researches that have transformed this discipline and aided its advancement. A number of latest researches have been included to keep the readers up-to-date with the global concepts in this area of study. Coherent flow of topics, student-friendly language and extensive use of examples make this book an invaluable source of knowledge.

This book was developed from a mere concept to drafts to chapters and finally compiled together as a complete text to benefit the readers across all nations. To ensure the quality of the content we instilled two significant steps in our procedure. The first was to appoint an editorial team that would verify the data and statistics provided in the book and also select the most appropriate and valuable contributions from the plentiful contributions we received from authors worldwide. The next step was to appoint an expert of the topic as the Editor-in-Chief, who would head the project and finally make the necessary amendments and modifications to make the text reader-friendly. I was then commissioned to examine all the material to present the topics in the most comprehensible and productive format.

I would like to take this opportunity to thank all the contributing authors who were supportive enough to contribute their time and knowledge to this project. I also wish to convey my regards to my family who have been extremely supportive during the entire project.

Editor

Driver Assistance System Based on Video Image Processing for Emergency Case in Tunnel

Huthaifa Ahmad Al_Issa

Department of Electrical and Electronics Engineering, Faculty of Engineering, Al-Balqa Applied University, Irbid Al-Huson, Jordan

Email address:

alissahu@yahoo.com

Abstract: This paper proposes architecture for detecting accidents system based on image processing techniques for emergency case in tunnel, as well as the technical challenges that had to be overcome to ensure that technology successfully operated under all conditions. The advantages of this method include such benefits as Non-use of sensors, low cost and easy setup and relatively good accuracy and speed. Because this method has been implemented using image processing and MATLAB software, production costs are low while achieving high speed and accuracy. Method presented in this research is simple and there is no need to use sensors that have been commonly used to detect traffic in the past. This research can be enhanced by helping out the driver assistance system, this is accomplished by informing the public traffic about accidents in specific areas so that they can avoid those routes.

Keywords: Driver Assistance System, Video Image Processing, Traffic Safety

1. Introduction

Video traffic detection is one of the most important applications for the combination of the two important toolboxes in MATLAB which are video processing toolbox and image processing toolbox. The main idea of the traffic detection application is to keep the controller aware of any new or emergency case in the traffic such as: car crash, water flood or any other unwanted case that would affect the traffic negatively. Once the emergency case was noticed and the alarm was triggered, the controller could handle that specific case as soon as possible [1, 2].

The image processing toolbox can help by some applications for image processing, analysis, visualization and algorithm development. These applications are provided as an extensive set of reference-standard algorithms and functions. Meanwhile, the video processing toolbox may help in analyzing the video and developing algorithms to read or detect specific parts in the video, such as: dark areas, light areas, borders and so on.

The video processing toolbox application can be used in both cases of video: offline and online videos, in the case of the offline video: the video is stored in the path of the MATLAB in the pc, where the user can upload the video to the program using a specific command to read and analyze the requested data. While in the case of the online video, the video will be broadcasted from a live video, similar to the procedure of this research, where the video should be coming from a live camera that is spotted somewhere in a tunnel to monitor the case of the traffic in order to be analyzed and decided whether the case is normal or not [3, 4].

The video traffic detection function has the following advantages: Allows analyzing the traffic, helps in handling the emergency cases in a short time, can be improved to detect the faces which may help to catch or track the criminals, moreover, surely it can help in traffic management.

2. Digital Image Processing

An image may be defined as a two-dimensional function $f = (x, y)$, where x and y are spatial (plane) coordinates, the amplitude of f at any pair of coordinates (x, y) is called the intensity or gray level of the image at that point. When x, y and the amplitude values of f are all finite, discrete quantities, we can call the image is as a digital image. The field of digital image processing refers to processing digital images by means of a digital computer. Note that the digital image is composed of a finite number of elements, each of which has a particular location and value. These elements are referred to as picture elements, image elements, and pixels. An image may be considered to contain sub-images sometimes referred to as regions–of–interest, and transformations [5].

Images based on radiation from the EM spectrum are the most familiar, especially images in the X-ray and visual bands of the spectrum. Electromagnetic waves can be conceptualized as propagating sinusoidal waves of varying wavelengths, or they can be thought of as a stream of mass less particles, each traveling in a wavelike pattern and moving at the speed of light. Each mass less particle contains a certain amount (or bundle) of energy. Each bundle of energy is called a *photon*. If spectral bands are grouped according to energy per photon, we obtain the spectrum as shown in figure 1, ranging from gamma rays (highest energy) at one end to radio waves (lowest energy) at the other. The bands are shown shaded to convey the fact that bands of the EM spectrum are not distinct but rather transition smoothly from one to the other [6].

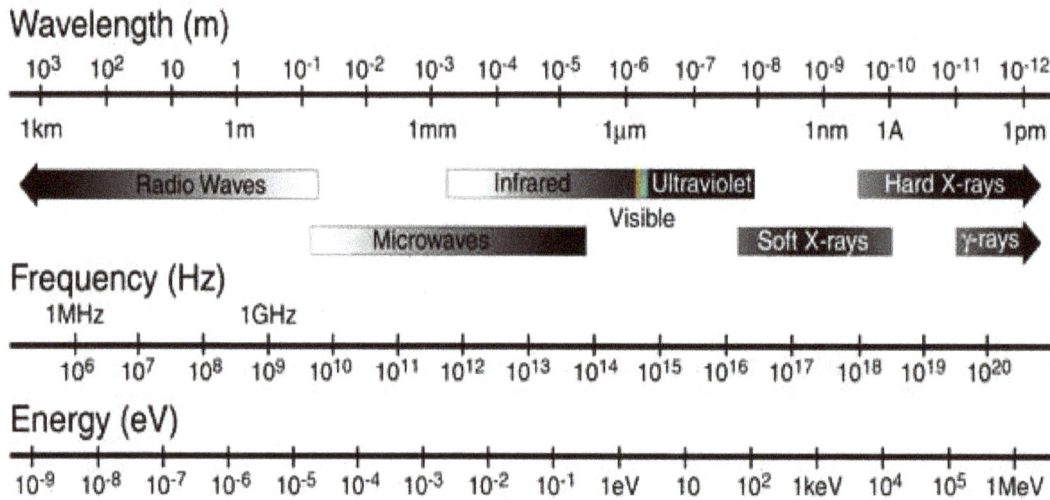

Figure 1. The electromagnetic spectrum arranged according to energy per photon.

3. Discussion of the Theory

Nowadays, tunnels handle large volumes of automobile traffic, sometimes moving at expressway speeds. In the narrow, confined environment of a tunnel, even minor incidents can cause deadly disasters, since critical, catastrophic incidents may occur even faster than on surface highways. An important objective of traffic management will be to identify potential sources of danger even before they reach the tunnel. In the face of the additional risks presented by subterranean accidents, every tunnel operator would like to be alerted early of any hazardous conditions. To pave the way to realizing this ideal, and the objective of this research is the traffic control in tunnel to ensuring the safety of the road users, and maximizing the availability of these structures [7].

All systems which are managed by traffic controller with its software that is capable of executing different logical, sequence, time, numeric arithmetic, regulative and communication functions. All subsystems, such as traffic lights, LED signalization, traffic count with inductive loops, emergency call system, video control, fire alarm, management of low voltage equipment and power supply, are connected to regional control center, where computers and supervisors are monitoring tunnels 24/7. Traffic safety is the key word. Function of Automatic Incident Detection (AID) is normally based on computer-based analysis of video image streams generated from cameras set up to view tunnel traffic [8].

Due to Needs a huge number of staff required to monitor each traffic area, and to the difficulties to describe the emergency case clearly to the operation center to handle the problem quickly, the need for automated detect and alarm system is essential. This research presents a method for estimating car accidents in the traffic using Image, this is done by using the camera images captured from the highway and videos taken are converted to the image sequences. The proposed algorithm expected to help in the traffic management field by applying the video processing functions in the traffic field, help in detecting the car crash cases in the tunnel, and help in detecting the water flood cases in the tunnel.

4. Methodology

In this research, we used the MATLAB software, we used a combination of two main toolboxes in MATLAB, video processing toolbox and image processing toolbox.

Image processing toolbox provides a comprehensive set of reference-standard algorithms, functions, and applications for image processing, analysis, visualization, and algorithm development. We can perform image enhancement, feature detection, noise reduction, image segmentation, geometric transformations, and image registration. Image processing toolbox supports a diverse set of image types, including high dynamic range, gig pixel resolution [9, 10].

Video processing toolbox is one of the main toolboxes in the image field in MATLAB, this toolbox offers the user many functions in order to analyze the videos and get the required data from those videos [11]. This type of processing is critical in systems that have live video or where the video data is so large that loading the entire set into the workspaces is inefficient. Computer vision system toolbox supports a stream processing architecture through system objects for use in MATLAB and blocks for use in Simulink.

The code was run in MATLAB it due some analysis in data from a film, the analysis includes finding number of frames, number of background frames, and due some process on dust, dirt, and brightness (based on night time, pixel segment with spatial temporal techniques) according to normal circumstances data, and focus on spots experienced notable changes. The code set a threshold on how much size a car can be recognized.

We set a threshold for the changes which are based on parameters, in order to reduce false alarms. Once an accident is recorded a peep alarm runs for three times every half a second. One drawback of the experiment that in order to monitor all the accidents a three seconds interval have to be for every two concessive accidents.

Video processing toolbox has many functions that help in analyzing the videos like:
- Video Reader:

Video reader is a function that is used to read the video data from a file into MATLAB workspace (in this research the file is a live cam).
- Implay:

The function IMPLAY is a function that opens a movie player for showing MATLAB image sequence (or image stacks), showing videos, and movies.
- Image Comparison: Imshowpair:

As this research requires a comparison between images, we need have a function to compare the images in order to analyze it. The function IMSHOWPAIR is a very good function to compare the differences between two different images.
- Alarming:

In this research we used alarming which is called beep, where beep is a sound that can be produced in MATLAB, so we used as alarm due to many reasons, one of them is our research needs to get a high speed response so the beep sound will keep running until the controller notices it and make a fast response. Another reason for choosing beep, as an alarm that its shorter to be generated and more effective tool to use in such cases.

The flow chart for this algorithm is describing below in figure 2.

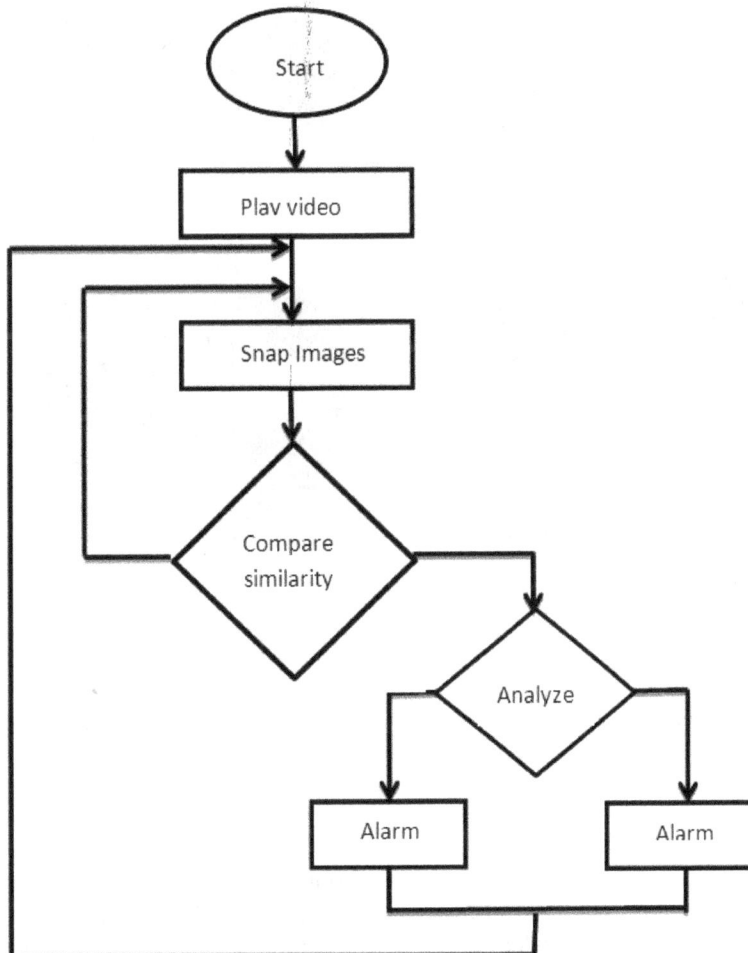

Figure 2. Flow chart describes the algorithm for this research.

5. Research Results

This program ensures the movement of cars and detects if accidents have occurred by seeking cars which are placed horizontally or the accumulation of dust and dirt. The program works in 4 second intervals in which the program checks for accidents by the above mentioned method. If there is more than one accident in each 4 second block, the number of accidents is shown above the second illustration, as shown in figure 3.

Figure 3. The code's generated images for a frame's accident recorded inside the tunnel.

This research can be enhanced by helping out the driver assistance system, this is accomplished by informing the public traffic about accidents in specific areas so that they can avoid those routes.

6. Conclusion

This paper presents a method for estimating car accidents in the traffic using image, this is done by using the camera images captured from the highway and videos taken are converted to the image sequences. Each image is processed separately and the advantages of this method include such benefits as Non-use of sensors, low cost and easy setup and relatively good accuracy and speed. Because this method has been implemented using image processing and MATLAB software, production costs are low while achieving high speed and accuracy. Method presented in this research is simple and there is no need to use sensors that have been commonly used to detect traffic in the past. One of the disadvantages of this method is extremely sincerity to the light this problem can be overcome by using specific camera filters during imaging processing or change in MATLAB code, this method can be used to detect rood accident.

References

[1] HunjaeYoo; Ukil Yang; KwanghoonSohn "Gradient-Enhancing Conversion for Illumination-Robust Lane Detection", *Intelligent Transportation Systems, IEEE Transactions on,* On page(s): 1083 - 1094 Volume: 14, Issue: 3, Sept. 2013

[2] Farrell, J.A. Givargis, T.D. Barth, M.J. "Realtime differential carrier phase GPS-aided INS".IEEE Transactions on Control Systems Technology,2000,8 (4):709-720.

[3] A. López, C. Cañero, J. Serrat, J. Saludes, F. Lumbreras, and T. Graf, "Detection of Lane Markings based on Ridgeness and RANSAC," IEEE Conference on Intelligent Transportation Systems, 2005, pp. 733-738.

[4] B. F. Wu, C. J. Chen, Y. P. Hsu, and M. W. Chung, "A DSP-based lane departure warning system," in Mathematical Methods and Computational Techniques in Electrical Engineering, Proc. 8th WSEASInt. Conf. Bucharest, 2006, pp. 240-245

[5] M. J. Jeng, P. C. Hsueh, C. W. Yeh, P. Y. Hsiao, C. H. Cheng, and L. B. Chang, "Real time mobile lane detection system on PXA255 embedded system," International Journal of Circuits, Systems and Signal Processing, vol. 1, no. 2, pp. 177-181, 2007. http://www.naun.org/multimedia/NAUN/circuitssystemssignal/cssp-31.pdf

[6] T. Al Smadi, Computing Simulation for Traffic Control over Two Intersections, Journal of Advanced Computer Science and Technology Research, Vol 1, No 1.pp. 10-24,2011.http://www.sign-ific-ance.co.uk/dsr/index.php/JACSTR/article/view/38

[7] Chen, L.; Li, Q.; Li, M.; Zhang, L.; Mao, Q. Design of a Multi-Sensor Cooperation Travel Environment Perception System for Autonomous Vehicle. Sensors 2012,12, 12386-12404. http://www.mdpi.com/1424-8220/12/9/12386

[8] Wang Y., Teoh E.K, Shen D. "Lane detection using B-snake".Int. Conf. Information Intelligent and Systems, Bethesda, MD, USA, 1999, pp. 438-443.

[9] Wang Y., Shen D, Teoh E.K. "Lane detection using spline model", Pattern Recognit. Lett., 2000, Vol. 21, no. 8, pp. 677-689.

[10] Yim Y.U., Oh S.Y. "Three-feature based automatic lane detection algorithm (TFALDA) for autonomous driving", IEEE Trans. Intell. Transp. Syst, 2003, Vol. 4 , no. 4,pp. 219-225.

[11] Mccall J.C, Trivedi M.M. "Video-based lane estimation and tracking for driver assistance: survey, system, and evaluation", IEEE Trans. Intell. Transp. Syst., 2006, Vol. 7, no. 1, pp. 20-37.

A Novel Hybrid Method for Face Recognition Based on 2d Wavelet and Singular Value Decomposition

Vahid Haji Hashemi[1, *], Abdorreza Alavi Gharahbagh[2]

[1]Computer Engineering, Faculty of Engineering, Kharazmi University of Tehran,Tehran, Iran
[2]Department of Electrical and Computer Engineering, Islamic Azad University, Shahrood, Iran

Email address:
hajihashemi.vahid@yahoo.com (V. H. Hashemi), R_alavi@iau-shahrood.ac.ir (A. A. Gharahbagh)

Abstract: An efficient face recognition system using eigen values of wavelet transform as feature vectors and radial basis function (RBF) neural network as classifier is presented. The face images are decomposed by 2-level two-dimensional (2-D) wavelet transformation. The wavelet coefficients obtained from the wavelet transformation are averaged for finding centers of features. In train process, four output of wavelet transform is analyzed and all eigenvalues of these images is obtained. At next step, the maximum 10 eigenvalues of wavelet sub images is stored as feature. Based on four sub images of wavelet transform and 10 eigenvalues of each sub image, the length of feature vector is 40. After obtaining features, in the train process for each person a center that has minimum Euclidean distance from all features is selected using RBF function. In fact the features are recognized by a RBF network. For a new input face image, firstly the feature vector is computed and then the distance (error) of this new vector with all centers of all persons is checked. The minimum distance is selected as target face. The proposed method on Essex face database and results showed that the proposed method provide better recognition rates with low computational complexity.

Keywords: Face Recognition, Singular Value Decomposition, SVD, Wavelet, Radial Basis Function, Neural Network

1. Introduction

Recognition of human faces is a very important task in many applications such as robotics, artificial intelligence, security systems etc. The wide category of face images such as its scene, brightness, lights, etc, are challenges for face detection algorithms. A face recognition system must be reliable and robust about all variable conditions of face images such as viewpoint, illumination, rotation, etc. The main tasks in the face recognition system are training with minimum face images and classification a new image based on its train process. Many researchers work about face recognition [1].In [1] face recognition methods classify to some groups. In the first group that named appearance based, total of face images or face objects are analyzed directly. Turk and Pent in [2] suggest using the eigenvalues in face recognition that is also popular than apparent base method. In [2] face matrix eigen values and vectors is computed and called Eigen faces, which are the principal components of face images. LDA[1]could be analyzed directly face images to extract the Fisher face [3] or analyzed the Eigen face to obtain a criteria for Eigen features of each face [4].Jain Yang et al. have developed a new technique by two-dimensional principal component analysis [2] for image representation. 2DPCA is based on 2D image matrices so the image matrix does not need to be transformed into a vector [5]. Rajagopalan proposeda system using multiple facial features extracted from the face [6].Image representation is a popular method in many image processing algorithms. Wavelet and FFT based feature extraction methods has many advantages. An appropriate wavelet transform is so robust regard to lighting changes and be capable of capturing substantial spatial features while has low computational complexity low. Bai-Ling Zhang use WaveletTransform and choosethe lowest resolution subband coefficients as face features [7]. Another technique is WPD[3] that build a compact and meaningful feature and is also used in face recognition method [8]. WPD

---

1 Linear Discriminant Analysis
2 2DPCA
3 Wavelet Packet Decomposition

is an extended form of the wavelet decomposition and includes multiple bases and different basis which result in better classification performance in compare to original wavelet [9].Ognian Boumbarov proposed a method that used wavelet packets for dimension reduction [10]. Vytautas Perlibakas [11] has presented a work on face recognition using both PCA and WPD. In the face recognition process, classifier plays an important role. Conventional classifiers such as SVM, boosting method and neural networks have been employed in different classification methods. [12] proposed a hybrid approach that used NFP [4] and NFS [5] classifiers for robust decision in the presence of wide face image variations. The mostly network topology that used in face recognition methods as classifier is RBFNN[6] due to its specifications and adaption to face features in comparison with other types of ANNs[13]. HaiGuo and Jing-ying Zhao have proposed Chinese minority script recognition using radial basis function network [14]. Many other researchers have implemented face recognition system using RBF neural network [15,16].Bicheng Li and Hujun Yin have presented a face recognition system using radial basis function neural network that its features is computed with wavelet transformation [17]. Ning Jin and DerongLiu have developed WBFNN[7]with sequential learning algorithm and showed it has better performance in compare to sequential learning algorithm ofRBF neural network [18].In this paper, a face recognition methodbased on wavelet transform in combination with eigen value of 2d wavelet coefficients matrix for extracting features and radial basis function neural network as classifier is proposed. The wavelet packet transform is used as preprocessing step. The eigen values of matrixes of wavelet packet coefficients are computed. Based on SVD[8] method the ten higher eigen values of each wavelet level is used as feature of image. The remainder of this paper is organized as follows. First, Section 2 describes preprocessing process. Section 3 presents the recognition process using RBFN. Section 4 describe results that showed improved correction accuracy andfinally, we conclude this paper and discuss our future work in Section 5.

2. Preprocessing and Feature Extraction

2.1. Preprocessing

Using the wavelet transformation, the image is mapped to a new time domain coordinates. This transform split the image in first level to four new sub images that the last matrix can mapped again in higher levels to four sub matrix. These new matrixes have many details about image and is so robust against noise and distortions. In the wavelet transformation, each new image has detailed coefficientsof the face that are decomposed as in Figure 1.

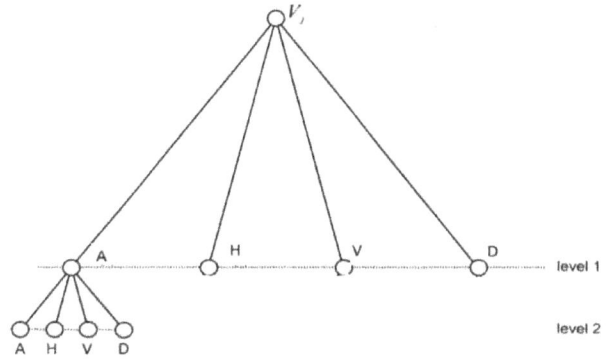

Fig. 1. *Wavelet transform structure.*

The face image V is decomposed in level 1 to four new images. In these new sub images A is a resized image of original image that include all details, H is an approximated gradient of image in horizontal direction, V is approximated gradient of image in vertical direction and D is include diagonal details. If algorithm need more details, wavelet in level2 decompose A to four new sub images again and this routine can repeat until the size of resized image stored in A, is lower than a specified threshold. Likewise, horizontal, vertical and diagonal details are decomposed into the new sub images such as original images. During the wavelet transformation, at level 2, only the A is decomposed into approximation and details. In the proposed method the wavelet packet transformation has been applied on the face images for only two levels and obtained matrixes are including a resized original image with size about 25% of original face image, two horizontal details, two vertical details, and two diagonal details. The obtained wavelet packets in the preprocessing step is used for finding reliable features at next step as an input for recognition.

2.2. Feature Extraction

In the feature extraction step, firstly the eigen values and eigen vectors of all wavelet sub matrixes is extracted. At preprocessing step we extract seven sub matrixes from a face image. All eigen values and eigen vectors of these sub matrixes is computed. The eigenvectors of a matrix are the non-zero vectors which, after being multiplied by the matrix, remain proportional to the original vector, any vector $\mathbf{x}$ that satisfies the equation:

$$\mathbf{Ax} = \lambda \mathbf{x},$$

where $\mathbf{A}$ is the wavelet sub image matrix, $\mathbf{x}$ is the eigenvector and λ is the associated eigenvalue. Eigenvectors are not unique for this reason usually pattern recognition methods use normalized eigen vectors that is a unique vector. If you consider a 2×2 matrix as a stretching, shearing or reflection transformation of the plane, you can see that the eigenvalues are the lines passing through the original matrix that are left unchanged by the transformation. In order to find the eigenvectors of a matrix we must start by

---

4nearestfeature plane

5 nearest feature space

6 radial basis function neural network

7 wavelet basis function neural network

8 Singular value decomposition

finding the eigenvalues. The only way for finding eigen values is finding the zero value for determinant of $\mathbf{A} - \lambda\mathbf{I}$. Once we have a set of eigenvalues we can substitute them back into the original equation to find the eigenvectors.

After finding all eigen values these values is sorted and ten maximum values is selected as the feature. In fact we summarize the total elements of a matrix in ten higher eigen values. In preprocessing step the wavelet transform implemented in level 2, split face image to seven submatrix. If we have select ten eigen value from each sub matrix we have a vector with length $7 \times 10 = 70$.

In train process, some face images of each person from different viewpoints, angles or scenes is selected and after preprocessing step the eigen values of wavelet sub matrixes from these images are extracted with above details as face features.

For eliminating noise or unwanted features from train process, after finding the features of a person face images, correlation coefficients between these vectors computed and if a vector is so different from other vectors or features is discard in train process. Another method that used for improving features performance is kmeans clustering method.

2.3. Kmeans for Noise Reduction

Kmeans clustering is a method of vector quantization, originally from signal processing that is popular for cluster analysis in image processing and data mining. Kmeansclustering aimsto map N observations or feature vector into k clusters or centers in which each observation belongs to the cluster with the nearest mean, distance or similarity.

This problem is computationally difficult however there are many efficient heuristic algorithms that are converge quickly to find centers. These method usually similar to the expectation-maximization algorithm for mixtures of Gaussian distributions however, kmeans clustering try to find clusters of comparable elements, while the expectation-maximization mechanism allows clusters to have different shapes. If the features is sparse or so different from each other and algorithm do not able to cluster all vectors in defined centers, the centers become diverge and algorithm missed centers. If a vector or features cause kmeans method to diverge, surely this feature is noise and should be discarded in train stage.

After analyzing all features accurately and eliminate weak and noisy features, in next step remaining features is used for training process.

3. RBFNN and Classification Step

The RBF network is a popular type of neural network including input, hidden and output layers with several forms of radial basis activation functions. Some of typical radial basis functions are Gaussian, Hardy Multiquadratic and Inverse Multiquadratic. The most common activation function is the Gaussian function defined by

$$\phi(x) = e^{-\frac{|x-\mu|^2}{2\sigma^2}} \quad \sigma > 0$$

Where σ is the width parameter, μ is the vector determining the center of basis function and x is the ddimensional input vector. In a RBF network, a neuron of the hidden layer is activated whenever the input vector is close enough to its center vector μ in comparison the other centers. There are several techniques and heuristic methods for optimize and fast training the basis functions parameters and determining the number of hidden neurons needed to best classification. In the proposed algorithm two training algorithms is tested: FS[9] and MM[10].

In FS method one neuron is allocated to each group of features belong to each individual and if different faces of a person are not close to each other, more than one neuron will be necessary. The second training method MM is a mixture density model, whose parameters μ and σ are to be optimized by maximum likelihood ratio action. In this method the K basis functions is used as an input to the model that is typically is much less than the total number of input data points. In the output layer, RBF classifier comprises one neuron to each person. The final output are linear function of the outputs of the neurons in the hidden layer. The final classification decision is based on the greatest output. With RBF networks, the regions of the input space associated to each person can present an arbitrary form and disjoint regions can be associated to the same person to very different angles, viewpoints or different facial expressions.

As a mathematical expression of the output layer:

$$f(\mathbf{x}) = \sum_{i=1}^{m} w_i \phi_i(\mathbf{x})$$

Where m is the number of hidden neurons, W_i are the weights connecting the hidden layer neurons and output layer neuron. The weights are adjusted using the following formula

$$w(t+1) = w(t) + \lambda(d_i - y_i)\varphi_j(x) \qquad i = 1, 2, ..., m$$

Where λ is a positive learning rate parameter and di is the desired output.

4. Simulation Results

The proposed method is tested on ORL face database. This standard database contains a set of face images taken between 1992 and 1994 at the Olivetti Research Lab in Cambridge, U.K, and include ten images for 40 individuals (totally 400 images). The images were taken with varying lighting, facial expressions (open and closed eyes, smiling or not smiling) and facial details (glasses or no glasses). All images has a dark homogeneous background with the person in an upright frontal position, with tolerance for some tilting and rotation of up to about 20 degrees. Scale varies about

---

9Forward selection
10Gaussian mixture model

10%. The original size of each image is 92x112 pixels, with 256 gray levels per pixel.

The proposed work has been carried out using Matlab 8.1. Figures 2 show the respective result of a face image after applying the wavelet transform in level 1.

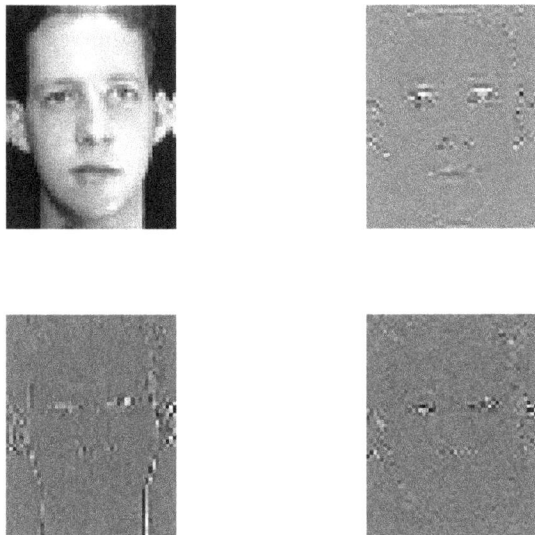

Fig. 2. All wavelet results of a face image in level 1.

Different wavelet filters such as Haar, Symlet, Daubechies and Coifletswith different orders is tested in proposed method. The Eigen values of face imagesafter wavelet transforms is analyzed by correlation coefficients and kmeans method given in section 2. Features obtained using this processapplied as input to RBFN network. In the training process 70% of face images (7 images for each person) is used and remaining 30% is used as test. The recognition rates are obtained by doing the training and testing process repeatedly for 20 times and averaged to eliminate randomness of training process in RBFN. The results such as recognition rate, and average epoch number obtained from proposed method are shown in Table 1. As shown in table 1, the Daubechies type of wavelet with order 8 give the best features in our algorithm.

For RBFN neural network as classifier, the number of hidden layers is set to 10 and number of inputs equal to Eigen values vector size is 70.

5. Conclusion

A face recognition algorithm using the main eigen values of 2D wavelet matrixes, kmeans and correlation coefficient as preprocessing method and RBFN network as classifier is proposed. The simulation results using a database with 250 face images were carried out to compare the performance of the RBF classifiers and extracted features with other methods. The results indicated that the proposed method reach its maximumperformance around 96% of recognition rate. The RBF classifiers showed a better performance regarding toits sensitiveness to the choice of the training and testing sets.

Table 1. recognition rate as % and average epoch number of iterations in proposed method.

Wavelet name	Recognition Rate %	Average Epoch number
Haar	95.57	470.64
Sym4	90.16	474.45
Sym8	80.00	462.41
Sym12	82.63	464.41
Sym16	86.05	443.20
Db4	82.23	458.20
Db8	96.00	477.02
Db12	86.63	464.06
Db16	85.56	478.93
Coif2	81.72	448.14
Coif4	83.46	446.59

The important unknown parameters such as the number of hidden layers in RBFN, number of eigen values selected from each wavelet matrix, level of wavelet transform and its filter type is set based on experimental results. Future work can be evaluate the robustness of ROI based methods in face recognition such as methods extract eyes, nose, lip and ears and analyze separately these elements.

References

[1] W. Zhao, R. Chellappa, A. Rosenfeld and P. J. Phillips, "Face Recognition: A Literature Survey," Technical Re- port CAR-TR-948, University of Maryland, College Park, 2000.

[2] M. Turk and A. Pentland, "Eigen faces for Recognition," Cognitive Neuroscience, Vol. 3, No. 1, 1991, pp. 71-86.

[3] P. Belhumeur, J. Hespanha and D. Kriegman, "Eigen faces vs Fisher Faces: Recognition Using Class Specific Linear Projection," IEEE Transactions on Pattern Analysis and Machine Intelligence, Vol. 20, No. 7, 1997, pp. 711-720.

[4] D. L. Swets and J. Weng, "Using Discriminant Eigen features for Image Retrieval," IEEE Transactions on Pattern Analysis and Machine Intelligence, Vol. 18, No. 8, 1996, pp. 831-836.

[5] J. Yang, D. Zhang, A. F. Frangi and J.-Y. Yang, "Two-Dimensional PCA: A New Approach to Appearance- Based Face Representation and Recognition," IEEE Transactions on Pattern Analysis and Machine Intelligence, Vol. 26, No. 1, 2004, pp. 131-138.

[6] A. N. Rajagopalan, K. S. Rao and Y. A. Kumar, "Face Recognition Using Multiple Facial Features," Pattern Recognition Letters, Vol. 28, No. 3, 2007, pp. 335-341.

[7] B.-L. Zhang, H. H. Zhang and S. Z. S. Ge, "Face Recognition by Applying Wavelet Sub band Representation and Kernel Associative Memory," IEEE Transactions on Neural Networks, Vol. 15, No. 1, 2005, pp. 166-177

[8] C. Garcia, G. Zikos and G. Tziritas, "Wavelet Packet Analysis for Face Recognition," Image and Vision Computing, Vol. 18, No. 4, 2000, pp. 289-297.

[9] J. Z. Xue, H. Zhang and C. X. Zheng, "Wavelet Packet Transform for Feature Extraction of EEG during Mental Tasks," Proceedings of the Second International Confer- ence on Machine Learning and Cybernetics, Vol. 1, 2003, pp. 360-363.

[10] O. Boumbarov, S. Sokolov and G. Gluhchev, "Combined Face Recognition Using Wavelet Packets and Radial Ba- sis Function Neural Network," International Conference on Computer Systems and Technologies—CompSysTech'07, Bulgaria, 14-15 June 2007, pp. v.4.1-v.4.7.

[11] V. Perlibakas, "Face Recognition Using Principal Component Analysis and Wavelet Packet Decomposition," Informatica, Vol. 15, No. 2, 2004, pp. 243-250.

[12] J.-T. Chien and C.-C. Wu, "Discriminant Wavelet faces and Nearest Feature Classifiers for Face Recognition," IEEE Transactions on Pattern analysis and Machine Intelligence, Vol. 24, No. 12, 2002, pp. 1644-1649.

[13] T. M. Mitchell, "Machine Learning," China Machine Press, Beijing, 2003.

[14] H. Guo and J.-Y. Zhao, "Chinese Minority Script Recognition Using Radial Basis Function Network," Journal of Computers, Vol. 5, No. 6, 2010, pp. 927-934.

[15] X.-Y. Jing, Y.-F. Yao, J.-Y. Yang and D. Zhang, "A Novel Face Recognition Approach Based on Kernel Dis- criminative Common Vectors (KDCV) Feature Extraction and RBF Neural Network," Neuro computing, Vol. 71, No. 13-15, 2008, pp. 3044-3048.

[16] M. J. Er, S. Q. Wu, J. W. Lu and H. L. Toh, "Face Recognition with Radial Basis Function (RBF) Neural Net- works," IEEE Transactions on Neural Networks, Vol. 13, No. 3, 2002, pp. 697-710.

[17] B. C. Li and H. J. Yin, "Face Recognition Using RBF Neural Networks and Wavelet Transform," Lecture Notes in Computer Science, Vol. 3497, 2005, pp. 105-111.

[18] N. Jin and D. R. Liu, "Wavelet Basis Function Neural Networks for Sequential Learning," IEEE Transactions on Neural Networks, Vol. 19, No. 3, 2008, pp. 523-528.

3

Static Heuristics Classifiers as Pre-Filter for Malware Target Recognition (MATR)

Anuj Lohani, Aditi Lohani, Jitendra Singh, Manish Bhardwaj

Dept. of Computer Science and Engineering, SRM University, NCR Campus, Modinagar, India

Email address:

lo.juna@gmail.com (A. Lohani), aditilohani@gmail.com (A. Lohani), jitendra.s@ncr.srmuniv.ac.in (J. Singh),
aapkaapna13@gmail.com (M. Bhardwaj)

Abstract: Now a day's malware are one of the major threats to computer information system. The current malware detection technologies have certain significant limitations on their part. Different organizations which deal with the protection of sensitive information may face the problem in identifying recent malware threats among millions and billions of benign executables using just signature-based antivirus systems. Currently for frontline defense against malware, signature-based antivirus products are used by organization.In the undergoing project, we proposed a detection approach by using static heuristics in MATR for malware in PE (portable executable) files. The project suggestslarger performance-based malware target recognition architecture that at present use only static heuristic features.Results of the experiments show that this architecture achieves an overall test accuracy of greater than 98% againstmalware set collected from various operational environments, while most antivirus provide detection accuracy of only 60% at their most sensitive configuration [1]. Implementations of this architecture enables benign executables to be classified successfully to some extent providing enhanced awareness of operators in hostile environments it also enable detection of unknown malware. We are to show the performance of Bagging and AdaBoostensemble.

Keywords: Malware, PE (Portable Executable), Bagging, AdaBoost (Adaptive Boosting)

1. Introduction

Computer networks of large organizations are difficult to secure and faces challenges due to their vast scale, scope, and complexity. The analysts must go thoroughly through a potentially overwhelmed set of data to discover new malware threats. As the data sets fails to be reduced to manageable extinct in large organizations, the attacker tools and malicious network traffic successfully hide in plain sight among millions of executable programs and billions of network connections making it difficult for humans to find without a form of automated assistance.

This field of malware detection has now been an active in computer security research area for decades. Advances in this area have not produced much significant solutions for this malware problem. The solutions must enhance human effectiveness to defend the network rather than replacing them.

Known threats are effectively detected and identified by signature-based antivirus and intrusion detection systems, but these could be nevertheless efficient with respect to new and unknown threats. For persons attacking these systems work as design constraints while designing new tools in order to avoid detection via system. [18]

An appreciable and useful data reduction might not be performed by these systems to reduce workload of analysts as their detections are typical binary yes or no outputs.

The research on malware detection provides a potentially viable method of data reduction of analyst workload. The heuristic analysis techniques are categorized into two distinct categories: static and dynamic. Static heuristics employs non-runtime indicators, such as structural anomalies, n-grams and program disassembly while on the other hand dynamic heuristics uses runtime indicators obtained in virtual environments, such as commercial sandbox applications. [1]

Both of these techniques complement one another(and even commercial antivirus products) hence neither static nor dynamic analysis is sufficient alone for providing full spectrum defense, with reduced effective scan and detection time against malware. Theoretically, sensitive and relatively fast static analysis methods can serve as pre-filters for slower dynamic analysis methods thus reducing the effective runtime scan performance of the overall system while producing a lower number of false positives than either method alone.[1] Contributions by this research are:

1) The MaTR architecture is extended which initially proposed, an operational model for self-discovery of malware in an organization with low effective scan times and low false positive rates through successive data reduction and analysis.
2) Provides generalization of the static analysis models trained from one or more current dataset.

These applications are challenging for research and commercial antivirus solutions, as they signify the prominent threats that were not consider prior for deployment by defensive system. Also the test environment considers various levels of sensitivity, simulation and organizational execution in unfriendly cyberspace environments to normal environment. [1]

2. Related Work

Recent static heuristic analysis research focuses on the occurrence in programs of n-grams, which are byte sequences of length. N-grams approach is appliedfor feature source which avoids the problems of generating pristine disassembly described by Moser [4], but non-instruction-based static heuristics like N-gram are not the only approach certainly available today. Researchers have also employed strings [2], [3], [5] and anomaly and structure data [2], [9], [6], [7] as features for malware detection classifiers. Presently, the scope of MaTR is strictly static heuristic analysis. The section describes related research work in static heuristic analysis of malware and detection evasion techniques. [1]

2.1. Static Approach

Static approach only checks executable binaries or assembly code without executing the codes. According to a survey done not long ago [9] the static approach attain a very high accuracy while maintaining low false positives while implying machine learning classifiers on static features in order to detect unknown malware. The advantage of this approach is that it does not add up overhead of execution time. [8]

Kolter [10] and Maloof improved Schulz's technique, by introducing byte n-grams rather than non-overlapping sequences. 500 n-grams are selected through measuring information gain and 4grams are employing as features. They applied decision trees; naïve instance based learners, support vector machines and TFIDF and, also improved some of these classification algorithms. Best result was achieved by boosted J48 at AUC, 0.996. [8]

2.2. N-Gram Features

The IBM research of Kephart, Tesauro, and Arnold provides the seminal research in n-gram analysis of malware [11]–[13]. These n-grams are byte sequences of length n that occur in the target, which theoretically represent program structural components and fragments of instructions and data.

They examine the use of n-grams in automatic signature extraction [11] for malware variants and generic detection [12], [13].

While searching for methods to automate signature extraction for new variants of known malware, Kephart et al. [11] discovered the utility of n-grams for generic malware detection. By determining the probability of finding specific n-grams in malicious and non-malicious programs, the authors fabricated a generic malware detection classifier.

Tesauro et al. [13] successfully used neural networks to detect boot sector viruses. They manipulated the decision threshold boundary to increase the cost associated with false positives as they cited that a single false positive reading likely affects thousands of systems. Despite significant computational and space constraints, and a small sample size for training and testing, they achieved a false positive rate (FPR) of less than 1% while detecting over 80% of unknown boot sector viruses.

3. Problem Description

Apart from performance issues MaTR has other issues to handle; these issues can be classified as problem other than performance heuristic based component. The section gives us the spread of problems, their possible solutions and the system process for MaTR.

3.1. Existing Problem

The problem within the current scenario of the PE's with malware and their detection can be described as follows:

1) Detection schemes which currently exist are based on an implicit assumption that each infected computer conducts scan over the Internet and propagate itself at the highest speed possible.
2) It has been demonstrated that the traffic volume of worm scan and the number of infected computers via worm show exponentially increasing patterns.
3) Worm may also make the use of traffic morphing and evasive scan technique to hide detection.
4) An analysis of a MaTR system presumes a decentralized deployment. A centralized deployment of a MaTR system makes it less vulnerable to adversary exfiltration and analysis.
5) The MaTR architecture may be susceptible to reverse engineering and exploitation.

3.2. Proposed Solution

The approachable solutions for the problem of malware detection that can be can be presented as follows:

1) Proposed Worm detection schemes are based on the global traffic monitor, scan and anomalous behavior in traffic. Other defense and worm detection methods such as sequential hypothesis testing applied for detecting worm-infected systems and payload-based worm signature detection are also present.
2) Use of a better performance-based static heuristic

classifier over MaTR architecture, to improve pre-filtering capability. Better detection of malware using both static and dynamic analysis, complementing each other.

3) A state-space feedback control model was presented which could control and detect spreading of worms or viruses. This is done by calculating the rate of new connections which the infected systems make.

4) In spite of different approaches mentioned above, we hold as an opinion that detection of wide scan anomalous behaviorcontinues to be a productive means against worms, and practically multifaceted defense is advantageous.

3.3. MaTR Process

The MaTR system uses a simple and straight forward method for detecting malware by using only high-level structural data of a program. While many commercial companies and researchersusesame structural data, none exclusively rely on this source of data and achieve the performance levels of MaTR. Fig. 1 shows the inputs and outputs of MaTR and illustrates its internal process. Inputs to MaTR are executable files, such as portable executable (PE) files common in the Microsoft Windows operating systems.

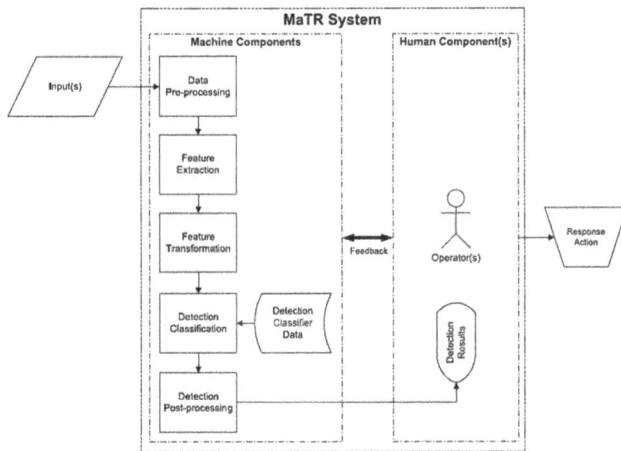

Figure 1. *MaTR system process [16].*

MaTR explicitly bounds themachine and human operator together within the overall system, a subtle yet significant distinction from other work that simply uses a computer to generate "solutions". In MaTR's architecture, the operator assertedly becomes a critical component receiving and providing feedback to the rest of the system and eventually initiating a response action. Since recognizing the operator's role allows for a more robust networkdefense. The discoveries of certain malware payloads (an area of future work) require distinct response. [16]

4. Methodology

MaTR architecture works on prioritizing efforts on malware detection from an organization perspective. The network of organizations can easily contain millions of programs, which is clearly infeasible to examine for a team of human analysts. Furthermore, if process behavior is observed using dynamic analysis for a 5 sec. period, it takes almost two months of time to record the behavior of a million processes, not including analysis time of the generated data. [1] The purpose of the MaTR architecture is to respond accurately to malware threats and enable organizations to efficiently discover threats. It accepts program files as input, after which the system makes predictions that enhance the effectiveness of malware analysts within the organization.

4.1. Static Analysis Component

The static component mostly targets malware defenses that result in program structural anomalies when compared to non-malware.

This component follows a standard machine learning process for malware detection using only non-instruction based static heuristics.

1) AdaboostMeta Classifier: Here static analysis component applies Adaboost meta classifier based on its performance on pilot studies, which can be observed asparallel findings of various researchers based onsimilar data sources. Boosting is a very different method to generate multiple predictions (function estimates) and combine them linearly. Boosting is a bias reduction technique, in contrast to bagging.

Boosting typically improves the performance of a single tree model. Reason for this is that we often cannot construct trees which are sufficiently large due to thinning out of observations in the terminal nodes. Boosting is then a device to come up with a more complex solution by taking linear combination of trees.Boosting is also very useful as a regularization technique for additive or interaction modeling.

4.2. Feature Set

A main difference betweenMaTR component and other commercial and research products are its feature set.MaTR component itself utilizes over 100 static heuristic features based on structural anomalies and structural information [1]. The feature set component in MaTRachieves high detection performance despite restricting its features exclusively to non-instruction based static heuristics. Other than following a mathematical model which are used to determine features, it utilizes anomaly and structural heuristic features commonly used by analysts [2], [6], [7] when examining samples to determine if they are indeed malicious.

To determine final feature set, MaTR component does not confide on complex computations of large samplings, thus it avert the selection steps overhead of a resource intensive feature. [1]

4.3. Experimental Details

The experiment uses a data sample of about 40137 malware, 10035 non-malware and 204 unknown sample instances. The data sample for malware came from VX Heavens dataset [17] and data sample for non-malware sample came from windows clean installs. The sample was remodelled to numeric value according to the experiment. The tool for the analysis was Weka 3.7 tool, mainly used for data mining.

The ROC curve generated during the classification process depicts a relationshipbetween true positive and falsepositive rate. The generalised role ofclassifiersin MaTRis to reduce the false positive rate, and to increase truepositive rate of malware detection.The two cardinal classifier used for the comparison was Adaboost and Bagging using decision tree J48 as classifier.

5. Result and Discussion

In this experiment after analysis from various sources AdaBoost was used against begged decision tree classifier, as boosting algorithm are also very efficient and work well with decision tree. The test was performed with both Bagging and AdaBoost using J48 decision tree algorithm. The plot for ROC (receiver operating characteristic)curve was almost the same for both ensemble classifier, but while classifying the malware the relative absolute error and time to test model both was lower for Adaboost.

First, we classified malware instance in the sample for both classifiers. Although samples were widely different, but for some unknown instance in sample the false positive rate for both classifiers was very low and the mean absolute error rate was also low considering the low percentage of correctly classified instances. The ROC for unknown instance this case was 0.9534. Some result is shown in Table 1.

Table 1. *Initial test result.*

	Bagging	Adaboost
Mean absolute error	0.2241	0.224
False-positive	0.237	0.237
Root mean square error	0.3347	0.3347

Then, we classified both malware and non- malware sample for both classifier.Results were almost similar in both cases.

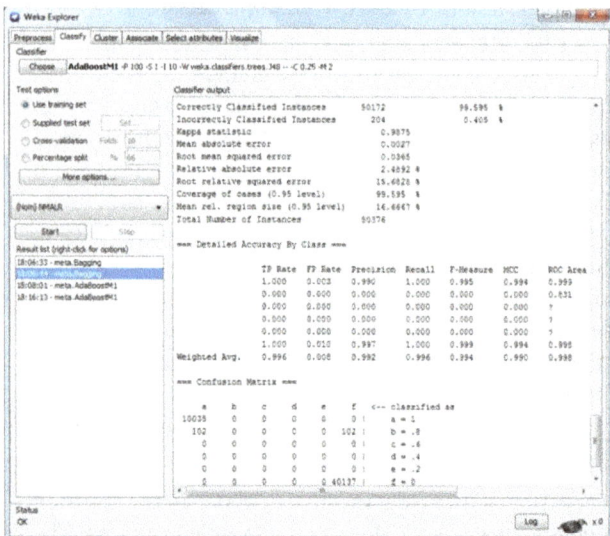

Figure 2. *Test result using Bagging on malware and non-malware.*

The Fig. 2 shows the test result for the bagging decision tree classifier in which the false positive rate was 0.008 and the value for the ROC curve was 0.998. The precision for the analysis was high at about 0.992 and relative absolute error was very low even for a large number of test instances.

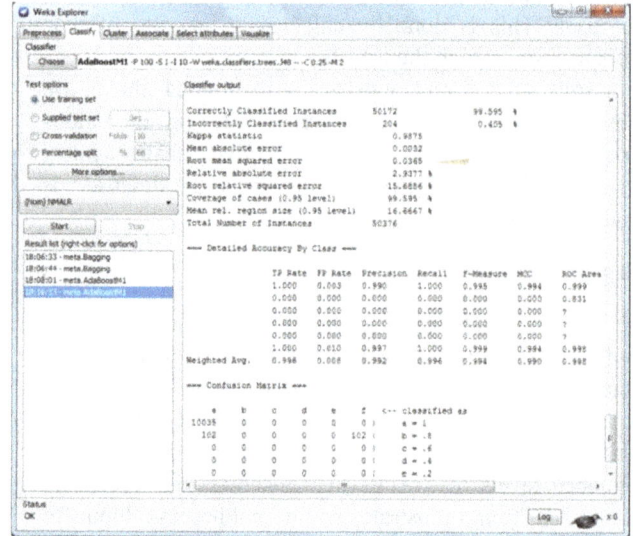

Figure 3. *Test result using AdaBoost on malware and non-malware.*

The test result for Fig. 3 shows that for the Adaboost classifier in which the false positive rate was 0.008 and the value for the ROC curve was 0.998. The precision for the analysis was also high at 0.992 and relative absolute error was very low for a large number of test instances, even for this classifier.

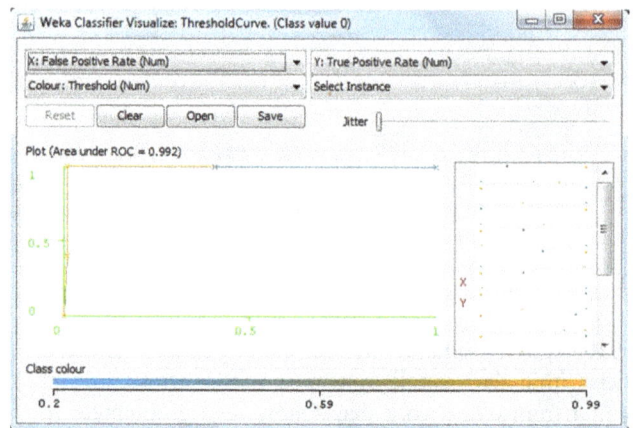

Figure 4. *ROC curve using AdaBoost.*

The Fig. 4 shows ROC curve using AdaBoost for one of the scenario. The result for the classification of malware and non-malware in caseof both Bagging and AdaBoost classification provided us with almost similar result, though in most cases generated test result in less time and with some ups and downs for other stats in both cases.

6. Conclusion and Future Work

We have proposed approach by using AdaBoost as static

heuristic classifier for MaTR. This approach happens to maintain precise detection of unknown malware, based on samples that are inspected earlier, while also maintaining low false positive rate. The indications of experimental results are that, the precision of the classification algorithms is 0.990 and above, and theoretically applying multi-level ensemble algorithms may improves classification accuracy.

6.1. Conclusion

As already seen in the experiments the ensemble classifier AdaBoost provide more or less same performance as that of Bagging Meta classifier with J48 in detecting malware with 0.45% more relative absolute error. Thoughtime to test model for AdaBoost was less in most comparison, but when weclassified only malware the relative absolute error and time to test model respectively0.02% and 1 sec lessforAdaBoost. Thus operational capability and accuracy was almost identical for both classifiers with some ups and downs for each.

6.2. Future Work

Examining performance characteristics of dynamic analysis methods to verify effective improvement in malware detection performance and analyze human-machine interface specifications. Given the capability of the MaTR architecture to provide focused information to the operator, generally accepted analysis processes may include redundant processes which humans can eliminate to increase overall performance further. Future investigations can also expand testing of other static heuristics classifiers as pre-filters for the MaTR architecture.

References

[1] T. E. Dube, R. A. Raines, M. R. Grimaila, K. W. Bauer, S. K. Rogers, "Malware Target Recognition of Unknown Threats," IEEE Systems Journal, 2013.

[2] P. Szor, "The Art of Computer Virus Research and Defense", IN: Addison-Wesley, 2005.

[3] M. Schultz, E. Eskin, E. Zadok, and S. Stolfo, "Data mining methods for detection of new malicious executables," in Proc. IEEE Symp. Security Privacy, May 2001, pp. 38–49.

[4] A. Moser, C. Kruegel, and E. Kirda, "Limits of static analysis for malware detection," in Proc. ACSAC, 2007, pp. 421–430.

[5] M. Christodorescu, N. Kidd, and W.-H. Goh, "String analysis for x86 binaries," ACM SIGSOFT Softw. Eng. Notes, vol. 31, no. 1, p. 95, 2006.

[6] N. Rafiq and Y. Mao, "Improving heuristics," Virus Bull., pp. 9–12, Aug. 2008.

[7] S. Treadwell and M. Zhou, "A heuristic approach for detection of obfuscated malware," in Proc. Intell. Security Inform., Jun. 2009, pp. 291–299.

[8] Jinrong Bai, Junfeng Wang, and Guozhong Zou, "A Malware Detection Scheme Based on Mining Format Information," The Scientific World Journal Volume 2014, Article ID 260905, 11 pages.

[9] A. Shabtai, R.Moskovitch, Y. Elovici, and C. Glezer, "Detection of malicious code by applying machine learning classifiers on static features: a state-of-the-art survey," Information Security Technical Report, vol. 14, no. 1, pp. 16–29, 2009.

[10] J. Z. KolterandM. A.Maloof, "Learning to detect and classify malicious executables in the wild," Journal of Machine Learning Research, vol. 7, pp. 2721–2744, 2006.

[11] J. O. Kephart and B. Arnold, "Automatic extraction of computer virus signatures," in Proc. 4th Virus Bull. Int. Conf., 1994, pp. 178–184.

[12] W. Arnold and G. Tesauro, "Automatically generated Win32 heuristic virus detection," in Proc. Virus Bull. Conf., Sep. 2000, pp. 51–60.

[13] G. Tesauro, J. Kephart, and G. Sorkin, "Neural networks for computer virus recognition," IEEE Expert, vol. 11, no. 4, pp. 5–6, Aug. 1996.

[14] T. E. Dube, R. A. Raines, S. K. Rogers, "Malware Target Recognition," United States Patent Application Publication [US 2012/0260342 A1], 2012.

[15] Symantec Corporation, "Understanding Heuristics: Symantec's Bloodhound Technology," Symantec White Paper Series, vol. XXXIV, no. 1, pp. 1–14, 1997.

[16] T. Dube, R. Raines, G. Peterson, K. Bauer, M. Grimaila, S. Rogers, "Malware target recognition via static heuristics," Elsevier computers & security 31 (2012) 137-1 47.

[17] VX Heavens. (2010, Apr. 15). Virus Collection [Online]. Available: vx.netlux.org/vl.php

[18] T. E. Dube, "A NOVEL MALWARE TARGET RECOGNITION ARCHITECTURE FOR ENHANCED CYBERSPACE SITUATION AWARENESS," Air Force Institute of Technology, AFIT/DCE/ENG/11-07, September 2011.

4

Audio Codecs Impact on Quality of VoIP Based on IEEE802.16e Considering Mobile IP Handover

Ali M. Alsahlany, Hayder S. Rashid

Department of Communication Engineering, Engineering Technical College / Najaf, Al-Furat Al-Awsat Technical University, Najaf, Iraq

Email address:

alialsahlany@yahoo.com (A. M. Alsahlany), hyder_hejajoo@yahoo.com (H. S. Rashid)

Abstract: A simulation model using OPNET tool is introduced for testing audio codecs impact on quality of VoIP based on IEEE802.16e with taking into account handover performance. Different parameters that indicate the quality of VoIP such as Throughput, MOS, End to End delay and traffic send and received. According to our simulation the acceptable MOS value is recorded for codecs G.711 and GSM-FR.

Keywords: Audio, Codecs, QoS, WiMAX, Handover

1. Introduction

In the recent years the Worldwide Interoperability for Microwave Access (WiMAX) placement is growing rapidly. IEEE 802.16e (Mobile WiMAX) becomes a prevalent technology for holding mobile clients because the base station (BS) of mobile WiMAX has the vital benefit to serve considerable coverage areas. During data transmission, mobile WiMAX permits the user to move freely. The major consideration of mobile WiMAX is that when the user is moving from one BS to another there must be no loss in data, i.e. during the handover which represents the method when a mobile node (MN) alters the serving BS [1]. Serving a large number of MNs practically demands an effective scheme of handover. Presently, a considerable delay in handover of mobile WiMAX is contributed to the overall end-to-end communication [2]. In the recent researches more attention is paid to increase the handover schemes efficiency. In this research, a comparison to study the performance of Mobile WiMAX handover among various audio codec schemes will be examined. The rest of this research will be structured as follows: the next section will object to the related works. Section 3 refers to the WiMAX handover procedure. In section 4, voice encoding schemes of utilizing VoIP over IEEE802.16e network will be clarified. Section 5 presents network model implementation. In Section 6 the simulation results will be discussed and finally, conclusion will be stated in section 7.

2. Related Works

Our study recognizes a number of earlier researches in the same course. The main contribution in [3] is to assess and examine the performance of several audio encoding schemes employing RSVP in a VoIP based on WLAN. The results were compared among three audio codecs namely; G.711, G.723.1 and G.729 utilizing the services of RSVP. On the other hand, the researchers in [4] examined mobile WiMAX performance using voice codecs under only two conditions and showed that the performance of audio codecs stay fixed for random way point and group mobility manner.

The researchers in [5] introduced an investigation to study the performance of different audio codecs based on various service classes such as rtPS, ertPS, nrtPS, Best Effort and NOAH routing protocols. While the authors in [6] introduced the performance evaluation of VoIP in wireless networks considering packet end-to-end delay and throughput without taking into account the mobility of nodes. In [7] the study examined the handover in order to delay characteristic and throughput involving UDP and TCP protocols for two types of fading models. An efficient technique of forced handover to progress capacity handling and performance of base station in [8] was introduced; however, the study did not mention audio performance during handover. In [9] the researchers provided an evaluation for different scenarios of WiMAX handover with several movements. While in [10] the authors presented an examination of handover impact on the speech performance in VoIP but audio codecs did not take into consideration. This study will provide an examination of the performance of WiMAX handover using four kinds of audio codecs namely; G.711, G.723.1, G.729A and GSM-FR.

3. IEEE 802.16e Handover Procedure

In general, due to weak signal received, cell capacity and radio channel condition the handover is occurring. The most important component before occurring handover process is the received signal strength. The MN will exchange to the adjacent BS if a significant drop in the received signal strength is observed [7, 8]. The procedure of handover can be made according to the following phases.

• Phase 1: Cell Reselection

Reselection of cell indicates to the procedure that MN is scanning and/or associating with one or more BSs according to fix handover suitability taking into account other performance issues. The MN might be used the acquired data of neighbor BS from a decoded data to arrange intervals of scanning or sleep intervals to scan the neighbor BS for the reason of evaluating the handover of MN to a potential target BS [8, 9].

• Phase 2: Handover Decision and Initiation

The decision to transfer the mobile station to the desired BS from the serving BS is known as initiation of a handover. Both BS and MN can be started this decision. Starting the actual handover requires sending a handover request that will activate a specific messages sequence of handover to be sent between MN and BS [8, 9].

• Phase 3: Synchronization

The synchronization of MN to its downlink channel is required to create the communication with the desired BS. In this stage, the MN starts receiving the uplink and downlink

transmission parameters. This process can be reduced if the MN earlier received data about the desired BS (during the acquisition of network topology) [8, 9].

• Phase 4: Ranging

The target BS and the MN must have handover ranging. Re-entry of network progress from the initial ranging steps in the network entry process negotiates basic capabilities, authentication stage, establishment stage, registration and optional network entry steps. Network re-entry may be reduced by target BS possession of information of MN, acquired from serving BS over the backbone network. Based on the amount of the acquired information, the target BS may choose to avoid one or several of network entry steps [8]. Regardless whether the serving BS received MN information or not the target BS will demand MN information from the backbone network [8, 9].

• Phase 5: Termination of Mobile Node

Termination of the MN context is expressed as the serving BS termination of the context of all connections belonging to the MN and the removal of the context associated with them [8, 9].

4. Voice Encoding Schemes

Voice codecs are used to convert an analog voice signal into digitally encoded version. The codecs may vary in sound quality, required bandwidth and computational requirements, etc. Table 1 shows the commonly used voice codecs with their algorithms, bit-rates and mean opinion score (MOS) [4]. Voice codec can be briefly defined as an algorithm used to code and decode voice stream. The voice quality experienced by the end-user depends on the performance of the codec [11].

Table 1. Features of the most common codecs.

IUT-T Codec	Algorithm	Frame (ms)	Bit Rate (kbps)	Bits per frame	Codec delay (ms)	Comments
G.711	Pulse code modulation (PCM)	0 .125	64	8	0.25	Delivers precise speech transmission. Very low processor requirements.
G.723.1	Algebraic Code Excited Linear Prediction (ACELP)	30	5.3	159	67.5	High compression with high quality audio. Can use with dial-up. Lot of processor power.
G.729A	Conjugate Structure Algebraic Code Excited Linear Prediction (CS_ACELP)	10	8	80	25	Excellent bandwidth utilization. Error tolerant.
GSM FR	Regular Pulse Excitation Long Term Predictor (RPE-LTP)	20	13	260	40	High compression ratio. Free and available in many hardware and software platforms. Same encoding is used in GSM cell phones

5. Network Model Description

The configuration of the suggested Mobile WiMAX network model will be done using OPNET environment. OPNET simulator is chosen to its global recognition in terms

of accuracy and reliability to get the results which are much closed to the real life. Also, it provides the necessary modules to study the performance of Mobile WiMAX (IEEE802.16e) with respect to the impact of audio codecs, mobile IP, handover and other mobility features.

The suggested model will be implemented using four

different scenarios to evaluate the best voice encoding schemes of utilizing VoIP over IEEE802.16e network. It is taken into account the impact of mobile IP and a handover mechanism which occur during the movement of MN across the cells. In the proposed simulation, all scenarios have the same structure and topology and the comparisons will be conducted using different values of parameters.

The first scenario will implement with voice encoding G.711 over mobile VoIP handover environment. The second scenario will employ with G.729A audio codec over mobile VoIP handover environment, while the third scenario will be examined using G.723.1 and finally, GSM audio codec over mobile VoIP handover environment will be used in scenario four. The network topology of the simulation environment is illustrated in Figure 1.

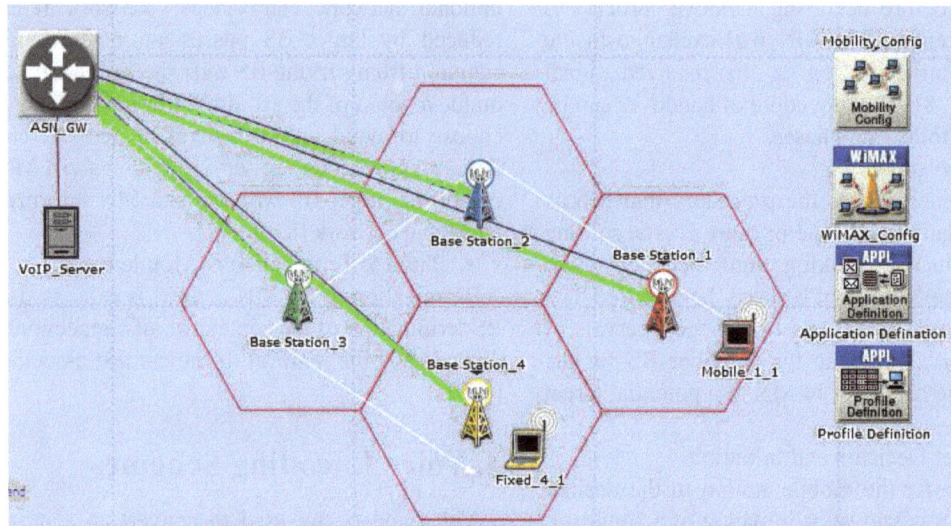

Figure 1. Network Topology of IEEE802.16e.

The scenario of the simulated network consists of four hexagonal cells of IEEE802.16e networks, i.e. single BS per cell. All BSs are connected to the core network (VoIP_Server) via Access Service Network Gateway (ASN-GW). The ASN-GW supports the mobility in IEEE802.16e network and the VoIP_Server handles the VoIP traffic. The researchers use two MNs with varying movements. The first one is (fixed_4_1) fixed and another node is (mobile_1_1) moved along with the trajectories indicated by a white color line around the cells. The green bidirectional dotted lines represent the generic routing encapsulation (GRE) tunnels. Each BS connected to the ASN-GW by using the PPP DS link which has a 1.544 Mbps capacity. The name of the link in OPNET model is 'point_to_point_link_adv'. The ASN-GW is connected to the VoIP_Server by using 1000BaseT Ethernet cable which has 1Gbps capacity.

The names of the ASN-GW and servers in OPNET model are 'ethernet4_slip8_gtwy' and 'ethernet_server', respectively. The VoIP node model is 'wimax_ss_wkstn_adv' and the BS (Foreign and Home agent) model is 'wimax_bs_router_adv'. The WiMAX parameters of MNs and BSs are given in Table (2) and Table (3) respectively.

Table 2. Mobile Node Configuration Parameters.

Application	VoIP
Antenna Gain	-1 dBi
Max Power Transfer	0.5 W
PHY Profile	Wireless OFDMA 20 MHz
Modulation and coding	Adaptive
Buffer Size	64 KB
Number of Transmitter	SISO

Table 3. Base Station Configuration Parameters.

Antenna Gain	15 dBi
Max Power Transfer	0.5 W
PHY Profile	Wireless OFDMA 20 MHz
Modulation and coding	Adaptive
Number of Transmitter	SISO

• Mobility Configurations

In our simulation model that shown in Figure (1) we assume that MN is carrying voice traffic which is ongoing between Mobile1_1 and VoIP_Server. When the simulation time is started, Mobile1_1 initiates the first session with Base station_1 (Home agent). After 9.18 min, the MN performs a handover from the current Home_agent spot where the session was generated to a Base station_2 (Foreign_agent_2) spot. This session will be continued until reaching Base Station_4 (Foreign_agent_4) which is passed through Base Station_3 (Foreign_agent_3). During this movement, the MN will change its Home agent and access address which imply that its IP address also will be changed. According to the handover process, the session at the Home agent will terminate, and the MN has to trigger the handover procedure. The new access router and the foreign BS have not information about the terminated session, so that MN has to register with the foreign BS and to establish IP connectivity with new access address and also reestablish session flow with VoIP_Server. Figure (2) shows an IP configuration of our simulation.

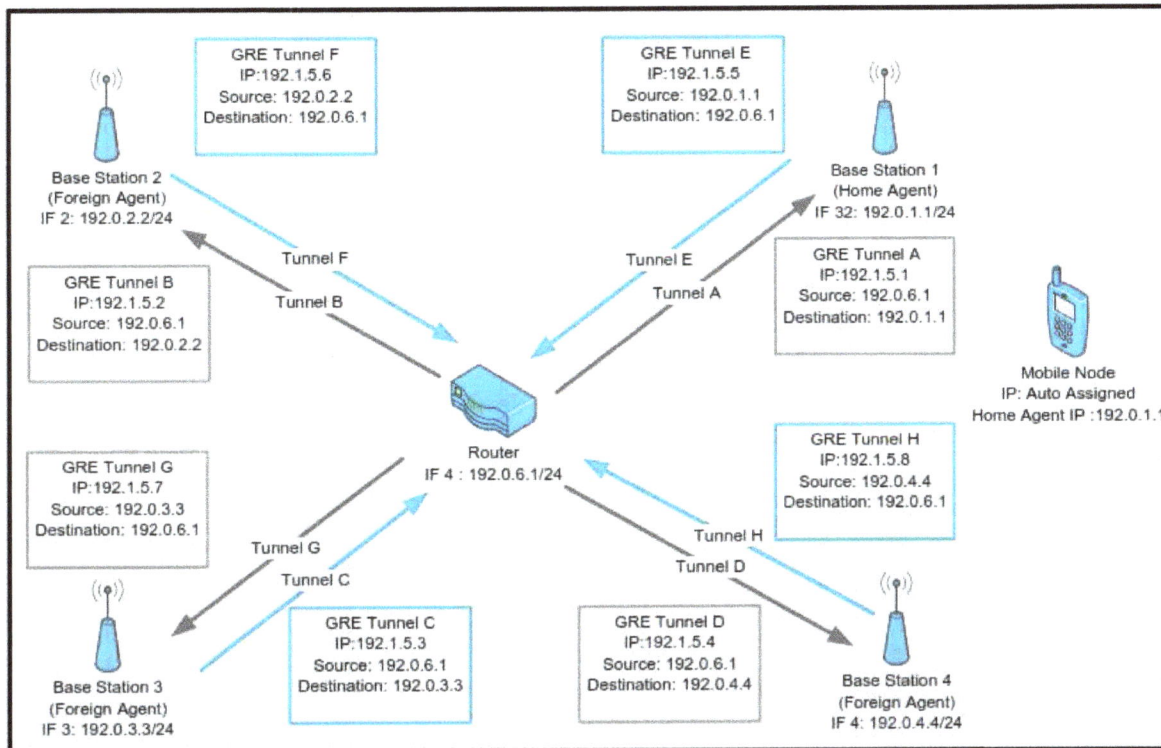

Figure 2. IP configuration for the proposed scenario.

The mobility Configuration defines a mobility profile for the wireless MNs in IEEE802.16e network. The mobility parameters of the simulation scenarios are given in Table (4) and Table (5) respectively.

Table 4. *Handover Parameters.*

MS Handover Retransmission Timer (ms)	**30**
Maximum Handover Request Retransmissions	6
Handover Threshold Hysteresis (dB)	0.4
Multi target Handover Threshold Hysterias (dB)	0
Maximum Handover Attempts per BS	3

Table 5. *Scanning Parameters.*

Scanning Threshold (dB)	**35**
Scan Duration (N) (Frames)	3
Interleaving Interval (P) (Frames)	255
Scan Iteration (T)	5
Maximum Scan Request Retransmissions	8
Ground Speed km/h	10

The application Definitions define traffic generation on the network. The VoIP application is set up as the application of the whole network.

6. Results of the Simulation

In this section, the results are obtained after implementing the IEEE802.16e network simulation and the statistics are collected by using OPNET Modeler under four different scenarios. The collected statistics from the simulation include Throughput (packet/sec), Mean Opinion Score (MOS), Voice packet end to end delay (sec), Voice traffic send and received (packet/sec). The results of the simulation potentially indicated to the degree of performance of the audio codecs over mobile IP handover environment based on voice communication.

The curves in Figure (3) correspond to the throughput of Mobile_1_1 node versus simulation time with the four audio codecs: G.711, G.723.1, G.729A and GSM-FR. According to the gained results, it is observed that the maximum rate of throughput is recorded for the codec G.929A with the rate of 195 packets /sec. On the other hand, G.711 and GSM-FR codecs have the lower throughput rates. The throughput in G.711 codec is ranged from 120 packets/sec to approximately 10 packets/sec, while GSM-FR codec is ranged from 20 packets/sec to roughly 65 packets/sec. The lowest rate belongs to codec G.723.1 which is oscillated from 5 packets/sec to about 23 packets/sec.

Additionally, it can be clearly seen that throughput of mobile1_1 node is reduced momentarily after 540 sec, 1320 sec and 2100 sec respectively; because when the MN moves outside the home agent coverage area and passes through boarder of foreign agent cell 1, 2 and 3 handover will occur. This reduction will continue until handover is completed. During the handover, the Mobile_1_1 node will receive signals from two base stations. When the setup of the new connection is completed, the transmission from/to the serving (Home Agent) BS will be ended, i.e. when the MN crosses a boarder of cells between the serving (Home Agent) BS and target (Foreign Agent) BS, the connection with the serving (Home Agent) BS will be finished. After that, a new connection with the target (Foreign Agent) BS will be established.

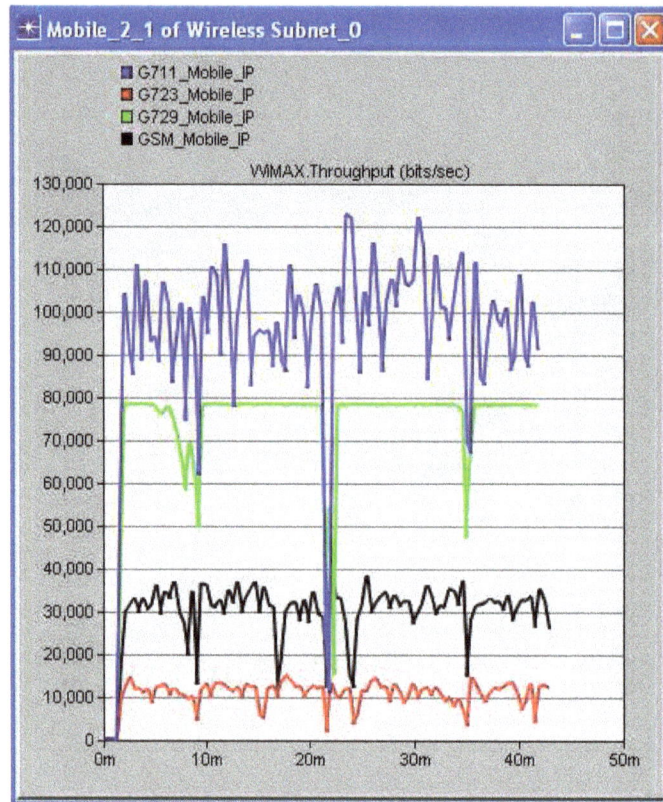

Figure 3. *Throughput of various audio codecs.*

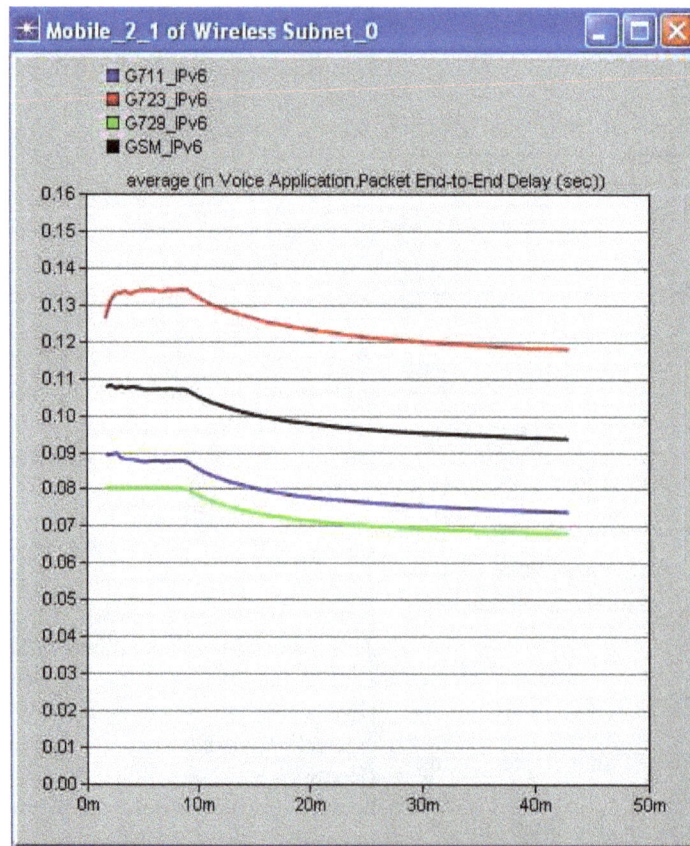

Figure 4. *packet End to End delay (sec) under various audio codecs.*

Figure (4) illustrates the obtained results between average end to end delay metric and simulation time. It can be seen that the average end-to-end delays of G.723.1 are larger than 0.11 sec which give a lower performance with respect to the other codecs. This result is due to small transfer rate 5.3 Kbps for G.723.1. On the other hand, the average end-to-end delays of GSM-FR and G.711 are larger than 0.09 second and 0.075 second, respectively, the reasons for that are the high bits per frame which are required more time to process them. While, G.729A has small delay with only 0.08 sec due to the high transfer rate and low packet size. These reasons make the G.729A provide the best performance of VoIP applications over IEEE802.16e network.

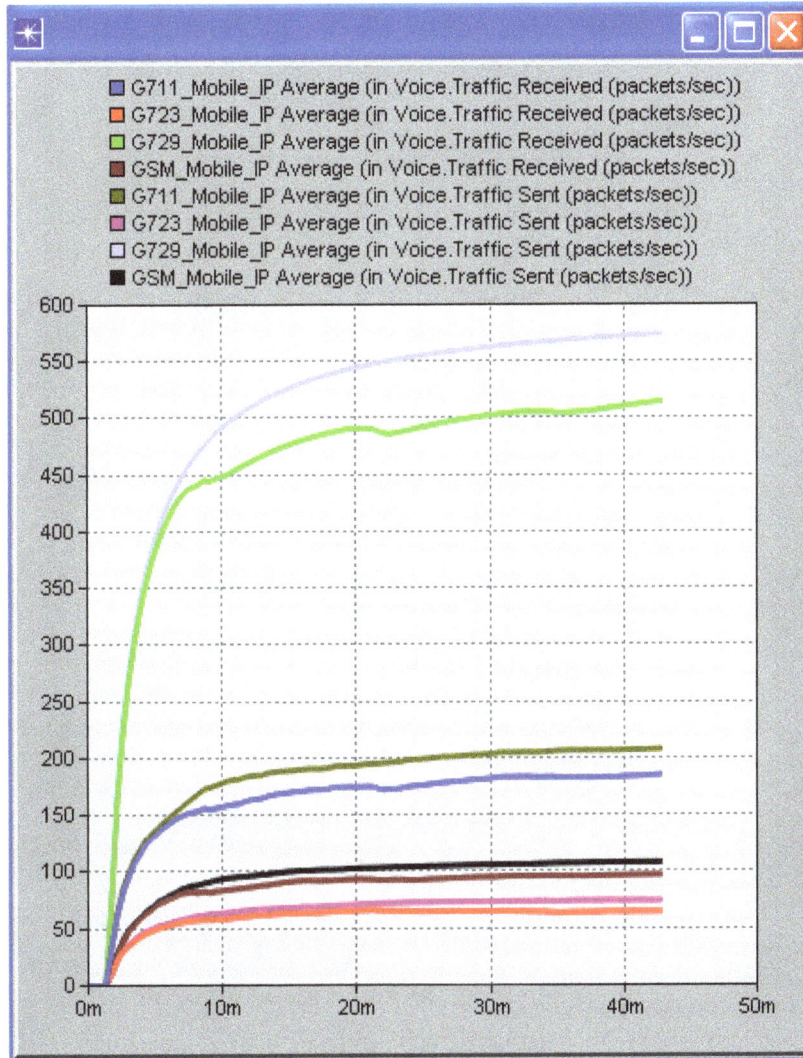

Figure 5. Traffic send and received.

Figure (5) represents the relationship between traffic received and traffic sent versus simulation time with different audio codecs. In order to examine network efficiency, the loss between two traffics (sent and received) must be less enough to make them identical to each other [13]. During the simulation time, the deviation of the traffic received to the traffic sent in G.711 and 729.A codecs is large comparing with GSM-FR and G723.1 codecs that have small deviation. The results indicate that codecs GSM-FR and G723.1 in IEEE802.16e network provide better efficiency since the path loss between traffic sent and received is less compared with other codecs.

Figure (6) shows the average MOS value for the four codecs. The MOS value gives a description of the perceived quality of receiving voice after being transmitted and compressed using codecs. According to [14], the acceptable MOS value in our results is recorded for codecs G.711 and GSM-FR with a value of 3.1 and 3 respectively. In the contrast, the MOS value of codecs G.723.1 and G.729A is less than 3 which mean that these values indicate that the quality of service is poor and cannot be acceptable if these codecs are used.

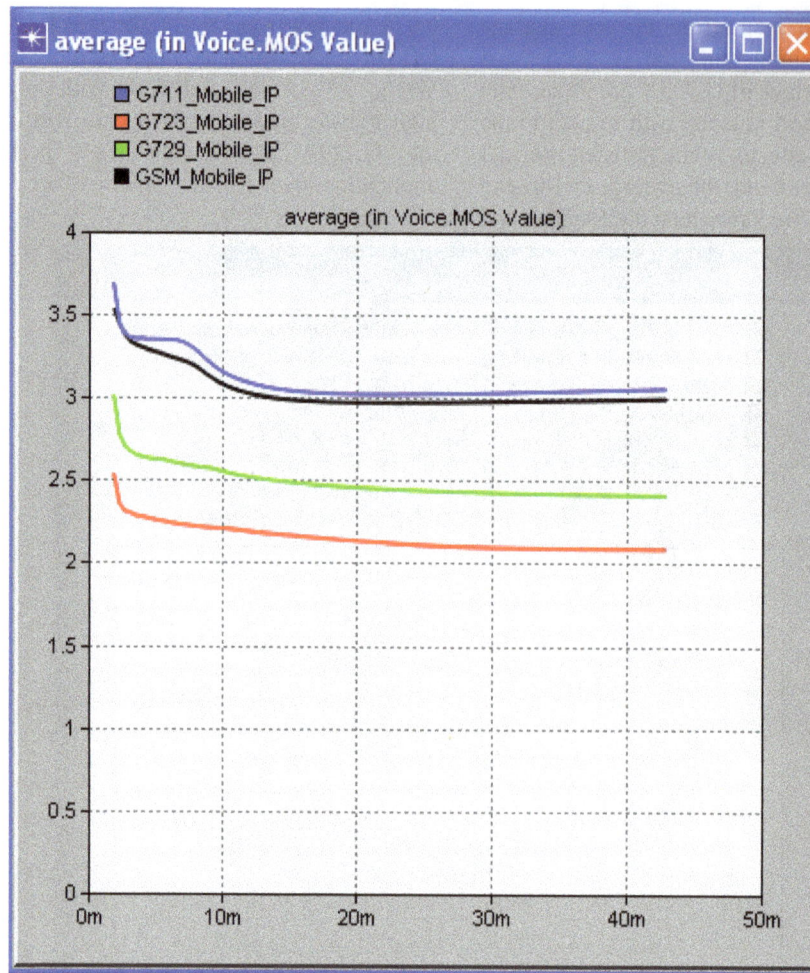

Figure 6. Average voice MOS under various codecs.

7. Conclusion

In this study, OPNET Modeler was occupied as a simulation tool for testing audio codecs impact on quality of VoIP based on IEEE802.16e. The proposed handover for IEEE802.16e was validated in various simulation scenarios using four audio codecs. The obtained results show that the acceptable MOS value was recorded for audio codecs G.711 with a value of 3.1, followed by GSM-FR codecs with a value of 3. Audio codecs G.723.1and GSM-FR recorded lower effect on handover, i.e. the best performance. Additionally, in traffics sent and received, they have small deviation. In terms of end to end delay, G.729A codec recorded a small delay comparing with other audio codecs.

References

[1]   Rambir Joon, Sandeep and Manveen Singh Chadha, ''Analysis of WIMAX Handover'', International Journal of Soft Computing and Engineering (IJSCE), pp. 476-479, Volume 2, Issue 3, July, 2012.

[2]   Bhaskar Ashoka, David Eyers, and Zhiyi Huang, ''Handover Delay in Mobile WiMAX: A Simulation Study'', 12th IEEE International conference on Parallel and Distributed Computing, Applications and Technologies (PDCAT), pp 305 – 312, 2011.

[3]   Priyanka Luthra, and Manju Sharma, ''Performance Evaluation of Audio Codecs using VoIP Traffic in Wireless LAN using RSVP'', International Journal of Computer Applications, pp. 15-21, Vol. 40, no. 7, February, 2012.

[4]   M. Tariq, M. Azad, and S. Rizvi, ''Effect of Mobility Patterns on VoIP QoS in Mobile WiMAX'', International Journal of Computer Science and Telecommunications, pp. 1-7, Volume 4, Issue 1, January, 2013.

[5]   Tarik ANOUARI, and Abdelkrim HAOI0, ''Comparative study and analysis of VoIP traffic over WiMAX using different service classes'', IEEE International conference on Next Generation Networks and Services (NGNS), pp. 87-93, 2-4 Dec., 2012.

[6]   Khamis AlAlawi, and Hussain Al-Aqrabi, ''Quality of service evaluation of VoIP over wireless networks'', 8th IEEE conference on GCC Conference and Exhibition (GCCCE), pp. 1-6, 1-4 Feb., 2015.

[7]   Mohd Pardi, Mohd Baba, and Muhammad Ibrahim, ''Analysis of handover performance in mobile WiMAX networks'', IEEE conference on Control and System Graduate Research Colloquium (ICSGRC), pp. 143 – 149, 27-28 June, 2011.

[8] Shahab Hussain, and et al, ''Handoff in mobile WiMAX: Forced handoff scheme with load balancing in mobile WiMAX networks'', 8th IEEE International Workshop on Performance and Management of Wireless and Mobile Networks, pp. 666 – 672, 22-25 Oct, 2012.

[9] Firas Al-Saedi, and Wafa A. Maddallah, ''Evaluation of Handover Process in WIMAX Networks'', International Journal of Computer Science Engineering and Technology (IJCSET), pp. 831-838, Vol 2, Issue 1, January, 2012.

[10] Zdenek Becvar, Pavel Mach, and Robert Bestak, ''Impact of Handover on VoIP Speech Quality in WiMAX Networks'', 8th International Conference on Networks, pp. 281 – 286, 1-6 March, 2009.

[11] S.Nithya, and M.S.K.Manikandan, ''Perform ace Analysis Of CODEC's With QoS Constrainsts In Voice Over Internet Protocol V6'', IEEE International Conference on Electronics and Communication Systems (ICECS), pp. 1-5, 13-14 Feb., 2014.

[12] Jamil Hamodi, and et al, ''Performance Study of Mobile TV over Mobile WiMAX Considering Different Modulation and Coding Techniques'', International Journal of Communications, Network and System Sciences, pp. 10-21, Vol.7, No.1, 2014.

[13] Ali M. Alsahlany, ''Performance Analysis of VOIP Trafic over Integrating Wireless LAN and WAN Using Different CODECS'', International Journal of Wireless & Mobile Networks (IJWMN), Vol. 6, No. 3, June, 2014.

[14] Shaffatul Islam, Mamunur Rashid, and Mohammed Tarique, ''Performance Analysis of WiMax/WiFi System under Different Codecs'', International Journal of Computer Applications, Vol. 18, No.6, March, 2011.

5

Efficient Sideband Noise Cancellation for Co-located Networks Using ANCT

Nosiri Onyebuchi Chikezie, Ezeh Gloria Nwabugo, Agubor Cosmos Kemdirim, Nkwachukwu Chukwuchekwa

Department of Electrical & Electronic Engineering, Federal University of Technology, Owerri, Nigeria

Email address:
buchinosiri@gmail.com (Nosiri O. C.), ugoezeh2002@yahoo.com (Ezeh G. N.), aguborcosy@yahoo.com (Agubor C. K.), nonwuchekwa2002@yahoo.com (Chukwuchekwa N.)

Abstract: An efficient noise cancellation technique for a co-located network was realized using Adaptive Noise Cancellation Technique (ANCT). The technique was developed as improved feature to the classical application of passive filters. The paper focused on achieving a theoretical perfect cancellation considering three essential parameters: the amplitude, phase and delay characteristics of the signal and noise at both the primary path and the reference path of the co-located system. An experimental test-bed of the ANC Architecture was developed using Matlab-Simulink block design which demonstrated the error signals before and after cancellation.

Keywords: Co-location, Noise, Interference, Cancellation, Adaptive

1. Introduction

Wireless communication networks are faced with interference challenges which require urgent attention for improved system performance. Co-location or infrastructure sharing has been an aspect of interest in the fields of wireless communication. Government armed forces in the United States were the first to adopt the strategy. They required different wireless platforms to share a small site because of the mobile nature of their equipment (e.g. battleships, aircraft and expeditionary fighting vehicles [1,2]. The strategy was introduced to reduce the incessant proliferation of cell towers which are capital intensive and could pose environment health hazard. The policy was globally adopted as a reliable means of reducing capital and operating expenditure to promote economic efficiency and improves network performance deployment for operators in telecommunication industry [3]. Telecommunication infrastructure sharing is of two types, namely; passive and active components sharing [4]. Active sharing involves sharing of electronic components and facilities such as base station, microwave radio equipment, switches, antennas and receivers while passive sharing involves sharing of non-electronic components and facilities such as towers, shelters, electric supply, easements and ducts.

The infrastructure sharing technique considered in this article is passive component sharing.

In a co-located setting, base station receivers are required to receive weak desired signals in the presence of high-power transmit signals, resulting to major interference [5]. When RF signal is amplified to form the transmit signal, a significant amount of emissions are generated outside the transmit band referred to as sideband noise emission [6]. The emissions are due to the non-linearity and noise generated inside the power amplifier, and may appear as a "skirt" or "shoulders" when observed through the power spectrum at the output of power amplifier. These emissions or undesired noise energy may fall within the pass bands of a co-located receiver, degrading the receiver sensitivity and increasing the noise floor level. These undesired noise energy also contribute to the Carrier to Interference (C/I) ratio degradation, reduction in the full utilization of the capacity and the coverage radius, thereby disfranchising the end users from enjoying their hard paid services. Therefore, the undesired spectral components are required to be optimally reduced to avoid the introduction of excessive noise in the receiver front end.

Interference between the transmitted signals and the received signals depends on factors such as: The number of active channels at the site; the interval between the working frequency ranges of the two systems; the spatial separation between the receiver and transmitter and the characteristics of the technology including base station equipment [7].

This work predominantly considered the application of Adaptive Noise Cancellation Technique (ANCT) to filter co-located high power transmit jammers and allow only the desired receive signal at the receiver. The technique generally comes into play as a measure to improve on the limitations of other industrial interference mitigation techniques for co-located networks such as passive filters.

The technique also operates on the principle of destructive interference between the primary path and the reference path correlated noise signals. It can theoretically provide optimum cancellation for all distortion signals. It is inherently and unconditionally stable, requires no prior knowledge of the environment and possess better power handling capabilities with higher frequency agility response.

1.1. Objective of the Study

The paper was focused on developing an ANCT towards achieving theoretical perfect error cancellation. It also required developing an expression that will characterize the quality with which information is transferred through the channel and the minimum required information carrying characteristics of the signal such that the information can be properly detected and recovered.

1.2. Significance of the Study

The article contributes to providing improved cellular network performance in a worst case scenario involving a co-located network. These include: enhanced capacity and coverage performance, guaranteed optimum network availability and rapid expansion of customer base services.

2. Literature Review

The application of active cancellation technique for

Transmitter/Receiver (Tx/Rx) feed-through in auxiliary transmitters as developed by [8], demonstrated how a flexible multiband front end involving an additional multi-band transmitter chain was used to cancel the transmitter interference signal. The cancellation signal was computed over the Transmission channel bandwidth in the baseband and then transformed up to radio frequency via an auxiliary transmitter. The design does not cancel the transmitter noise in the receiver band because the transmitter noise and the auxiliary transmitter noise introduced are not correlated, instead of cancelling the transmitter noise, the technique introduced additional noise components to the receiver. Also, the introduction of auxiliary transmitter adds more cost to the design.

Single-loop adaptive cancellation system for co-locating a global positioning system (GPS) receiver unit within mobile communication user equipments was developed by [9] and [10]. The cancellation system coupled out a sample of the interfering mobile communication transmit signal. The compensation path signal was then adaptively gain-phase adjusted such that it is equal in magnitude and 180^0 out of phase to the interference path signal when coupled back in the receive path of the GPS unit. This mitigates the interfering mobile communication transmit signal before it reaches the GPS receiver. A variable attenuator and a phase shifter can only cancel narrow-band noise signals. Accurate delay match for compensation path was deficient in the design.

3. Methodology

ANCT was applied as a technique to sample the undesired spectral components that interfere with the receiver front end. The primary antenna picks up the desired signal, S_{Pri} with the jamming signals X_{Pri}. The reference antenna was directed to pick only the jamming signal X_{Ref}, illustrated in Fig. 1

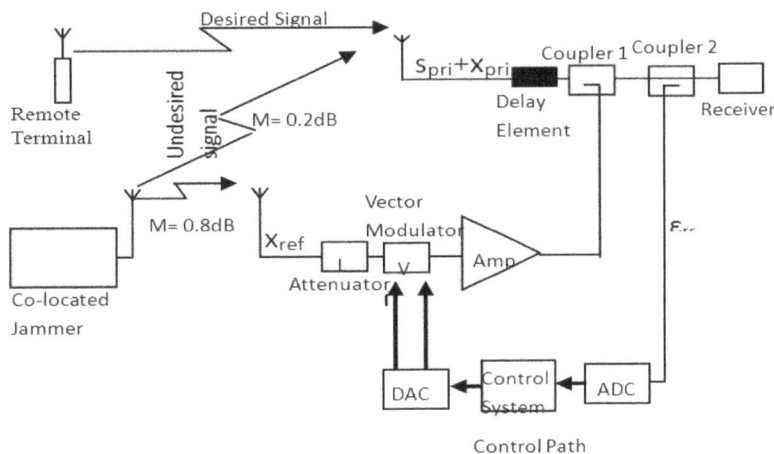

Figure 1. Conceptual Adaptive Cancellation System for a co-located.

The interfering signal was sampled using a reference antenna oriented in the direction of arrival of the interfering signal or directly from the source such as the output of the

power transmitter. The sampled signal is used to cancel by vector addition the interference present in the primary antenna path of the affected receiver before the interference enters the receiver channel.

The ANCT signal controller consists of two components at the cancellation path: The vector modulator and amplifier. The vector modulator was necessary to modify the noise signal such that it matches its complement in the primary path. An attenuator in the vector modulator was required when the amplitude noise signals picked on the reference antenna was sufficiently larger than that on the primary. The amplifier was necessary to compensate for the insertion loss generated by the vector modulator (I and Q) and provides the amplification needed to cancel the noise correlated signals at the primary path. The vector modulator works by adjusting the I and Q components of the noise signal. These modulators act as an actuator used to adjust the cancellation signal in the cancellation signal path, usually described as the weight variables of the adaptive systems. The reference path and the primary path have different delay characteristics; the reason was that each path uses different components. The primary path consists of two directional couplers and a delay line, exhibit fairly active component characteristics with low insertion loss while the reference path contains both passive and active RF components, namely attenuator, vector modulator and amplifier.

The ANCT was implemented using MATLAB-Simulink test bed as shown in Fig. 2. Let the signal generator A, generates the desired signal and represented as

$$S_{GA}(t) = V_{Spri} \sin(2\pi f_c t + \varphi_{Spri}) \qquad (1)$$

where $S_{GA} = S_{pri}$,

- f_c is the instantaneous frequency,
- φ_{Spri} is the initial phase of the signal,
- V_{Spri} is the amplitude of the desired signal at the primary path.

Let the signal generator B generates the interfering signal and represented as

$$X_{GB} = V_{GB} \cos(2\pi f_c - \beta_c I_{GB} + \varphi_{GB}) \qquad (2)$$

where

- V_{GB} is the amplitude of the undesired signal, generated by the signal generator B,
- β_c is the phase shift constant at f_c for a particular transmission line,
- I_{GB} is the length of the transmission path the signal passes before being input to the cancellation coupler,
- φ_{GB} is the initial phase of the signal.

Since the cancelation signals are operating at the same frequency, therefore the reference signal and primary cancellation signal are represented to be a two single tone sinusoidal signal.

Let the power splitter be represented as S_p, from the design, the amplitude of the power splitter was structured in the ratio of 0.2:0.8 at the primary and reference paths. This was prioritized due to the design placement of the antenna.

$$\therefore S_p = X_{GB}(t) = 0.2 X_A(t) + 0.8 X_B(t) \qquad (3)$$

where X_A represents the cancellation signal at the primary path and X_B represents the interfering (jamming) signal at the reference path.

$$\therefore X_A(t) = X_{Pri}(t) = V_{X Pri} \cos(2\pi f_c t - \beta c I_{X Pri} + \varphi_{X Pri}) \qquad (4)$$

$$X_B(t) = X_{Rref}(t) = V_{X Ref} \cos(2\pi f_c t - \beta c I_{X Ref} + \varphi_{X Ref}) \qquad (5)$$

Coupler A was represented as C_A. It is the summation point for the uncorrelated signals at the primary path.

$$\therefore C_A(t) = X_A(t) + S_{GA}(t) \qquad (6)$$

where $X_A = X_{Pri}$ and $S_{GA} = S_{Pri}$

$$C_A(t) = X_{Pri}(t) + S_{Pri}(t) = V_{X Pri} \cos(2\pi f_c t - \beta c I_{X Pri} + \varphi_{X Pri}) + V_{S Pri} \sin(2\pi f_c t + \varphi_{S Pri}) \qquad (7)$$

At the input of the Vector Modulator the interfering signal was represented as:

$$X_{Ref}(t) = V_{X Ref} \cos(2\pi f_c t - \beta c I_{X Ref} + \varphi_{X Ref}) \qquad (8)$$

The expected output of the vector modulator is

$$X_{0Ref}(t) = V_{X Ref}(t) \cos\left\{2\pi f_c t - \beta c I_{X Ref} + (\varphi_{X Ref} + 180^0)\right\} \qquad (9)$$

At the Coupler 1, which is the summation point or cancellation point?

$$X_R = X_{Ref} - X_{Pri} = e_{rr} \qquad (10)$$

Where X_R is the residual noise signal .

The magnitude and phase characteristics of the error signal depends on the phase error, amplitude imbalance and delay mismatch between the primary signal path and reference signal path at the point of cancellation.

Let Δ_V represent the difference in amplitude imbalance θ_{err} represent the phase error and Δ_L represent the delay mismatch

$$\Delta_V = V_{X Ref} - V_{X Pri},$$

$$\therefore V_{X Ref} = \Delta_V + V_{X Pri} \qquad (11)$$

$$\theta_{err} = (\varphi_{X Ref} - \varphi_{X Pri}) - 180^0,$$

$$\therefore \varphi_{X\,\mathrm{Re}f} = \theta_{err} + \varphi_{X\,\mathrm{Pri}} + 180^0 \qquad (12)$$

$$\therefore I_{X\,\mathrm{Re}f} = \Delta_L + I_{X\,\mathrm{Pri}} \qquad (13)$$

$$\Delta_L = I_{X\,\mathrm{Re}f} - I_{X\,\mathrm{Pri}},$$

Substituting equations 11, 12 and 13 into equation 8. Therefore,

$$X_{\mathrm{Re}f} = (\Delta_V + V_{X\,\mathrm{Pri}})\cos(2\pi f_c - \beta c(\Delta_L + I_{X\,\mathrm{Pri}}) + \theta_{err} + \varphi_{X\,\mathrm{Pri}} + 180^0 \qquad (14)$$

Hence, ε_{rr} is the outcome of the vector addition of the primary signal and the reference signal at the cancellation point, which takes place at coupler B.

$$\therefore \varepsilon_{rr} = \varepsilon_{Xref} + \varepsilon_{X\,\mathrm{Pri}} \qquad (15)$$

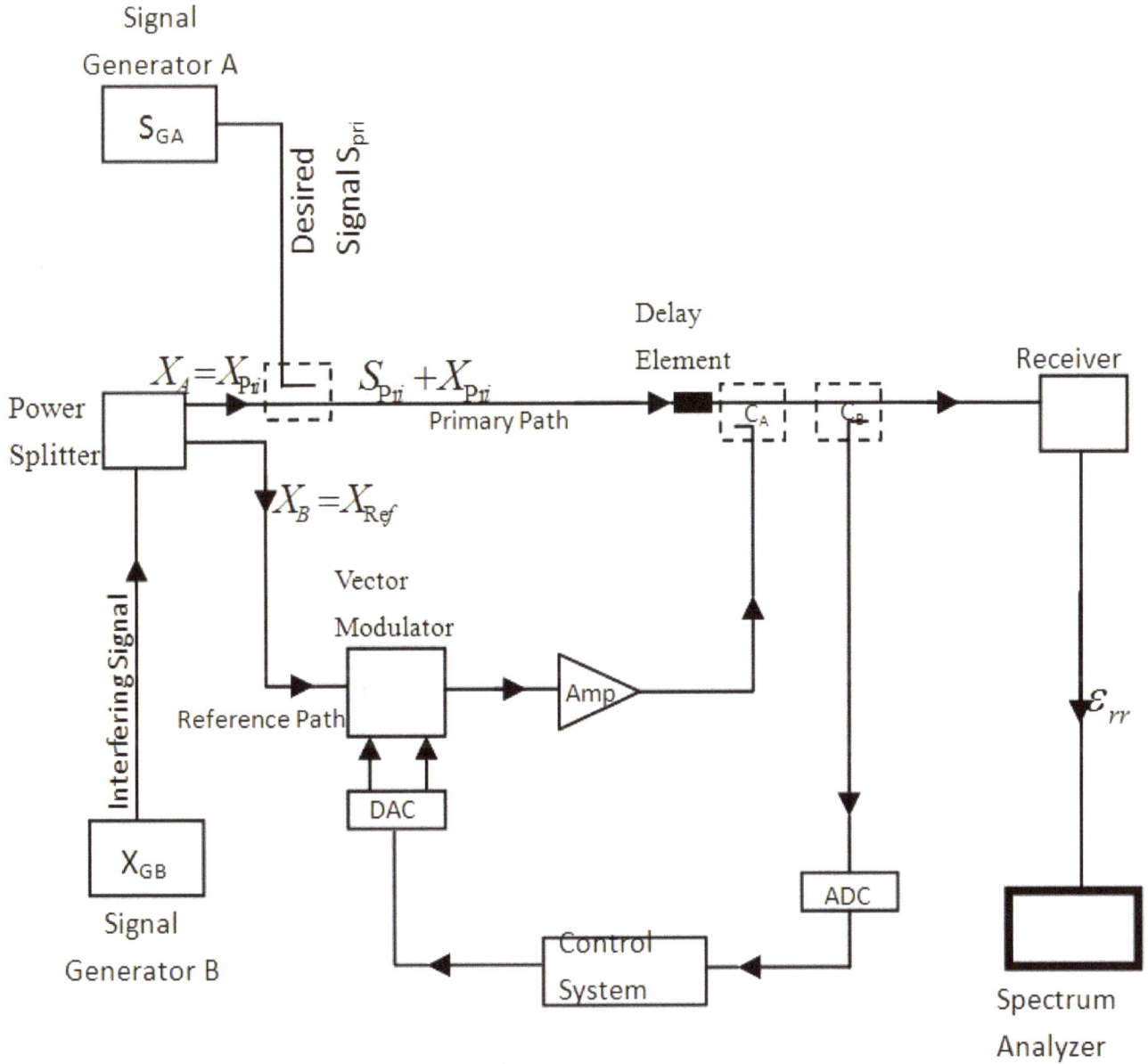

Figure 2. *Implementation Test-Bed of the ANCT.*

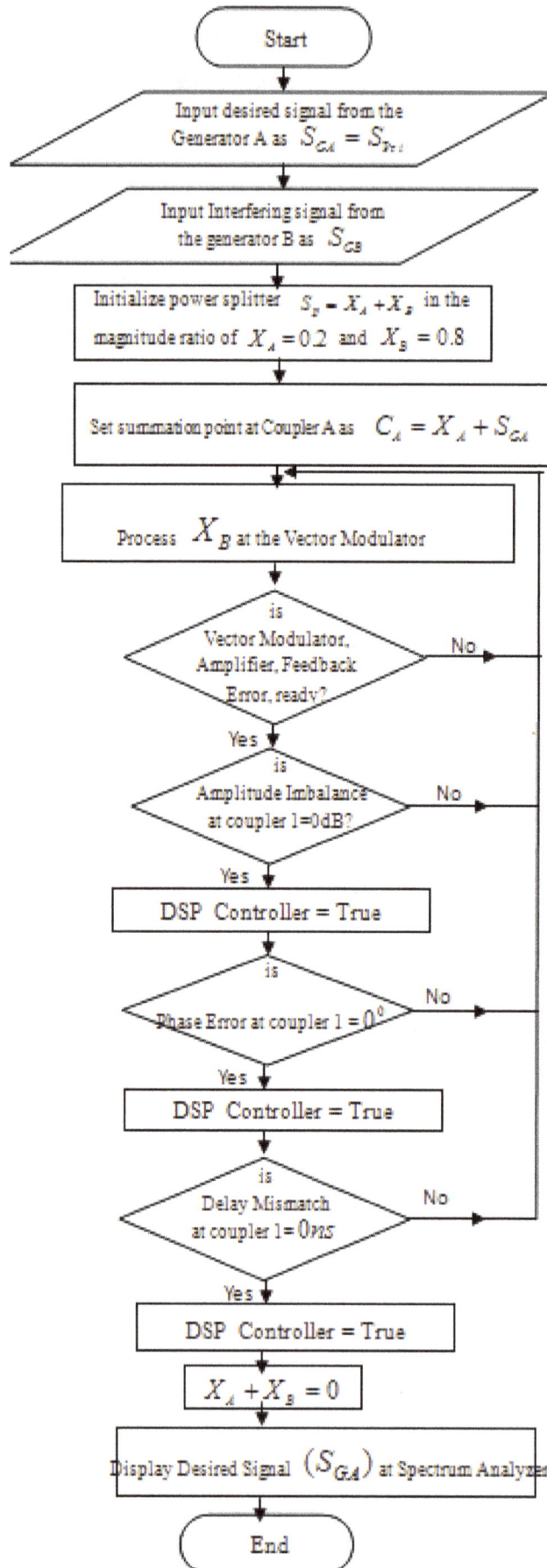

Figure 3. *Steps for implementing the ANC system represented in a flowchart algorithm.*

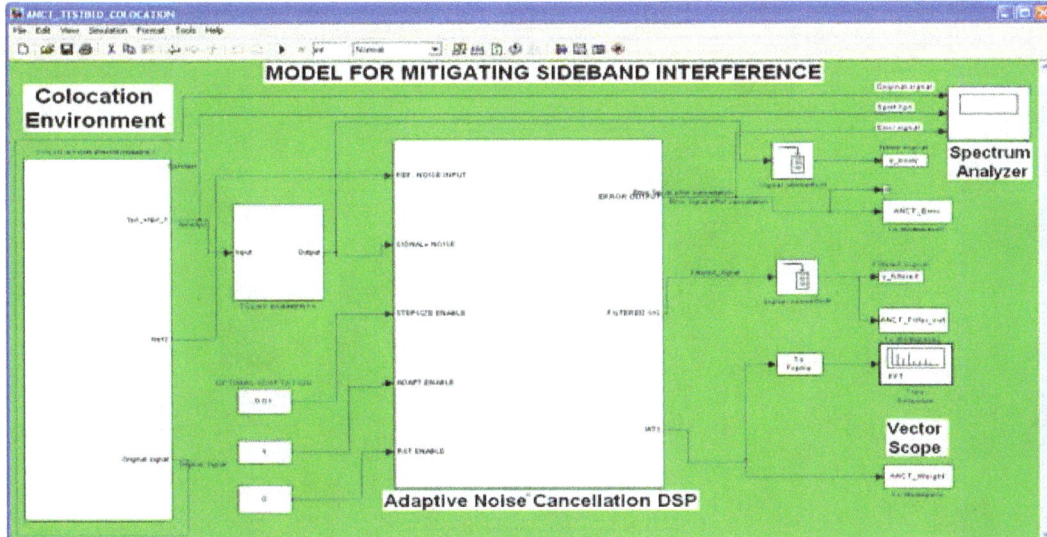

Figure 4. *Matlab-Simulink Test-bed for the Adaptive Noise Cancellation System.*

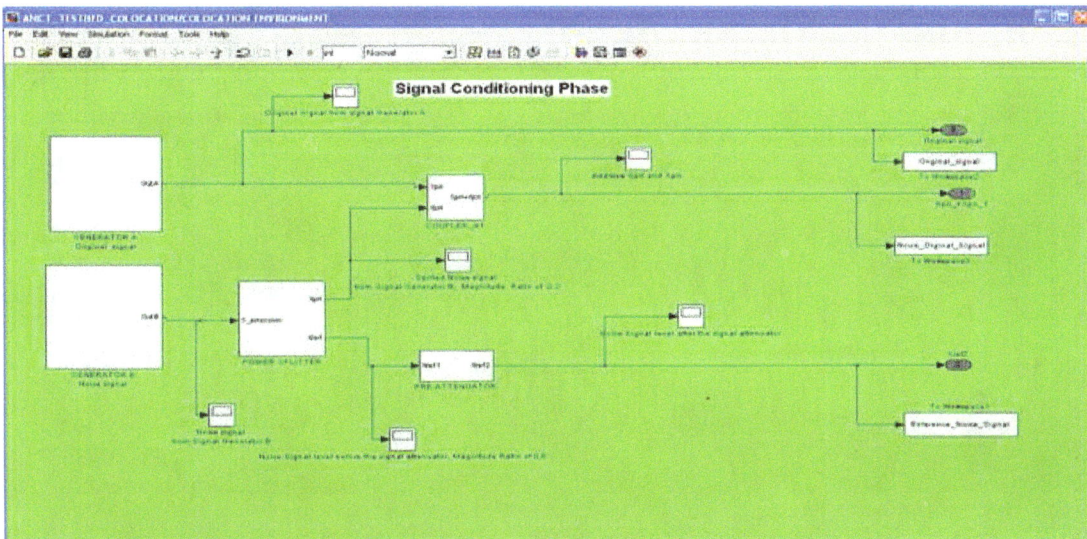

Figure 5. *Signal Conditioning Phase.*

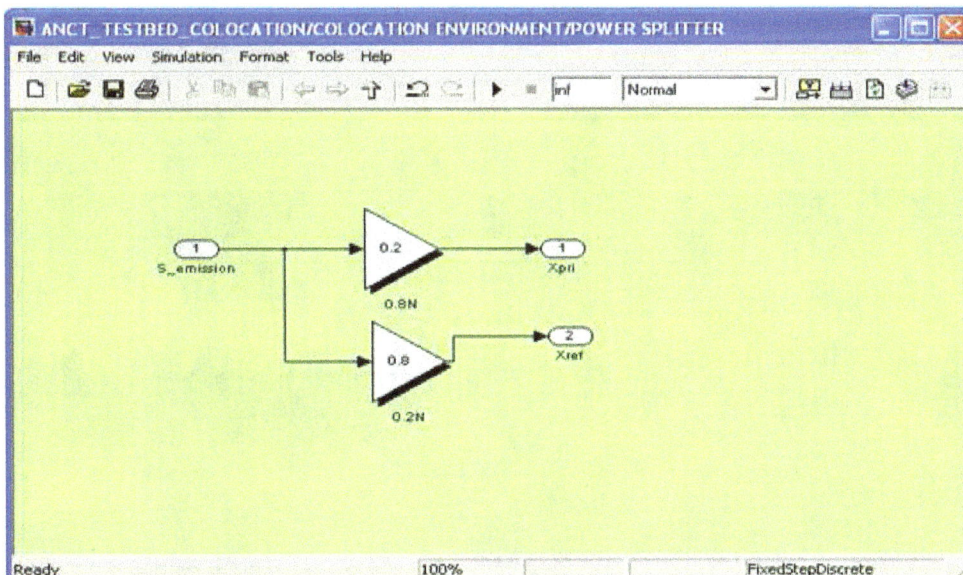

Figure 6. *Power Splitter.*

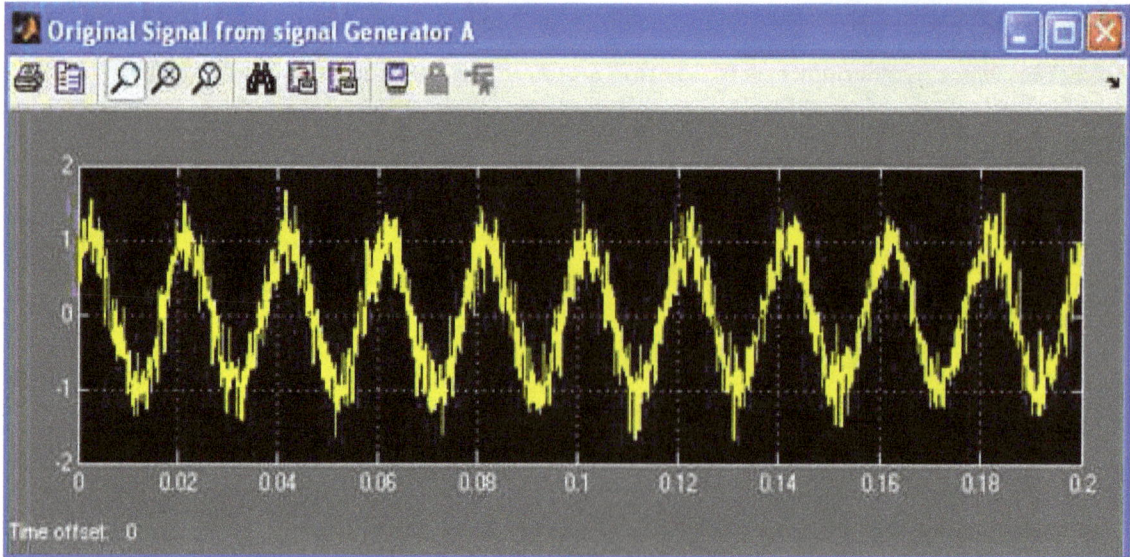

Figure 7. Original Signal from Signal Generator A.

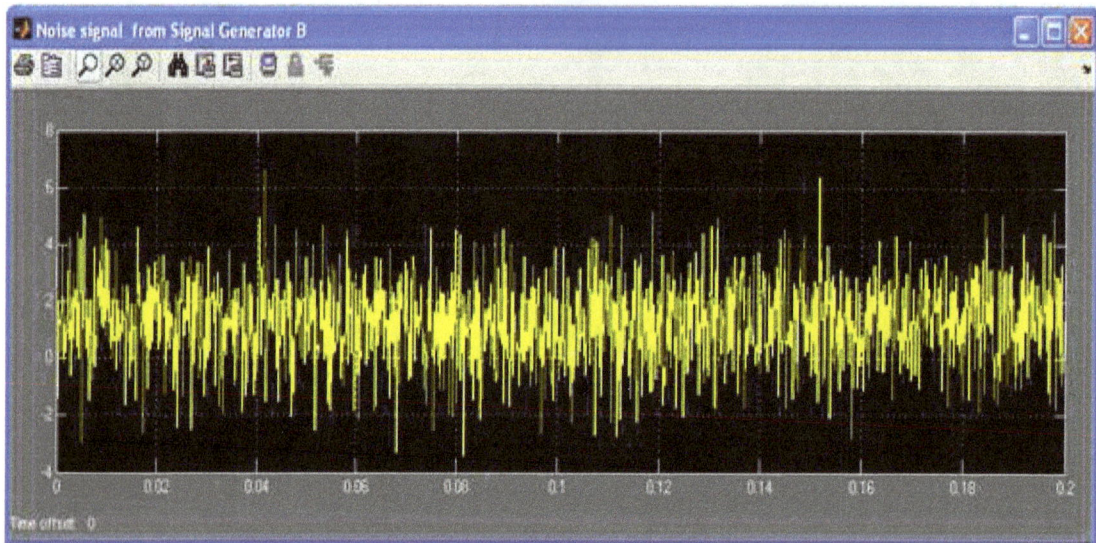

Figure 8. Noise Signal from Signal Generator B.

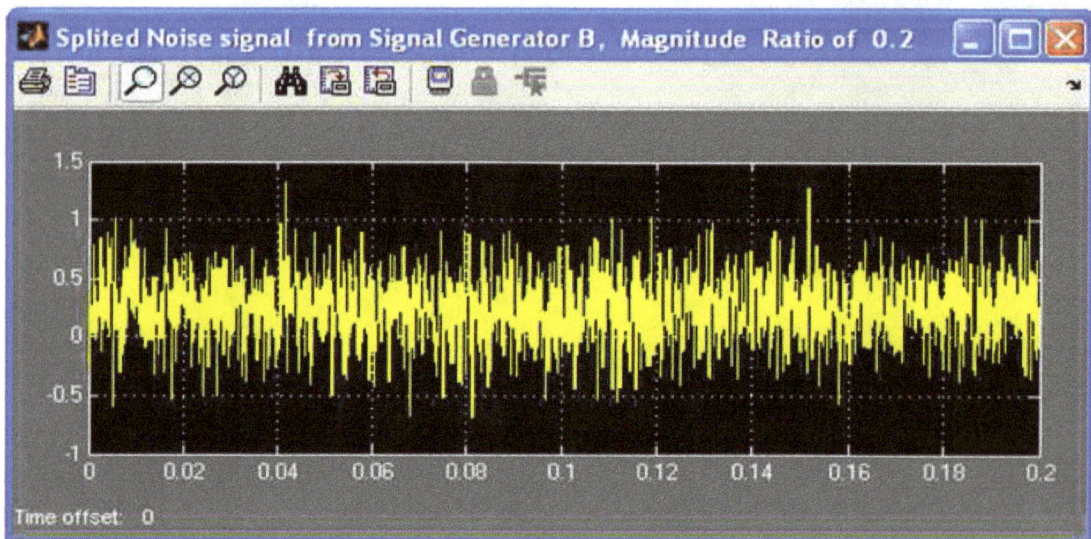

Figure 9. Splitted Noise signal from Signal Generator B, Magnitude Ratio 0.2.

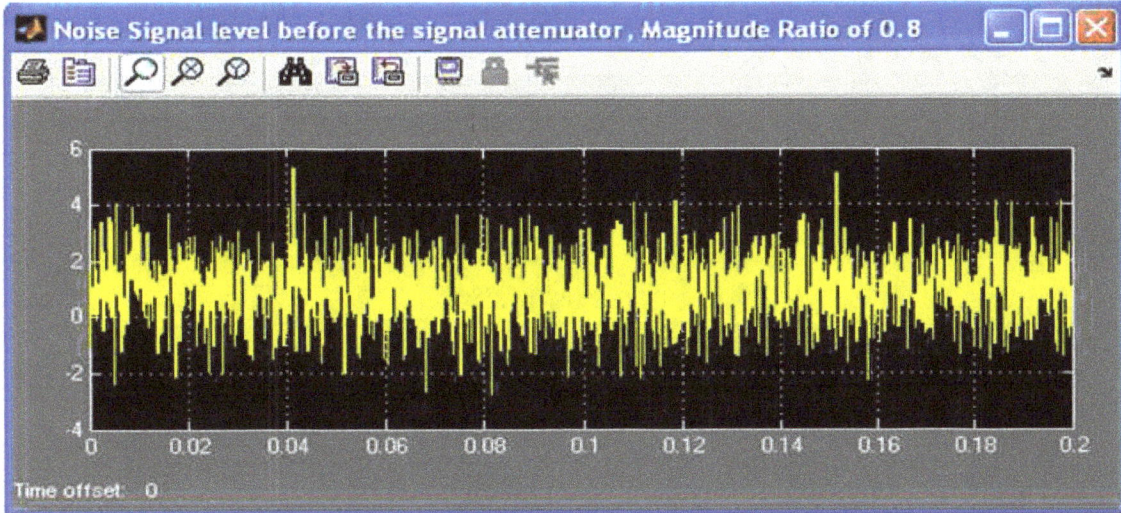

Figure 10. *Noise signal level before the signal attenuator, Magnitude Ratio 0.8.*

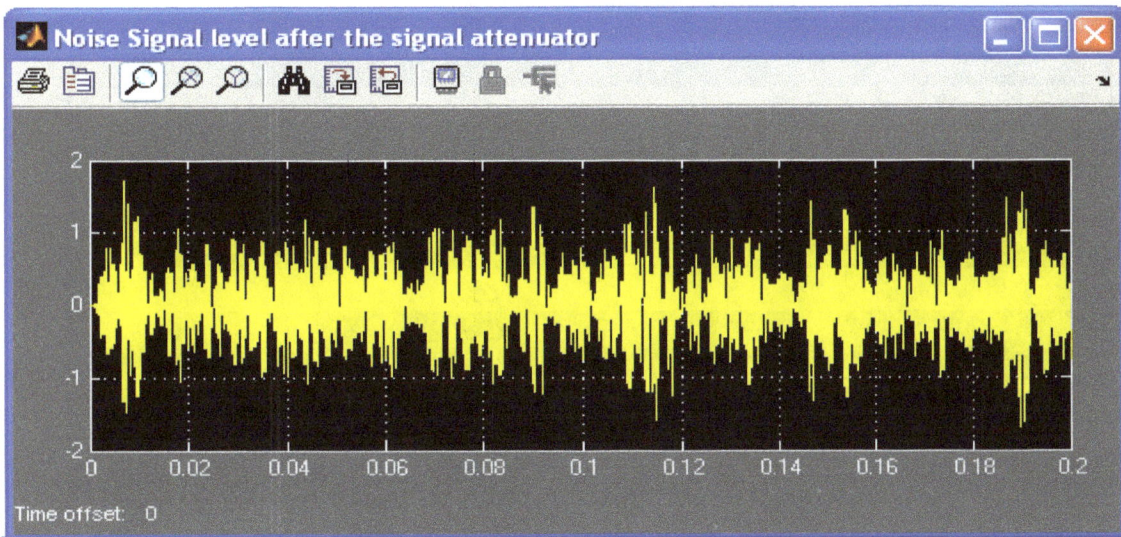

Figure 11. *Noise signal level after the signal attenuator.*

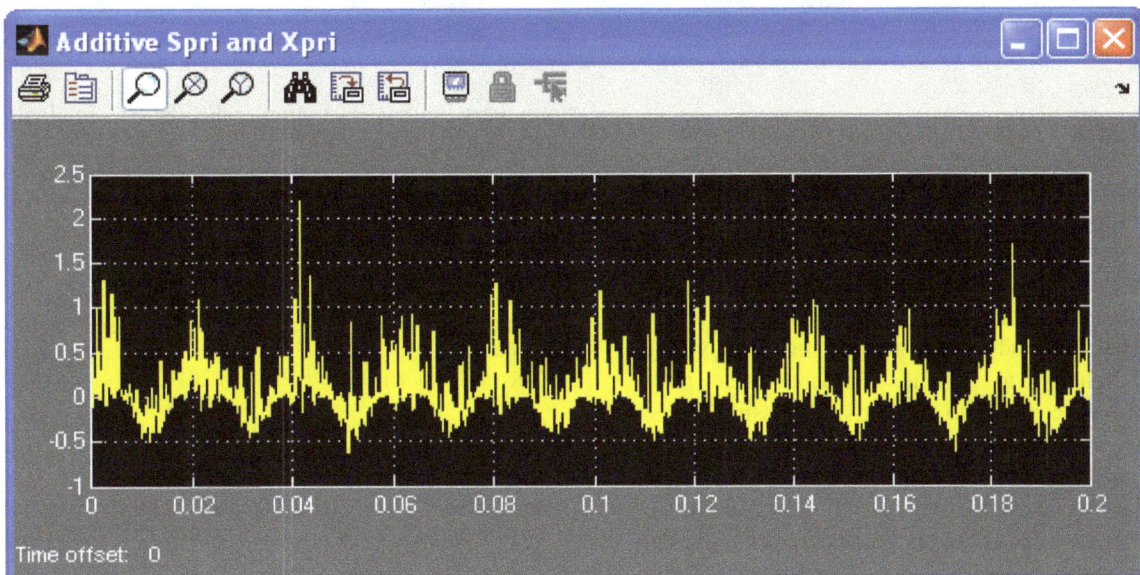

Figure 12. *Additive Spri and Xpri.*

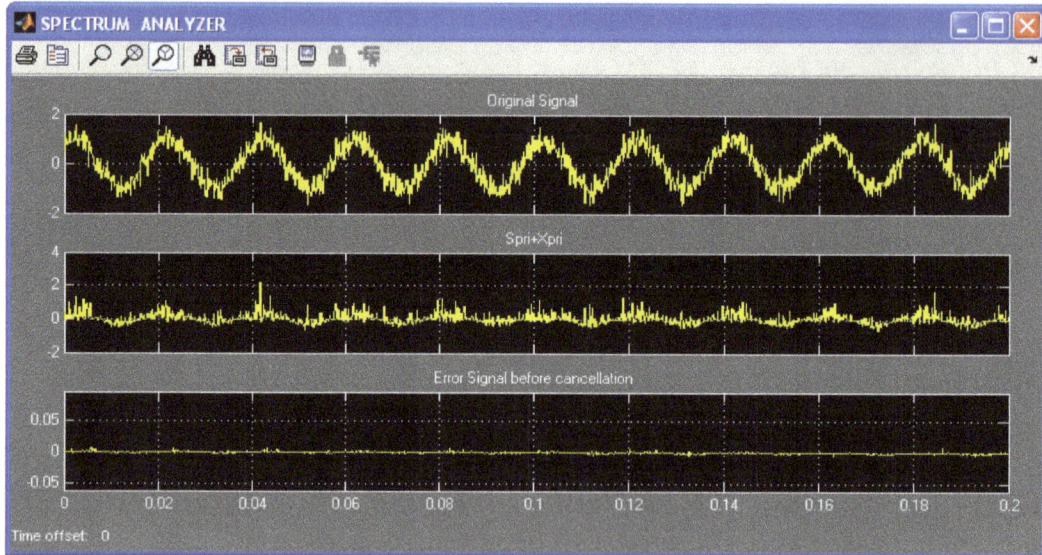

Figure 13. *Error Signal before cancellation.*

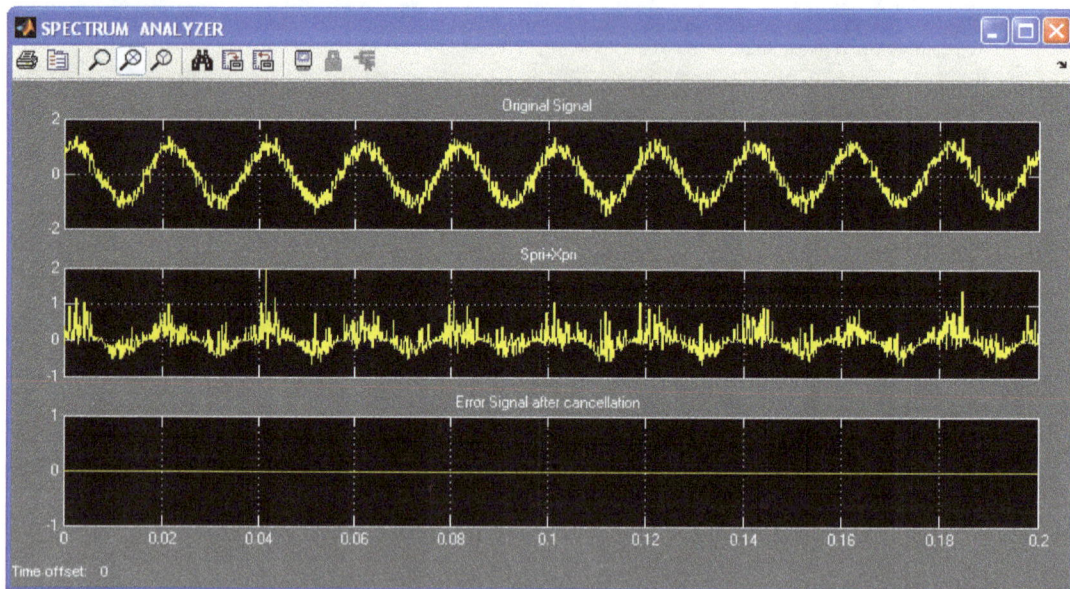

Figure 14. *Error Signal after cancellation.*

4. Conclusion

Adaptive Noise Cancellation Technique was once a weapon against intentional jammers, now a weapon against unintentional jammers. It has been shown capable of eliminating in-band interference over a wider bandwidth and over a changing environment without appreciably affecting the desired signal. The study was primarily aimed at providing an improved solution to mitigate sideband noise in a co-located network compare to the classical passive filters. The technique was primarily focused on achieving perfect theoretical cancellation where the amplitude imbalance, phase error and delay mismatch are dynamically adjusted by the weight variables towards achieving perfect cancellation. An experimental test-bed set-up was performed using Matlab-simulink environment to realize the error cancellation performances before and after cancellation, implemented using a developed algorithm.

References

[1]   German F., Annamalai K., Young M., Miller M.C., "Simulations and Data Management for Co-site Interference Prediction", *in Proc. IEEE International symposium on Electromagnetic compatibility,* July 2010.

[2]   Demirkiran I., Weiner D.D., Drozd A. and Kasperovich I., "Knowledge-based Approach to Interference Mitigation for EMC of Transceivers on Unmanned Aircraft," *in Proc. IEEE International Symposium on Electromagnetic Compatibility,* July, 2010.

[3]   Louay A. C., Bahjat E., Ghassan H., Mohamad M., "Telecom Infrastructure Sharing: Regulatory Enablers and Economic Benefits," *Booz Allen Hamilton,* 2007.

[4]   Bala-Gbogbo E., "Telecom Industry Operators Opt for Infrastructure Sharing.", 2009.

[5]   Razavi B., "RF Microelectronics". *Upper Saddle River, NJ: Prentice Hall*, 1998,pp. 11-53.106.

[6]   A. Roussel "Feedforward interference cancellation system applied to the 800MHz CDMA cellular band". *A Master of applied science in Electrical Engineering. Ottawa-Carleton Institute*. Pg 6, 2003.

[7]   Infrastructure sharing and collocation services License by Nigerian Communications Commission Section 32 of the *Nigerian Communications Act*, 2003.

[8]   Charu S. and Monika B. "Active cancellation concept of the Tx/Rx feedthrough applying an auxiliary transmitter", *International Journal of Computer science Issues*, 2011

[9]   Thomas H., Kodim W., Gloeckler R., Dingfelder H., "Transmitter Leakage Cancellation Circuit for Co-Located GPS Receiver," *European Patent 1 091497*, 11 April 2001.

[10] McConnell R. J., Tso R., "Method and Apparatus for Reducing GPS Receiver Jamming during Transmission in a Wireless Receiver," *U.S. Patent 6 961 019*, 1 Nov 2005.

6

Enhanced Privacy-Preserving Multi-Keyword Ranked Search over Encrypted Cloud Data

Pourush, Naresh Sharma, Manish Bhardwaj

Department of Computer Science and Engineering, SRM University, NCR Campus, Modinagar, India

Email address:

pulkittyagi1991@gmail.com (Pourush), nrssharma@gmail.com (N. Sharma), aapkaapna13@gmail.com (M. Bhardwaj)

Abstract: To protect the privacy, sensitive information has to be encrypted before outsourcing to the cloud. Thus the effective data uct keyword search. Related works on searchable encryption emphasis on single keyword based search or Boolean keyword based search, and hardly work on sorting the search results. Our work focuses on realizing secure semantic search through query keyword semantic extension. We mix-ups and used architecture of two clouds, explicitly private cloud and public cloud. The search process is distributed into two steps. The leading step develops the question keyword upon warehoused database in the private cloud. The subsequent step uses the drawn-out query keywords set to recover the index on public cloud. Finally the matched files are resumed in order. Complete security analysis shows that our explanation is privacy-preserving and secure. Trial evaluation determines the efficiency and effectiveness of the scheme.

Keywords: Motion Detection, Background Modeling (BM), Block Based, Human Detection, Wavelet Threshold Algorithm, Confusion Matrix

1. Introduction

We are existing in an exceedingly organized environment, where enormous measures of information are warehoused in confined, yet not basically trusted servers. There are various protection issues concerning to getting to information on such servers; two of them can just be perceived: affectability of

i. Keywords sent in questions and

ii. The information recuperated; both need to be concealed.

A related convention, Private Information Retrieval (PIR) empowers the client to get to public or private databases without uncovering which information he is extricating. Since protection is of an incredible concern, PIR conventions have been widely considered previously.

In today's information technology scene, clients that need high warehousing and processing power have a tendency to out- source their information and administrations to clouds. Clouds empower clients to remotely store and access their information by bringing down the expense of hardware possession while giving strong and quick administrations. The significance and need of protection saving pursuit procedures are significantly more claimed in the cloud applications. Because of the way that extensive organizations that work people in public clouds like Google or Amazon may get to the delicate information and hunt examples, concealing the question and the recovered information has extraordinary imperativeness in guaranteeing the protection and security of those utilizing cloud administrations.

This research paper concentrates on to the arrangement of multi-keyword ranked search encrypted (MRSE) over cloud information while protecting strict framework perceptive security in the cloud computing ideal model. The query to be tended to here is, given an arrangement of keywords, how would we use as ranking framework to secure cloud information storage and access?

To actuate the ranked search for successful use of outsourced cloud information under the previously stated model, the framework ought to be proposed by considering the security contemplations too. The framework is required to give the accompanying security and execution ensures as follows:

* Multi-keyword Ranked Search: To plan search plans which permit multi-keyword query and give result similarity ranking to authoritative information recovery, as opposed to returning undifferentiated results.

* Privacy-Preserving: To keep the cloud server from taking in extra data from the dataset and the record, and to meet the essential security necessities

- Efficiency: Ranked search should ensure privacy and computation overhead and also low communication.

2. Literature Review

The objective of this literature review is to summarize the data utilization and security issues of various searching techniques in the encrypted cloud data.

Qin Liu proposed this search that gives keyword security, semantic secure and information protection and public key encryption. Here, CSP is included in halfway decipherment by lessening the computational overhead and correspondence in deciphering for clients. The clients submit the keyword trapdoor encrypted by clients' private key to CS safely and recover the scrambled reports. Cong Wang proposed this search which understands transforming overhead, information and keyword protection, least correspondence and processing overhead. The executive construct file alongside the keyword recurrence based pertinence scores for records. Client demand "w" to CS with unrestricted "k" as Tw utilizing the private key. The CS seeks the record with scores and sends scrambled document focused around ranked grouping.

Wenhai Sun proposed this analysis that gives resemblance based query element ranking, keyword security, Index and Query privacy and Query Unlink ability. The encrypted record is assembled by vector space model supporting disjunctive and conjunctive document search. The searchable list is invented utilizing Multidimensional B tree. Holder makes encrypted inquiry vector $\bar{Q}$ for record keyword set. Client gets encrypted question vector of W from holder which is given to CS. Presently CS pursuits list by MD algorithm and compares at cosine measure of record and question vector and returns top k encoded records to client.

J. Baek proposed this strategy, in which CS makes its own particular public-private key pair. Sender encrypts all documents, keyword utilizing servers' and clients' public key before outsourcing. Client demands keyword trapdoor Tw to CS utilizing its private key. CS checks the Tw utilizing servers' private key and returns encrypted record. H. S. Rhee proposed this pursuit in which the outsourcing is carried out as SCF-PEKS. Client demands Tw to CS encoded with servers' open key and clients' private key. CS checks Tw utilizing servers' private key and returns encrypted record matching the keyword. Here the untouchable can't perform KGA without server's private key.

PengXu proposed this search, in which client makes fuzzy keyword trapdoor Tw and definite catchphrase trapdoor Kw for W. Client demands Tw to CS. At that point CS checks Tw with fluffy keyword file and sends superset of matching cipher messages by Fuzztest calculation that is executed by CS. The client process Exacttest calculation for checking ciphertexts with Kw and recover the encrypted records. Ning proposed this search for known cipher content model and foundation display over encoded information giving low calculation and correspondence overhead. The direction matching is picked for multi-keyword search for. They

utilized internal item likeness to quantitatively assess comparability for ranking records. The disadvantage is that MRSE have little standard deviation which debilitates keyword protection.

Wenhai Sun proposed Verifiable that gives multi-keyword seeks Privacy-Preserving Multi-keyword Text Search by similitude pursuit based result ranking. Owner outsources encoded report $\check{D}$ utilizing vector space model and confirmed secure list tree manufactured utilizing Multidimensional B-tree encrypted utilizing RSA and SHA-1. Client submits W to holder and gets encoded question vector $\bar{Q}$ for W. The question $\bar{Q}$ alongside inquiry parameter k is given to CS. Presently CS looks $\bar{Q}$ utilizing MD algorithm and thinks about cosine measure of $\bar{Q}$ and $\check{D}$ and returns top k encrypted records to client. At that point client looks this base tree utilizing the same inquiry calculation as CS and checks the question results.

3. Problem Statement

The issue in recovering the important records is that clients may not need archives, which they demand, to be uncovered, since their substance may be sensitive and they are typically specifically identified with query terms in their inquiries. In our plan, the server can furnish a proportional payback records to the client.

In the proposed approach, we require the information manager or its delegate that does not stratagem with the server, to be dynamic. The use of a dynamic agent for the information manager is a typical approach that is cognizant with earlier works. As clarified in proposed architecture, the information holder encodes archives with a symmetric-key encryption technique utilizing an alternate secret key for each one record. The server ought not to have the capacity to decrypt those ciphers since this would suggest that the server takes in the substance of the archive the client demands for. Accordingly, the information manager encodes the symmetric-keys with an open key encryption strategy, which has blinding capacity, and stores the encoded symmetric-keys in the server. In cryptography, blinding is a procedure, whereby a specialists can register a cryptographic capacity (e.g. marking and unscrambling), without knowing either the genuine info or the genuine yield of the capacity. We pick the RSA as the general population key encryption, which assistance blinding.

Accept that the client asks for the archive R. He gets the RSA encryption of the symmetric-key (sk), in particular RSAe(sk), where e means the general public key of the information holder. The client does not know the private key of the RSA (i.e. d), in this way he needs the information holder to perform the unscrambling of sk without indicating y = RSAe(sk), which would uncover the report he recovers. The client utilizes the blinding strategy and interfaces with the information holder for unscrambling the RSA encryption without taking in the private key d. Firstly, y is blinded by an irregular blinding variable c picked by the client as $z = c^e y$ mod N, where N is the RSA modulus. At that point, the client sends the blinded result z to the information manager, who unscrambles it utilizing his private key and gives back where its due (ž = zd mod N) over to the client. At last the client unblinds the result utilizing the blinding variable as sk = ž c-1 mod N. The information manager can't figure out which mystery key it is decoding following the ciphertext is blinded, thus arbitrary looking.

4. Existing Technology

The protection definition for analysis routines in the related text is that the server must study only the query items. We further tighten the protection over this general security definition and build a set of security necessities for privacy-preserving search protocols. A multi-keyword look system must give the accompanying client and information security properties (First instincts and afterward formal definitions are given):

1. Data Privacy: No one however the client can take in the genuine recovered information.
2. Index Privacy: The search file or the inquiry list does not release any data about the comparing keywords.
3. Trapdoor Privacy: Given one trapdoor for a set of keywords, the server can't produce an alternate legitimate trapdoor.
4. Non-Impersonation: No one can imitate an authentic client.

Definition 1- (Data Privacy) A multi-keyword search approach has information protection, if there is no polynomial time foe A that, given the recovered encrypted information and the comparing encoded key, realizes any data about the information.

Definition 2- (Index Privacy) A multi-keyword search approach has record protection, if for all polynomial time foes A that, given two distinctive keyword records L1 and L2 and a list $I_b$produced from the magic word list L_b where b $\in_R$ {0, 1}, the playing point of An in discovering b is insignificant. The playing point of A is irrefutably the estimation of the distinction between its prosperity likelihood and 1/2.

Definition 3- (Trapdoor Privacy) A multi-keyword search approach has trapdoor security, if for all polynomial time enemies A that, given a legitimate trapdoor for a set of keywords, A can't create a substantial trapdoor for its subset.

Definition 4- (Non-Impersonation) A multi-keyword search approach has non-mimic property, if there is no foe A that can imitate a genuine client U with likelihood more noteworthy than $\in$ where $\in$ is the likelihood of breaking the signature scheme

5. Proposed Methodology

Under the above system, we propose two answers for searching on encrypted information, in particular APKS and APKS+. We make novel utilization of a late cryptographic primitive; various hierarchical predicate encryption (HPE), which offers assignment of search capacities. Both of our

answers empower effective multi-dimensional questions with correspondence, subset and a class of basic extent inquiries. Since the PKC-based SE plans experiences a keyword reference assault that uncovers the basic magic words in an inquiry to the server, in APKS+ we upgrade the question security by keeping that sort of attack with the assistance of extra intermediary servers. To the best of our insight, the APKS+ plan is the first to accomplish productive multi-dimensional extent question, ability appointment and inquiry protection at the same time.

• Architecture:

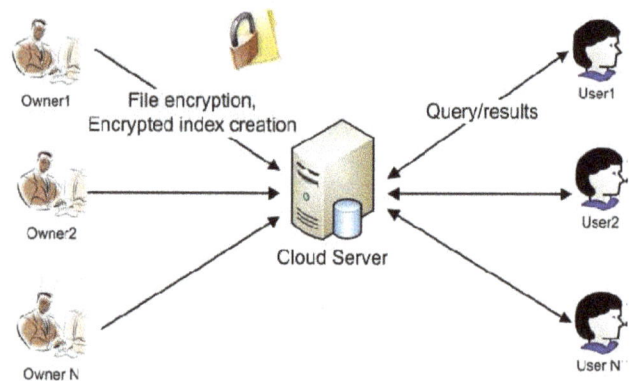

Fig. 1. *Architecture of proposed work.*

• Modules: The substances in the framework are: information managers/clients, trusted powers, and the cloud server.

1. Data Owner: In this paper, information manager depicts somebody who claims the data, e.g., a patient encrypts her information and needs them to be put away in the cloud server while protecting her information. The framework ought to permit numerous owners to encrypt and help information, while empowering a substantial number of clients to search over different managers' information. In accomplishing this, the framework ought to have high versatility, i.e., low key administration overhead. Additionally, productivity ought to be adequate for every inquiry operation from a client's perspective.

2. Cloud Server: The cloud server stores the encrypted information helped by various owners in a database and performs look for the clients.

3. User: The "clients" for the most part allude to the individuals who can perform search for required data over the encrypted database. At the point when a client asks for an ability for inquiry ^ Q from a LTA, the LTA checks whether a client either really has the quality worth set W basic the ^ , or is "qualified" for those qualities. One approach to accomplish this is to keep up a database of attribute qualities for all clients in the LTA's neighboring space. On the other hand, the LTA can issue to every client in its area a set of authorizations confirming the client's characteristic values, and confirms those qualifications upon a solicitation for capacity. With a specific end goal to

demonstrate its approval on ability, a TA/LTA can issue a signature that will be identity-based in light of every capacity it created/assigned. The server needs to confirm that a got ability has a legitimate signature from an enrolled LTA before performing quest for a client.

Age	Sex	Region	Provider
60	Male	Boston	Hospital A
Z_1	Z_2	Z_3	Z_4

Age 1	Age 2	Age 3	Age 4	Sex	Region 1	Region 2	Region 3	Provider
0-100	31-60	51-60	60	Male	MA	East MA	Boston	Hospital A
$Z_{1,1}$	$Z_{1,2}$	$Z_{1,3}$	$Z_{1,4}$	Z_2	$Z_{3,1}$	$Z_{3,2}$	$Z_{3,3}$	Z_4

Fig. 2. Index conversion- Age and region are hierarchical fields.

Fig. 3. Query processing and search using ranked multi keywords.

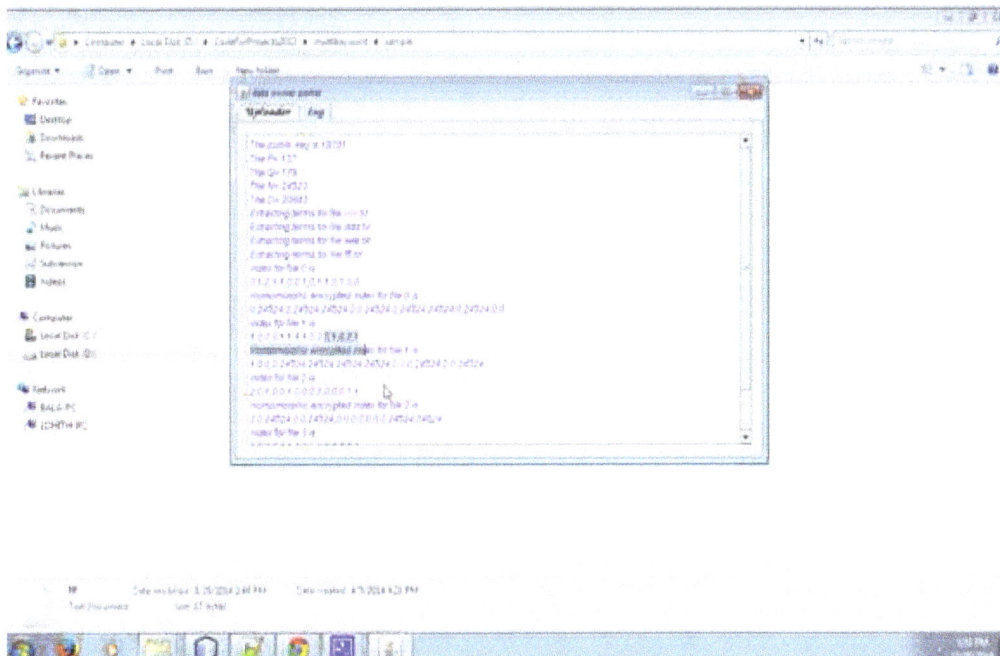

4. Encrypted Index Generation and query privacy-Index and Query Privacy: The essential security objective is to keep the cloud server from realizing any helpful data about the encrypted reports, records, and the clients' questions, with the exception of what can be acquired from the query items. List protection refers to confidentiality of the record, while inquiry security ensures clients' inquiries.

5. Search: Multi-dimensional Keyword Search: The framework ought to support multi-dimensional keyword search for usefulness, in particular, we need to help conjunctions among distinctive measurements where in each one measurement there can be different keywords (counting balance, subset and extent inquiries). The normal hunt handling time on single scrambled list under diverse n values It can be seen that the inquiry is much speedier than encryption and is straight to n, since it just takes $n + 3$ matching operations.

6. Results

"Direction matching", i.e., whatever number matches as would be prudent, is a proficient similitude measure between such multi-keyword semantics to refine the result importance, and has been broadly utilized as a part of the plaintext data recovery (IR) group.

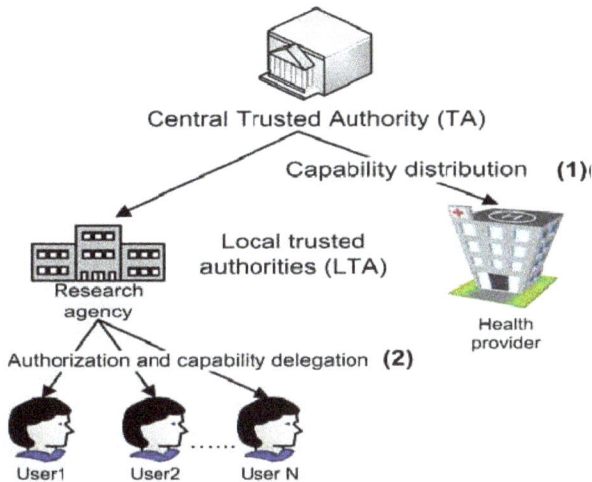

Fig. 4. Profile of different data users at current time.

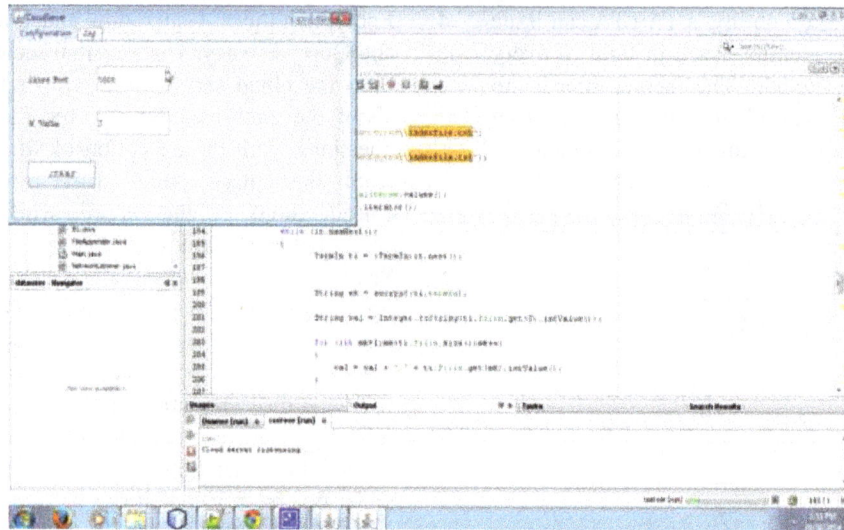

Fig. 5. *Log file calling means index generation.*

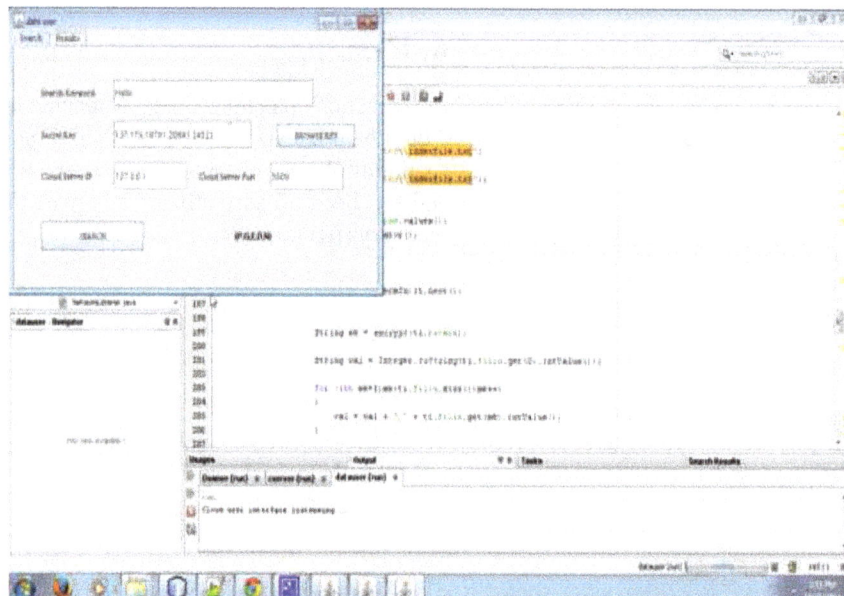

Fig. 6. *Key generation for a particular profile.*

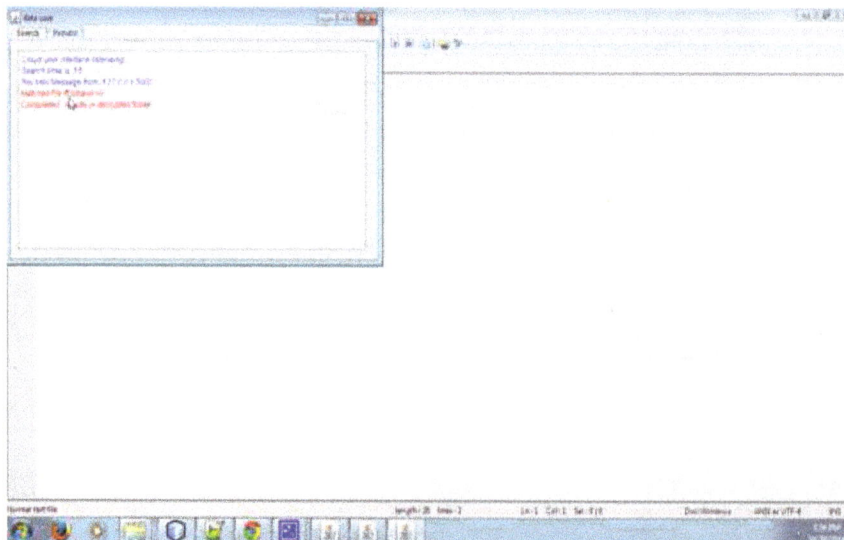

Fig. 7. *Encrypted profile search.*

Fig.4 shows the key pair of public and private are generated and then the using the keys the file data changed into cipher text shown by fig.5. The keyword is indexed for

key protection in fig6. The results of user queries are search and ranked shown by fig 7 and fig 8.

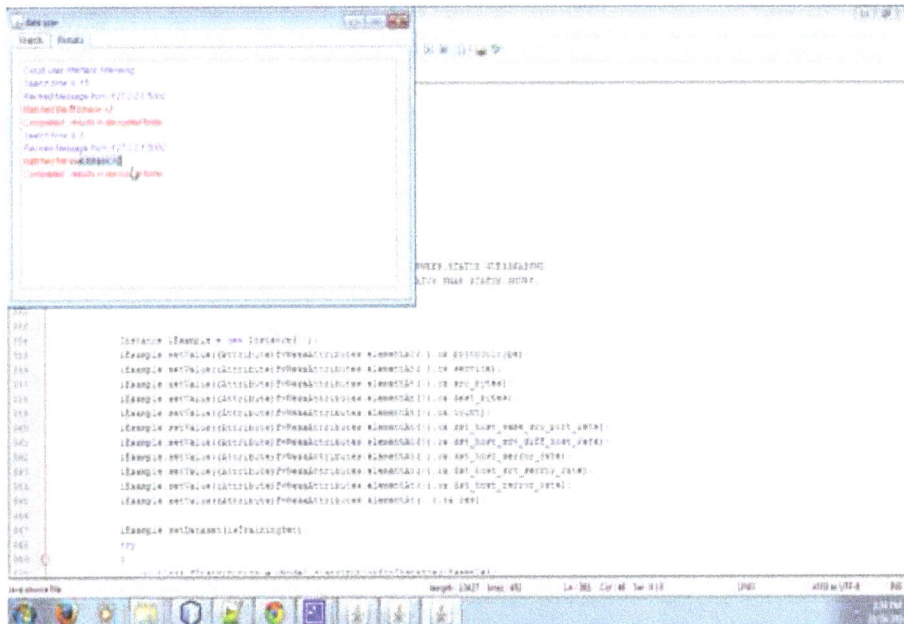

Fig. 8. *Decrypted profile search.*

In any case, how to apply it in the encrypted cloud information search framework remains an extremely difficult assignment as a result of immutable security and protection problems, including different strict necessities like the information protection, the file protection, the keyword protection, and numerous others.

7. Conclusion

Ranked Search can likewise richly eliminate unnecessary system traffic by sending back just the most important information, which is exceedingly attractive in the "pay-as-you- utilization" cloud standard. For protection assurance, such ranking operation, notwithstanding, ought not to release any keyword related data. To put it plainly, none of existing Boolean keyword searchable encryption plans help different essential words ranked inquiry over scrambled cloud information while saving security as we propose to investigate. Our proposed work recognizes the new approach and architecture algorithm to upgrade the execution of multi keyword ranked inquiry.

Firstly, the client side of proposed framework will be actualized on cell phones running Android and ios working frameworks since the potential application situation imagines that clients acquire to the information anyplace and whenever. Also besides, the proposed technique will be tried on a genuine dataset with a specific end goal to think about the execution of our ranking strategy with the ranking strategies utilized as a part of plain datasets that don't include any security or privacy-preserving techniques.

References

[1]   Oxford dictionaries, the oec: (June 2011)."Facts about the language."http://oxforddictionaries.com/page/oecfactslanguage/the-oec-facts-about-the-language.

[2]   D. Boneh, E. Kushilevitz, R. Ostrovsky, and W. Skeith. (2007). "Public key encryption that allows pirqueries"; In Advances in Cryptology - CRYPTO, volume 4622 of Lecture Notes in Computer Science, pages 50{67. Springer Berlin / Heidelberg, 2007.

[3]   N. Cao, C. Wang, M. Li, K. Ren, and W. Lou. (2011). "Privacy-preserving multi-keyword ranked search over encrypted cloud data". In IEEE INFOCOM.

[4]   Y.-C. Chang and M. Mitzenmacher. (2005). "Privacy Preserving Keyword Searches on Remote Encrypted Data". In Applied Cryptography and Network Security, pages 442{455. Springer.

[5]   D. Chaum. (1982). "Blind signatures for untraceable payments". In Advances in Cryptology: Proceedings of CRYPTO'82, pages 199{203.

[6]   B. Chor, E. Kushilevitz, O. Goldreich, and M. Sudan. (November 1998). "Private information retrieval". J. ACM, 45:965{981.

[7]   L. E. Dickson. (2003). "Linear Groups with an Exposition of Galois Field Theory". Dover Publications, New York.

[8]   M. J. Freedman, Y. Ishai, B. Pinkas, and O. Reingold. (2005). "Keyword search and oblivious pseudorandom functions". In Theory of Cryptography Conference -TCC 2005, pages 303{324.

[9]   J. Groth, A. Kiayias, and H. Lipmaa. (2010). "Multi-query computationally-private information retrieval with constant communication rate". In PKC, pages 107{123.

[10] W. Ogata and K. Kurosawa. (2004). "Oblivious keyword search". In Journal of Complexity, Vol.20, pages 356{371.

[11] J. T. Trostle and A. Parrish. (2010). "Efficient computationally private information retrieval from anonymity or trapdoor groups". In ISC'10, pages 114{128.

[12] L. M. Vaquero, L. Rodero-Merino, J. Caceres, and M. Lindner. (December 2008). "A break in the clouds: towards a cloud definition". SIGCOMM Computer. Commun. Rev., 39:50{55.

[13] C. Wang, N. Cao, J. Li, K. Ren, and W. Lou. (2010). "Secure ranked keyword search over encrypted cloud data". In ICDCS'10, pages 253{262.

[14] P. Wang, H. Wang, and J. Pieprzyk. (2009). "An efficient scheme of common secure indices for conjunctive keyword-based retrieval on encrypted data". In Information Security Applications, Lecture Notes in Computer Science, pages 145{159. Springer.

[15] J. Zobel and A. Mo_at. (1998). "Exploring the similarity space". SIGIR FORUM, 32:18{34.

7

ICT Policy Outcomes for National Development: The Place of Knowledge Integration and Management in Nigerian Higher Education

Chinyere Onyemaechi Agabi[1], Comfort Nkogho Agbor[2], Nwachukwu Prince Ololube[1]

[1]Department of Educational Foundations and Management, Faculty of Education, Ignatius Ajuru University of Education, Port Harcourt, Rivers State, Nigeria

[2]Department of Environmental Education, Faculty of Education, University of Calabar, Calabar, Cross River State, Nigeria

Email address:
chinyereagabi@yahoo.com (C. O. Agabi), commyspaco@yahoo.com (C. N. Agbor) ololubeprince@yahoo.com (N. P. Ololube)

Abstract: The prologues of Information Communication Technology (ICT) usage including its integration and diffusion kicked off a new era in educational processes and has fundamentally changed the conventional methods of teaching and learning in higher education institutions (HEIs) around the world, and have transformed contemporary processes of teaching and learning experiences of both lecturers and students. The debates here are made with reference to (1) the contexts of ICT and knowledge integration (2) the challenges of ICT usage and knowledge integration, and (3) ICT policy outcomes and national development. A qualitative research method was adopted; the use of document and observation were indispensable part of the methods for data gathering. The study found that the lofty hopes, keenness and enthusiasm for ICT and knowledge integration and management are obstructed as the nation is faced with inadequacies in essential ICT infrastructures and services such as telecommunication services, electricity, incompetent ICT personnel, inadequate funding, poor economic situation, poverty, high ICT literacy rate and so on. However, there is an ongoing moves and development to ensure effective ICT knowledge integration and management in education resources in Nigeria and Africa higher education institutions. This novel study recommends that higher education should become expansive, positive and proactive actors in ICT knowledge integration and management in teaching, learning and research for academics, non-academics to foster admirable academic environment aimed at meeting national development. This learned debate has implication for education practitioners, curriculum developers and designers, policy makers, planners and the government.

Keywords: ICT, Knowledge Integration, Management, Diffusion, Higher Education, Policies, National Development

1. Introduction

Information Communication Technology (ICT) is a complex network of connections that presents an increasingly expanding collection of innovative services that have key financial consequences in the processes of knowledge integration and management in higher education and maybe adjudged to be very important for national development (Ifinedo & Ololube, 2007). The processes that lead to knowledge integration according to Ololube, Kpolovie, Amaele, Amanchukwu and Briggs (2013) is the use of computer-based (Ololube, 2009) and web-based learning technological (Ifinedo, 2006) tools adopted by institutions of higher education in organising and the processing of information and communication needs of such organisations.

Promoting and developing a knowledge society through computer-based and web-based learning technological tools is one of the strategic plans that is increasingly adopted in recent times by knowledge advancement institutions around the world who wish to progress and support economic growth for national development (Ololube, Ubogu & Egbezor, 2007). In particular, higher education institutions worldwide are saddled with the responsibility of growing in importance as agencies for the development of knowledge (Akuegwu, Anijaobi-Idem & Ekanem, 2011). As a result, a number of countries, particularly those in the west have been proactive towards developing strategies to drive efforts aimed at providing institutions of higher education the opportunity to achieve knowledge parity for their citizens (Department of Education and Skills, 2011). However, despite advances in ICT, higher education institutions in the Third World are masqueraded with complex problems especially in their academic programmes in reaching the goals of promoting the development of knowledgeable society (Johnson, 2007).

Global efforts at proactively creating knowledge society through ICT include components of computer hardware and software, network connectivity and several other devices that are essential to achieving effective knowledge management. The components include but are not limited to audio, camera, e-mail, facsimile (Fax), internet, intranet, main-frame computer, minicomputer, micro-computer, photography, teleconference, video, websites, word processing computer etc. (Nwafor, 2005). According to Lopez (2003) ICT has provided innovative opportunities for teaching, learning, research and administration in higher education.

The ICT literacy rate in many institutions of higher learning in the Third World has been a craze of major concerns and has transformed how we see other counties in the global contexts (Mac-Ikemenjima, 2005). ICT literacy rate in this context is the competence (familiarity, skills and ability) of students, academic and non-academic staff to identify, explore actual fact, presents and diffuse information in order to assemble knowledge and develop the spirit of learning to be critical and self-critical, and learn to create knowledge significant to a theme of study (Ololube, 2012). To be able to create knowledge is the ability to generate new knowledge on the basis of one's own experiences and constructing knowledge where none is available (Mohanan, 2005). This has given rise to progresses in our conducts in life because ICT has transformed and impacted on educational methodologies in higher education globally (Pena-Bandalaria, 2007; Richardson et al., 2015). However, it is pertinent to note that the revolution taking place in higher education is not widespread and there is the need to strengthen ICT penetrations to reach a larger percentage of students, faculty and non-faculty in our institutions of higher education (Ifinedo, 2005, 2006). Therefore, knowledge integration and management approaches (Bellinger et al., 2004) (see figure 1) are extremely essential to guarantee successful national development of Nigeria's economy and society. These approaches involve the process of including novel information into an existing body of knowledge using multiple approaches. The processes entail the determination of how new information and the existing knowledge interrelate to each other, how existing body of knowledge should be made to other to accommodate new information, and how new information should be structured to reflect existing knowledge (Cárdenas, Al-Jibouri, Halman & van Tol, 2013).

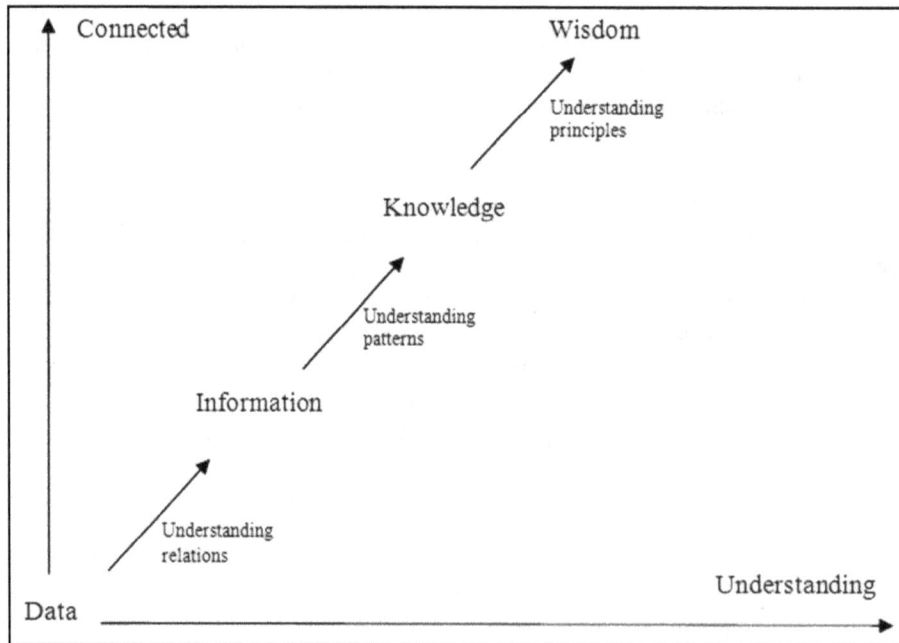

Figure 1: Knowledge hierarchy (Bellinger et al., 2004).

The knowledge hierarchies presented above can be explained thus:
- Data: incoherent collection of facts about a phenomenon that has little basic interest;
- Information: evolves when relationships between the connecting facts are established and understood;
- Knowledge: this is when patterns of relationships are determined, identified and understood;

- Wisdom: judging to produce understanding where there was no prior understanding;
- Understanding: recognition and logical synthesis of novel knowledge from existing knowledge.

The academic landscape of Nigeria's higher education includes traditional f2f teaching-learning and the distance education processes, along its research processes. Additionally, it includes the certificate and degree programmes and courses, theses, disseminations and academic publications; the libraries and information services; higher education administration and management. The 6th edition of the Nigerian national policy on education (FRN, 2013) perceived higher education as post-secondary education system which is given in universities, Polytechnics, Colleges of Education and Monotechnics, as well as those organisations offering correspondence courses. The Federal Republic of Nigeria (FRN, 2013) through its national policy on education stated that the goals of higher education shall be:

- To acquire both physical intellectual skills which will enable individuals to be self-reliant and useful members of society;
- To contribute to nationwide development through high level germane manpower training;
- To expand and inculcate proper value for the survival of the individual and society;
- To enlarge the intellectual capacity capability of individuals to understand and appreciate local and external environments;
- To forge and cement national unity; and
- To encourage and encourage scholarship and community service;
- To encourage national and international understanding and interaction.

The Federal Republic of Nigeria through its higher education institutions shall pursue these goals to make for favourable contributions to national development. And ICT is seen as an indispensable constituent towards the move for knowledge integration (Oyovwe-Tinuoye & Adogbeji, 2013). Nonetheless, ICT diffusion, use and its penetration into research, teaching and learning in higher education are dependent upon the resources availability, which are heavily dependent on government policies and the political environment (Ololube, 2013). However, the policies and political atmosphere has necessitated gaps in addressing innovative ideas, free enterprise, development problems, strategic planning (Meek, et al., 2009) and the digital divide in Nigerian academic landscape especially in higher education. Scholars (e.g., Mac-Ikemenjima, 2005; Ifinedo, 2005; Ololube, 2009; Cárdenas, Al-Jibouri, Halman & van Tol, 2013) belief that ICT has a lot to offer towards knowledge integration, creation and management. However, the Nigeria government has failed to deliver on that front in line with international best practices, because a great deal of time and efforts has been and is still endowed in developing policy guidelines, instead of effective implementation of ICT programmes aimed at elevating and positioning Nigeria in the global academic community (Ololube et al., 2013).

According to professional colleagues and in line with an academic work entitled "Active Learning Application of Technology Tools and Services to Increase Student Achievement in Online and Blended Learning Environments in Higher Education Institutions" by Ololube (2015), this learned research however revealed 'significant improvement in the use of ICT tools and blended learning methods to accomplish effective and active academic performance in students and academics. The impact of ICT and blended learning in higher education institutions is evidenced in the changing instructional strategies to improved student academic attainments, which results from more active interactive learning processes'. Nevertheless, the perceptions of the respondents are that the Nigerian government's policy on higher education to enhance ICT usage and integration is still struggling to yield greater positive results as achieved in the west. The study contends that Nigeria's usual place of paying lip service or little attention to empowering higher education must stop if Nigeria is to make headway in its knowledge integration agenda.

To this end, the objective of this paper is to understand perceptions concerning the role of ICT in knowledge integration and management in Nigeria. Despite studies confirming the importance of knowledge integration and management to national and global development (e.g., Agbeja & Fajemisin, 2008; Sridharan, Tretiakov, & Kinshuk, 2004), and the perception among some professionals in Nigeria who say ICT is not related to knowledge integration and management. Rather they see knowledge integration to be tied to intelligence, available resources, interest in academic work, and other personality traits as factors that propels knowledge integration and management. As a result, there is need to theoretically ascertain if the factors mentioned by colleagues are responsible for knowledge integration in Nigeria higher education environment. This study attempts to offer new perspectives into the varied factors that support the use of ICT for knowledge integration and management. This study will further address the intellectual gap in understanding the role of ICT in knowledge integration.

2. Literature Review

2.1. The Contexts of ICT, Knowledge Integration and Management

As humans we are challenged with the need to resolving our daily problem in every life and in academics. The knowledge we possess and the capability to make the most of the appropriate information in our daily lives require meeting some certain task. Taking tactical decision requires sharing of knowledge among colleagues and during teaching and learning processes. Basically, knowledge refers to the understanding an individual has about information and the integration and management of knowledge is a comprehensive process, which includes all the processes that allow for knowledge capitalisation in higher education (Oladejo & Osofisan, 2011). Information communication technology (ICT) literacy has been recognised to influence the search for relevant information to solve daily and academic problems (Ifinedo & Ololube, 2007).

The acceptance and use of ICT to boost knowledge integration and management processes is valuable for higher education in positive ways—enhancing and enabling and in negative ways—blocking and frustrating methods. However,

ICT can increase the knowledge content of the teaching-learning methods and the type of students who graduate from higher education institutions. It can make possible as well as hinder the processes of knowledge integration, acquisition, diffusion, relevance and preservation. The acceptance and use of ICT to enhance and facilitate knowledge integration and management has brought to the centre state the burning need for new policies and methodology towards achieving quality higher education around the world (Omona, van der Weide, & Lubega, 2010).

ICT fundamentally refers to the application of technology in communication to influence the teachers and learners acquisition of knowledge. It facilitate the achievement of educational goals, benefits the advancement of critical literacy including technological literacy among teachers and students using computer-based teaching and learning methods (CBTL) (Ifinedo, 2005; Ololube, 2009). CBTL improves methods of teaching and learning and makes them more efficient and even more effective; it makes teaching and learning extremely interesting to both the teachers and the learners, it deepens knowledge acquisition for the teachers and learners, but has superior impact on learners. It increases the ability of learners and adds value to knowledge (Ololube et al., 2013). Thus, CBTL improves the quality of information, knowledge integration and management.

2.2. The Challenges of ICT Usage and Knowledge Integration

The most recent Global Information Technology Report (GITR) (2015) featured the latest Networked Readiness Index (NRI). NRI assessed the indicators, policies and institutions that facilitates and enables a country to fully influence ICTs for improved competitiveness and knowledge integration. The timing and release of this report is germane when many economies around the world are struggling to make sure that economic growth is equitable and provides benefits for their entire populace. As shown in the GITR (2015) report, ICTs act as a means of social development and transformation by improving access to basic services, enhanced connectivity, creating employment opportunities and knowledge societies. In this year's NRI ranking, Nigeria dropped seven places to rank 119th out of 143 countries. NRI for 2014 was 112th out of 148 countries, while that of 2013 was 113th out of 144 countries. The low ranking is due to several factors as indicated in table 1 below.

Table 2 presents a detailed analysis of the indicators that necessitated Nigeria to be listed amongst the ICT low rank countries. The most important indicators to this study are the 3rd, 4th, 5th, 6th, and the 10th pillars. Indices in the 3rd pillar—infrastructure revealed that the electricity production per Kwh is very low to boost and power ICT tools and services. Nigeria rank 125 out of the 143 countries surveyed as a result of the poor electricity supply in the country. The mobile network coverage per percentage of the population showed that Nigeria ranked 116 out of the 143 countries investigated. What this depicts is that the mobile network coverage for Nigeria is low. Information on the International Internet

bandwidth per kb for each user show that Nigeria ranked very low, in a position of 130 out of the 143 countries evaluated. Same is true of the data for secure Internet servers/million of the population. The revelation is that Nigeria ranked 119 of the 143 countries examined.

Table 1: 2015 Nigeria's Networked Readiness Index.

Group	Indicators	Global Ranking / 143 Countries
A.	Environment Subindex	120
	1st pillar: Political and regulatory environment	116
	2nd pillar: Business and innovation environment	111
B.	Readiness Subindex	123
	3rd pillar: Infrastructure	121
	4th pillar: Affordability	104
	5th pillar: Skills	135
C.	Usage Subindex	104
	6th pillar: Individual usage	114
	7th pillar: Business usage	79
	8th pillar: Government usage	95
D.	Impact Subindex	104
	9th pillar: Economic impacts	81
	10th pillar: Social impacts	116

Source: Global Information Technology Report for Nigeria (2015, p. 213)

Table 2: Detailed Analysis of the 5th Pillar.

Indicators	Global Ranking/143 Countries
3rd Pillar: Infrastructure	
Electricity production, kWh/capita	125
Mobile network coverage, % pop.	116
Int'l Internet bandwidth, kb/s per user	130
Secure Internet servers/million pop.	119
4th pillar: Affordability	
Prepaid mobile cellular tariffs, PPP $/min	35
Fixed broadband Internet tariffs, PPP $/month	118
Internet & telephony competition, 0–2 (best)	1
5th Pillar: Skills	
Quality of educational system	121
Quality of math & science education	132
Secondary education gross enrollment rate, %	125
Adult literacy rate, %	108
6th Pillar: Individual Usage	
Mobile phone subscriptions/100 pop	87
Individuals using Internet, %	119
Households w/personal computer,	112
Fixed broadband Internet subs/100 pop	140
Mobile broadband subs/100 pop	98
Use of virtual social networks	82
10th pillar: Social Impacts	
Impact of ICTs on access to basic service	123
Internet access in schools	111
ICT use & gov't efficiency	119
E-Participation Index, 0–1 (best)	88

Source: Global Information Technology Report for Nigeria (2015, p. 213)

The NRI index for the 4th pillar—affordability indicates that prepaid mobile cellular tariffs for PPP $/minute shows a great improvement. This means that the population is willing to pay as high as one United State Dollar per minute tariff. Nigeria ranked 118 for its fixed broadband Internet tariffs for PPP $/month. The result revealed that the fixed broadband

Internet tariffs for PPP $/month is low. On Internet and telephony competition Nigeria was ranked 1in a 0–2 (best) rating. This means that Nigeria is comparatively competing moderately fine with its counterparts in the west because Nigeria is one of the fastest emergent telecommunications market in the world.

The 5th NRI indices revealed astonishing results when Nigeria ranked 121st of the 143 countries surveyed based on quality of educational system. This is in the same way true when no Nigerian university was ranked amongst the 1000 best universities in the world. Nigeria was ranked 132nd of the 143 countries examined on quality of math and science education. This result did no portray Nigeria in good light. On secondary education gross enrollment rate per percentage of the population placed Nigeria at a distant 125th position. The adult literacy rate per percentage of the population of Nigeria positioned her in 108th out of the 143 countries surveyed. This low ranking could be deduced that the low ranking is as a result of Nigeria's low ICT penetration.

The Readiness Subindex: 3rd pillar: Infrastructure, 4th pillar: Affordability and 5th pillar: Skills are influence by vandalisation of ICT infrastructure, insecurity and scarce electrical power supply for operations, local participation in key ICT areas, too little skills and enabling environment (Okwuke, 2013) and poverty (Ololube, 2013).

Data for the NRI for the 6th pillar—individual ICT usage revealed that mobile phone subscriptions/100 population placed Nigeria in the 87th position, slightly above half of all the 143 countries surveyed. Individuals using Internet per percent of the population depicts that Nigeria ranked 119th of the 143 countries. Information on households/personal computer revealed that Nigeria was ranked 112th. The fixed broadband Internet subscription/100 population is 140th which means Nigeria is among the last four countries that are low in fixed broadband Internet subscription. Mobile broadband subscription/100 population placed Nigeria in 98th position while on the use of virtual social networks; Nigeria was moderately placed at 82nd position.

The NRI index for the 10th pillar—social impacts portray Nigeria very low in the NRI. Nigeria was ranked 123rd of the 143 countries surveyed on the impact of ICTs on access to basic service. The report rated Nigeria low on Internet access in schools at 111th position, and ICT use and government efficiency in 119th place, while the E-Participation index rated on 0–1 (best), Nigeria was rank 88th. These demining results are nothing to write home about considering Nigeria's position as a major player in world economy and Nigeria's ambition to position itself to be one of the top 20 countries in the world by 2020.

A comparative analysis of three years NRI for selected countries in sub-Saharan Africa and the west revealed that among the sub-Saharan African countries, Nigeria 119th, Gambia 108th, Ghana101st, Senegal 106th, and South Africa 75th, was found by the report to have low NRI, meaning that all the indicators for improved ICT penetration are not well developed compared to the countries in the west who topped the list, Germany, Finland, USA, UK, and Sweden ranked

13th, 2nd, 7th, 8th, and 3rd respectively in the 2015 NRI. (See table 3 and figure 2 for detail). The overall results indicated that given the existing ICT penetration rate, it may take African countries and indeed Nigeria over 100 years to catch up with the west in her drive for enhanced knowledge integration and management.

Table 3: *Networked Readiness Index for Selected Countries.*

Country/Year	2013	2014	2015
Nigeria	113	112	119
Gambia	98	107	108
Ghana	95	96	101
Senegal	107	114	106
South Africa	70	70	75
Germany	13	12	13
Finland	1	1	2
USA	9	7	7
UK	7	9	8
Sweden	3	3	3

Source: Global Information Technology Report for Nigeria (2015)

2.3. ICT Policy Outcomes for National Development

The Federal Executive Council of Nigeria approved a national ICT policy in 2007 with the creation of the National Information Technology Development Agency (NITDA). NITDA was charged with the responsibility of developing and implementing ICT policies (NITDA, 2015). However, voices from several quarters (e.g., Okwuke, 2013; Osuagwu, 2015) are of the view that NITDA's mandate is long overdue for review due to advances in global ICT use, integration and diffusion. NITDA interventions in promoting inclusive development are in policy areas like ICT for Development (ICT4D), National ICT Policy, National Software Policy (NSP), Local Content Guidelines, National e-Government Master Plan, States ICT Policy/Strategic Plan and Open Data Development Initiative (see figure 3) aimed to develop innovative ICT policies to drive nationwide development. NITDA has failed in its ICT action plan and roadmap for the nation, this is evidence from the Global Information Technology Report for Nigeria (2015) because ICT4D and other policies have failed to impel ICT facilities, use and integration in the country.

On another front, the former president of Nigeria (Dr. Goodluck Ebele Jonathan) in 2011 appointed a Minister into the new Nigerian Ministry of Communication Technology. The President's after consultations with ICT stakeholders, set up an Adhoc Committee to harmonize the diverse policies for the different sectors in the ICT landscape including telecommunications. The committee was charged with the responsibility of balancing and harmonizing all existing guiding principle and policies in the ICT sector into a single ICT policy.

Notwithstanding the efforts by the Adhoc Committee and NITDA to drive nationwide development following multiple interventions, facts shows that the inadequate or the nonattendance and deficiencies of well thought policies and the existing situation in the ICT penetration in Nigeria has widened the digital gap between Nigeria a developing countries and the developed west (Okwuke, 2013). Countries with enhanced and increased access to ICT and those who apply ICT extensive in inclusive manner are able to get hold of the advantages of globalization. Conversely countries with laughable ICT infrastructure end up not joining the race for globalization. The truth remains that ICTs provide efficiency in global educational management systems, increase faculty and students' productivity and opens up novel opportunities that drives growth, development and improvement in every sector of the economy, particularly in knowledge integration and management.

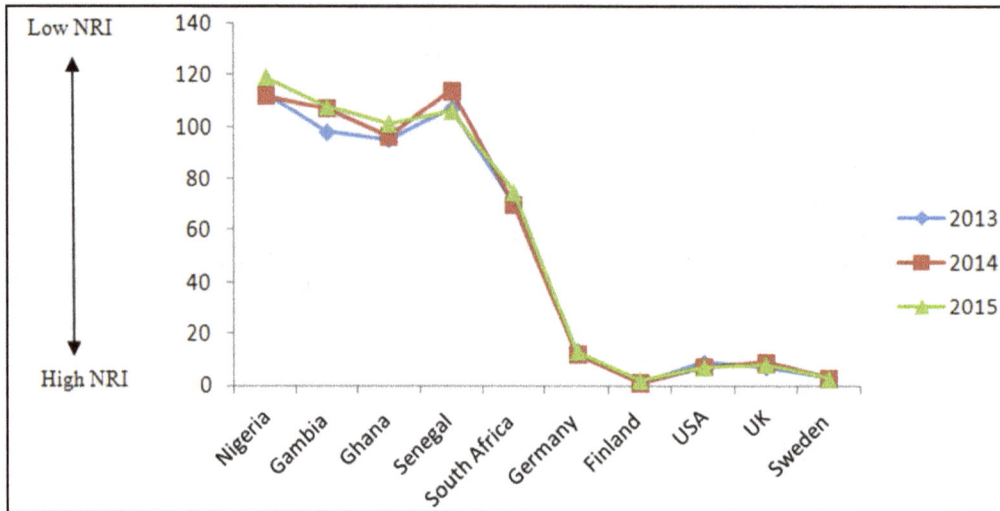

Figure 2: *Line representation of some selected countries NRI for 2013, 2014 and 2015.*

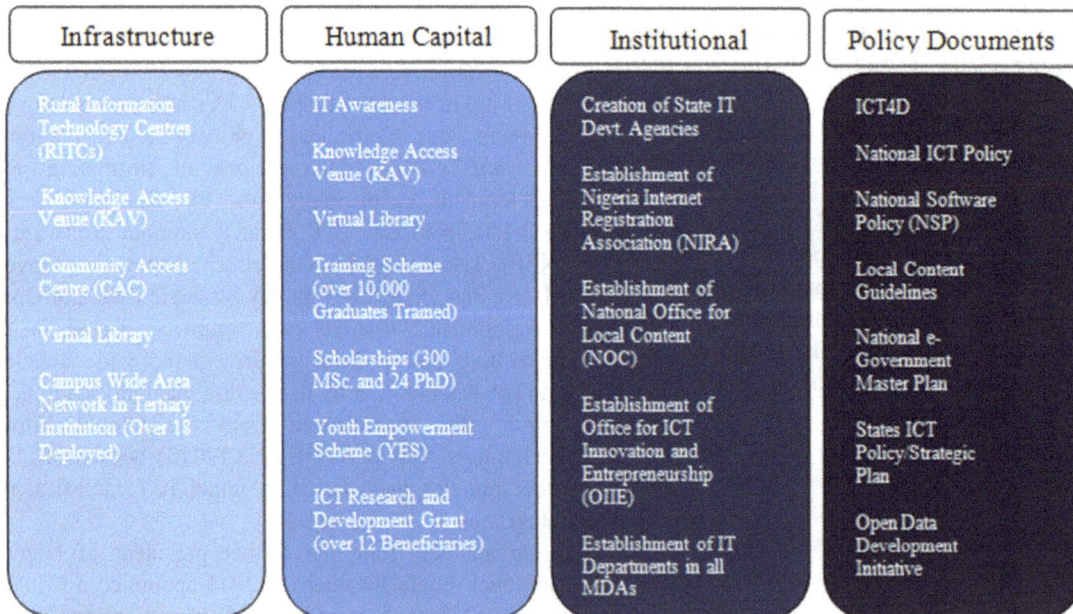

Source: National Information Technology Development Agency (NITDA) (2015)

Figure 3: *NITDA interventions in promoting inclusive development.*

3. Discussion/Conclusions

In this paper, it is evidence that ICT enhances knowledge integration and management in higher education. The several outstanding issues that will bridge the current existing gaps between the requirements and the challenges facing the acceptance and use of ICT to enhance knowledge integration and management in higher education was addressed. This work highlights the relationships between ICT and knowledge integration and management in higher education systems.

This unique study realises that in spite of the unparalleled mobile penetration and the substantial market prospects and potentials, Nigeria is far from being near the digital age if plausible and credible policy measures are not immediately established, because ICT services and products are changing by the day. The effort put so far does not guarantee success;

what matters is the implementation and the speed of incorporating new ICT trends into Nigeria's higher education (Osuagwu, 2015).

According to the Global Information Technology Report (2015) on NRI, less than 40% of the world population enjoys access to the Internet in spite of the fact that more than half of the population now own a mobile phone. Africa is ranked as one of the highest in mobile phone penetration but the report found that the lack of access to Internet is depriving many African higher education institutions the opportunity to take full advantage of blended learning and web-based learning and services to enhance knowledge integration and management for development to strive.

At this point, African and indeed Nigerian higher education industries can make massive strides in connecting more of their students, academic and non-academic staff to bring about positive knowledge integration by improving its Networked Readiness Indices. Government policies and the political and regulatory environment will show commitment to bring the benefit of ICTs to our knowledge institutions. Africa and indeed Nigeria needs to prioritise ICT usage, knowledge, skills, and development if it has to benefit from the new knowledge economy and the competences needed to stimulate national development.

The 2015 Global Information Technology report suggests that investment in ICT and education infrastructure; including creating and promoting an enabling environment and competition through sound higher education policies and regulation will correct the imbalance in Nigeria's academic landscape in Higher education. Therefore, genuine policies and strategies that will strengthen digital transformation are direly needed at this stage of our national development.

Not until Nigeria formulates enabling policies, regulations and laws that will improve access to ICT infrastructure in all spheres of life. The building of an ever-present broadband infrastructure, supporting device ownership for all Nigerians, encouraging local content development, stimulating transparency, efficiency and productivity in government and citizen engagement, Nigeria's aim of multiple policies will go to waste.

References

[1] Agbeja, O., & Fajemisin, D. O. (2008). Knowledge Management: Strategy for Corporate Survival and Sustainable Global Development. *Journal of Knowledge Management Practice, 9*(2). Retrieved August 28, 2015 from http://www.tlainc.com/articl157.htm.

[2] Akuegwu, B. A., Anijaobi-Idem, F. N., & Ekanem, E. E. (2011). Higher Institution Students' Access to Information and Communications Technology in Nigeria: Management Imperatives for Labour Market Preparations. *European Journal of Business and Management, 3*(11), 29-40.

[3] Bellinger, G., Castro, D., & Mills, A. (2004) Data, Information, Knowledge and Wisdom. Retrieved July 30, 2015 from http://www.systems–thinking.org/dikw/dikw.htm.

[4] Cárdenas, I., Al-Jibouri, S. H. S., Halman, J. I. M., & van Tol, F. A. (2013) Capturing and integrating knowledge for managing risks in tunnel works. *Risk Analysis, 33*(1), 92–108.

[5] Department of Education and Skills (2011). National Strategy for Higher Education to 2030 - Report of the Strategy Group. Retrieved August 18, 2015 from http://www.hea.ie/sites/default/files/national_strategy_for_higher_education_2030.pdf.

[6] Federal Republic of Nigeria (FRN) (2013). *National Policy on Education 6th edition*. Lagos: NERC Press.

[7] Global Information Technology Report (2015). Networked Readiness Index. Retrieved August 30, 2015 from http://reports.weforum.org/global-information-technology-report-2015/downloads/.

[8] Ifinedo, P. & Ololube, N. P. (2007). A Discourse on the Problems, Prospects, and Progress of Distance Education in a Developing Country. In E. P. Bailey (Ed.) *Focus on Distance Education Developments* (pp. 183-194). New York, NY: Nova Science Publishers.

[9] Ifinedo, P. (2005). Measuring Africa's e-readiness in the global networked economy: A nine country data analysis. *International Journal of Education and Development using Information and Communication Technology, 1*(1), 53-71.

[10] Ifinedo, P. (2006). Acceptance and Continuance Intention of Web-Based Learning Technologies (WLT) among University Students in a Baltic Country. *The Journal of Information Systems in Developing Countries, 23*(6), 1-20.

[11] Johnson, S. O. (2007). Enhancing quality in higher education through information and communication technology in Nigeria. In J. B. Babalola, G. O. Akpa, A. O. Ayeni, & S. O. Adedeji (Eds.) *Access, Equity and Quality in Higher Education* (pp. 505-512). Ibadan: NAEAP Publication.

[12] Lopez, V. (2003). An exploration of the use of information technologies in the college classroom. *College Quarterly, 6*(1). Retrieved August 15, 2015 from http://www.collegequarterly.ca/2003-volo6-num01-fall/lopes.html.

[13] Mac-Ikemenjima, D. (2005). *e-Education in Nigeria: Challenges and Prospects*. Paper Presented at the 8th UN ICT Task Force Meeting April 13-15, 2005 Dublin, Ireland.

[14] Meek, V. L., Teichler, U., & Kearney, M. (Eds.) (2009). Higher Education, Research and Innovation: Changing Dynamics. Report on the UNESCO Forum on Higher Education, Research and Knowledge 2001-2009. Kassel: International Centre for Higher Education Research.

[15] Mohanan, K. P. (2005) 'Assessing qualities of teaching in higher education', *Centre for Development of Teaching and Learning (CDTL)*. Retrieved September 12, 2015 from http://www.cdtl.nus.edu.sg/publications/assess/who.htm.

[16] NITDA (2015). NITDA Interventions in promoting Inclusive Development. Retrieved September 12, 2015 from http://www.ncs.org.ng/wp-content/uploads/2015/08/NITDA-Presentation.pdf.

[17] Nwafor, S. O. (2005). Information Technology: A modern tool for the administration of universities in Rivers State. *Nigerian Journal of Educational Administration and Planning, 5*(2), 184-188.

[18] Okwuke, E. (2013). New ICT policy as catalyst for national development. Retrieved August 2, 2015 from http://dailyindependentnig.com/2013/04/new-ict-policy-as-catalyst-for-national-development/.

[19] Oladejo, B. O., & Osofisan, A. O. (2011). A Conceptual Framework for Knowledge Integration in the Context of Decision Making Progress. *African Journal of Computer & ICT, 4*(2), 25-32.

[20] Ololube, N. P. (2009) Computer communication and ICT attitude and anxiety among higher education students. In A. Cartelli & M. Palma (Eds). *Encyclopedia of Information and Communication Technology*, (pp. 100-105). Hershey, PA: Information Science Reference DOI: 10.4018/978-1-59904-845-1.ch014.

[21] Ololube, N. P. (2012). *Sociology of education and society: an interactive approach*. Owerri: SpringField Publishers.

[22] Ololube, N. P. (2013). Evaluating the usage and integration of ITs and ISs in teacher education programs in a sprouting nation. *Mediterranean Journal of Social Sciences, 4*(16), 63-72. DOI:10.5901/mjss.2013.v4n16p63.

[23] Ololube, N. P. (2015). Active Learning Application of Technology Tools and Services to Increase Student Achievement in Online and Blended Learning Environments in Higher Education Institutions. In N. P. Ololube, P. J. Kpolovie, & L. N. Makewa (Eds.), *Handbook of Research on Enhancing Teacher Education with Advanced Instructional Technologies* (pp. 109-129). DOI: 10.4018/978-1-4666-8162-0.ch006.

[24] Ololube, N. P., Kpolovie, P. J., Amaele, S., Amanchukwu, R. N., & Briggs, T. (2013). Digital Natives and Digital Immigrants: A study of Information Technology and Information Systems (IT/IS) Usage between Students and Faculty of Nigerian Universities. *International Journal of Information and Communication Technology Education, 9*(3), 42-64. DOI: 10.4018/jicte.2013070104.

[25] Ololube, N. P., Ubogu, A. E., & Egbezor, D. E. (2007). ICT and Distance Education Programs in a Sub-Saharan African Country: A Theoretical Perspective. *Journal of Information Technology Impact, 7*(3), 181-194.

[26] Omona, W., van der Weide, T., & Lubega, J. (2010). Using ICT to Enhance Knowledge Management in Higher Education: A Conceptual Framework and Research Agenda. *International Journal of Education and Development using Information and Communication Technology, 6*(4), 83-101.

[27] Osuagwu, P. (2015). Nigeria drops in Networked Readiness ranking. Retrieved August 31, 2015 from http://www.vanguardngr.com/2015/06/nigeria-drops-in-networked-readiness-ranking/.

[28] Oyovwe-Tinuoye, G., & Adogbeji, B. O. (2013). Information Communication Technologies (ICT) as an Enhancing Tool in Quality Education for Transformation of Individual and the Nation. *International Journal of Academic Research in Business and Social Sciences, 3*(4), 21-32.

[29] Pena-Bandalaria, M. D. (2007). Impact of ICTs on Open and Distance Learning in a Developing Country Setting: The Philippine experience. *International Review of Research in Open and Distance Learning, 8*(1), 1-15.

[30] Richardson, J. C., Koehler, A. A., Besser, E. D., Caskurlu, S., Lim, J., & Mueller, C. M. (2015). Conceptualizing and Investigating Instructor Presence in Online Learning Environments. *International Review of Research in Open and Distributed Learning, 16*(3), 256-297.

[31] Sridharan, B., Tretiakov, A., & Kinshuk, S. (2004), Application of Ontology to Knowledge Management in Web based Learning. In L. Kinshuk, C. K. Sutinen, E. Sampson, D. Aedo, L. Uden, and E. Kahkonen (Eds.), Proceedings of the 4th IEEE International Conference on Advanced Learning Technologies (pp. 663–665), August 30–Sept 1, 2004, Joensuu, Finland. Los Alamitos, CA: IEEE Computer Society.

8

Performance Analysis of CPU-GPU Cluster Architectures

Ho Khanh Lam

Faculty of Information Technology, Hung Yen University of Technology and Education, Hung Yen, Vietnam

Email address:

lamhokhanh@gmail.com

Abstract: High performance computing (HPC) encompasses advanced computation over parallel processing, enabling faster execution of highly compute intensive tasks such as climate research, molecular modeling, physical simulations, cryptanalysis, geophysical modeling, automotive and aerospace design, financial modeling, data mining and more. High performance simulations require the most efficient compute platforms. The execution time of a given simulation depends upon many factors, such as the number of CPU/GPU cores and their utilization factor and the interconnect performance, efficiency, and scalability. CPU and GPU clusters are one of the most progressive branches in a field of parallel computing and data processing nowadays. GPUs have become increasingly common in supercomputing, serving as accelerators or "co-processors" in every node CPU-GPU to help CPUs get work done faster. In this paper I use the Multiclass Closed Product-Form Queueing Network (MCPFQN) and Mean Value Analysis (MVA) to analyze effects of the CPU-GPU cluster interconnect on the performance of computer systems.

Keywords: CPU-GPU Clusters, Performance, Multiclass Product Form Queueing Network

1. Introduction

Efficient high performance computing systems require high bandwidth, low latency connections between thousands of multiprocessor nodes, as well as high speed storage.

Parallel computing using accelerators has gained widespread research attention in the past few years. In particular, using GPUs (Graphics Processing Units) for general purpose computing (GPGPU) has brought forth several success stories with respect to time taken, cost, power, and other metrics. . In the most recent list of the world's fastest 500 supercomputers, 53 systems used co-processors and 38 of these used Nvidia chips. The second and sixth most powerful supercomputers used Nvidia chips along side CPUs. Intel still dominates, providing processors for 82.4 percent of Top 500 systems. As the price/performance of GPUs has improved, a number of petaflop supercomputers such as Tianhe-I and Nebulae have started to rely on them. However, other systems such as the K computer continue to use conventional processors such as SPARC-based designs and the overall applicability of GPUs in general purpose high performance computing applications has been the subject of debate, in that while a GPU may be tuned to score well on specific benchmarks its overall applicability to everyday algorithms may be limited unless significant effort is spent to tune the application towards it. China Tianhe-1A, an upgraded

supercomputer in 2010, was equipped with every node of 14,336 Xeon X5670 processors and 7,168 NVIDIA Tesla M2050 GPGPUs. It has a theoretical peak performance of 4.701 petaflops. NVIDIA suggests that it would have taken "50,000 CPUs and twice as much floor space to deliver the same performance using CPUs alone." The current heterogeneous system consumes 4.04 megawatts compared to over 12 megawatts had it been built only with CPUs.

The overall most efficient GPU processing model determined thus far is Single Instruction Multiple Data (SIMD). The SIMD model has been leveraged to great advantage in traditional vector processor/supercomputer designs. As displayed in figure 1, the SIMD GPU is nominally organized as an assembly of 'N' distinct multiprocessors ($N_{MP/TPC}$) in every thread processing cluster (TPC). Each multiprocessor per TPC consists of 'M' cores - distinct thread processors. ($M_{TP/MP}$). There are some TPC in one GPU ($N_{TPC/GPU}$). Total number of thread processors (core) per GPU is:

$$N_{Thread/GPU} = N_{TPC/GPU}N_{MP/TPC}M_{TP/MP} \qquad (1)$$

SIMD-Cores share an Instructiuon Unit with other cores in a multiprocessor. Multiprocessors have local registers (splits to local memory), local L1 cache, shared L2 cache, constant cache, texture cache, and shared memory. Constant/texture cache are readonly and have faster access than shared memory. Global/Device memory for GPU (shared for all multiprocessors) is DRAM (DDR3 or DDR5). Threads are organized into blocks, which are organized into a *grid*. A multiprocessor executes one block at a time. A *warp* is the set of threads executed in parallel. The number of thread per a warp may be same as the number of thread processor cores. Programming model of GPU is heterogeneous Computing, where GPU and CPU execute different types of code. CPU runs the main program, sending tasks to the GPU in the form of kernel functions, and multiple kernel functions may be declared and called, but only one kernel may be called at a time. Nvidia developed Compute Unifed Device Architecture (CUDA), a simple language for GPU computing allows a programmer to use the C programming language to code algorithms for executions on the GPU. Using CUDA or OpenCL programming toolkits many real-world applications can be easily implemented and run significantly faster than on multi-processor or multi-core systems. The CUDA framework allows to lower the CPU load and send the computations to the GPU which by construction has a highly parallel architecture, thus accelerating the processing of large amounts of data arrays using an identical set instructions (called kernel) for each element in one data array. In this way the processor is only busy with supplying the data and the kernels to the GPU and then collecting the results. Although the transfer between CPU (host) and GPU (device) might appear as a bottleneck due to the slow speed of the bridge bus. One could obtain very good results if it exposes as much data parallelism as possible in the algorithm and also takes care of mapping it to the hardware as eficiently as possible in such a way that the transfers are minimized.

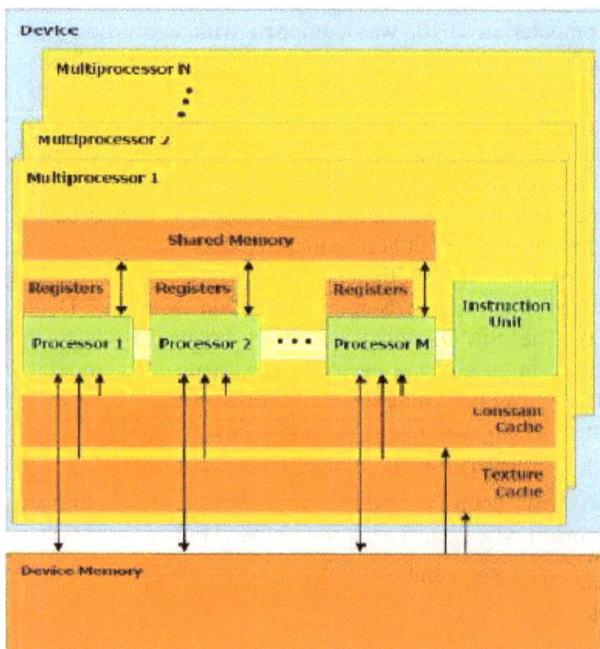

Figure 1. GPU hardware.

2. Choice of the CPU-GPU Cluster

Typical cluster consists of homogenous Central Processing

Units (CPUs). A new model for parallel computing based on using CPUs and GPUs together to perform a general purpose scientific and engineering computing was developed in the last years, and used to solve complex scientific and engineering problems. Craving for more computational power led to the idea of plunging the CUDA framework in a Message Passing Interface (MPI) environment, thus arising the concept of cluser of GPUs. Inside the cluster each node of the MPI network would pass most of intensive parallel tasks to the GPU, off-loading the CPU which is free to handle the network communication between nodes. A cluster is a computer system comprising two or more computers ("nodes") connected with a high-speed network. Cluster computers can achieve higher availability, reliability, and scalability than is possible with an individual computer. There are three principal components used in a GPU cluster: host nodes, GPUs, and interconnect.

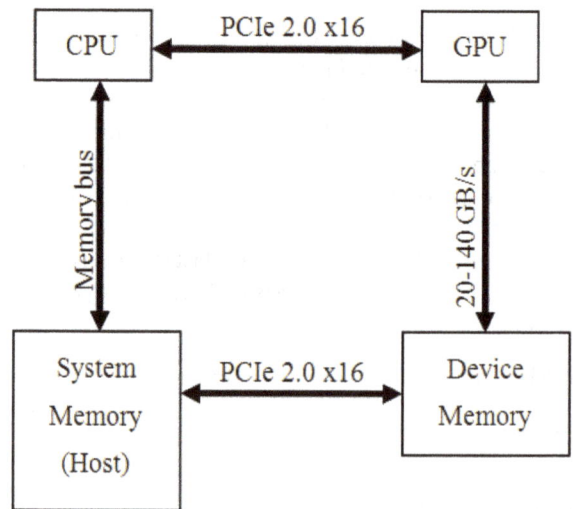

Figure 2. CPU/GPU interconnect.

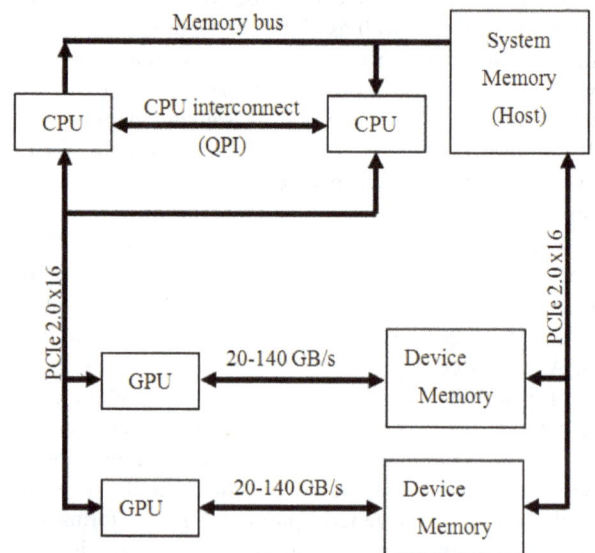

Figure 3. 2CPU/2GPU interconnect.

Since the expectation is for the GPUs to carry out a substantial portion of the calculations, host memory, PCIe bus, and network interconnect performance characteristics need to be matched with the GPU performance in order to maintain a well-balanced system. In particular, high-end GPUs, such as the NVIDIA Tesla, require full-bandwidth PCIe Gen 2 x16 slots that do not degrade to x8 speeds when multiple GPUs are used. Also, InfiniBand QDR interconnect is highly desirable to match the GPU-to-host bandwidth. Host memory also needs to at least match the amount of memory on the GPUs in order to enable their full utilization, and a one-to-one ratio of CPU cores to GPUs may be desirable from the software development perspective as it greatly simplifies the development of MPI-based applications.

The figures 2 and 3 show typical CPU/GPU interconnect with 1 and 2 CPUs and 2GPU. The figure 4 shows a CPU/GPU desktop supercomputer organized in clusters.

The System interconnect (cluster interconnect) is 1GbE/10GbE Ethernet or InfiniBand switch. Every node in a cluster consist of one/or two multicore-multithreading CPU chips on two sockets with high speed CPU interconnect, e.g. used Intel QuickPath Interconnect – QPI with a clock rate of 3.2 GHz yields a data rate of 25.6 GB/s = (3.2 GHz x 2 bits/Hz (double data rate) x 20 (QPI link width) x (64/80) (data bits/flit bits) x 2 (unidirectional send and receive operating simultaneously))/8 (bits/byte), or AMD HyperTransport bus with a clock rate 3.2 GHz yields a data rate of 25.6 GB/s = (3.2 GHz x 2 bits/Hz (double data rate) x 32 bits/s))/8 bits/byte; two/or three multiprocessor-multithreading GPU installed on PCI Express 2 x16 slots (for all sorts of graphics cards, can access a total bandwidth of 8 GB/s). The Cluster Interconnect is QDR (40 Gbps) InfiniBand connectivity on each node in one cluster. An Infiniband link is a serial link operating at one of five data rates: single data rate (SDR) switch chips have a latency of 200 ns (12X: 24 Gbit/s), double date rate (DDR) switch chips have a latency of 140 ns (12X: 48 Gbit/s and quad data rate (QDR) switch chips have a latency of 100 ns (12X: 96 Gbit/s), fourteen data rate (FDR-10, FDR) (12X: 163.64 Gbit/s), and enhanced data rated (EDR) (12X: 300 Gbit/s). Larger systems with 12X links are typically used for cluster and supercomputer interconnects and for inter-switch connections. For cluster interconnect, the InfiniBand uses a switched fabric topology, as opposed to a hierarchical switched network like traditional Ethernet architectures, although emerging Ethernet fabric architectures propose many benefits which could see Ethernet replace InfiniBand. Most of the cluster interconnect topologies are Fat-Tree, 2D-mesh/2D-Torus or 3D-Torus, butterfly (Clos) as well.

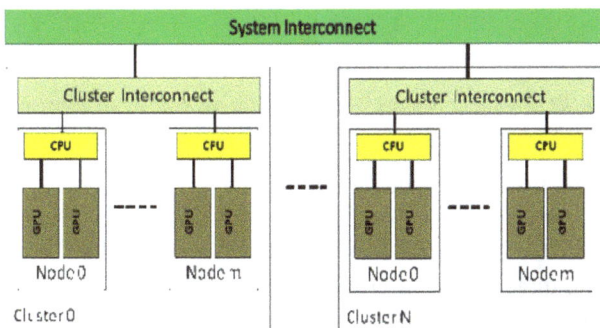

Figure 4. *CPU/GPU supercomputer organized in clusters.*

In this context, multiprocessor operation is defined modulo an ensemble of threads scheduled (by hardware scheduler) and managed as a single entity across thread processing clusters. In this manner, shared-memory access, SIMD instruction fetch and execution, and cache operations are maximally synchronized. Memory is organized hierarchically *Global -> Device -> Shared* where Global/Device memory transactions are understood as mediated by high-speed bus transactions (PCIe, HyperTransport, or QPI). A key subtlety associated with the CPU/GPU processing architecture is GPU processing is effectively *non-blocking*. Thus, CPU processing may continue as soon as a work-unit has been written to the GPU transaction buffer. Host (CPU) processing and GPU processing may be overlapped as displayed in figure 1. In principle, GPU work unit assembly/disassembly and I/O at the GPU transaction buffer may to large extent be hidden. In such case, GPU performance will effectively dominate system performance. As might be expected, optimal GPU processing gain is achieved at an I/O constraint boundary whereby thread processors never stall due to lack of data.

At an application level, the maximum achievable speedup is governed by Amdahl's Law; any acceleration ('*A*') due to thread parallelization will critically depend upon: (1) the fraction of code than can be parallelized ('*P*'), (2) the degree of parallelization ('*N*'), and (3) any overhead associated with parallelization. Thus, expected acceleration is modeled by:

$$A = \frac{1}{(1-P) + P/N} \qquad (2)$$

A key consideration is the limiting case:

$$\lim_{N \to \infty} \left(\frac{1}{(1-P) + P/N} = \frac{1}{1-P} \right) \qquad (3)$$

This indicates a theoretical maximum acceleration for the complete application. However, CPU code pipelining, (i.e. overlap with GPU processing), must also be factored into any calculation for '*P*'; pipelining effectively parallelizes CPU and GPU code segments reducing the non-parallelized code fraction (1-*P*). Thus, under circumstances where decrease is sufficient to claim $\frac{P}{N} \gg (1-P)$, Amdahl's Law then becomes:

$$A \cong \frac{1}{\frac{P}{N}} = \frac{N}{P} \cong N \qquad (4)$$

CPU/GPU-based desktop supercomputing model is cluster architecture (figure 1), that characteristic acceleration values approaching a limit:

$$A_{Cluster} \cong N_{Node/CL} N_{Thread/GPU} N_{GPU} \qquad (5)$$

$N_{Node/CL}$ - number of nodes per cluster,

N_{GPU} - number of GPUs in the cluster .

Total acceleration of the system with N_{CL} clusters:

$$A_{Supercomputer} \cong (N_{Node} N_{Thread/GPU} N_{GPU}) N_{CL} \qquad (6)$$

3. The MCPFQN of the CPU/GPU Node

Queueing networks [1] consisting of several service stations are more suitable for representing the structure of many systems with a large number of resources than models consisting of a single service station. In a queueing network (QN) at least two service stations are connected to each other. A station, i.e., a node, in the network represents a resource in the real system. Jobs in principle can be transferred between any two nodes of the network; in particular, a job can be directly returned to the node it has just left. The QN is called open when jobs can enter the network from outside and jobs can also leave the network. Jobs can arrive from outside the network at every node and depart from the network from any node. The QN is said to be closed when jobs can neither enter nor leave the network. The number of jobs in a closed network is constant. Jobs can be different in their service times and in their routing probabilities. Jobs with same service times and routing probabilities belong to one class. So QNs can be single class or multiclass networks (MCQN). It is also possible that a job changes its class when it moves from one node to another. If the QN contains both open and closed classes, then it is said to be the mixed network. Behaviours of many queueing system models can be described using Continuous –time Markov chains (CTMCs). The QNs that have an unambiguous solution of the local balance equations are called product-form queueing networks (PFQNs). The term product-form of open and closed queueing networks with exponentially distributed interarrival and service times. The queueing discipline at all stations was assumed to be FCFS. As the most important result for the queueing theory, it is shown that for these networks the solution for the steady-state probabilities can be expressed as a product of factors describing the state of each node. This solution is called product-form solution. For the performance analysis of CPU/GPU architectures, we can use MCPFQN with fixed number of jobs, which are threads.

Figure 5 shows the p_{MC21} MCPFQN model of the CPU/GPU architecture in figure 3. 2 CPUs are 2-core multithreading: CPU 1 server node (include private L1 and L2 caches) with: CPU core 11 and CPU core 21, CPU 2 server node: CPU core 21 and CPU core 22; every threads are modeled by thread processor with service rate μ_T. CPUint node with service rate μ_{int} is the CPU interconnect (e.g. Intel QPI or AMD HyperTransport bus). 2 GPUs (GPU 1 and GPU 2 nodes include caches, have local thread processors with service rate μ_{TP}) are installed on the PCIe 2.0 x 16 slots (node CPU-GPU int with service rate μ_{PCi}) of CPU mainboard. The Memory bus (via MCH chip of the chipset) is the node Memory bus with service rate μ_{MB}. Memory modules are nodes System Memory with service rate μ_M. Every GPU connect to the local Device Memory modules (with service rate μ_{DM}) via interconnect bus (nodes GPU1-Device Memory, GPU2-Device Memory) with service rate μ_{GM}.

The connection between System memory and Device Memories may be PCIe 2.0 x16 or Infiniband (node DM-SM int with service rate μ_{MG}). Nodes CPU and GPU are type M/G/m-PS with service disciplines: FCFS, all other nodes are type M/M/m – FCFS.

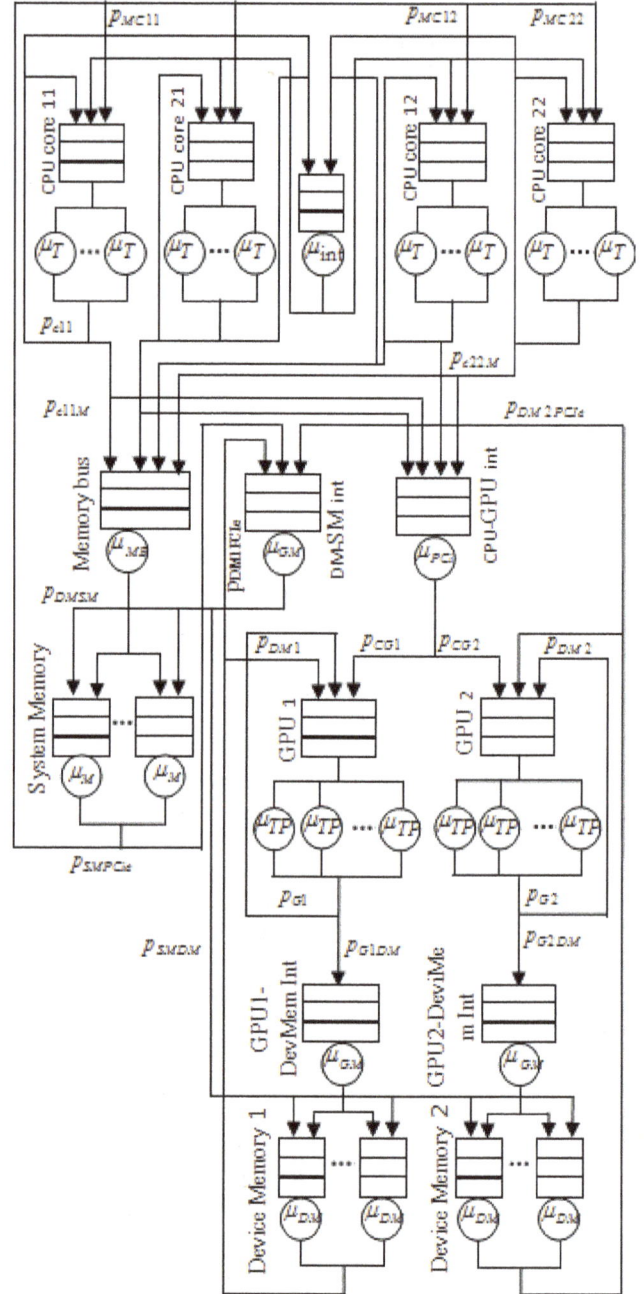

Figure 5. *Closed queue network of 2CPU/2GPU desktop supercomputer model in figure 3.*

Assume that: GPUs and CPUs (include private L1 and L2 caches) work in 2GHz have average service time for every tread is 0.5ns ($\mu_T = \mu_{TP} = 0.5ns$). From every CPU there are 4 routing directions with probabilities (their sum = 1): core loopback ($p_{c11}, p_{c21}, p_{c12}, p_{c22}$), to other core in one CPU (p_{c11c21}, p_{c12c22}), to other CPU

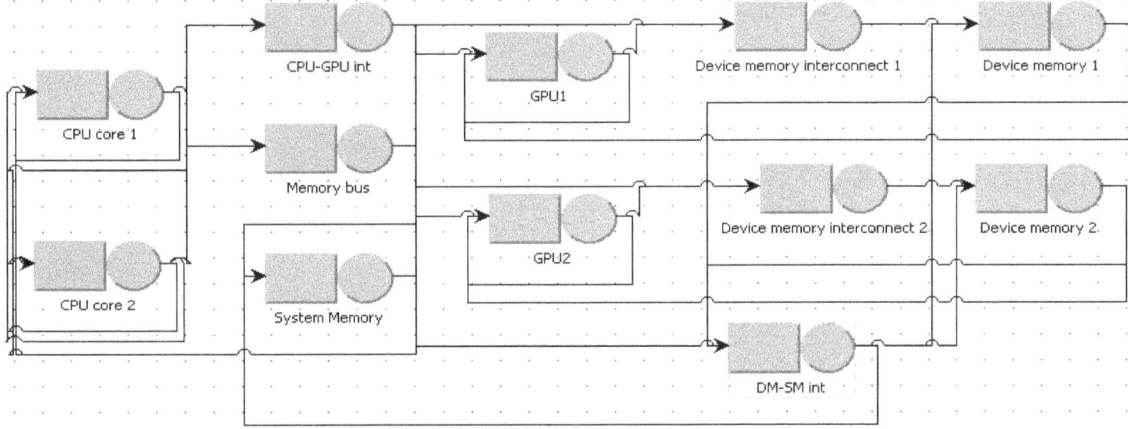

Figure 6. *MCPFQN of the CPU-2-core-4-thread/2GPU-448-core desktop architecture. $(p_{c11int}, p_{c21int}, p_{c12int}, p_{c22int})$ and to System memory $(p_{c11M}, p_{c21M}, p_{c12M}, p_{c22M})$.*

From every GPU there are 2 routing flows: GPU loopback (p_{G1}, p_{G2}) and to Device memories (p_{G1DM}, p_{G2DM}).

From CPUs to GPUs (via node CPU-GPU int) there are 2 flows: to GPU1 (p_{CG1}), to GPU2 (p_{CG2}).

From GPU Device memories there are 2 routing flows: loopback to GPU (p_{DM1}, p_{DM2}) to bus connect (node DM-SM int) to System memory $(p_{DM1PCIe}, p_{DM2PCIe})$.

From node DM-SM int there are 2 routing flows: to System memory (p_{DMSM}), to Device memory (p_{SMDM}).

From System memory there are 2 routing flows: to CPU cores $(p_{MC11}, p_{Mc12}, p_{MC21}, p_{MC22})$.

For QN, most important performance measures are:

- Marginal probabilities $\pi_i(k)$ for closed queueing networks, that ith node in the state $S_i = k$, is :

$$\pi_i(k) = \sum_{\substack{\sum_{j=1}^{N} S_j = K \\ \& S_i = k}} \pi(k_1, ..., k_N) \qquad (7)$$

- Utilization ρ_{ir} of the ith node with respect to jobs of the rth class is:

$$\rho_{ir} = \frac{1}{m_i} \sum_{\substack{all\ states\ k \\ with\ k_r > 0}} \pi_i(k) \frac{k_{ir}}{k_i} \min(m_i, k_i), \ k_i = \sum_{r=1}^{R} k_{ir} \qquad (8)$$

And if the service rates are independent on the load:

$$\rho_{ir} = \frac{\lambda_{ir}}{m_i \mu_{ir}}. \qquad (9)$$

Throughput λ_{ir} is the rate at which jobs of the rth class are services and leave the ith node:

$$\lambda_{ir} = \sum_{\substack{all\ states\ k \\ with\ k_r > 0}} \pi_i(k) \frac{k_{ir}}{k_i} \mu_i(k_i) \qquad (10)$$

Or if the service rates are independent on the load:
$\lambda_{ir} = m_i . \rho_{ir} . \mu_{ir}$.

- System throughput λ_r of jobs of the rth class:

$$\lambda_r = \frac{\lambda_{ir}}{e_{ir}} \qquad (11)$$

Mean number of Jobs (or customer number) $\overline{K}_{ir}$ of the rth class at the ith node is:

$$\overline{K}_{ir} = \sum_{\substack{all\ states\ k \\ with\ k_r > 0}} k_r . \pi_i(k) \qquad (12)$$

Little's theorem can also be used here: $\overline{K}_{ir} = \lambda_{ir} . \overline{T}_{ir}$

Mean Response Time $\overline{T}_{ir}$ of the jobs of the rth class at the ith can also be determined using Little's theorem:

$$\overline{T}_{ir} = \frac{\overline{K}_{ir}}{\lambda_{ir}} \qquad (13)$$

- Mean Waiting Time $\overline{W}_{ir}$: if the service rates are load-independent, then the mean waiting time is given:

$$\overline{W}_{ir} = \overline{T}_{ir} - \frac{1}{\mu_{ir}} \qquad (14)$$

For analysis, performance parameters are: Number of customers, Response Time, Utilization, Throughput, and System Throughput in relation with service times at interconnect networks (e.g. PCIe 2.0 x16) between CPU and GPU (node CPU-GPU int), and between System memory and Device memories (node DM-SM int). The analysis is made for architectures: 2-core/4-threading CPU, 2 GPUs with 448 thread cores in each, number of jobs: 448, and exponential distribution service time, $f(t) = \lambda e^{-\lambda t}$ of interconnects (CPU-GPU int) and DM-SM int in figure 6) for three cases: i) mean: 0.25, $\lambda = 4$; ii) mean: 1, $\lambda = 1$; iii) mean: 4, $\lambda = 0.25$. Results of Performance parameters are given in figures 7. For the performane analysis, I take only response times and system throughput of three cases.

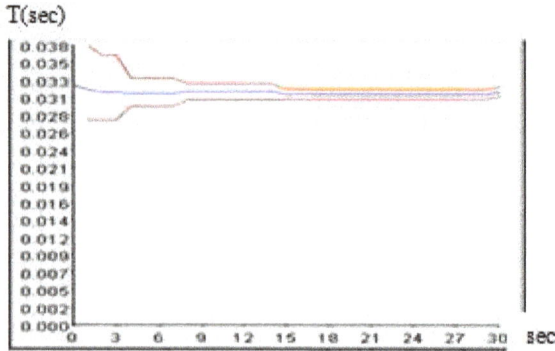

Figure 7(a). *System Response Time T (seconds): mean: 4167.3293, min:4047.4249, max:4287.2338 in case i) Secvice Time distribution of CPU-CPU int=DM-SM int: mean:0.25. $\lambda = 4$.*

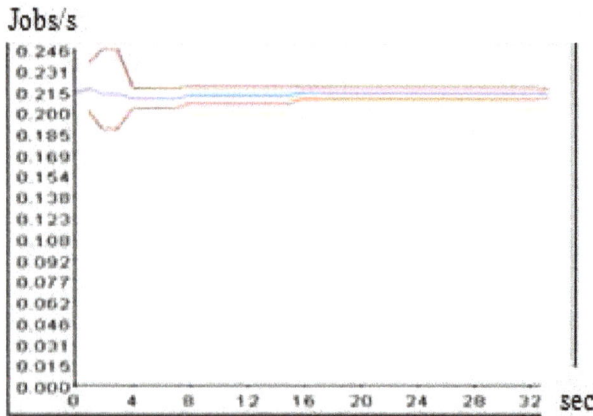

Figure 7(b). *System Throughput (jobs/sec): mean: 0.2143, min:0.2108, max:0.2178 in case i) Secvice Time distribution of CPU-CPU int=DM-SM int: mean:0.25. $\lambda = 4$.*

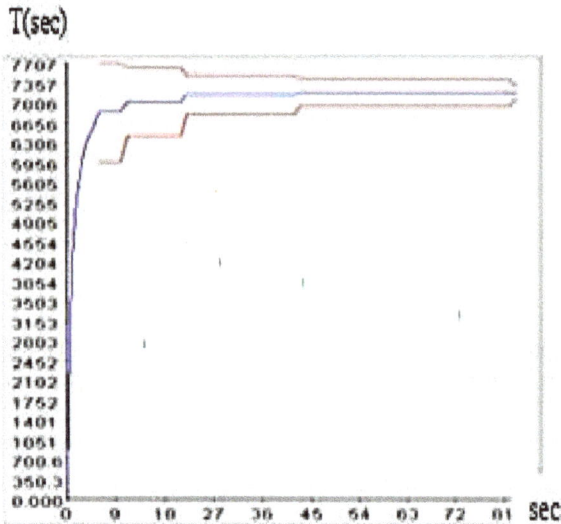

Figure 7(c). *System Response Time T (seconds): mean: 7188.6417, min:7073.0428, max:7304.2406 in case i) Secvice Time distribution of CPU-CPU int=DM-SM int: mean:1. $\lambda = 1$.*

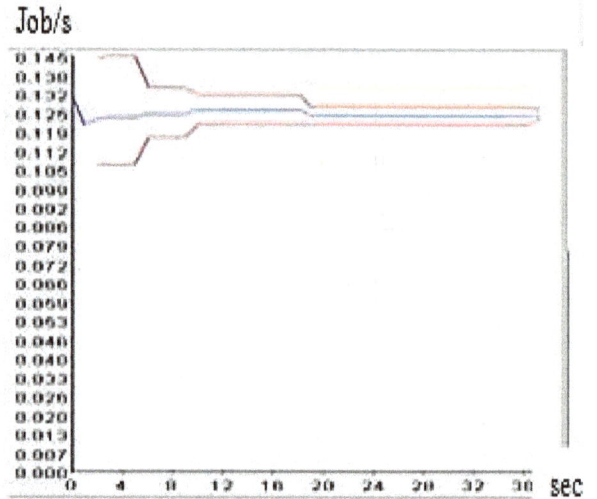

Figure 7(d). *System Throughput (jobs/sec): mean: 0.1244, min:0.1223, max:0.1265 in case i) Secvice Time distribution of CPU-CPU int=DM-SM int: mean:1, $\lambda = 1$.*

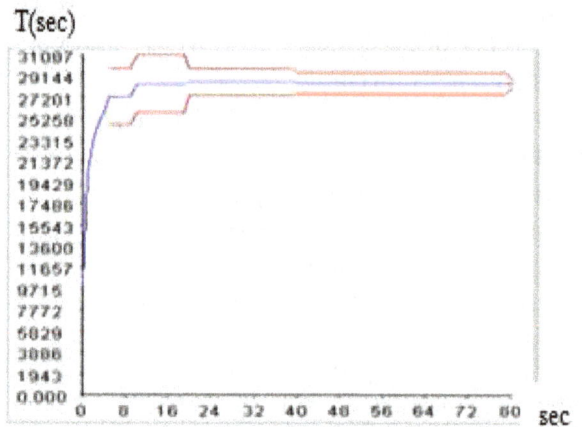

Figure 7(e). *System Response Time: mean: 2.853E4, min:2.807E4, max:2.900E4 in case iii) Secvice Time CPU-CPU int=DM-SM int: mean:4, $\lambda = 0.25$*

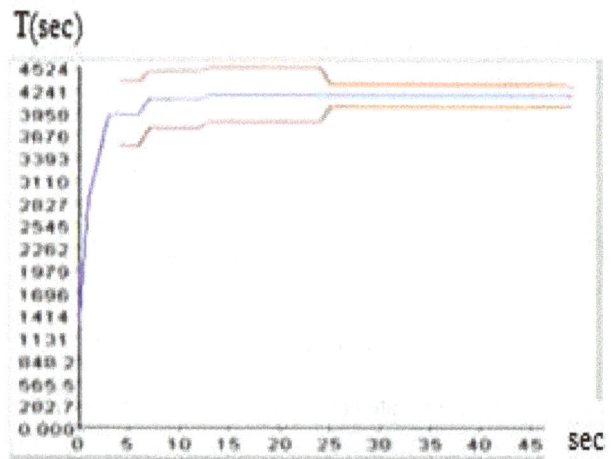

Figure 7(f). *System Throughput: mean: 0.0313, min:0.0307, max:0.0320 in case iii) Secvice Time distribution of CPU-CPU int=DM-SM int: mean:4, $\lambda = 0.25$*

Table 1. *Avarage service times of queue nodes.*

	Class 1	Class 2	Class 3	Class 4	Class 5
CPU core 1	1.0	0.5	0.25	0.15	0.1
CPU core 2	1.0	0.5	0.25	0.15	0.1
GPU 1	1.0	0.5	0.25	0.15	0.1
GPU 2	1.0	0.5	0.25	0.15	0.1
CPU-GPU int	4.0	2.0	1.0	0.5	0.25
GPU1-DevMem	2.0	1.0	0.5	0.25	0.2
GPU2-DevMem	2.0	1.0	0.5	0.25	0.2
Device memory 1	40.0	35.0	30.0	25.0	20.0
Device memory 2	40.0	35.0	30.0	25.0	20.0
Memory bus	2.0	1.0	0.5	0.25	0.2
System memory	40.0	35.0	30.0	25.0	20.0
DM-SM int	4.0	2.0	1.0	0.5	0.25

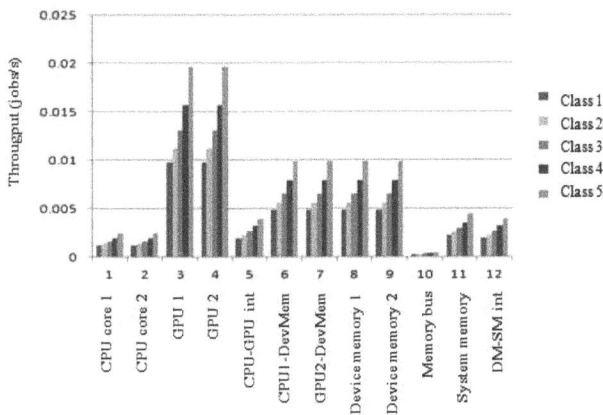

Figure 8. *Average service times for the MVA algorithm.*

Table 2. *Throughput of the CPU-GPU node.*

	Class 1	Class 2	Class 3	Class 4	Class 5
CPU core 1	0.0012	0.0014	0.0016	0.0020	0.0025
CPU core 2	0.0012	0.0014	0.0016	0.0020	0.0025
GPU 1	0.0098	0.0112	0.0131	0.0158	0.0197
GPU 2	0.0098	0.0112	0.0131	0.0158	0.0197
CPU-GPU int	0.0020	0.0023	0.0026	0.0032	0.0039
GPU1-DevMem	0.0049	0.0056	0.0066	0.0079	0.0098
GPU2-DevMem	0.0049	0.0056	0.0066	0.0079	0.0098
Device memory 1	0.0049	0.0056	0.0066	0.0079	0.0098
Device memory 2	0.0049	0.0056	0.0066	0.0079	0.0098
Memory bus	2.5E-4	2.8E-4	3.3E-4	4.0E-4	4.9E-4
System memory	0.00227	0.0026	0.0030	0.0035	0.0044
DM-SM int	0.00201	0.0023	0.0026	0.0032	0.0039

Based on the MCPFQN in the figure 6, we can use Mean Analysis (MVA) algorithm [1] for calculating performance measures. For this case, I define CPU-GPU node: CPU core 1 and CPU core 2 by 6-core, GPU 1 and GPU 2 with 512 cores. Average service times (ns) of all nodes and five classes are listed in table 1 and in figure 8. The throughput (jobs/s) of the CPU-GPU node with the initial setting 12 jobs (number of customers) in CPU core 1 and CPU core 2 nodes for all classes are listed in table 2 and in figure 9.

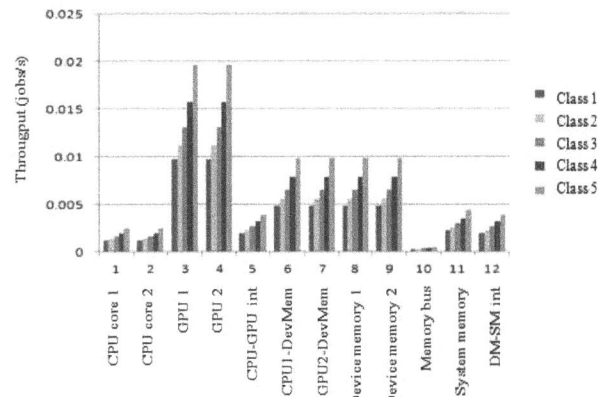

Figure 9. *Throughput of CPU-GPU node is taken by the MVA algorithm.*

4. Conclusion

The performance rerults in figures 7 and figure 9 of MCPFQN model of the CPU-GPU node: system response times T (sec) and system throughput (jobs/s) show that: The changes of interconnect service times affect deeply system performance parameters: system response time and throughput that belong to service times of CPU-GPU interconnect and GPU device memory-CPU system memory interconnect, in fixed average service rates of all other nodes. Clearly, more service time of interconnects, more system response time and less system throughput. Using MCPFQN, we can evaluate system performance depend on any other node or complet CPU+GPU desktop supercomputers with many cluster and diffrent cluster-to-cluster interconnect network topologies and delays.

References

[1] Gunter Bolch, Stefan Greiner, Hermann de Meer, Kishor S.Trivedi: Queueing Networks and Markov Chains; A John Wiley & sons, Inc., Publication.

[2] Peng Wang, NVIDIA. Fundamental Optimizations in CUDA, GPU technology conference.

[3] Bryan Schauer. "Multicore Processors-A Necessity", 9/2008.

[4] John Mellor-Crummey, Department of Computer Science Rice University: Caching for Chip Multiprocessor, 8/2009.

[5] Sarah Bird, …University of Texas at Austin, IBM Austin: Performance Characterization of SPEC CPU Benchmarks on Intel's Core Microarchitecture based processor. 2006.

[6] R.Ubal, J.Sahuquillo,..:Multi2Sim. A Simulation Framework to Evaluate Multicore-Multithreaded Processors, 2006.

[7] W.M.Zuberek. Performance eqivalence in the simulation of Multiprocessor systems, 2002.

[8] Scott.T.Lentenegger, Mary K.Vernon. A mean-Value performance Analysis of a New Multiprocessor Architecture, 12/1988.

[9] Angel Vassilev Nikolov, National University of Lesotho, 180, Roma, May,2009. Model of a shared Memory Multiprocessor.

[10] Susmit Biswas, Diana Franklin,,,2008. Multi-Execution. Multicore Caching for Data- Similar Executions.

[11] Intel Multicore microprocessor technology. http://www.Intel.ccom/.

[12] Rafael H. Saavedra-Barrera, David E.Culler. An analytical solution for a markov chain modelling multithreaded execution.

[13] Michael R.Marty, University of Wisconsin-madison, 2008. Cache coherence techniques for multicore processors.

[14] Richard Mcdougall and James Laudon. Multi-core microprocessors are here.

[15] Avinatan Hassidim, 16/09/2009. Cache replacement Policies for Multicore processors.

9

On the Development Robust and Fast Algorithm of Action and Identity Recognition

Khawlah Hussein Alhamzah[1], Tianjiang Wang[2]

[1]Computer Science Department, Basrah University, Basrah, Iraq
[2]School of Computer Science, HUST University, Wuhan, China

Email address:
Khawlahussein@yahoo.com (K. H. Alhamzah)

Abstract: Human action recognition and surveillance applications are playing a key important in the present days and took an increasing interest in modern. Since most previous methods strictly limited to action classification in different scenarios and not take attention to human identity that makes an action at the same time. We present a novel and fast algorithm to recognize action and identity in a single framework. We assumed one person makes one action in a video. To identify and training the owner of the video to the classifier, we proposed the watermark embedded as 2-D wavelet transform as binary image, which is contains identity information in the training video. We used these wavelet coefficients as identity descriptors. To represent feature motion representation, we used motion energy image (MEI) and motion history image (MHI) as temporal template of the human actions and Zernike moments to extract shape features of the action from MEI and MHI. In this research, a set of Zernike moment based feature vectors is proposed for human action recognition, which is capture the global properties of an object rather than the local ones. We have composed two different feature vectors by evaluating the variance values of lower order Zernike moments in the four-dimensional Zernike moment space with encouraging experimental results. It has discriminative information that is suitable for classification, especially on related actions, such as running and jogging, that is most previous researches fail to classify them even human vision HVS. Nearest neighbor classifier is used for action and identity categorization. The result of these experiments suggests that this method has a high recognition rate in both action and identity accuracy on KTH data sets.

Keywords: Action Recognition, Digital Watermarking, 2-D Wavelet Transform, MEI and MHI, 2-D Zernike Moments, Nearest Neighbor Classifier

1. Introduction

Human action recognition is an important field in computer vision. The security and surveillance are playing a key role in the present day and took an increasing interest in modern society, especially in light of the challenges facing the world. Therefore, biometrics, such as face, fingerprint and iris cannot be used unless the distance is close and if there is cooperation by the user [1-3]. Action recognition and identity from distance platforms such as surveillance camera is a difficult task. However, previous techniques that heavily relied on distance image face significant difficulties. On the other hand, object identity recognition and object tracking can be closely related problems. In a tracking scenario objects can be represented by their shapes and appearances. However, in tracking, objects are usually considered to have small displacements between observations; Therefore, most tracking techniques such as Mean Shift and Kalman filter-based tracking make use of this information and search for the tracked objects within small spatial variation limits. Such techniques have proved their efficiency in continuous scenes where disappearances and clutters are minor. However, in the cases of long occlusions, tracking performance considerably decays and it even becomes totally inefficient when discontinuities are inherent in the video. Most of researches focus on gait recognition, which is define the style of people that they walk, but this is face difficulties, since people may change their walking depend on external situations. In this research, the feature vectors based Zernike moments are proposed for an action recognition system. From statistics view, we select four lower order Zernike moments with the highest variance values to form a set of moment feature vectors in the four dimensional Zernike moment space. To test the proposed algorithm, we used a set of KTH video frames. The rest of the paper is organized as follows: In the next section, we present an overview of the related works. Section three; we describe the proposed method of single framework. Section four illustrates the results and discusses the experiments. Finally, section five presents the conclusions.

2. Related Works

Over the past two decades, many studies and searches of action recognition in different scenarios. Some of the current state-of-the-art solutions are either inaccurate or computationally intensive while others require human intervention. Several techniques have been developed in the recent years mainly based on the use of local descriptions of the video frames [1-3]. Following the success of SIFT [4] and HOG in image and scene recognition and classification, several methods of space-time extensions have been proposed. Videos are represented as a collection of descriptors or bag of visual features. Space-time descriptors represent the appearance and the motion features. Laptev et al. [5] have defined a descriptor as a concatenation of histograms of oriented 2D gradients and histograms of optical flow. In order to reduce the computation burden extensions of SURF have been presented in [6]. Scovanner et al. [7] Extended the SIFT to three-dimensional gradients normalizing 3D orientations bins. Klaser et al. [8] proposed to exploit 3D pixel gradients developing a technique based on Platonic solids. Finally Ballan et al. [9] developed an efficient descriptor, describing the spatial and temporal components and creating separated histograms of 3D gradient orientations. Moussa et al. [10] proposed an enhanced method for human action recognition by using fine tuning to reduce the number of interest point detectors that produced from SIFT algorithm, they get high accuracy of action recognition. Khawlah et al. [11] proposed identity as watermark with SIFT descriptors. However all of these descriptors are extremely high-dimensional, in terms of time and space complexity and often retain redundant information. Bobick and Davis [12] present a human action recognition algorithm with two temporal templates, named motion energy image (MEI) and motion history image (MHI).In the same time, researchers have exploited moments and invariant moments in pattern recognition [12]. Moments are scalar quantities used to characterize a function and to capture its significant features and they have been widely used for hundreds of years in statistics for description of the shape of a probability density function.

The first attempt by Hu [13], which is used moments for image processing and analysis of seven order moments, that is robust to scaling, transformation. The recent papers in [14] Khan et al. used Hu invariant moments and neural networks to train and recognize different actions, the accuracy is high. Besides Hu moment invariants are sensitive to noise, the recognition rate is descending sharply in the noisy environment. Okay et al. [15] used 3D Zernike moments, they obtain 3D volumetric data (voxel data), and used the surface information contained in the dataset. In [16] their model, four lower orders of Zernike moments with the highest variance

values are utilized to form a feature vector in a four dimensional Zernike moments to recognize a Chinese characters. In [16] their system used Hu invariant moments to identify aircraft by moment invariants. The Hu moments are more redundancy information in higher order of moments which are not suitable for recognition. In [18] their model used geometric Hu moments to drive invariant moments for gray images. Not much work has been reported in the literature of action and identity recognition. Zhou Zhang et al. [19] address the two problems for recognizing actions and identity. They propose an action and identity representation based on computing rich descriptors from ASIFT key point trajectories which capture more global spatial and temporal information. And they used the object categories depicted using a model developed in the statistical text literature: Latent Dirichlet Allocation (LDA), and combined action recognition and identification in a whole framework. They used trajectory extraction, which is may be not always useful for action recognition. Their method requires more computation and time for recognition and identification. They were separating two problems, at first; they classify action, and then second, assign the identity that belongs to the action. Their procedures take more time to identify the human that makes an action.

3. Methodology for Extracting Features of Action and Identity

We want to develop computer vision algorithms that go beyond learning one task at time, and improve the learning features in terms of speed and accuracy. In this paper, we show that our identity and action approach to human action recognition allows us to do a better job in the recognition task and provides many new abilities

3.1. Identify the Human by Using a Watermark

Based on the author's method [10], we develop the watermarked method, which the watermark can be embedded into each frame of the video. Digital watermarking [20] is the process of embedding or hiding digital information called watermark into a multimedia product, and then the embedded data can later be extracted or detected from the watermarked product. In this paper, we want to recognize the action from the video frames and also we want to know the owner of the video, therefore, the classifier are training for both action and hidden identity at the same time, without any computational cost in terms of time and space. So we present a single framework for recognize action and identity, as shown in Fig. 1a and Fig. 1b. We assumed that there is a single person performing only one action in the video like KTH dataset. The person identity embedded as a logo binary watermark at each video clip. This watermark based on 2-D wavelet transforms which decomposes a video frame into sub-frames, 3 details and 1 approximation. The 2-D wavelet transform separates the frequency band of a frame into a lower resolution approximation sub-band (LL) as well as horizontal (HL), vertical (LH) and diagonal (HH) detail components.

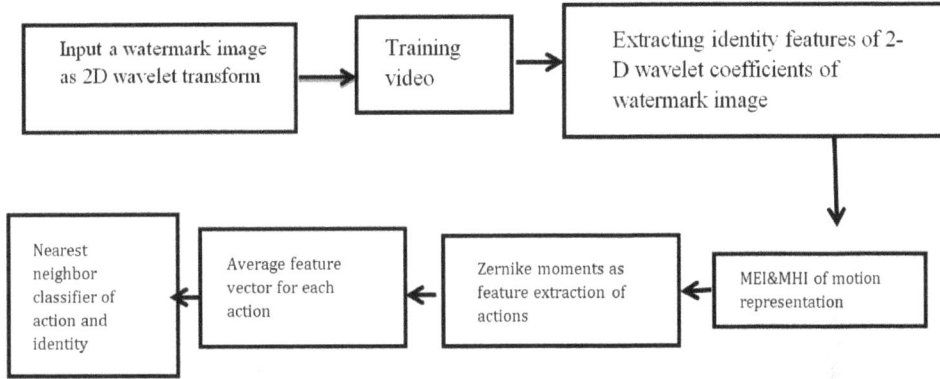

Figure 1a. *Single framework of our proposed algorithm for action and hidden identity classification for training video.*

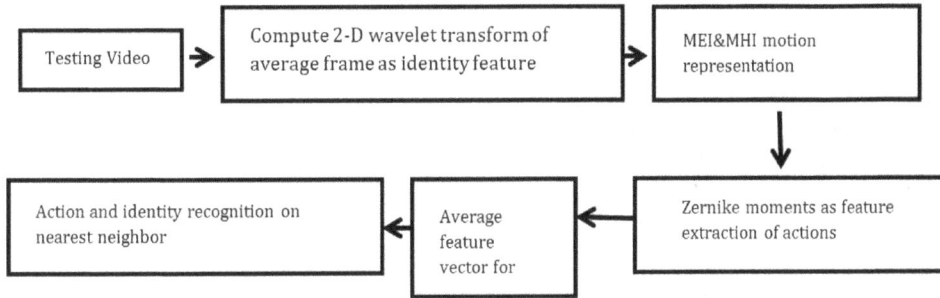

Figure 1b. *Single framework of our proposed algorithm for action and identity recognition for testing video.*

3.2. Embedding Procedure

We assumed the same procedure for embedding watermark in previous methods as follows [20, 21]:

Step 1: Convert the n × n binary watermark logo into a vector $W = \{w_1, w_2, ..., w_{n \times n}\}$ of '0's and '1's.. Out of the four sub bands, only the three high resolution detail sub bands {LH, HL, HH} are selected.

Step 2: Pseudorandom sequence is used to embed the zero watermark bit in the selected sub-band as PN sequence of sub band matrix is generated for uniformly distributed.

Step 3: A watermark amplification factor K is embedding with the PN sequence in the selected DWT sub band. If we denote W_i as coefficients matrix of the selected sub band, then embedding is done according to the equations as follows:

If the watermark bit is 0

$$L_{i,uv} = W_{i,uv} + K.PN_{uv}\ where\ W \in \{LH, HL, HH\} \quad (1)$$

Otherwise,

$$L_{i,uv} = W_{i,uv} \quad (2)$$

At this step, the coefficients of wavelet transform are describing feature vector identity of the owner's human video that makes an action. Watermarking in the DWT domain can be split into the two procedures: embedding of the watermark and extraction of the watermark. More details about embedding watermark in video by using DWT can be found in [21]. The quality of the video is the same as original as shown in Fig. 2 and Fig.3 respectively.

Figure 2. *Frames with watermark embedding as 2-D wavelet in the KTH dataset (not affect the quality).*

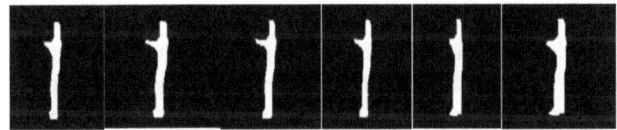

Figure 3. *Example of frames that embedded a watermark of boxing action in KTH data set.*

3.3. Temporal Templates of Human Action

There are many different types of methods to present human actions. These methods mainly can be divided into two kinds: model-based and appearance based. Appearance-based features are commonly used in action recognition since they can be easily and robustly extracted from video. We assumed motion energy image (MEI) and motion history image (MHI) in [2, 12] as the representation of human action.

A. Motion-Energy Images (MEI)

MEI is a cumulative motion image of a sequence of binary images. Let I(x, y, t) be an image sequence and let

D(x, y, t) be a binary image sequence indicating regions of motion; the MEI $E\tau$ (x, y, t) can be defined as [2, 12]:

$$E(\tau) = \cup_{i=0}^{\tau-1} D(x,y,t-i) \quad (3)$$

B. Motion-History Images (MHI)

To describe how the image is moving, we form a MHI. MHI. $H^\tau(x, y, t)$ can be defined as [2, 12]:

$$H^\tau(x, y, t) = \begin{Bmatrix} \tau \ if \ D(x, y, t) = 1 \\ \max(0, H^\tau(x, y, t-1) - 1) \ otherwise \end{Bmatrix} \quad (4)$$

Here τ is time of action, presented by the length of sequence of binary images as shown in Fig.4.

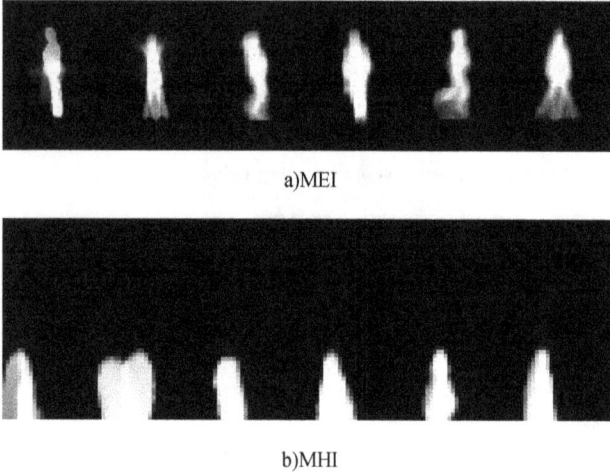

a)MEI

b)MHI

Figure 4. *MEI and MHI in KTH dataset.*

3.4. Extracted Features Using Zernike Moments

For action recognition we use moments, such that moments have been used in image processing and classification. Since Hu introduced them in his groundbreaking publication on moment invariants [22]. Hu used geometric moments and showed that they can be made to be translation and scale invariant. Since then more powerful moment techniques have been developed [23]. A notable example is Teague's work on Zernike Moments (ZM); he was the first to use the Zernike polynomials (ZP) as basis functions for the moments [24]. The use of ZP's as a basis function is theoretically beneficial because they are orthogonal polynomials which allows for maximum separation of data points, given that it reduces information redundancy between the moments, this is very useful in recognition task [25, 26]. Their orthogonal properties make them simpler to use during the reconstruction process as well. Furthermore, the magnitudes of ZM's are rotationally invariant, which is crucial for certain image processing applications, such as classifying shapes that are not aligned. Moments are common in statistics and physics, however, they are often referred to as variance or mean. Hu's moments [13] contain much redundant information about a character's shape. Nowadays, Zernike moments are becoming popular for shape recognition. Zernike moments have been also proposed in action recognition as holistic features in [28] to describe the human silhouettes. We used four lower orders of Zernike moments with the highest variance. Zernike polynomials [29] are a set of complex polynomials in which a complete orthogonal set is formed over the interior of the unit circle. The Zernike basis is a set of complete and orthogonal

functions on the unit disk defined as [16, 30]:

$$V_{nm}(x, y) = R_{nm}(p)e^{jm\theta} \quad (5)$$

Where $= \sqrt{x^2 + y^2}$, $\theta = tan^{-1}\left(\frac{y}{x}\right)$, n is non negative integer and m can be positive, negative, or zero

The radial polynomials of Zernike polynomials are:

$$R_{nm}(\rho) = \sum_{k=0}^{(n-m)/2}(-1)^k \frac{(n-k)!}{k!\left(\frac{n-m}{2}-k\right)!\left(\frac{n-m}{2-k}\right)!} \rho^{(n-2k)}, \quad (6)$$

From (3), we can find that V_{nm} is a complex number when $m \neq 0$. It makes Zernike moments different from other moments with real numbers only, we can use several different types of values from the same A_{nm}.

Where ρ is the radial distance $0 \ll \rho \leq 1$, and R_{nm} are the radial polynomials. Zernike polynomials have the property of being limited to a range of -1 to $+1$, i.e. $|Z_n^m(\rho, \varphi)| \ll 1$.

The complex Zernike moments of order n with repetition m, A_{nm}, is defined as [16]:

$$A_{nm} = \frac{n+1}{\pi} \iint_{x^2+y^2 \leq 1}^1 f(x, y)V_{nm}^*(x, y)dx \, dy, \quad (7)$$

Where * represents complex conjugate. To compute the Zernike moments of an image, the center of the image is taken as the origin point of the unit circle and only pixels whose centers are located on the unit disk are used in computation. The Zernike moments have several advanced fundamental properties, such as being invariant to rotations and reflections. If an image is rotated σ degree, unlike geometric moments their invariants can be calculated independently to arbitrary high orders without having to recalculate low order invariants, the Zernike moments of the rotated image are:

$$A_{n,m}^\alpha = A_{n,m}e^{-jm\alpha}, \quad (8)$$

From Eq. (6), the magnitudes of A_{nm} are rotationally invariant.

These orthogonal properties also allow one to evaluate up to what order to calculate these moments to get a good descriptor for a given database. Another reason for using OG moments frequently mentioned in literature is for better image reconstruction [28, 29]. Therefore, the image reconstruction from OG moments can easily compute. We use these values of Zernike moments as action descriptors for the feature vectors, which is contain characteristic information about the contents of each frame in a numerical form. Furthermore, a framework is capable of recognizing the hidden features of identity features and recognizing action features for each defined human action.

4. Classification

The nearest neighbor is used in machine learning and pattern recognition. It obtains class membership for some testing feature descriptor based on its nearest neighbor from training feature descriptors in feature space. The testing is classified by a majority vote of its K nearest neighbors. The KNN [31] will go over all training samples to find the nearest neighbor for testing sample, therefore, it takes long time, if the training samples are numerous, but in our research, the training data are small. In the KNN, three parameters are used. First parameter, K is set up to number of voting members. Second, distance metric type, we used Euclidean (squared difference). Third, the rule for selecting estimated class for testing sample is set up into: nearest neighbor. The KNN classifier is calculated the distances (d) between testing sample (x) and each training sample (m) provided from Eq. 6, where d is a distance metric, x is a testing sample, m are training samples, j = [1, 2, …, N], N is a number of training samples [32]

$$D(x, m) = \arg \min_j \{d(x, m_j)\} \qquad (9)$$

Where d is Euclidean distance in this research. To classify identity human, the 2-D wavelet coefficients of watermarked image used as feature identity vectors, of video frames of KTH dataset. We divide each video clips of each action into small clips of 0.5 second (13 frames) to reduce the high dimensions of the video, and minimize the complexity time for training. We suggest that it is enough to recognize action and identity, which is discriminate the shape of the motion of video frames as shown in Fig. 4. We split the video frames into training frames and testing frames. We compute real and imaginary parts of Zernike moments of video frames of all actions, and then perform average feature vector of each action. Nearest neighbor classifier used to classify action and identity. The frames are selected which embedded a watermark in video training to learn the features of videos belonging to different classes and labeling is done with action and identity in this step. For each class a number of frames having different styles of the same action performed by same person are taken for training.

5. Results and Discussion

In our algorithm, a one watermark embedded as binary image into frames, to identify the human identity. We assumed one person makes one action in each video clip. If we have more complex data sets such as multiple people do actions in same video, we can embedded each watermark for each human appears in the video, this is our future works. The binary human identity is of size 24 x 24; embedded in the training video. This requires a feature representation to describe the human identity as feature vectors of 2-D coefficients of wavelet watermarked image to use for identity classifier. We firstly calculate MEI and MHI of each action using the number of frame for each action as parameter. Secondly calculate the feature vectors of every MEI and MHI. To calculate the Zernike moments of a frame, the center of the frame is taken as the origin point of the unit circle and only pixels whose centers are located on the unit disk are used in computation [29], we calculated real part values, and magnitudes of 4 lowest orders Zernike moments and highest variance for the 6 actions of KTH dataset of feature vector of every MHI and MHI. The results are shown in Table 1, 2, and 3 respectively. Next, select label some actions with hidden

identity, using the constrained nearest neighbor to find minimum Euclidean distance similarity. Then use the same method to extract the feature vectors of test actions. Finally apply the nearest neighbor classifier to recognize test actions and identity. Since, the computation of moments requires that the image is of binary nature [16]. Thus, we not need any additional cost computation, because MHI and MEI temporal template representation provide binary representation of actions. For identity classifier, wavelet transformation provides both frequency and spatial description of an image. Temporal information retained in the transformation process. The confusion matrix in table 5, and table 6 respectively, show high accuracy of action and identity, which is better than [19]. Their work is similar to our paper, but different in methodology.

Table 1. Variance values of the 10 lowest orders of Zernike moments Amn calculated by their magnitudes.

	N=0	N=1	N=2	N=3
|m|=0	0.00340		0.0083	
|m|=1		0.00037		
|m|=2			0.00021	
|m|=3				0.00043
	N=4	N=5		
|m|=0				
|m|=1		0.00073		
|m|=2	0.00072			
|m|=3		0.00071		
|m|=4	0.00176			
|m|=5		0.00083		

Table 2. Variance values of the 10 lowest orders of Zernike moments Amn calculated by their real parts.

	N=0	N=1	N=2	N=3
|m|=0	0.00040		0.00216	
|m|=1		0.00042		
|m|=2			0.00206	
|m|=3				0.00435
	N=4	N=5		
|m|=0				
|m|=1		0.00135		
|m|=2	0.00211			
|m|=3				
|m|=4		0.00159		
|m|=5	0.00173	0.00197		

Based on our selection, the following two Zernike moment feature vectors are computed:

Vreal [f1 = A02; f2 = A22; f3 = A33; f4 = A24]; (10)

Vmagnitude [f1 = A00; f2 = A02; f3 = A15; f4 = A55]; (11)

Tabel 3. Selected four highest Zernike moment values from each part.

	F1	F2	F3	F4
Real part	A02	A22	A33	A24
Magnitude part	A00	A02	A15	A55

When the Zernike moment feature vectors *Vreal* and *Vmagnitude*, as defined in (7) and (8), are applied, we only need to consider the real part or magnitude of A_{nm}:

$$d(f_i, f_j) = \sqrt{(f_{1i} - f_{1j})^2 + (f_{2i} - f_{2j})^2 + (f_{3i} - f_{3j})^2 + (f_{4i} - f_{4j})^2} \quad (12)$$

Where i and j denote any pair of human actions of training and testing of video frames respectively. From the result of accuracy recognition in Table. 4. Khan et al. [14], and Moussa et al. [10], propose methods of action recognition with performing quite high recognition rate, because in [14], they used whole parts of seven Hu invariant moments, and in [10] used normalization after feature detection of interest points, this means more cost in terms of computation time, their method, time is 14.446, although the accuracy is high. While our method use four feature vectors, so our method performs faster than [10] even though calculating more feature vectors.

Table 4. *Descriptor complexity (in time) comparison together with accuracy (the value results of other methods taken from their papers).*

Method	Size	Computation time	Accuracy
Our method	54	0.0334 s	96%
Hu Moments [14]	-	-	97%
Moussa et al. [10]	-	14.446	97%
Pyramid Zernike 3D [11]	84	0.0300 s	91.30%
Gradient + PCA[33]	100	0.0060 s	81.17%
3D SIFT [31]	640	0.8210 s	82.60%
3D Gradn[32]	432	0.0400 s	90.38%
HOG 3D [7]	380	0.0020 s	91.40%
SURF 3D [6]	384	0.0005 s	84.26%

The difficulties of other similar related methods as in [19] are to classify between jogging and running, they used SIFT descriptor for identity representation and ASIFT descriptor for action representation in expensive computing, their method of action accuracy is 81.7 and for identity accuracy is 58.8. As experimental results of our method in recognition accuracy of action and identity are high and the computation time for extracted features 0.033, as shown in Table 4. Unlike the current state of the art approaches such that 3D Zernike, SIFT, HOG, MHI, etc., which is need more computation time to extract Spatio and temporal features. We expect that the property of discriminative Zernike moments and unique coefficients of wavelet values for identity features would increase the efficiency of the proposed action and identity recognition.

Table 5. *Confusion matrix of action accuracy is 96.5%.*

	walk	Jog	run	box	hand-wave	hand-clap
walk	98	2.0	0.0	0.0	0.0	0.0
jog	3.1	93	3.9	0.0	0.0	0.0
run	2.0	3.0.	95	0.0	0.0	0.0
box	0.0	0.0	0.0	100	0.0	0.0
hand-wave	0.0	0.0	0.0	0.0	95	5.0
hand-clap	0.0	0.0	0.0	2.0	0.0	98

Table 6. *Confusion matrix of identity accuracy is 89.7%.*

	John	Wang	Ming	said	Hasan	Zane
John	90	5	5	0	0	0
Wang	0	88.4	7.2	4.4	0	0
Ming	0	0	93.4	0	3.5	3.1
Said	0	9.0	0	87.0	0	4.0
Hasan	5.0	5.0	0	0	90.0	0
Zane	3.0	0	0	5.0	2.6	89.4

6. Conclusion

In this paper, we have proposed a novel, robust and fast algorithm for action and identity recognition. We identify the human identity by embedding a watermark as binary image of 2-D wavelet transform in the video frames, then extract features identity as wavelet coefficients. To represent the motion features, we use temporal templates MEI and MHI representations. Zernike moments based feature extraction method which captures global features is proposed for human action description. Zernike moment has discriminative features than Hu moments, which are necessary to classify related actions such as jogging and running, which most previous methods fail to classify them. The advantage of using Zernike moments than Hu moments is orthogonal properties, so no redundancy of data, and robust to translation, rotation and scaling. The nearest neighbor classifier can be train to classify different actions of Zernike feature vectors and wavelet coefficients of human identity with higher accuracy in action and identity, without any expensive computation for identifying human identity. The experimental results show that the described system is simple and robust to few illumination changes while dealing with indoor and outdoor actions and only using very small number of labeled training data examples. There is a room to improve and extend this framework to recognize actions in more and complex data sets such that Hollywood data sets.

Acknowledgements

We would like to thank the anonymous reviewers and editors for their constructive comments and suggestions that help to improve the quality of this manuscript.

References

[1] A. Wahi,. S.S., P.P. A Comparative Study for Handwritten Tamil Character Recognition using Wavelet Transform and Zernike Moments. International Journal of Open Information Technologies ISSN: 2307-8162 vol. 2, no. 4, 2014.

[2] Yanan Lu et al. A Human Action Recognition Method Based on Tchebichef Moment Invariants and Temporal Templates, 4th International Conference on Intelligent Human-Machine Systems and Cybernetics, 2012.

[3] Omar O., Ramin M., M. Shah. Human Identity Recognition in Aerial Images. IEEE Conference on CVPR, CA, 2010.

[4] Lowe, D.G.: Distinctive image features from scale-invariant key points. International Journal of Computer Vision 60(2), 2004

[5] I. Laptev, M. M., C. Schmid, and B. Rozenfeld. Learning Realistic Human Actions from Movies. Proc. IEEE Conf. Computer Vision and Pattern Recognition, 2008.

[6] Willems, G., Tuytelaars, T., Van Gool, L. An efficient dense and scale-invariant spatio-temporal interest point detector. In: Proc. of ECCV. (2008)

[7] Scovanner, P., Ali, S., Shah, M.: "A 3-Dimensional SIFT Descriptor and its Application to Action Recognition". In: Proc. of ACM Multimedia. (2007)

[8] A. Klaser, M. Marszałek, and C. Schmid. A spatio-temporal descriptor based on 3D-gradients. In: Proc. of BMVC, 2008.

[9] Ballan, L., Bertini, M., Del Bimbo, A., Seidenari, L., Serra, G. Recognizing human actions by fusing spatio-temporal appearance and motion descriptors. In: Proc. Of ICIP. (2009)

[10] Mona M. Moussa, Elsayed Hamayed, Magda B., Heba A. Enhanced method for human action recognition. Elsevier, Journal of Advanced Research, University of Cairo, 2013.

[11] Khawlah Hussein Ali and T. Wang. Recognition of Human Action and Identification Based on SIFT and Watermark. Springer International Publishing Switzerland 2014.

[12] Bobick, A.F., Davis, J.W.: The Recognition of Human Movement Using Temporal Templates. J. IEEE Trans. on Pattern Analysis and Machine Intelligence 25, 257–267, 2001.

[13] H. Ming-Kuei. Visual pattern recognition by moment invariants. IRE Transactions on InformationTheory, vol. 8, no. 2, pp. 179–187, 1962.

[14] Y. D. Khan, Nabeel S., S. F. Adnan A., and M. K. Mahmood. An Efficient Algorithm for Recognition of Human Actions. Hindawi Publishing Corporation, the Scientific World Journal Volume 2014.

[15] Okay Ank and A.Semih Bingo. Human Action Recognition Using 3D Zernike Moments. 978-1-4799-3866-7/14, IEEE. (2014)

[16] T. Wang and Simon Liao. Chinese Character Recognition by Zernike Moments. Conference, ICALIP 2014978-1-4799-3903, IEEE 2014.

[17] S. A. Dudani, Kenneth J., and Robert B. Mcghee. Aircraft Identification by Moment Invariants. IEEE Transaction on computers, 2009

[18] Y. Wang1, X. Wang, Bin Zhang. A Novel Form of Affine Moment Invariants of Grayscale Images. Elektro Technika, ISSN 1392-1215, VOL. 19, NO. 1, 2013

[19] Zhuo Zhang, Jia Liu. Recognizing Human Action and Identity Based on Affine-SIFT. IEEE Symposium on Electrical & Electronics Engineering (EEESYM). 2012

[20] Hai Tao 1, Li Chongmin,, Jasni Mohamad Zain1, Ahmed N. Abdalla, Robust Image Watermarking Theories and Techniques: A Review, Malaysia, Vol. 12, February 2014.

[21] Nataša Terzija, Markus R. Kerstin Luck, Walter G. Digital Image Watermarking Using Discrete Wavelet Transform Performance Comparison of Error Correction Codes. From Proceeding (364) Visualization, Imaging, and Image Processing, 2002.

[22] Andrzej Sluzek. Shape Identification Using New Moment-based Descriptors. Conference on computer technology, 1994

[23] S. S. Reddi, "Radial and Angular Moment Invariants for Image Identification", IEEE Transaction on Pattern Analysis and Machine Intelligence, vol. PAMI-3, no. 2, march 1981.

[24] L. Shao, L. Ji, Y. Liu, and J. Zhang, "Human action segmentation and recognition via motion and shape analysis," *Pattern Recognition Letters*, vol. 33, no. 4, pp. 438–445, 2012.

[25] Z. Jiang, Z. Lin, and L. Davis, "Recognizing human actions by learning and matching shape-motion prototype trees," *IEEE Transactions on Pattern Analysis and Machine Intelligence*, vol. 34, no. 3, pp. 533–547, 2012.

[26] J. Flusser, B. Zitova, and T. Suk. Moments and Moment Invariants in Pattern Recognition. JohnWiley & Sons, New York, NY, USA, 2009.

[27] S. Daniel Madan Raja, A. Shanmugam. Zernike Moments Based War Scene Classification Using ANN and SVM: A Comparative Study. Journal of Information & Computational Science 8: 2 (2011).

[28] Sun, X., Chen, M., Hauptmann, A. Action recognition via local descriptors and holistic features. In: Proc. of Workshop on CVPR for Human communicative Behavior analysis (CVPR4HB). (2009).

[29] Boiman, O., Shechtman, E., Irani, M. In defense of nearest-neighbor based image classification. In: Proc of. CVPR. (2008)

[30] Margarita N. F., Lakhmi C. Jain, Editors. Computer Vision in Control Systems-2: Innovations in Practice. Springer International Publishing Switzerland 2015.

[31] H. Jos'e Antonio Mart'ın, M. Santos, and J. de Lope. Orthogonal variant moments features in image analysis. Information Sciences, vol. 180, no. 6, pp. 846–860, 2010.

[32] Hejin Yuan and Cuiru Wang. A Human Action Recognition Algorithm Based on Semi-supervised Kmeans Clustering. Transactions on Edutainment VI, LNCS 6758, pp. 227–236, 2011.

[33] Dollar, P., Rabaud, V., Cottrell, G., B., S. Behavior Recognition via Sparse Spatio-Temporal Features. In: Proc. of VSPETS. (2005).

10

Portable Weather System for Measure and Monitoring Temperature, Relative Humidity, and Pressure, Based on Bluetooth Communication

Edgar Manuel Cano Cruz, Juan Gabriel Ruiz Ruiz

Department of Computer Science, University of the Istmo, Ixtepec, Mexico

Email address:

ie.edgarcano@gmail.com (E. M. C. Cruz)

Abstract: The system also proposes a wireless connectivity by using the Bluetooth communication standard providing of a reliable, portable and a low-cost tool for industry where it is necessary to have an environmental control to carry out critical processes. The weather system consist of an embedded system to the development of multimedia applications based on the PIC32 microcontroller, and development is performed using the SPIES methodology for the construct embedded systems. In this paper the design of a portable system that allows monitoring of four climatic variables (temperature, relative humidity, pressure and altitude). The purpose of the system is to serve as an auxiliary tool to make decisions subsystems for environmental control in different areas.

Keywords: Embedded System, Bluetooth, Weather System, Sensors

1. Introduction

At present, the monitoring of climate variables such as temperature, relative humidity, atmospheric pressure and altitude are of great importance [1], because in places like farms, greenhouses, computer centers and hospitals is needed monitoring systems that allow better control of environmental conditions. The main activities of the Mexican Meteorological Service (SMN) are the acquisition, processing and diffusion of weather data at local and national levels. SMN has 188 surface automatic weather stations (surface AWS) installed throughout Mexico and 10 installed in Oaxaca state (see Figure 1).

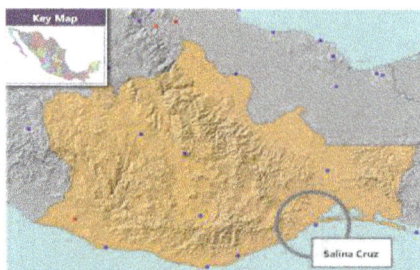

Figure 1. Surface automatic.

The AWS station is composed of sensors to receive and send weather information from where they are installed (see Figure 2).

Figure 2. Surface automatic weather station.

The sensors can meassure parameters such as: Wind direction, wind direction burst, wind speed, wind speed burst, average ambient temperature, relative humidity, atmospheric pressure, precipitation y radiation.

However, one of the main shortcomings with surface and synoptic AWS stations is that their radio action is limited to only 5 km on flat terrain. Also, there is no official reference for areas or zones they do not cover. Therefore, other alternatives to weather stations are necessary for the acquisition of meteorological variables [6]. Additional complications with these stations are related to the availability of information provided by sensors towards the monitoring stations. The stations usually have technical complications, for example, in the SMN list of stations [13], 26 out of 188 surface AWS are out of operation. At the time of this research, SMN had only one surface AWS station out of operation, which was located in Coatzacoalcos, Veracruz [8]. Moreover, most of the weather stations described here are not low-cost and need special facilities. Their cost ranges from $25,000 USD to $35, 000 USD [12].

This paper proposes the design and development of a weather which is embedded, flexible, and mobile with wireless capacity and an easy-to-install system. Trying to successfully accomplish all these project requirements is challenging research.

This paper is organized as follows. The second section provides a background of projects that focus on measuring weather variables and the motivation of our research. The third section summarizes the design and development of the system proposes. The fourth section presents our results. Finally, the fifth and sixth sections present the discussion and conclusions of the research, respectively.

2. Related Work

There are several projects that focus on the measuring of temperature and relative humidity to control the automatization of industrial processes. In "Monitoring System of Humidity and Temperature in Biological Collections by using Free Software Tools", Vargas [15] presents a system for the automation of the preventive maintenance of biological collections using the low-cost sensor TEMPerHUM and connecting the system to a database via an Ethernet connection. Similarly, "Control System of Humidity and Temperature for Greenhouses" [3] presents some ways to accurately control humidity, temperature, lighting, ventilation and other relevant variables of greenhouses. This work develops a control system based on an SHT11 sensor with actuator and control modules. Communication between modules is transmitted wirelessly by using a transceiver operating at 2.4 GHz. PIC Microcontrollers of 16F8xx and 18F4xx series made by Microchip are used in all modules.

Another example is "Automatic Control of Agricultural Irrigation with Capacitive Sensors of Soil Moisture. Applications in Grapevine and Olive" [11] which describes the features of a capacitive humidity sensor of soils developed

in the Institute of Automation (INAUT), where an automatic irrigation closed-loop control was developed.

The control system acts on the pump and irrigation valves. It identifies when it is appropriate to irrigate and how long the irrigation should last.

Another drawback is that only the manufacturer distributes all the hardware components. Updates to the system are not easy to carry out and most of them are arbitrary. A very convenient point is related to the easy setup and installation of these kinds of systems. Indeed no installation server is required.

Unlike the aforementioned research, this paper proposes a system to measure the average temperature and relative humidity with wireless communication capacity under the Bluetooth standard. It is built with an electronic board for the development of embedded multimedia systems with a touch screen to take measurements in various sites.

The proposed system is highly flexible and open to further expansion.

3. Design and Development Portable Weather System

The weather system was implemented with the help of the methodology for the development of embedded systems SPIES, which is detailed in [17]. SPIES uses the Unified Modeling Language (UML) to represent the phases that make up the methodology with an iterative top-down approach. It ensures that the system is tested at each phase and not only at the end. Hence, the system improves continuously over time (see Figure 3).

SPIES is extracted from process areas and specific practices in CMMI-DEV v1.2 level 2 [16], which are simplified for embedded software demands.

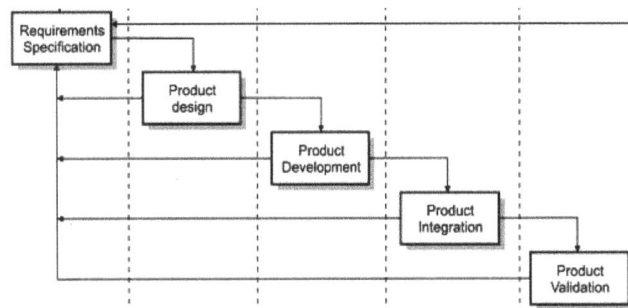

Figure 3. SPIES methodology.

3.1. Requirements Specification

The weather system is divided in three subsystems: main control, weather measure control, and Bluetooth controller (at the level of electronic components) and two communication interfaces: GUI Touch and Serial Interface. Table 1 describes each module of our proposed system.

Table 1. Modules of the system.

Column1	Column2
Main control	It is responsible for initializing and prioritizing of processor tasks.
Weather control	This module is responsible for calibrating and controlling the sensors in the system. It also computes the relative humidity, average temperature, pressure, and altitude.
Bluetooth control	This subsystem is responsible for calibrating and handling Bluetooth communication. It looks for Bluetooth devices in its vicinity to establish communication and data transfer
GUI control	This subsystem organize all functionalities about Graphical user interface.

3.2. Product Design

In this phase, the sensors for the product development are choose. The structure diagram of the system is depicted in Figure 4. It shows the interaction and the static functionality of the proposed system.

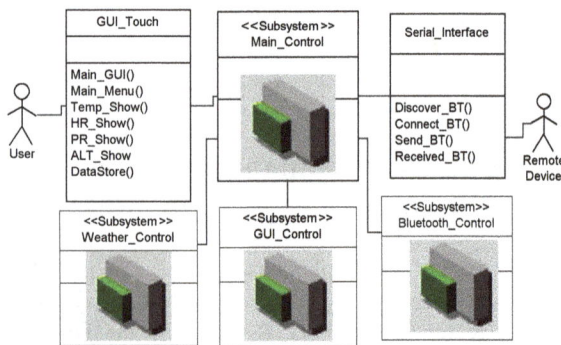

Figure 4. Structure diagram of the system proposed.

Due to the structural configuration of the system, each module can be tested individually during the system development.

The SHT11 Click sensor was selected to measure temperature and relative humidity (see Figure 5a). To measure pressure and altitude we chosen the Altitude Click sensors (see Figure 5b), both sensor of the mikroelektronika signature [9].

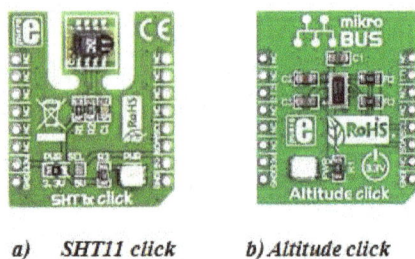

a) SHT11 click *b) Altitude click*

Figure 5. Senssor of the weather system.

The SHT11 Click sensor includes a digital humidity and temperature sensor SHT11. A unique capacitive sensor element is used to measure relative humidity while the temperature is measured by a band-gap sensor (SENSIRION, 2015).

The Altitude click features a MEMS pressure sensor MPL3115A2, which provides accurate pressure/altitude (20-bit). Resolution is down to 30 cm (1.5 Pa) (freescale, 2015). The MPL3115A2 contains automatic internal data

processing with data acquisition and compensation as show in Figure 6. The MPL3115A2 sensor outputs are digitized by a high resolution 24-bit ADC.

3.3. Product Development

The system functionality was programmed under mikroC compiler for PIC32 microcontrollers [4] to be executed on a mikromedia card. During the assembly of the system, a rechargeable battery is connected to the rear side of the mikromedia card. The expansion card and the Bluetooth antenna are also connected.

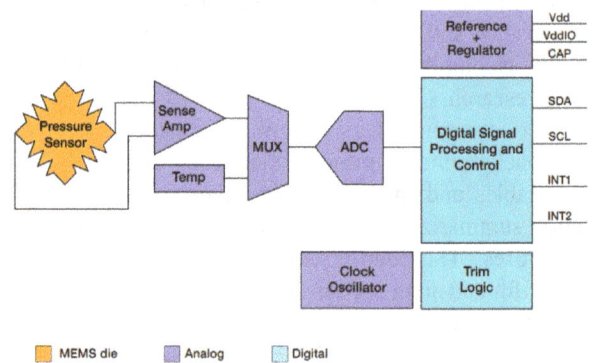

Figure 6. MPL3115A2 Pressure Sensor [11].

The sensor STH11 operates under a dedicated protocol designed by SENSIRION (see Figure 7).

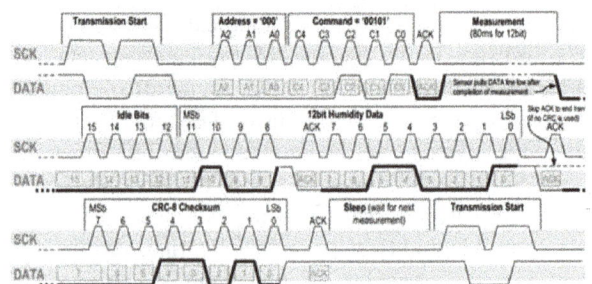

Figure 7. Sensirion protocol [10].

The bus is composed of two serial lines: serial clock input (SCK) for signal synchronization; and serial data line (DATA) for reading/writing to an SHT11 sensor. The SHT11 resolution is 14 bits for temperature and 12 bits for humidity. However, resolution can be adjusted to 12 and 8 bits respectively. In our weather system, resolutions by default were used to ensure a more accurate analog-to-digital conversion. Meanwhile, the pressure and altitude sensor MPL3115A2, the data can be accessed through an I2C/SPI interface making the device particularly suitable for direct interfacing with a

microcontroller; and the altitude sensor MPL3115A2.

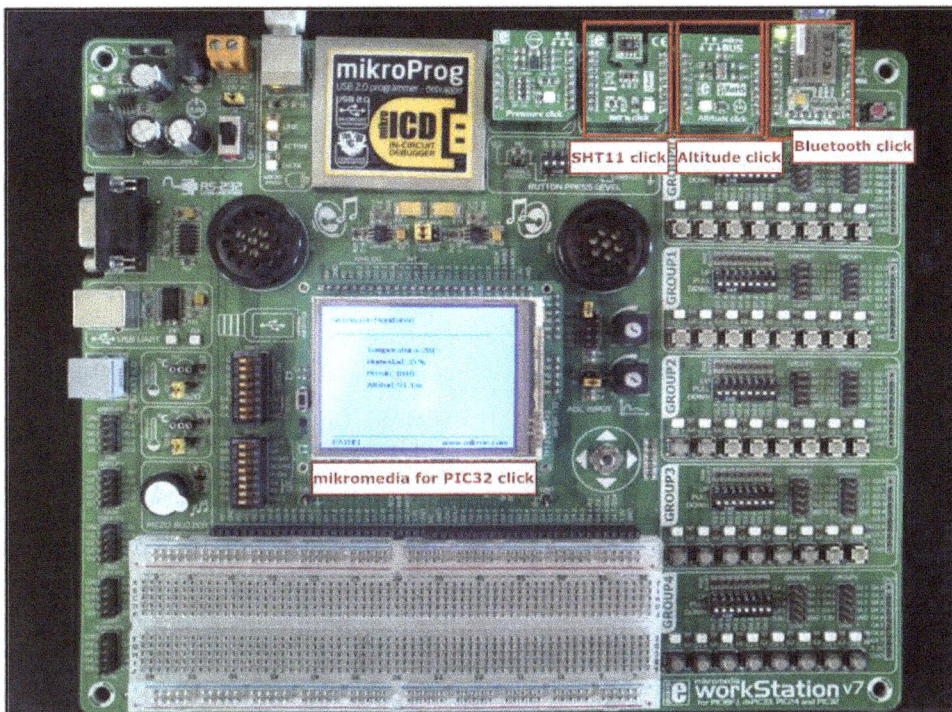

Figure 8. *Weather system integration.*

3.4. Product Integration

In the integration of the weather system a mikromedia for PIC32 card was selected (see Figure 8 and Figure 9). Mikromedia card is a compact design with different embedded peripherals for multimedia application development. The main features Mikromedia card specified in [4].

4. Experiment Results (Product Validation)

The system functionality was construct with the mikromedia for PIC32 on a workstation v7 under mikroC compiler for PIC32 microcontrollers [4].

During the assembly of the system, a rechargeable battery is connected to the rear side of the mikromedia card. The expansion card and the Bluetooth antenna are also connected.

Figure 9. *Weather system interface data on mikromedia.*

In order to validate and compare our results and, official data from the surface AWS - SMN Salina Cruz was considered. The following weather variables were used:

- *Temp.* Average temperature (in °C) over a 10- minute period. Samples are taken each minute.
- *RH.* Average relative humidity (in percentage) over a 10-minute period. Samples are taken each minute.
- *PB.* Average Pressure (in hectopascal) over a 10- minute period. Samples are taken each minute.

SMN website use UTC (Universal Time Coordinated), which is based in zero hour GMT (Greenwich Mean Time). In this way, equation (1) should be used to make a time conversion [5].

$$\text{Local Time} = Z + DZ \qquad (1)$$

Table 2 shows weather variables Temp, RH and PB from the SMN webpage, taken on April 20th, 2015 from 11:00 Z hrs to 14:00 Z hrs (every 15 minutes).

The weather system was programmed in the same way to measure and store samples of temperature, relative humidity, pressure, and altitude every minute. Afterwards, an average was computed over a 10- minute period resulting in Temp, RH, and PB.

Every two hours, computed averages were sent by a Bluetooth connection to a PC where they were displayed and stored. Our results were very close to those found by SMN.

Correlation coefficients are: (AWS Temp, WS Temp) 0.976 ($p<0.05$), (AWS RH, WS RH) 0.892 ($p<0.05$), and (AWS PB, WS PB) 0.927 ($p<0.05$).

Table 2. Temperature (Temp), relative humidity (RH), and Pressure measures from AWS and weather system proposed (WS).

Time	Temp (°C)		RH (%)		PB (pa)	
	AWS	WS	AWS	WS	AWS	WS
11:00	23.5	24.2	27.0	26.2	98	99
11:15	25.8	26.5	27.1	26.3	97	99
11:30	21.8	23.4	27.1	26.3	96	99
11:45	25.8	27.5	27.0	26.5	95	98
12:00	26.4	28.6	26.9	26.6	91	95
12:15	27.5	28.9	26.9	26.7	92	95
12:30	26.9	28.4	26.9	26.7	94	97
12:45	26.4	28.2	27.0	26.5	93	96
13:00	26.9	28.5	27.2	29.0	92	95
13:15	26.4	28.6	27.4	29.4	93	95
13:30	26.4	28.4	27.7	29.5	92	94
13:45	28.1	30.1	27.8	29.6	92	94
14:00	29.8	31.2	28.0	30.0	89	94

5. Conclusions

This paper proposes the development of a portable weather system to measure relative humidity, temperature, pressure and altitude. Its importance is that it provides a reliable and practical tool to measure temperature with a very simple device. The core component of the system, the SHT11 sensor and Pressure click, measures variables whereas Bluetooth click device establishes the communication between weather and a personal computer where this parameters are displayed and stored. A strength of this kind of system is that its installation and implementation are quick and simple because server installation is not needed.

In future research, we hope to develop the weather system by using the ARDUINO UNO card to replace the current mikromedia card, as mentioned and lower the cost of our system. We should also consider the design of a protective case for outdoor use of the device without any problems, decreasing the risks of it malfunctioning. In the very near future, a webpage could also be built in order to display the weather system data in a real-time graphical interface. Replacing the Bluetooth communication by Wi-Fi is also under consideration in order to transfer data to a cloud.

References

[1] Official webpage of humidity and temperature sensors APC (2011). Retrieved by 2014, from http://www.apc.com/.

[2] Official webpage of Arduino firm (2014). Retrieved by 2014, from http://arduino.cc/en/Main/arduinoBoardUno

[3] Hernández, R. Control System of Humidity and Temperature for Greenhouses (2004). Universidad Pedagógica y Tecnológica de Colombia (UPTC).

[4] Mikro Elektronika. Official webpage of Mikro Elektronika firm (2014). Retrieved by 2014, from http://www.mikroe.com/.

[5] OMM. Official webpage of World Meteorological Organization (2014). Retrieved by 2014, from http://www.wmo.int/pages/index_es.html.

[6] Prieto, R. Diagnóstico de las capacidades, fortalezas y necesidades para la observación, monitoreo, pronóstico y prevención del tiempo y el clima ante la variabilidad y el cambio climático en México (2008). INECC.

[7] SEMAR. Official webpage of Mexican Navy Secretariat (2013). Retrieved by August 2014, from http://meteorologia.semar.gob.mx/index.php.

[8] SEMAR. Information of surface automatic weather systems (AWS) in Coatzacoalcos, Veracruz (2013). Retrieved by 2014, from http://meteorologia.semar.gob.mx/datos_emas/coatza.htm.

[9] SENSIRION. Official webpage of SHT11 device (2014). Retrieved by 2014, from http://www.sensirion.com/en/products/humiditytemperature/ humidity-sensor-sht11/.

[10] SENSIRION. Reference manual of SHT11 sensor (2011).

[11] Shugurensky, C., & Caprano, F. (2011). Automatic Control of Agricultural Irrigation with Capacitive Sensors of Soil Moisture. Applications in Grapevine and Olive. Institute of Automation (INAUT).

[12] SMN. Official web page of Mexican Weather Service (SMN), Mexico (2010). Retrieved by August 2014.

[13] SMN. Weather stations of Mexican Weather Service (2010) (SMN). Retrieved by 2014, from http://smn.cna.gob.mx/emas/catalogoa.html.

[14] Telemetría. Official webpage of Telemetría firm (2014). Retrieved by 2014, from http://www.telemetria.com.mx/monitoreo-de-datacenter.html

[15] Vargas, H. Monitoring System of Humidity and Temperature in Biological Collections by using Free Software Tools (2011). Instituto Nacional de Biodiversidad.

[16] CMMI for Development (CMMI-DEV v1.2), 2006. Software Engineering Institute, Carnegie Mellon University.

[17] García, I., & Cano, E. Designing and implementing a constructionist approach for improving the teaching–learning process in the embedded systems and wireless communications areas, 2011 Computer Applications in Engineering Education.

11

Enhancement LTE System Based on DWT and Four STBC Transmit Antennas in Multichannel Models

Laith Ali Abdul-Rahaim

Electrical Engineering Department, Babylon University, Babil, Iraq

Email address:

drlaithanzy@yahoo.com

Abstract: In this paper Enhancement detail for the two main applications of Long Term Evolution (LTE) these are fixed and Mobile LTE. Fixed LTE will send the data from a single point-to-multipoint like user's houses and companies. While full mobility achieved by Mobile LTE to cellular networks. This done at very high broadband data rate comparted with other broadband networks like WiMax and Wi-Fi. The two types of LTE above are used in planning of a proper network which offers better throughput wireless broadband connectivity with lower cost. This work present a new proposed structures for LTE based on Space Time Block Coding (STBC-LTE) and Discrete Wavelets Transform (DWT) as multicarrier. The purpose of these new proposed structures is to improve the performance of bit error rate (BER) compared with the conventional STBC-LTE that use fast Fourier transform (FFT) as multicarrier. In addition, the new proposed structures with more than three transmits antennas and DWT was used in first time in LTE systems to enhance spectral efficiency and supports BER performance. The proposed STBC- LTE systems have been examined under different channel models like "AWGN, flat fading, and three types of multipath selective fading channel models (Extended Pedestrian, Extended Vehicular and Extended Typical Urban)". The simulation results achieved in this work show that STBC- LTE based on DWT with four transmits antennas given best between other conventional STBC-LTE based on FFT systems. All these LTE systems models were built using MATLAB 2014a to permit various system parameters to be changed and tested like signal to noise ratio (SNR) and maximum Doppler shift and type of channel and channel parameters of the system.

Keywords: LTE, Multipath Fading Channels, DWT, STBC, OFDM

1. Introduction

With the Large increase of wireless networks and multimedia applications such as audio streaming, mobile TV, interactive gaming, video and Internet browsing, the mobile communication technology needs to meet different requirements of mobile data, mobile calculations and mobile multimedia operations. In order to prepare this increasing in mobile data usage and the new applications of multimedia, LTE and (LTE-A advance) technologies have been specified by the 3GPP as the emerging technologies of mobile communication for the next generation wireless broadband mobile networks.

The LTE system is designed to be a packet-based system containing less network elements, which improves the system capacity and coverage, and provides high performance in terms of high data rates, low access latency, flexible bandwidth operation and seamless integration with other existing wireless communication systems [1]. The LTE-A system specified by the 3GPP LTE Release 10 enhances the existing LTE systems to support much higher data usage, lower latencies and better spectral efficiency [2]. In addition, both of the LTE and LTEA systems support "flat IP connectivity, full interworking with heterogeneous wireless access networks and many new types of base stations such as pico/femto base stations and relay nodes in a macro-cellular network". Due to the introduction of the new characteristics, it incurs a lot of new challenges in security issue with the design architectures of the LTE and LTE-A systems.

The Long-Term Evolution LTE is the newest expansion in the systems of 3GPP [3,4]. In 2004, the beginning of the first study in LTE systems, at this time the LTE was refers to the prospect evolution of UMTS. However, the word "LTE" bewitched many researchers, and so it became the favorite name between all new wireless communication systems. "Evolved UTRAN (E-UTRAN)" is name of radio access network in LTE systems as compared with the UTRAN in UMTS. The upgraded 3GPP radio access network from GSM to UMTS needed a little change in core network. However, when the radio access network was upgraded from UMTS to LTE, also would enhanced the core network. "System Architecture Evolution (SAE)" is targeting enhancement the standardization work of the core network architecture. And "the Evolved Packet Core (EPC)" term means the evolved core network. SAE based on an all-IP network and supports the GPP radio access networks and non-3GPP radio access networks like CDMA2000, Wi-Fi and WIMAX. However EPC's support the non-3GPP radio access networks and enables the operations of previous non3GPP radio access networks to espouse LTE as their future radio access network of these networks. The term "Evolved Packet System (EPS)" means E-UTRAN and EPC combination [4, 5].

The description to the LTE systems was been defined in Release 8 as the 1st idea of the standard of LTE system that including the basal functionality to perform it as a wireless system. LTE Release 9 makes a major addition by adding "the Multimedia Broadcast/Multicast Service (MBMS)", which provide LTE broadcast and multicast services. The new features that added in Release 10 of LTE will support "Carrier Aggregation (CA)" that encloses multiple LTE carriers used to provide a high speed rate. This additional will increase LTE coverage; and support access from many machine-type devices which called Machine-Type Communication (MTC). In beginning of 2012, "LTE Release 11 standardization" was been published and the Major works LTE Release 11 are the improvement of "MBMS and CA" [6, 7]. At the World Radio communication Conference (WRC07), it was pointed out that there is a strong need to add spectrum for mobile systems due to the expected large increase in next 15 years. The WRC07 defined "new bands several for *IMT-Advanced: 450 MHz band; UHF band (698–960 MHz); 2.3GHz band; C-band (3400–4200 MHz)*" [8,9].

2. The Proposed LTE Structure

The LTE can implement using Discrete Fourier transform (DFT) or fast Fourier transforms (FFT) as complex exponential functions. So it can be replaced by or discrete wavelets transform (DWT). This replacement will decrease the interference level. From many research, it's found that uses of discrete orthonormal wavelets will reduce the ICI and ISI because the DWT will Strengthens the orthogonality between the subcarrier [10]. The simulation results in [11, 12, 13] show this idea of replacement FFT by DWT in some multicarrier system and calculate the BER performance with these orthogonal bases. The simulations of LTE system with new transform have shown the dependence of channel on the performance of DWT and FFT. The main idea for using DWT in LTE system is the excellent spectral containment wavelet filters properties over Fourier filters. Under certain channel conditions, it has been found that DWT based LTE

does outperform better than FFT based LTE. The implementations of LTE in practice today have been done by using FFT and its inverse operation IFFT (or DWT and its inverse operation IDWT) to represent multicarrier modulation and demodulation. The Intersymbol interference (ISI) can be eliminated almost completely by adding a guard time interval in each packet of LTE frame and this will causes a lose about 25%-40% from data rate and this is one of the disadvantage of FFT-LTE. So the uses of DWT instead of DFT will increase the orthogonality between the subcarrier of the LTE packet and will be combat the narrowband interference and so no need to adding a guard time interval [14, 15].

The second proposed idea to LTE system is adding space time blocks coding (STBC) to the system. The STBC reduce the effect of multipath frequency selective Multipath fading channel .The aims this paper are designing a wireless communication system with least bit error rate (BER) for high data rate to fix stationary nodes and mobile users under multichannel models. These ideas will be implemented in LTE system by adding STBC with more than two antennas and using DFT or DWT [1,16,17].

The proposed STBC-LTE transceiver is shown in Fig. (1). All the type of space-time block codes with three transmitters or more has a coding rate of 1/2, to satisfy orthogonality condition. The space-time block code for four transmits antennas N = 4, with input symbols (S_1, S_2, S_3, S_4), the output will be over T = 8 symbol periods, thus the coding rate R =1/2 [18, 19]. At a given symbol period, four antennas transmitted four signals simultaneously. At time slot T0, transmitted signal from first transmitter (T_{x1}) is denoted by S_1, the signal from second transmitter (T_{x2}) by S_2 and the signal from third transmitter (T_{x3}) by S_3 and the signal from fourth transmitter (T_{x4}) by S_4. This process will go on in the same manner for each time slot until transmitting the last row of Table 1. This table has a rate of (1/2) and is used as STBC encoder to transmit any complex signal constellations [20,21]. For the four transmit and one receive antenna system, the channel coefficients are modeled by a complex multiplicative distortions, h_1 for the first transmit antenna, h_2 for the second transmit antenna and h_3 for the third transmit antenna. h_4 for the fourth transmit antenna[22]. Since some models used in this work are time varying and frequency selective for wide band mobile communication systems, so a dynamic estimation of channel is necessary to compensate LTE signal [4]..There are two types of channel estimations, block type and comb-type pilot channel estimation as shown in [15]. After pilot-carrier (training sequence) is generated as a bipolar sequence {±1}, the receiver previously knows this sequence. So the system can estimate the channel transfer function $h_1(t)$,$h_2(t)$,$h_3(t)$ and $h_4(t)$. The inverse of these channels also will be calculated. The channel transfer function estimation and the inverse of it are applied to each LTE packet to reduce the channel effects and bit errors rate BER, much like equalization [12]

Table (1). STBC mapping for four transmit antennas using complex signals

| Time slot | Four transmit antennas | | | |
| | Three transmit antennas | | | |
	T_{x1}	T_{x2}	T_{x3}	T_{x4}
Slot T0	S_1	S_2	S_3	S_4
Slot T1	$-S_2$	S_1	$-S_4$	S_3
Slot T2	$-S_3$	S_4	S_1	$-S_2$
Slot T3	$-S_4$	$-S_3$	S_2	S_1
Slot T4	S_1^*	S_2^*	S_3^*	S_4^*
Slot T5	$-S_2^*$	S_1^*	S_4^*	S_3^*
Slot T6	$-S_3^*$	S_4^*	S_1^*	$-S_2^*$
Slot T7	$-S_4^*$	$-S_3^*$	S_2^*	S_1^*

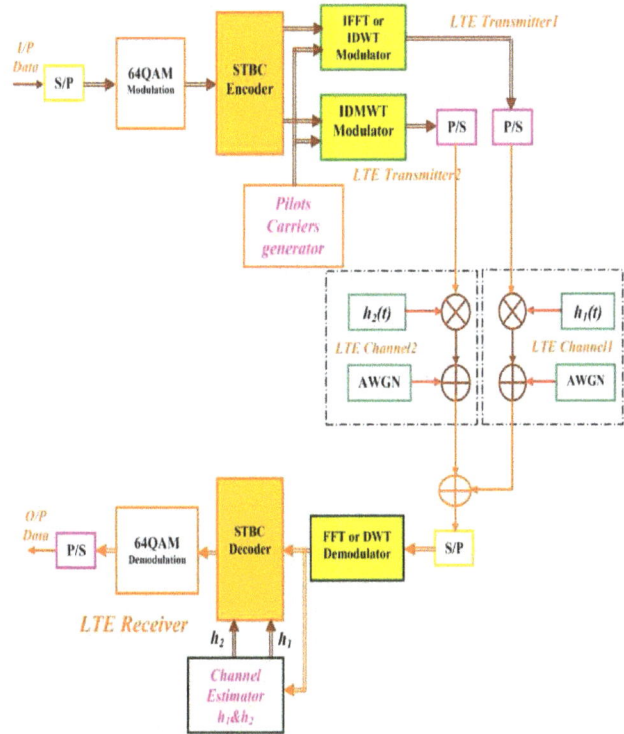

Fig. (1). Proposed Structure of STBC- LTE system

3. The Specific LTE Channel Models

The "Third Generation Partnership Project" 3GPP "Technical Recommendation" (TR) [3] defines three different type of multipath fading channel models: "the Extended Pedestrian A (EPA), Extended Vehicular A (EVA), and Extended Typical Urban (ETU)". All these channel-modeling functions will be used in this work examined the effect of these models. The higher-mobility profiles will not be used as "the closed-loop spatial-multiplexing mode". It is applicable to high data-rate and low-mobility scenarios only. These models enable the system to evaluate the performance of the proposed LTE transceiver in multichannel conditions reference. The model of any multipath fading channel can be defined by delay profiles and its relative power vectors. The maximum Doppler shift (*MDS*) or Doppler frequency must define with data rate in the channel model. The delay profiles

of these models of channel define at low, medium, and high delay spread environment, respectively corresponding to (5, 70, or 300 Hz) as the maximum Doppler shift as shown in table (2) that clarify the channel delay profile of each model with values of tap delay (in nanoseconds) and relative power (in decibels).

In multi input multi output (MIMO) scenario of transmission, the spatial correlations between transmit and receive antennas are important factor that affect directly to the overall transceiver connection performance.

Table (2). LTE channel models (EPA, EVA, ETU) and delay profiles [3]

Channel model	Excess tap delay (ns)	Relative power (dB)
Extended Pedestrian A(EPA)	[0, 30, 70, 90, 110, 190, 410]	[0, −1, −2, −3, −8, −17.2, −20.8]
Extended Vehicular A(EVA)	[0, 30, 150, 310, 370, 710, 1090, 1730, 2510]	[0, −1.5, −1.4, −3.6, −0.6, −9.1 ,−7, −12, −16.9]
Extended Typical Urban (ETU)	[0, 50, 120, 200, 230, 500, 1600, 2300, 5000]	[−1, −1, −1,0,0,0, −3 ,−5, −7]

4. Proposed STBC-LTE Systems Simulation Results

In this section the proposed STBC-LTE DWT and FFT systems had been simulated using MATLAB 2014a. The BER performance of the proposed LTE system considered in different type of channel models mentioned above , the AWGN channel, the flat fading channel, and the selective multipath fading channel. The carrier frequency used in all these scenario was 2.3GHz to fixed and mobile proposed LTE system with three values of *MDS* (5Hz represented a mobility speed of 2km/hr, 70Hz to speed 30 km/hr, and 300Hz to speed 120 km/hr) to used LTE channels models Path Loss (Extended Pedestrian, Extended Vehicular, Extended Typical Urban) that was mentioned in section (3) for selective Multipath fading channel. Table (3) shows the parameters used to simulate the proposed LTE systems.

Table (3). Parameter of Simulation LTE system

Parameter	Fixed LTE OFDM-PHY	Mobile LTE Scalable OFDMA-PHY			
Multicarrier size FFT or DWT	256	128	512	1024	2048
Number data used as subcarriers	96	64	180	360	720
Types of Modulation	64QAM				
Cyclic prefix or guard band (Tg/Tb)	1/16				
Bandwidth of Channel (MHz)	20	20	20	20	20

4.1. STBC-LTE Performance in AWGN Channel

As shown in the table (3), only the size of 256-subcarrier is used to (FFT or DWT) for fixed STBC-LTE. Also only AWGN model will be used to represent the channel model. The After simulation to all STBC-LTE system and collected the result of each system in Fig.(2). Also to compare the performance of these system, the BER=10^{-4} will be taken as a level of comparison. This figure shows that in FFT system reach the comparison level at SNR 38 dB for 1 antenna and SNR decreasing to 32dB in 4 antenna while in DWT system the SNR is about 18dB in 1 antenna and decreasing to 8dB in 4 antenna, so a gain of 6dB due to use STBC in FFT systems and 10dB in DWT system because of multiple antennas in transmitter side will enhance the spectral efficiency of LTE system and improve better error rate and these benefits come with no extra cost in power and little lost in bandwidth. Also from the same figure, it is shown clearly that the proposed DWT based STBC-LTE is much better than the FFT based STBC-LTE with a gain of 20dB, so with using of 4 transmitters with STBC-LTE, again of 30dB can be achieved. These results confirm that the DWT orthogonal base is more significant than the FFT orthogonal bases.

The size of (128,512,1024,2048) to (FFT or DWT) For mobile LTE used are as shown in table (3) .Generally the performance of BER decrease with the increasing of the subcarriers size as shown in fig.(3) , fig.(4) , fig.(5) and fig.(6). And also the idea of using STBC will decrease the BER and enhance the performance of system. And using of DWT instead of FFT will enhance the performances of proposed system.

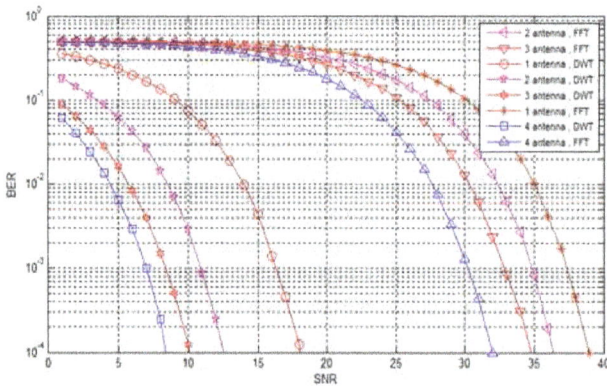

Fig. (2). *SNR Versus BER to Fixed STBC-LTE-256 subcarriers in AWGN channel model*

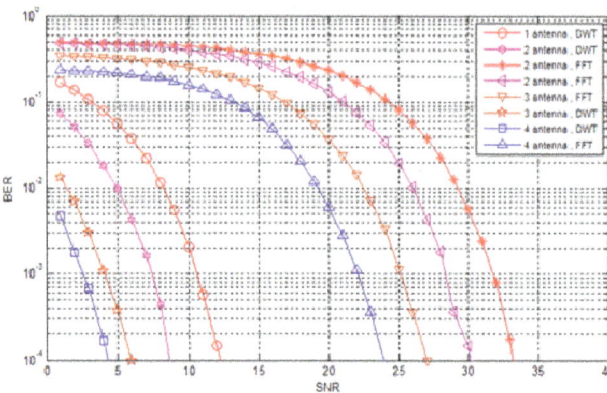

Fig. (3). *SNR Versus BER to Mobile STBC- LTE-128 subcarriers in AWGN channel model*

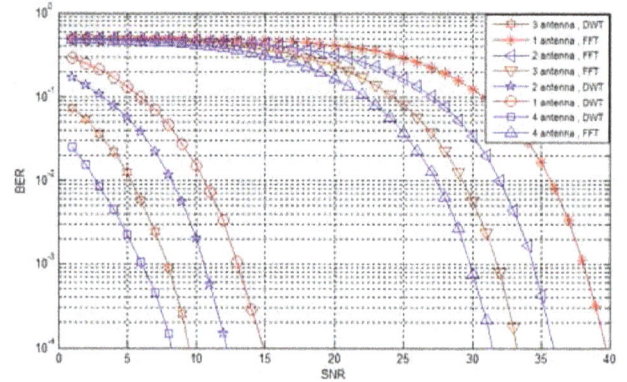

Fig. (4). *SNR Versus BER to Mobile STBC- LTE-512 subcarriers in AWGN channel model*

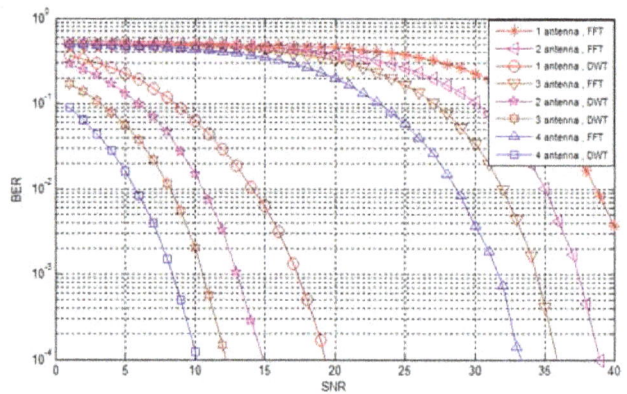

Fig. (5). *SNR Versus BER to Mobile STBC- LTE-1024 subcarriers in AWGN channel model.*

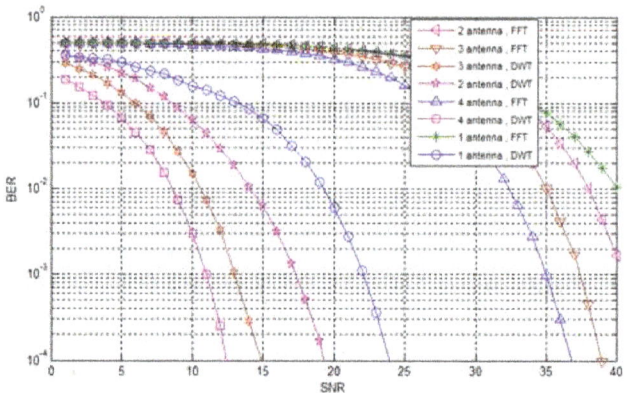

Fig. (6). *SNR Versus BER to Mobile STBC- LTE-2048 subcarriers in AWGN channel model*

4.2. STBC-LTE Performance in Flat Fading Channel

The same proposed STBC-LTE will examine by flat fading model in addition to AWGN. So the transmitted signals will be suffer from a constant attenuation and linear phase distortion through MIMO channels. The model that will represent flat fading channel will be the Rayleigh's distribution model. For fixed STBC-LTE system, the proposed system with STBC and DWT still performs better than the STBC-LTE based on FFT as shown in fig.(7). This figure shows that in FFT system reach the comparison level at SNR 39 dB for 1 antenna and SNR decreasing to 33dB in

4 antenna, while in DWT system the SNR is about 19dB in 1 antenna and decreasing to 9dB in 4 antenna, so a gain of 6dB due to use STBC in FFT systems and 10dB in DWT system due to multiple antennas used in transmitters to enhance the system spectral efficiency and supports BER performance. These benefits come with no extra cost in power and little lost in in data rate and bandwidth. Also from the same figure, it is shown clearly that the proposed STBC-LTE based on DWT is much better than the STBC-LTE based on FFT with a gain of 20dB, so with using of 4 transmitters with STBC-LTE, again of 30dB can be achieved.

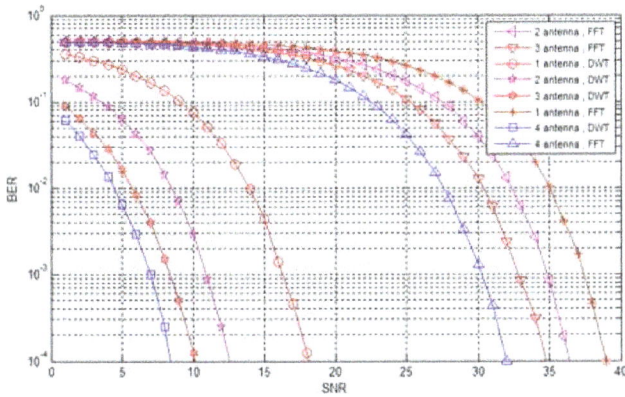

Fig. (7). SNR Versus BER to Fixed STBC-LTE-256 subcarriers in model of flat fading Channel

The simulation had been done in values of the Doppler frequencies (*MDS*) of 5Hz, 70Hz and 300Hz for mobile system that correspond to speed between the transmitter and receiver about 2, 30 and 120km/h respectively. In all subcarriers sizes, a smaller effect in BER performance appears in *MDS* =5Hz while the larger effect appears in high Doppler frequency *MDS* =300Hz as shown in fig.(8),(9) and (10) for subcarrier 128. And fig.(11),(12) and (13) for subcarrier 512. And fig.(14),(15) and (16) for subcarrier 1024. And fig.(17),(18) and (19) for subcarrier 2048. Also it is clear from all simulation results that proposed system STBC-LTE with DWT perform better performance the STBC-LTE based on FFT in all Doppler frequencies used in model of flat fading Channel.

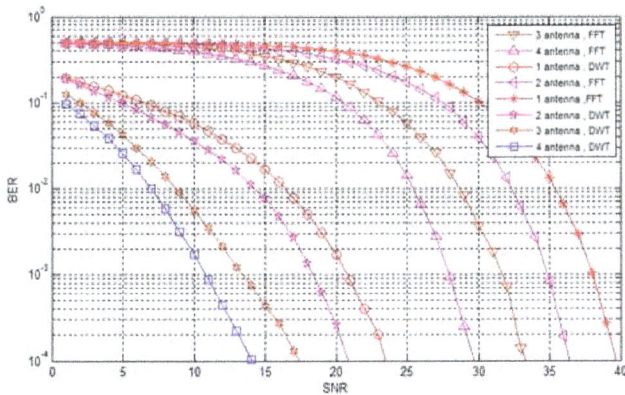

Fig. (8). SNR Versus BER to Mobile STBC-LTE-128 subcarriers in model of flat fading Channel-MDS=5 Hz

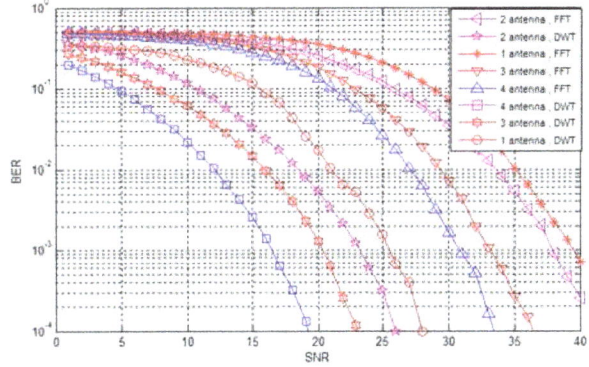

Fig. (9). SNR Versus BER to Mobile STBC-LTE-128 subcarriers in model of flat fading Channel - MDS=70Hz

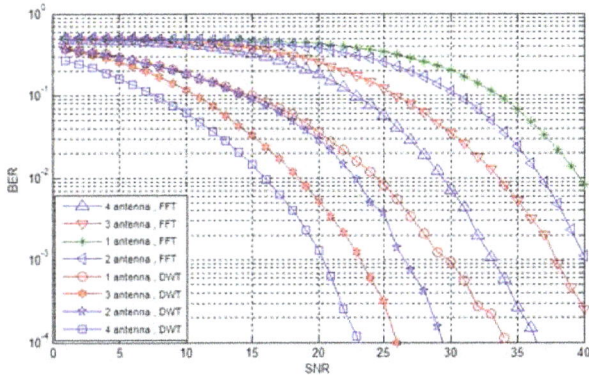

Fig. (10). SNR Versus BER to Mobile STBC-LTE-128 subcarriers in model of flat fading Channel - MDS=300Hz

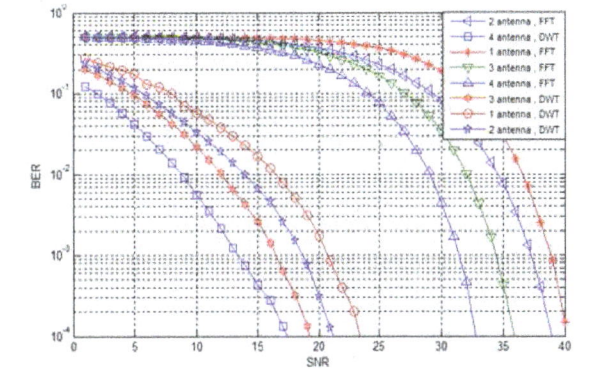

Fig. (11). SNR Versus BER to Mobile STBC-LTE-512 subcarriers in model of flat fading Channel - MDS=5Hz

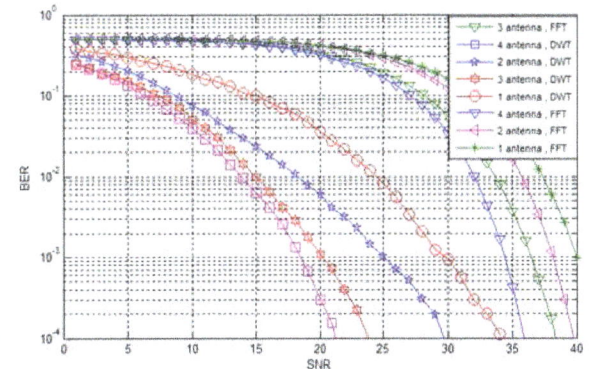

Fig. (12). SNR Versus BER to Mobile STBC-LTE-512 subcarriers in model of flat fading Channel - MDS=70Hz

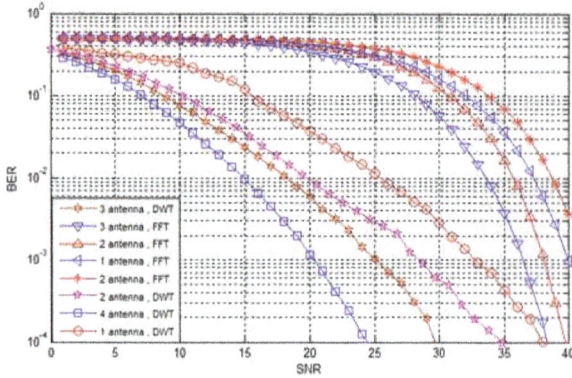

Fig. (13). *SNR Versus BER to Mobile STBC-LTE-512 subcarriers in model of flat fading Channel - MDS=300Hz*

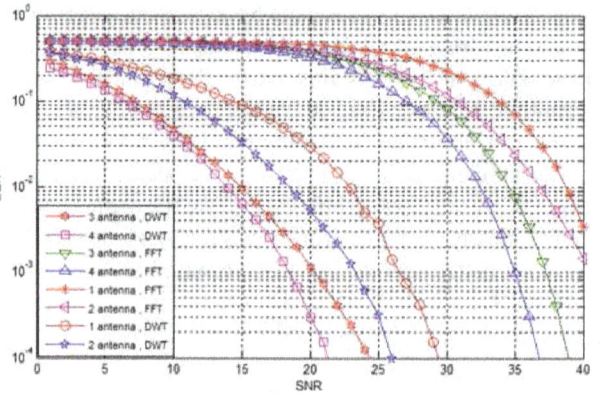

Fig. (14). *SNR Versus BER to Mobile STBC-LTE-1024 subcarriers in model of flat fading Channel - MDS=5Hz*

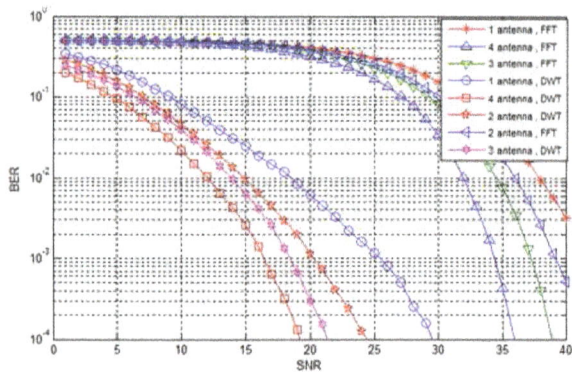

Fig. (15). *SNR Versus BER to Mobile STBC-LTE-1024 subcarriers in model of flat fading Channel - MDS=70Hz*

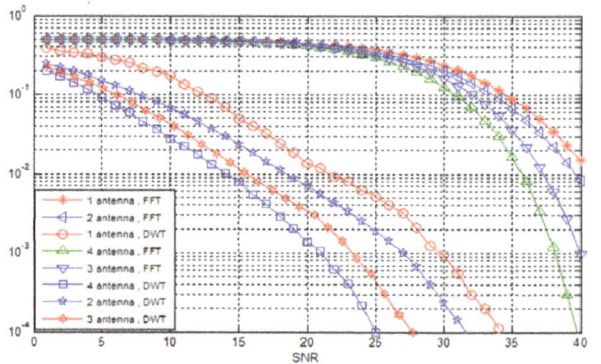

Fig. (16). *SNR Versus BER to Mobile STBC-LTE-1024 subcarriers in model of flat fading Channel - MDS=300Hz*

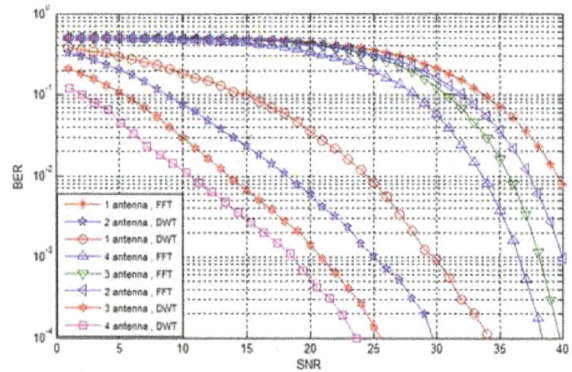

Fig. (17). *SNR Versus BER to Mobile STBC-LTE-2048 subcarriers in model of flat fading Channel - MDS=5Hz*

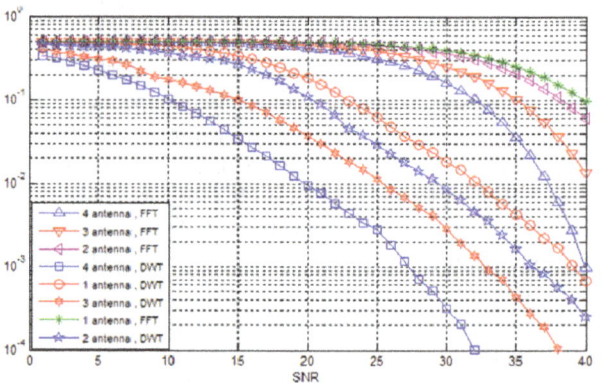

Fig. (18). *SNR Versus BER to Mobile STBC-LTE-2048 subcarriers in model of flat fading Channel - MDS=70Hz*

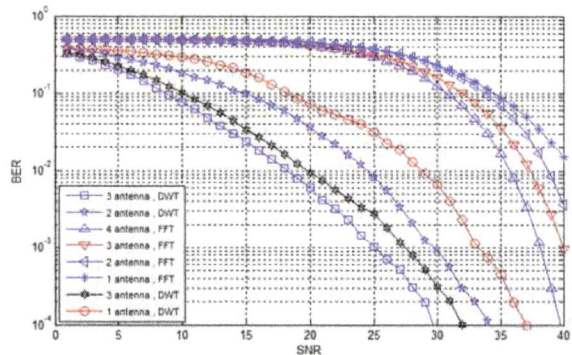

Fig. (19). *SNR Versus BER to Mobile STBC-LTE-2048 subcarriers in model of flat fading Channel - MDS=300Hz*

4.3. STBC-LTE Performance in Multipath Fading Channel

The same proposed STBC-LTE system shown in fig.(1) will examine by multipath selective Fading Channel model in addition to AWGN. So the transmitted signals will be suffer from multipath a constant attenuation and nonlinear phase distortion through MIMO channels. The model that will represent selective flat fading channel will be the Rayleigh's distribution model. All these model would been examining in three models of multipath fading channel: "the Extended Pedestrian A (EPA), Extended Vehicular A (EVA), and Extended Typical Urban (ETU)". In these types of channel, the transmitted signals frequency components are affected by uncorrelated changes corresponding to multipath. The Line of Sight (LOS) is one of them and the others paths are the reflected paths.

4.3.1. Extended Pedestrian A (EPA) Channel Model

The model of this channel was shown in table (1) first raw. The simulation had been done to all scenario of STBC-LTE with DWT and FFT and the simulation result shown in fig.(20) to fig.(26). In fixed STBC-LTE system it is noted that the proposed STBC-LTE based on DWT performs better than STBC-LTE based on FFT. The SNR at a BER 10^{-4} is about 19 dB for STBC-LTE based on DWT with 4 antennas and about 37 dB for STBC-LTE based on FFT with also using 4 antennas. A gain of about 18 dB has been achieved by using DWT way over the using FFT system and this value is much need in the communications systems to save transmitted power and increase data rate.

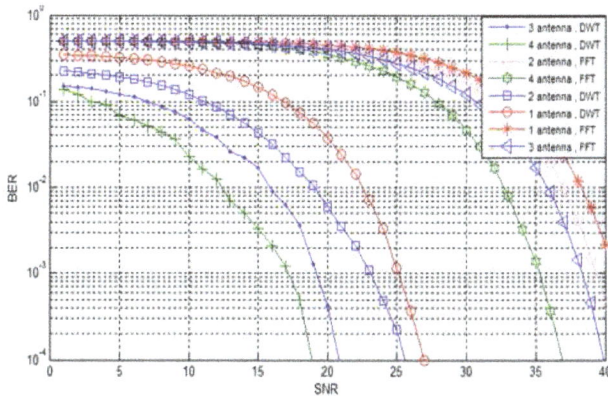

Fig. (20). *SNR Versus BER to Fixed STBC-LTE-256 subcarriers in Extended Pedestrian A (EPA) Channel model*

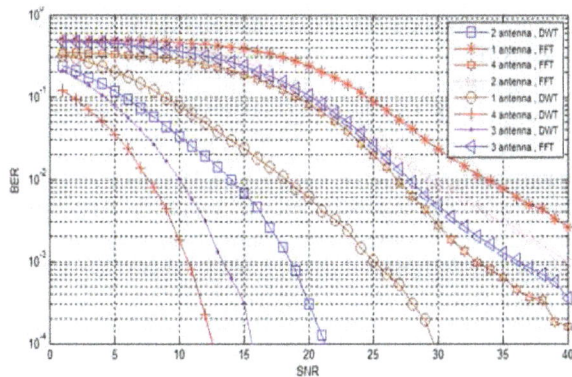

Fig. (21). *SNR Versus BER to Mobile STBC-LTE-128 subcarriers in Extended Pedestrian A (EPA) Channel model - MDS=5Hz*

For mobile STBC-LTE system, the results of 128 and 1024 subcarriers (the size of DWT or FFT) only have been found in three cases of *MDS* (5Hz, 70 Hz and 300 Hz). It can be seen from fig. (21) and fig. (22) (for small *MDS*=5Hz) that the proposed STBC-LTE based on DWT still performs better than STBC-LTE based on FFT and the system of STBC still gives good results for small and large subcarriers (128 and 1024 respectively). The SNR at a BER 10^{-4} is about 12 dB for 4 antennas in proposed system of 128 and about 17 dB in 1024, while it's not reach the desired value in STBC-LTE

based on FFT with 128 and 1024 subcarriers. In addition, a wide improvement span is obtained for all values of SNR in these systems.

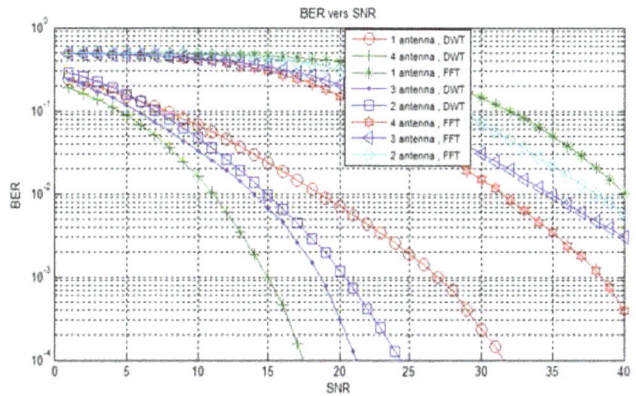

Fig. (22). *SNR Versus BER to Mobile STBC-LTE-1024 subcarriers in Extended Pedestrian A (EPA) Channel model - MDS=5Hz*

It can be seen from figures (23), (25) for 70 Hz and (24) ,(26) for 300 Hz that the STBC- LTE based on DWT is performing better than the STBC-LTE based on FFT but without STBC, LTE based on DWT is better because of the effect of STBC is eliminated in high Doppler frequency larger than 50 Hz and the same for other systems .

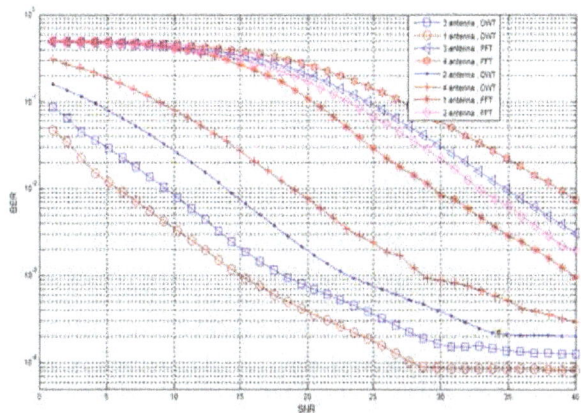

Fig. (23). *SNR Versus BER TO Mobile STBC-LTE-128 subcarriers in Extended Pedestrian A (EPA) Channel model - MDS=70Hz*

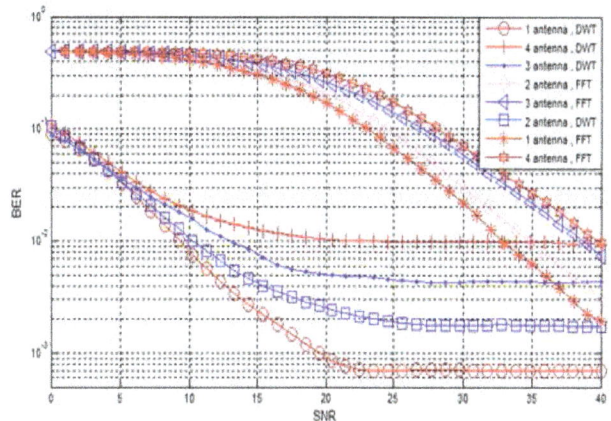

Fig. (24). *SNR Versus BER to Mobile STBC-LTE-128 subcarriers in Extended Pedestrian A (EPA) Channel model - MDS=300Hz*

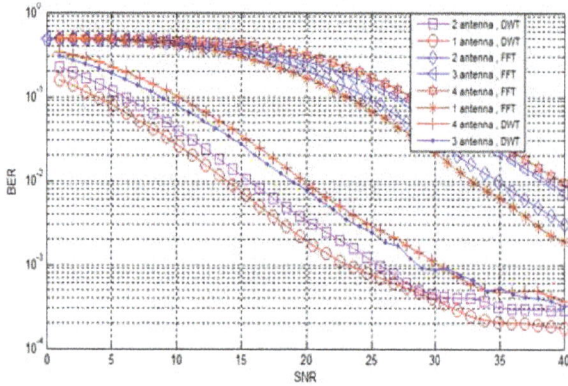

Fig. (25). *SNR Versus BER to Mobile STBC-LTE-1024 subcarriers in Extended Pedestrian A (EPA) Channel model - MDS=70Hz*

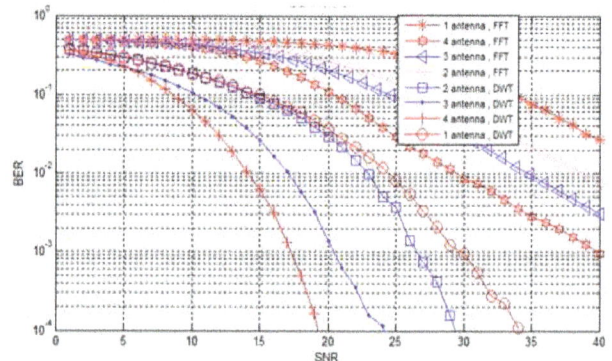

Fig. (26). *SNR Versus BER to Mobile STBC-LTE-1024 subcarriers in Extended Pedestrian A (EPA) Channel model - MDS=300Hz*

4.3.2. Extended Vehicular A (EVA) Channel Model

In this section the results of Extended Vehicular A (EVA) channel model will be simulated according to the table (1) second raw. For fixed system, It is clear from Fig.(27) that BER performance of STBC-LTE based on DWT is better than the system of STBC - LTE based on FFT. The SNR at a BER=10^{-4} is about 21dB for 4 antennas at proposed system and cannot been reached for STBC-LTE based on FFT system and this will give losses in the gain about 20 dB for proposed system against STBC-LTE with FFT system when compared with Extended Vehicular A (EVA) channel model.

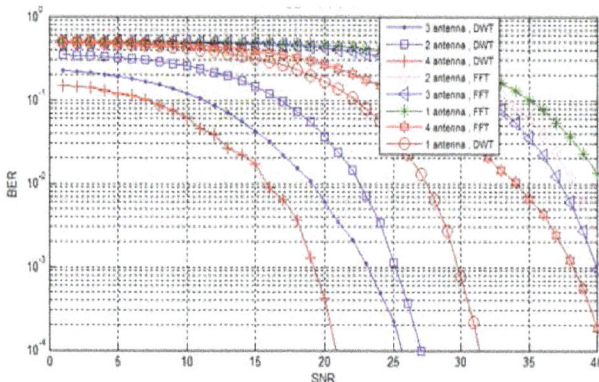

Fig. (27). *SNR Versus BER to Fixed STBC-LTE-256 subcarriers in Extended Vehicular A (EVA) Channel model*

Also for mobile system, the results have simulated only for 128 and 1024 (the size of DWT or FFT) in three cases of *MDS*=(5Hz, 70 Hz and 300 Hz).

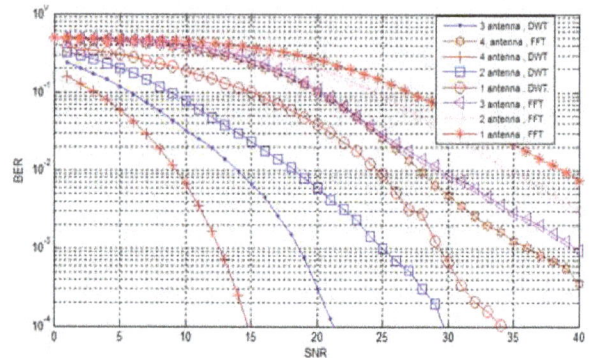

Fig. (28). *SNR Versus BER to Mobile STBC- LTE-128 subcarriers in Extended Vehicular A (EVA) Channel model - MDS=5Hz*

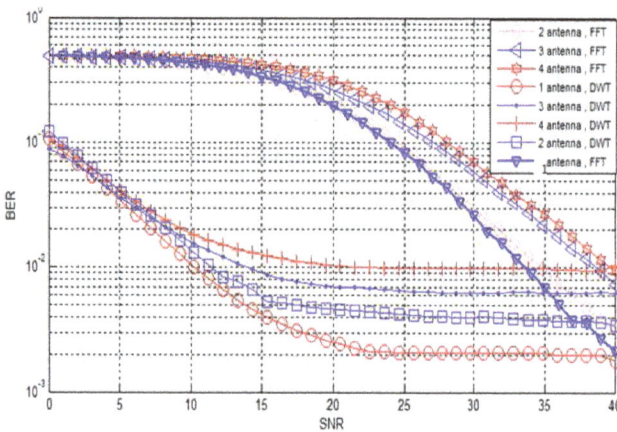

Fig. (29). *SNR Versus BER to Mobile STBC-LTE-1024 subcarriers in Extended Vehicular A (EVA) Channel model - MDS=5Hz*

From figure (28) and (29), it can be seen that the proposed DWT based STBC-LTE still performs better than FFT based STBC-LTE and the system of STBC still gives good results. The SNR at a BER 10^{-4} is about 15 dB for 4 antennas at proposed system of 128 and about 19 dB for 1024, while it is cannot been reach in FFT based STBC-LTE system of 128 and 1024, this means losses in gain is about 20 dB for the proposed STBC-LTE system and cannot been reach for the traditional system in comparison with the Extended Vehicular A (EVA) channel model at MDS of 5Hz.

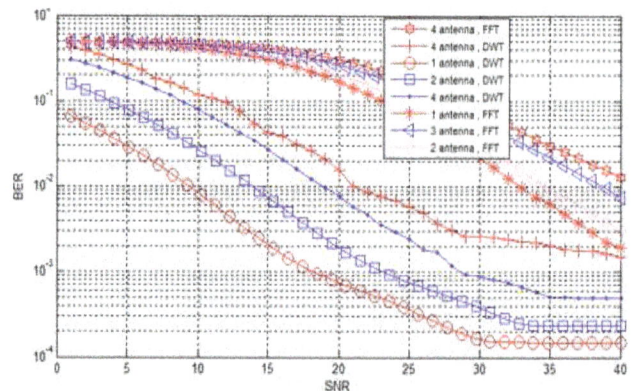

Fig. (30). *SNR Versus BER to Mobile STBC-LTE in Extended Vehicular A (EVA) Channel model -128 subcarriers- MDS=70Hz*

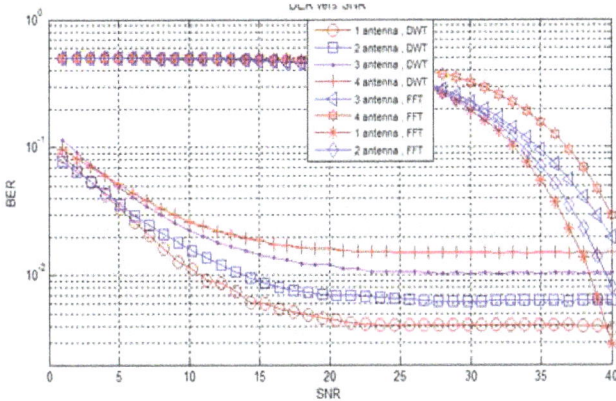

Fig. (31). *SNR Versus BER to Mobile STBC- LTE in Extended Vehicular A (EVA) Channel model -128 subcarriers- MDS=300Hz*

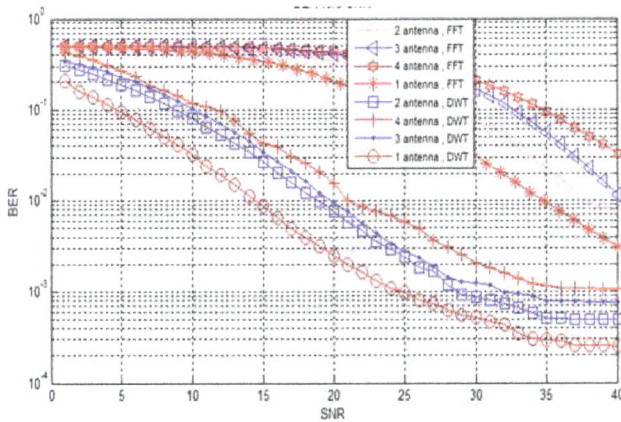

Fig. (32). *SNR Versus BER to Mobile STBC-LTE-1024 subcarriers in Extended Vehicular A (EVA) Channel model - MDS=70Hz*

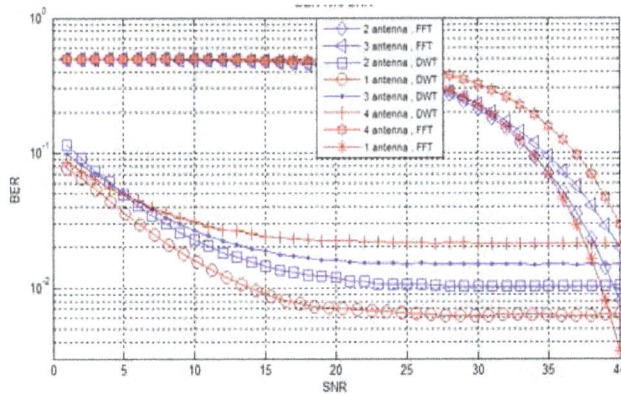

Fig. (33). *SNR Versus BER to Mobile STBC-LTE-1024 subcarriers in Extended Vehicular A (EVA) Channel model - MDS=300Hz*

It can be seen from figures above that the losses will be increased for both systems due to Doppler Effect.

4.3.3. Extended Typical Urban (ETU) Channel Model

In this section the results of Extended Typical Urban channel model for vehicular test environment will be achieved. In this case the results will be worse than two other channel models because there are six cases with higher relative delay. For fixed LTE, It can see that the proposed DWT based STBC-LTE still performs better than FFT based STBC-LTE. The SNR at a BER 10^{-4} is about 21 dB for 4

antennas at the proposed system and non at FFT based STBC-LTE.

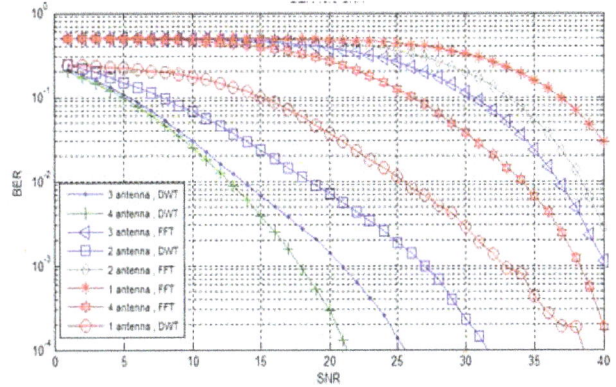

Fig. (34). *SNR Versus BER to Fixed STBC-LTE-256 subcarriers in Extended Typical Urban (ETU) Channel model*

For mobile system, the effect of MDS will appear and will directly affect on the system of STBC in the case of MDS higher than 50 Hz and the BER will increase as the Doppler frequency increases in both models. This will lead to results worse than the results of Extended Typical Urban (ETU) Channel model as display in the below figures.

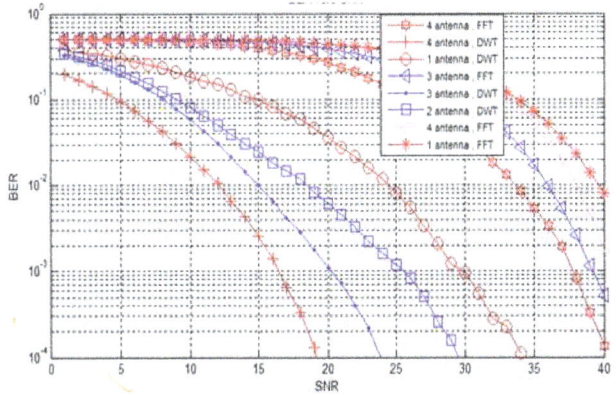

Fig. (35). *SNR Versus BER to Mobile STBC- LTE-128 subcarriers in Extended Typical Urban (ETU) Channel model -MDS=5Hz*

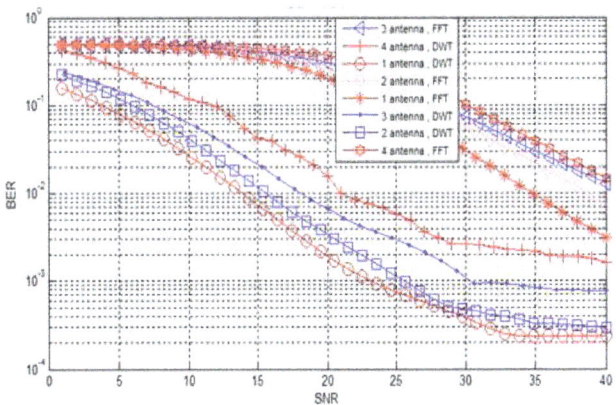

Fig. (36). *SNR Versus BER to Mobile STBC-LTE-128 subcarriers in Extended Typical Urban (ETU) Channel model -MDS=70Hz*

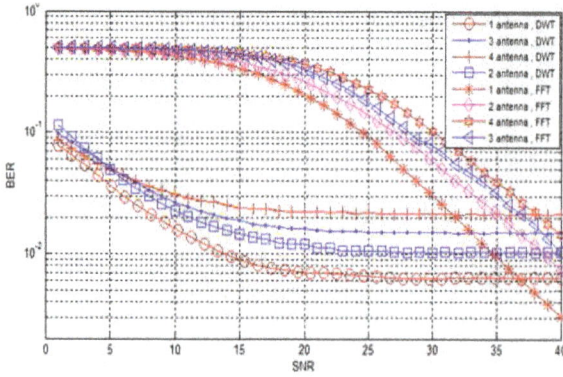

Fig. (37). *SNR Versus BER to Mobile STBC-LTE-128 subcarriers in Extended Typical Urban (ETU) Channel model -MDS=300Hz*

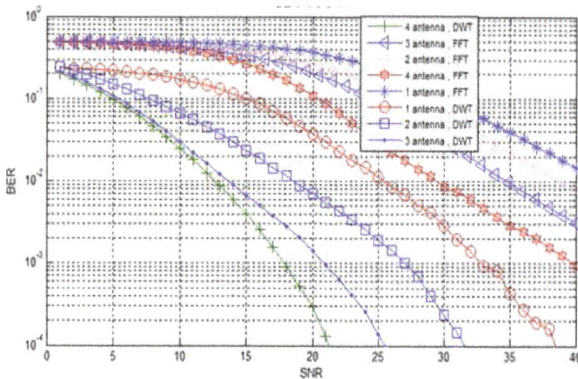

Fig. (38). *SNR Versus BER to Mobile STBC-LTE-1024 subcarriers in Extended Typical Urban (ETU) Channel model -MDS=5Hz*

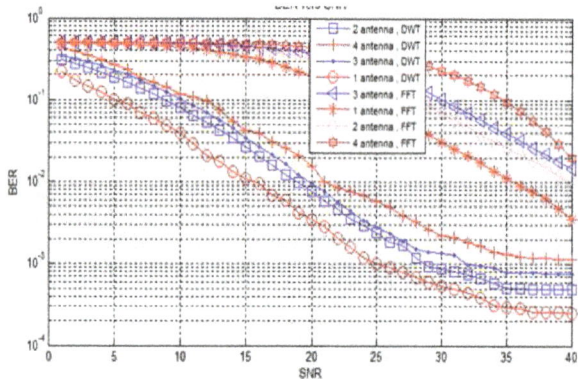

Fig. (39). *SNR Versus BER to Mobile STBC-LTE-1024 subcarriers in Channel model -MDS=70Hz*

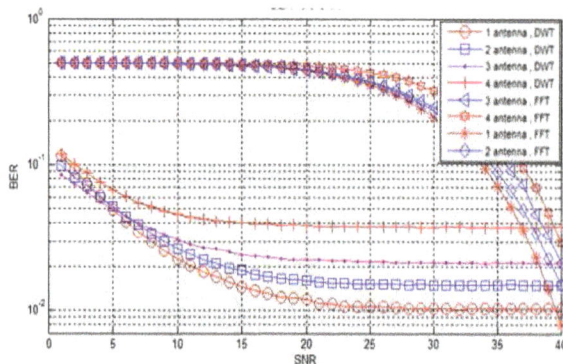

Fig. (40). *SNR Versus BER to Mobile STBC-LTE-1024 subcarriers in Extended Typical Urban (ETU) Channel model -MDS=300Hz*

All the results present in paper computed after testing the system used transfer over 1M symbols.

4.4. Discussion

The BER performance of AWGN channel as show from the simulation results is the best of all channels as it has the lowest bit error rate (BER) and power losses with using 64QAM modulation schemes. The amount of noise occurs in the BER to this channel type is quite slighter than flat fading channels models. Simulations proved that the accumulation STBC and DWT improved the bit error rates BER performance than accumulation STBC and FFT in LTE system. The simulation results show that the worst scenario to all of the STBC- LTE systems, in terms of performance, occurs when the system parameter is changed. The STBC-LTE with FFT shows that it requires a longer time adapting to noise and performs poorly to system parameter changes. It can be concluded from the comparison of the performance results of new STBC-LTE based DWT structure with the STBC-LTE based FFT that for the same model it gives a robust implementation and still performs better BER performance than FFT in all values of the Doppler frequencies model of flat fading Channel.

In the case of selective multipath fading channel, the simulation done in three cases of propagation losses model according to LTE channels models. Therefore it clear from the simulation results that DWT based STBC-LTE performs better than FFT based STBC-LTE ,but STBC advantage will be eliminated in high Doppler shift like at 70 Hz and 300 Hz because of each antenna attenuate each other and the losses in Extended Typical Urban (ETU) channel model is the worst case among other channel models Extended Pedestrian A (EPA), Extended Vehicular A (EVA), because of the combined effects of signal level attenuation in vehicular channel model due to the free space propagation, reflection, diffraction and scattering is more than those occurring in the other two channel models .

5. Conclusion

In this paper, first the companion of DWT and STBC in LTE structure has been designed and simulated for the first time. These simulations confirm successful operation and these structures are possibility of implementation. Also simulation results demonstrate BER performance enhancement that could achieve by combining the DWT and STBC techniques with very little decoding complexity to LTE system. As a result, the following points can be concluded:

1. It is clear that proposed STBC-LTE structure achieves much lower BER in AWGN, flat fading channel and selective Multipath fading channel. Therefore, this structure can be considered as an alternative to the conventional LTE system.

2. It is well known that the worst scenario to all of the STBC- LTE systems, in terms of performance, occurs when the system parameter is changed. The conventional scheme with FFT shows that it requires a longer time adapting to

noise and performs poorly to system parameter changes. It can be concluded from the comparison of the performance of this new structure with the FFT that for the same model it gives a robust implementation.

3. In selective Multipath fading channels models, the simulation results are presented by isolating individual propagation effects, to discover which channel parameters have the most significant impact on the performance. In Doppler shift, it is seen that DWT based STBC- LTE still performs well better than FFT based STBC- LTE, but STBC advantage will be eliminated or lost in high Doppler shift above 50 Hz because of each antenna attenuate each other and the losses in Extended Typical Urban (ETU) channel model is the worst case among other channel models Extended Pedestrian A (EPA), Extended Vehicular A (EVA),.

References

[1] Baig S. and Mughal M. J., 2007 "A Frequency Domain Equalizer in Discrete Wavelet Packet Multitone Transceiver for In-Home PLC LANS", IEEE International Symposium on Power Line Communications and its Applications, March 2007, PISA, Italy.

[2] Fazelk and S. Kaiser-2002, "Multi-Carrier and Spread Spectrum Systems". 1st Edition, John Wiley & Sons.

[3] Kitming, Tommy C.-2002, "Hybrids OFDM- CDMA: A Comparison of MC/DS-CDMA, MC-CDMA and OFCDM", Dept. of Electrical and Electronic, Adelaide University, SA 5005, Australia.

[4] Koga H., N. Kodama, and T. Konishi-2003, "High-speed power line communication system based on wavelet OFDM," in Proc. IEEE ISPLC 2003, Kyoto, Japan, May, pp. 226-231.

[5] Keita I., Daisuke U., and Satoshi D., 2007 "Performance Evaluation of Wavelet OFDM Using ASCET" IEEE.

[6] You-L. C. and Shiao-L. T., 2012 "A Low-Latency Scanning with Association Mechanism for Real-Time Communication in Mobile WiMAX", IEEE Transactions On Wireless Communications, Accepted For Publication.

[7] Qinghua Shi, Yong Liang Guan, Yi Gong and Choi Look Law, 2009 "Receiver Design for Multicarrier CDMA Using Frequency-Domain Oversampling", IEEE Transactions on Wireless Communications, Vol. 8, No. 5.

[8] Roberto C. and García A., 2009 "Joint Channel and Phase Noise Compensation for OFDM in Fast-Fading Multipath Applications", IEEE Transactions on Vehicular Technology, Vol. 58, No. 2.

[9] Sobia B., Gohar N.D., Fazal R., 2005 "An efficient wavelet based MC-CDMA transceiver for wireless communications", IBCAST

[10] Shun-Te Tseng and James S. Lehnert, 2009 "Windowing for Multicarrier CDMA Systems", IEEE Transactions on Communications, Vol. 57, No. 10

[11] Third Generation Partnership Project (3GPP), Dec. 2004 "Universal Mobile Telecommunications System (UMTS); Deployment aspects," (release 6), 3GPP. TS 25.943, version 6.0.0.

[12] Vasily S., Peter N., Gilbert S., Pankaj T., and Christopher H.-1999, "The Application of Multiwavelet Filterbank to Image Processing", IEEE Transactions on Image Processing, Vol. 8, No. 4, pp(548-563),April.

[13] Yeen, Linnartz J-P and Fettweis G. -1993, "Multicarrier CDMA in Indoor Wireless Radio Networks". Proc. of IEEE PIMRC 1993,Yokohama, Japan, Sept. , pp.109-13

[14] Yu-Wei Lin, Hsuan-Yu Liu, and Chen-Yi Lee-2005 "A 1-GS/s FFT/IFFT Processor for UWB Applications" IEEE Journal of Solid-State Circuits, Vol. 40, No.8, pp.1726-1735.

[15] Yuan D., Zhang H., Jiang M. and Dalei Wu-2004 "Research of DFT-OFDM and DWT-OFDM on Different Transmission Scenarios." Proceedings of the 2nd International Conference on Information Technology for Application (ICITA), pp. 31–33.

[16] Zhang H., D. Yuan, M. Jiang and Dalei Wu-2004 "Research of DFT-OFDM and DWT-OFDM on Different Transmission Scenarios." Proceedings of the 2nd International Conference on Information Technology for Application (ICITA).

[17] Gupta A., Chandavarkar B. R.,2012, "An Efficient Bandwidth Management Algorithm for WiMAX (IEEE 802.16) Wireless Network EBM Allocation Algorithm." IEEE Industrial and Information Systems (ICIIS).

[18] H. Zarrinkoub "Understanding LTE with MATLAB : from mathematical foundation to simulation, performance evaluation andimplementation" John Wiley & Sons, Ltd,2014

[19] P. Tong, R. C. de Lamare, and A. Schmeink, "Adaptive distributed space-time coding based on adjustable code matrices for cooperative MIMO relaying systems," IEEE Trans. Commun., vol. 61, no. 7, pp. 2692–2703, Jul. 2013.

[20] Third Generation Partnership Project (3GPP), "Evolved Universal Terrestrial Radio Access (E-UTRA); Physical Channels and Modulation," standard specifications TS 36.211, 2009, available on www.3gpp.org, accessed on February 10, 2014.

[21] M. Suryanegara, and M.Asvial, "In Searching for 4G Mobile Service Applications: The Case of Indonesia Market," Telecommunications Journal of Australia, Vol.63, No.2, 2013.

[22] His-LuChao,Chia-kai Chang,Chia-Iung Liu, "A novel channel-aware frequency-domain scheduling in LTE uplink", Wireless Communication and Networking Conference (WCNC),2013 IEEE

12

Modeling of Congestion and Traffic Control Techniques in ATM Networks

Bourdilllon Odianonsen Omijeh, Philip Ogah

Department of Electronic & Computer Engineering, University of Port Harcourt, Port Harcourt, Nigeria

Email address:

bourdillon.omijeh@uniport.edu.ng (B. O. Omijeh), omijehb@yahoo.com (B. O. Omijeh), ogan.pd@gmail.com (P. Ogah)

Abstract: In this Paper, Computer – based Simulation models for effective Congestion control and Traffic management in Asynchronous Transfer Mode (ATM) network have been developed providing a basis for monitoring ATM networks performance for traffic and congestion control purposes ,providing a system with a reduce short -term congestion in ATM networks, and enhancing a fair operation of networks in spite of the challenges in designing ATM traffic management system to make maximal use of network resources. An IDCC scheme was implemented, applying IDCC methodology to the ATM Network. Using analysis performance, limits were created for robust controlled network behaviour, as dictated by reference values of the desired queue length. By tightly controlling output of the controller, the overall network performance was adjusted and also controlled. A simulation tool, MATLAB/SIMULINK, was used for this purpose. An improvement was observed in the delay performance of ATM networks. The results were obtained by running several simulations and populating a table with the outcome over a number of simulation runs. The effectiveness of the congestion control techniques was tested by analysing the dynamic performance of the model through variation of some parameters. The performance of this model proved to be efficient if applied in the ATM network of today.

Kewords: ATM, IDCC, Traffic, Congestion, Model, Matlab/Simulink

1. Introduction

Congestion is seen as a serious "situation in Communication Networks because too many packets are present in a part of the subnet, performance degrades. Congestion in a network may occur when the load on the network *(i.e.* the number of packets sent to the network) is greater than the capacity of the network *(i.e.* the number of packets a network can handle)"[1].Therefore in an actual sense, Congestion occurs when resource demands exceed the capacity. As users come and go, so do the packets they send. "During a session, each host may transmit one or more packet flows to the other host a flow is commonly considered to be a series of closely spaced packets in one direction between a specific pair of hosts" [2]

The main aim of this paper is to develop an effective congestion Control and traffic management model in ATM Networks.

2. Related Works

"ATM congestion can be defined as a state of network elements in which the network is not able to guarantee the negotiated Network Performance objectives for the already established connections Congestion can be caused by unpredictable statistical fluctuations of traffic flows Since an ATM network supports a large number of bursty traffic sources, statistical multiplexing can be used to gain bandwidth efficiency, allowing more traffic sources to share the bandwidth. But if a large number of traffic sources become active simultaneously, severe network congestion can result" [15].

"In ATM networks, such circuits are called virtual circuits (VCs) The connections allow the network to guarantee the quality of service by limiting the number of VCs. Typically, a user declares key service requirements at the time of connection set up, declares the traffic parameters and may agree to control these parameters dynamically as demanded by the network Asynchronous Transfer Mode (ATM) networks fulfill high-rate transmissions and heterogeneous requirements with very high rates and different types of services" [3].

"On the other hand, they are less predictable networks

These "types of traffic patterns imposed on ATM networks, as well as the transmission characteristics of those networks, differ markedly from those of other switching networks. One reason is due to the applications that may generate very different traffic patterns and they require different network services. Another reason for this is that trusted statistics from traffic characteristics and connection establishment patterns are not given"[4].

"ATM networks also guarantee the engaged Quality of Service(QoS), so when a connection is accepted, the network must fulfill all the agreements reached with the user. In effect, the network and the subscriber inter into a traffic contract. But due to the statistical multiplexing of the traffic, an inherent feature of ATM, it is possible to have congestion even though all connections carry out their contracts" [5].

"On the one hand, networks need to serve all user requests for data transmission, which are often unpredictable and bursty with regard to transmission starting time, rate, and size.

On the other hand, any physical resource in the network has a finite capacity, and must be managed for sharing among different transmissions"[4].

2.1. Mathematically Speaking

"Network protocols which use aggressive retransmissions to compensate for packet loss tend to keep systems in a state of network congestion, even after the initial load has been reduced to a level which would not normally have induced network congestion. The stable state with low throughput is known as congestive collapse "[6][17].

2.2. Congestion Control

"Congestion control mechanisms are however divided into two categories: one category prevents the congestion from happening (Preventive congestion control) and the other category removes congestion after it has taken place'[1] (Reactive congestion control).

2.3. Traffic Management

"Monitoring ATM networks performance for traffic and congestion control purposes leads to difficulty because there are some parameters whose values must be tuned. These parameters are not only based on users' traffic descriptors but on the different traffic control policies implemented"[4].

In ATM network, traffic policing is located at the access point to the User Network Interface (UNI). What traffic policing does is to check the validity of VPI and VCI (Virtual Channel Identifier) values and monitor the cells arriving at the network and determines whether it violates any parameters. In ATM network, two most popular methods exist to control the peak rate, mean rate and different load States. These are the Leaky Bucket method and Window Method otherwise referred to as the Usage parameter Control (UPC) according to Uyles Black (1995). The UPC detects any nonconforming sources and the necessary actions it takes are as follows:

i.   To dropping violating cell
ii. To delay violating cells in queue
iii. To mark violating cells and treat them accordingly
iv. To control traffic by informing the source to take action when violation occurs.

"One of the earliest documents that mention the term 'congestion collapse' is by John Nagle; here, it is described as a stable condition of degraded performance that stems from unnecessary packet retransmissions. Nowadays, it is, however, more common to refer to 'congestion collapse' when a condition occurs where *increasing* sender rates *reduces* the total throughput of a network" .[7]"The existence of such a condition was already acknowledged in [8] which even uses the word 'collapse' once to describe the behaviour of a throughput curve"[19]. "Therefore, even if a network adopts a strategy of congestion avoidance, congestion recovery schemes would still be required to retain throughput in the case of abrupt changes in a network that may cause congestion" [15].

"Congestion control mechanisms are divided into two categories, namely: Open Loop and Closed Loop. In this open loop, policies are used to prevent the congestion before it happens. Congestio control is handled either by the source or by the destination while in the closed loop the congestion is removed after it happens" [10]

2.4. Traffic Modeling

A key element in simulating or analyzing communication networks is traffic modeling. "Traffic models reflect our best knowledge of traffic behavior" [11]. "Traffic is easier to characterize at sources than within the network because flows of traffic mix together randomly within the network. When flows contend for limited bandwidth and buffer space, their interactions can be complex to model"[12-14].

In order to obtain a successful performance evaluation, a clear understanding of the following is crucial: (i) The selection of a suitable random traffic model and (ii) The pattern or nature of the traffic in the target system.

3. Simulated Design of a Congestion and Traffic Management Model

The Design Stages involved in the work include: (i) Developing a dynamic Fluid Flow Model: This has the ability to capture the essential dynamics of the system. (ii) Designing of inflow to the switch: This is helps in regulating the flow of traffic at the source (consumer/user end) into the ATM network (iii) Designing of a simple non-linear congestion controller: This model attempts to keep a queue buffer length close to a reference value without knowledge or measurement of the flow in for both available (best effort) and guaranteed sources. (iv) Design of the queuing state model in the buffer of our ATM design.(v) Designing of Linear Controller Model Subsystem: For the implementation of congestion control by an Integrated dynamic congestion controller (IDCC).

3.1. Congestion Control in an ATM Network

The block diagram of the congestion control in an ATM Network is represented in Figure 1.

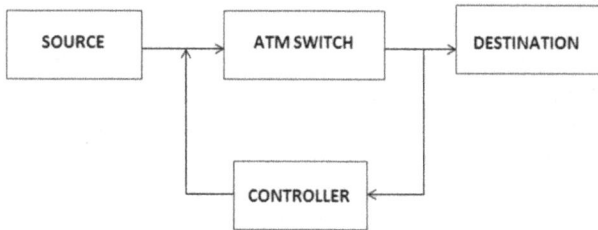

Figure 1. Block Diagram of Congestion Control System.

Source: This originates the traffic that transverses the links and ATM switches of an ATM network. This traffic which could be in form of data, voice or video. ABR traffic source (which is a model of real time services) is represented as λ^{ABR} (t).

ATM Switch: This "accepts the incoming cell from an ATM source or another ATM switch. It then reads and updates the cell-header information and switches the cell to an output interface toward its destination. The basic operation of an ATM switch is easy: The cell is received across a link on a known VCI (Virtual channel identifier) or VPI (Virtual path identifier) value"[19]. For a dynamic fluid flow model of ABR "traffic through the buffer of an ATM switch and assuming the link has a First-In-First-Out (FIFO) service discipline, the following standard assumptions are made: packets arrive according to a Poisson process, packet transmission time is proportional to packet length, and that the packets are exponentially distributed with mean length 1"[16].

The queue state model is given in (1)

$$x(t) = \frac{-x(t)}{1+x(t)} C(t) + \lambda ABR(t) \tag{1}$$

For several numbers of switches (1) can further be extended to (2).

$$x_i(t) = -C_i(t)\left[\frac{X_i(t)}{1+X_i(t)}\right] + \gamma_i ABR(t)\ i=2,\dots,M. \tag{2}$$

Where
$C_{i(t)}$ = bandwidth (capacity, cell service rate)allocated to the ABR traffic at node i

$x_{i(t)}$ = state of the queue (i.e. ensemble average of number of cells in the queue) at node i,

$\lambda^{ABR}_{(t)}$ = arrival rate due to ABR traffic
$\gamma_i^{ABR}_{(t)}$ = ABR traffic entering node i, arriving from the previous node, delayed by a deterministic amount $\tau_i - 1$ due to the transmission propagation and is further represented in Equation(3)

$$\gamma_i^{ABR}(t) = \left[C_{i-1}(t-\tau_{i}-1)\frac{x_i ABR-1(t-\tau_i-1)}{1+x_{i-1}(t-\tau_i-1)}\right] \tag{3}$$

Where
x_I^{ABR} (t) is the ensemble average of the number of cell places in the buffer occupied by ABR traffic.

3.2. Integrated Dynamic Congestion Controller

The controller used in the work is the IDCC (Integrated Dynamic Congestion Controller) as shown in Figure 2.

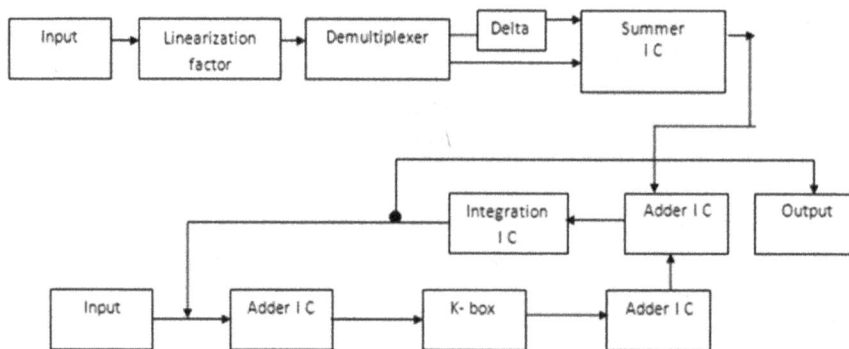

Figure 2. Block Diagram of the IDCC Architecture.

Input: The controller input is usually a control signal denoted by (4).

$$Input1 = C_{max} - C_{p(t)} \tag{4}$$

Where, C_{max} = maximum rate at which packets is transmitted

$C_{p(t)}$ = control signal determined by the congestion controller.

Linearization factor ($P_{i(x)}$): This helps to simplify our design procedure and reduce the model to a linear one as

shown in (5)

$$P_{i(X)} = \begin{cases} (x_p + 1)/x_p & x_p >= 0.2 \\ 0 & x_p < 0.2 \end{cases} \tag{5}$$

Where x_p = state of the queue or queue length.
Demultiplexer: This actually helps to separate combined signal into separate single signal component.
Delta (Δ): This denotes model uncertainties.
Summer IC: This circuitry is similar to the multiplexer, it tends to add two different signals together.

K-box: This box is comprised of what is known as the controller factor (K) as shown in (6)
Where

$$K = K_c * Kc_{\_bar} \qquad (6)$$

$$Kc = x/(x+1) \qquad (7)$$

Integration IC: This circuitry is responsible for producing outputs that are integral of its input at the current time step. The following equation represents the output of the block y as a function of its input u and an initial condition y_0, where y and u are vector functions of the current simulation time t.

$$y(t) = \int_{t_0}^{t} u(t)dt + y_0$$

The selected control strategy for the switch is developed using (2) as follows:
Let

$$x\_bar = x - x\text{ref} \quad \dot{x}\_bar = \dot{x} \qquad (8)$$

Then from Equation (3.2), we have:

$$\dot{X}\_bar(t) = -C(t) [x(t)/1 + x_{(t)}] + \lambda^{ABR}(t) \qquad (9)$$

The control input is given by (10) below

$$C = \rho[(1+x(t))/x_{(t)}] [\alpha\, x\_bar + k] \qquad (10)$$

$$\rho = \begin{cases} 0 & \text{if } x \le 0.01 \\ (100/99)\,x - (100/99 - 1) & 0.01 < x \le 1 \\ 1 & x > 1 \end{cases} \qquad (11)$$

$$K\_dot = P_r[\delta\, x_{bar}] \qquad 0 \le k(0) \le \check{K}$$

$$P_r[\delta x_{bar}] = \begin{cases} \delta\, x & \text{if } (0 \le k(0) \le \check{K}) \\ 0 & \text{else} \end{cases} \qquad (12)$$

$A, \delta > 0$ and $\check{K} >= \lambda^{ABR}(t)$ are design constants.

$$\dot{X}\_bar(t) = -\rho\,\delta\, x - \rho\, k_o + \lambda_1^{ABR}$$

Where $k_o = \lambda_1^{ABR}$ $\check{K} = k + k_o$ and $k_d = P_r[\delta\, x\_bar]$
For this research work, the following Lyapunov's function for analysis was proposed

$$V_{dot} = \frac{X\_bar^2}{2} + \check{K}^2/2\delta_1$$

The time derivative of V from (10) and (11) is

$$V_{dot} = \rho\,\alpha\, x\_bar^2 - \rho\, k\, x\_bar + x\_bar(\lambda^{ABR} - \rho\, k*) + P_r[x\_bar]$$
$$= -\rho\,\alpha\, x\_bar^2 + x\_bar(\lambda^{ABR} - \rho\, k*) + (P_r[x\_bar] - \rho\, x... \qquad (13)$$

Equation (13) can also be write as,

$$V_{dot} \begin{cases} x\_bar(\lambda ABR) + P_r[x\_bar] & \text{if } 0 \le x \le 1 \\ (x-1)\,\alpha\, x\_bar2 + x\_bar(\lambda ABR - (x-1)\,k*) + (P_r[x\_bar] - (x-1)\,x\_bar) & \text{if } 1 \le x \le 2x \\ -\alpha\, x\_bar2 + x\_bar(\lambda ABR - k*) + (P_r[x\_bar] - x\_bar) & \text{if } x\_bar > 2 \end{cases} \qquad (14)$$

Examining the properties of V_{dot} we have
$V_{dot} = 0$ for $x\_bar = 0$, and $V_{dot} < 0$ for $x\_bar \ne 0$
From Lyapunov theory and additional arguments it can be deduced that x_{bar} is bounded and $x_{bar(t)} \to 0$ as $t \to \infty$. With the above choice of C we can confidentially state that the maximum possible value C_{max} of C satisfies

$$C, max \le 2[\alpha(x,_{buff} - x^{ref}) + k\_bar] \qquad (15)$$

Where α and k_bar are design constants which provide some flexibility in meeting possible constraints.

3.3. Algorithm for Computing Process in a Congestion Controller

From (2), we have that the control strategy of our IDCC controller is given by,

$$x(t) = -C(t) \left[\frac{x(t)}{1+x(t)} \right] + \lambda^{ABR}(t), \, i=2...\, M$$

The control input described in (3) is given by,

$$C = \rho\, [(1+x(t))/x_{(t)}] [\alpha\, x\_bar + k]$$

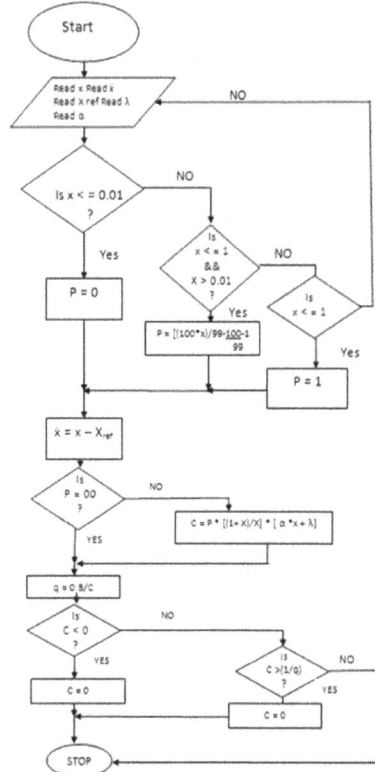

Figure 3. Algorithm for computing process in congestion controller.

From Figure3 q= cell service time and x (t), α , x_bar are kept constant.

The meanings of the symbols are treated in (2).The flow process is shown in Figure 4.

3.3.1. Computation Process of the Controller

This process is shown in Figure 3.

The process of setting the reference value for an available traffic is elucidated in Fig 4 below.

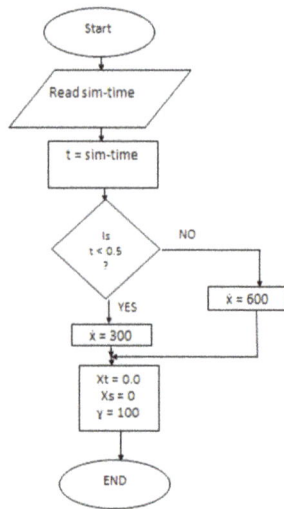

Figure 4. *A flowchart of reference value setting in Available traffic.*

Where, sim-time = simulation time
Xt = total queue length
Xs = Reading at start time
y = ABR traffic entering a switching node

3.3.2. Setting Reference Value for Abr Traffic

DESTINATION: This serves as the final end point for the traffic that transverses the links and ATM switches of an ATM network.

3.4. Modeling of Dynamic Fluid Flow Model and Non-Linear Congestion Controller

The Matlab model of congestion control in ATM network is better understood by representing a dynamic fluid flow model on Matlab/simulink as shown in Figure 5 developed in this work.

3.4.1. Dynamic Fluid Flow Model

This model is made up of the source(s), a flow controlled subsystem, transmission delay, a switch and non-linear controller subsystem and the destination.

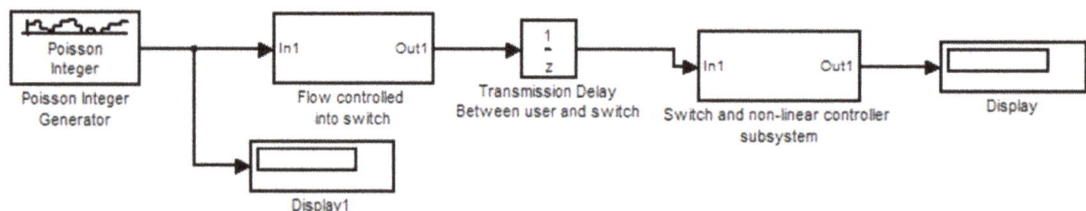

Figure 5. *Computer-based Design of Dynamic Fluid Flow Model.*

3.4.2. Poisson Integer Generator

The Poisson Integer Generator block models the ABR traffic that transverses through the ATM network.it generates random integers using a Poisson distribution.

$$\lambda^{k} \exp(-\lambda)/(k!) \qquad (16)$$

where; λ = positive number known as the Poisson parameter, exp = base of the natural logarithm.

3.4.3. Non-Linear Congestion Controller

The model as designed in this work is seen in Figure 6.

The flow control subsystem helps in regulating the flow of traffic at the source (consumer/user end) into the ATM network. ABR traffic passes through point 1 to Add1 and also through the limiter to Add2.The output of the two Add blocks were passed to the switches. The switches looks at the output of the Add block to determine if it's going to pass the third or the first input. The output of the switch 1 and switch 2 are being fed into another set of Add block, the Add3 and Add4

block. The Add3 block subtracts the output of Add5 block from the output of switch2. The Add4 block sums the output of Add5 block and switch1 block. The output of the Add3 block is passed into the integrator block that outputs the integral of its input at the current time step .This output is being fed back into the Add5 block to be added to the ABR traffic from the Poisson integer generator. Now, the output of the Add4 which forms the controlled ABR traffic is now passed to the ATM switch subsystem.

3.4.4. Transmission Delay

This deals with delay in transmission link between the ABR traffic source generator(consumer) and ATM switch. The Unit Delay block delays controlled ABR traffic by sample period of -1 seconds. The delay block is used to model transmission delay and it is Equivalent to the z^{-1} discrete-time operator. The block accepts the ABR traffic input and generates one output. If the input is a vector, all of its elements are delayed by the same sample period.

The value of ABR traffic due to delay is stated in (2).

3.4.5. Switch and Non-Linear Controller Subsystem

This subsystem model and its derivative are given in (1) (8) (9) & (14).

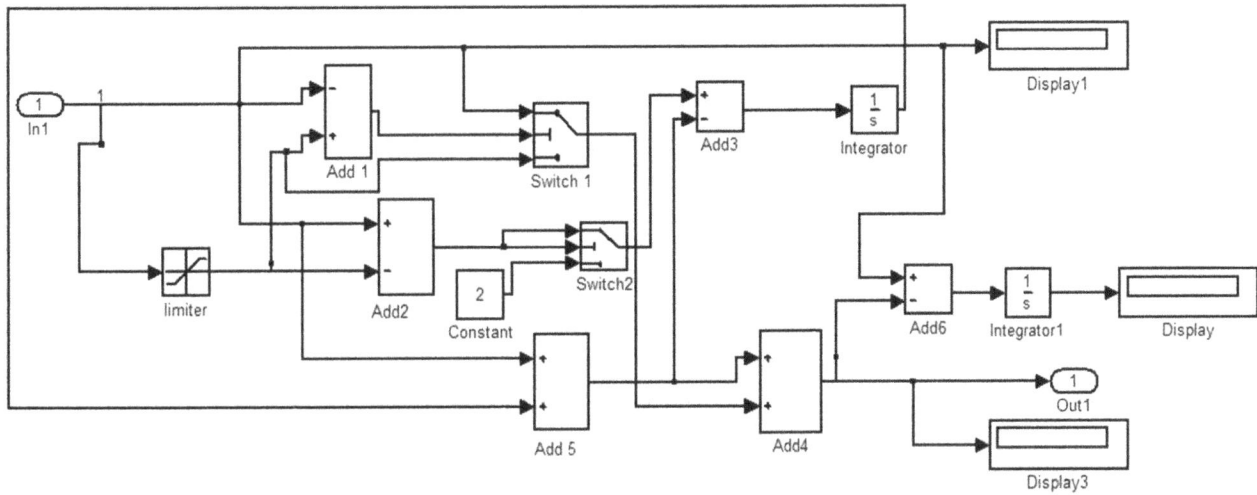

Figure 6. *Non-Linear Congestion Controller Model.*

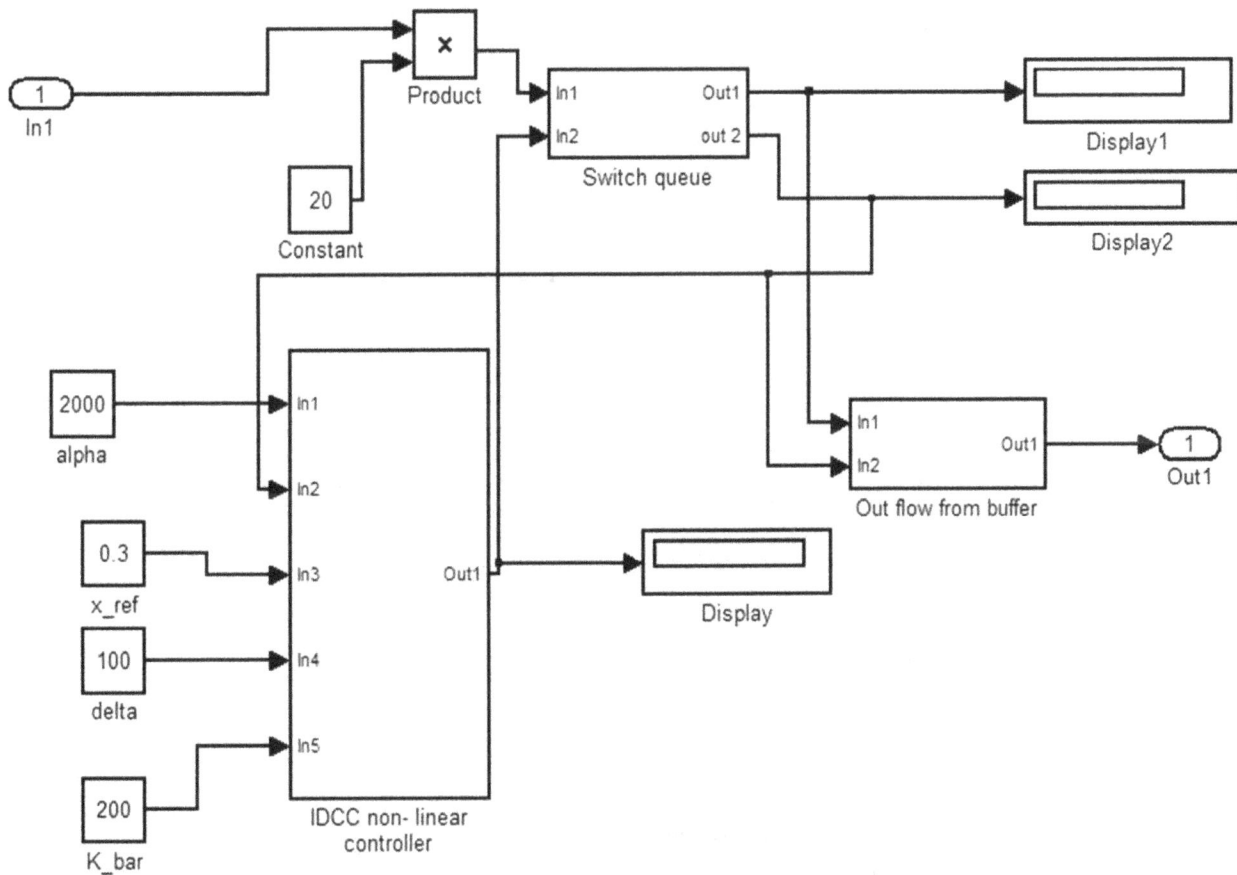

Figure 7. *Designed Switch and Non-Linear Controller Subsystem.*

This subsystem is divided into:
 i. ATM Switch subsystem
 ii. IDCC (Integrated dynamic congestion controller) subsystem
 iii. Out flow from buffer subsystem.

3.4.6. ATM Switch Subsystem

The model for this ATM switch is given in (1).

The ATM switch subsystem as seen in Fig 3.9, is made up of various systems/ blocks that work together, according to Equation(1), inorder to achieve its switching function. The

Cserver (finite buffer) is the bandwidth (capacity, cell service rate) allocated to the ABR traffic at the ATM switch and x_{buff} is the maximum possible queue size for the ABR traffic. The

relationship between C_{max}, α, x_{buff}, x^{ref}, k_bar is given and defined in (14).

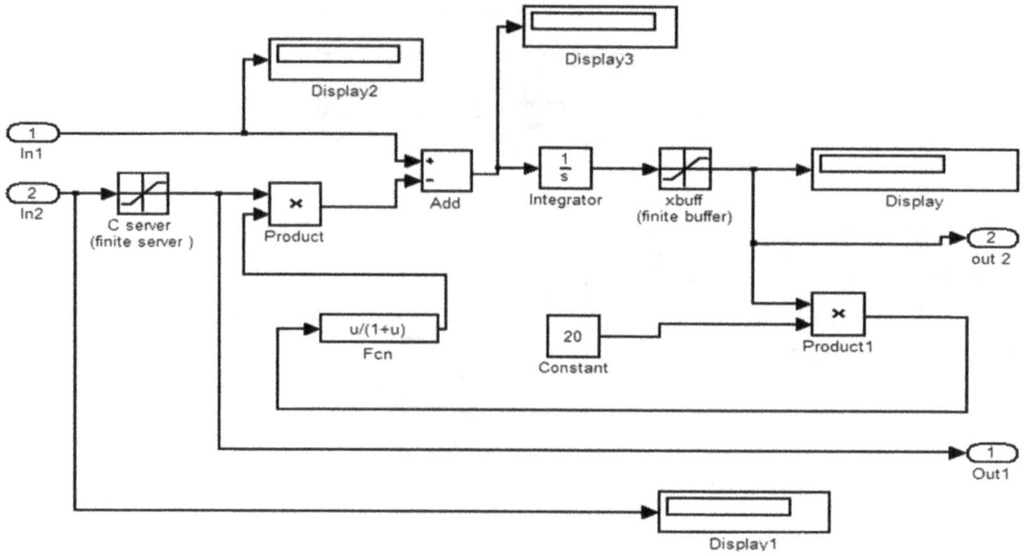

Figure 8. ATM Switch Subsystem Model.

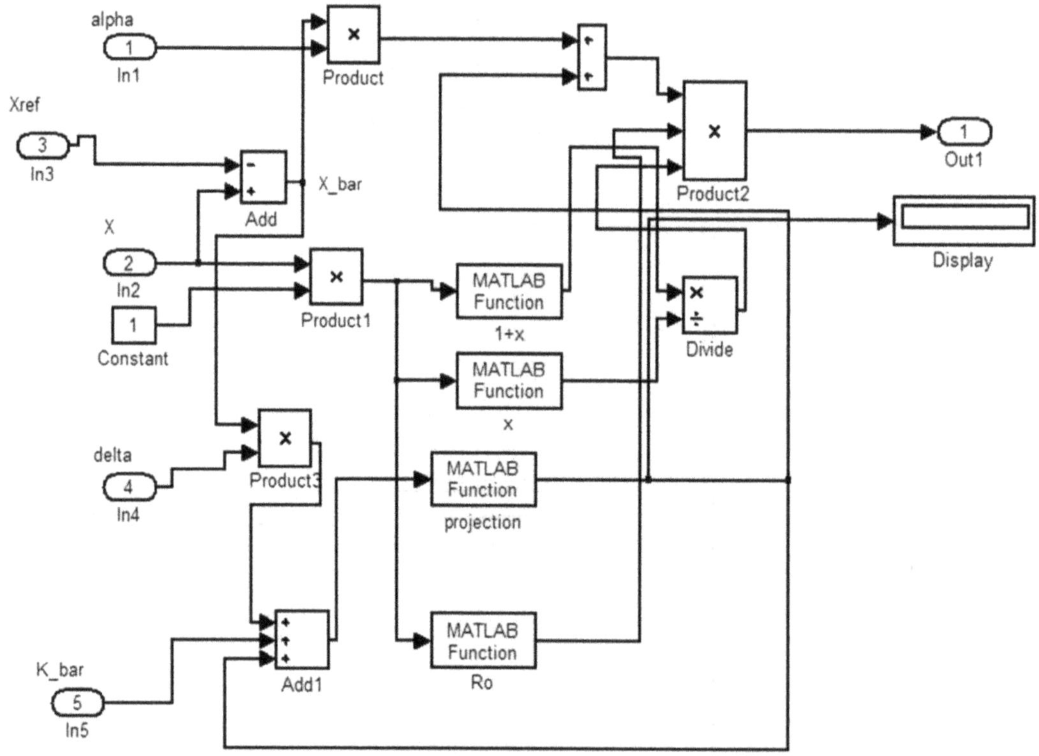

Figure 9. IDCC Model.

3.4.7. IDCC (Integrated Dynamic Congestion Controller) Subsystem

The IDCC controller is based on the algorithm shown in Equation (4) below.

It is made up of five (5) input namely: $\ddot{x}$, x_{ref}, α, δ and $\check{k}$ and one output (C) that is fed back into the ATM switch.

a. $\ddot{x}$ (represented by X_bar in Fig 3.10) is given by the

difference between X and X_{ref} as stated in (8).

b. X_{ref} is a constant set by the designer or network operator for adjusting the ATM switch queue state.

c. α, δ and $\check{k}$ (represented by alpha, delta and K_bar in Figure 9) are the design constants for the non-linear controller.

The $\frac{1+x}{x}$ part of (10) is represented by the MATLAB function in Figure 8 and ρ represented by a MATLAB function Ro in Figure 9.

3.4.8. Out Flow from Buffer

Table 1. ABR sources.

Parameter	Values	Specification
Lambda	10	Poisson generator
Initial seed	43	Poisson generator
Sample time	0.005	Poisson generator

This MATLAB model in Figure 10 represents the outflow of the ATM switch to the next node which can be either a destination or another ATM switch.

$$\text{Flowout} = \frac{x}{1+x} * C \qquad (17)$$

$\frac{x}{1+x}$ = MATLAB Fcn in Fig .10 above
C = the output of cell service rate.

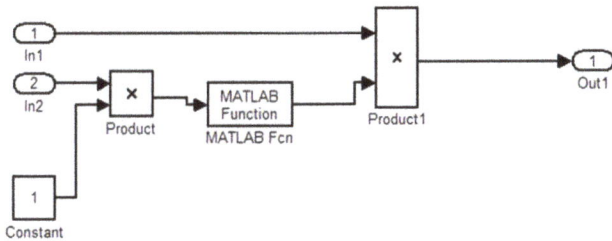

Figure 10. BUFFER Model.

Parameters adopted for this simulation are stated in Table 1, 2, 3 and 4.

Table 2. Switch queue.

Parameter	Values	Specification
Upper limit	112040	C server
Lower limit	0	C server
Initial condition	0	Integrator
Upper limit	128	Xbuff
Lower limit	-128	Xbuff
Expression	u/(1 + u)	Fcn

Table 3. ATM/flow controlled into switch.

Parameter	Values	Specification
upper limit	2	limiter
Lower limit	-0.5	limiter
Threshold	0	switch1
Threshold	0	switch2
initial condition	0	Integrator
initial condition	0	Integrator1

Table 4. IDCC.

Parameter	Values	Specification
alpha(α)	2000	constant
Xref	0.2	constant
delta(δ)	100	constant
K_bar	200	constant

4. Results and Discussion

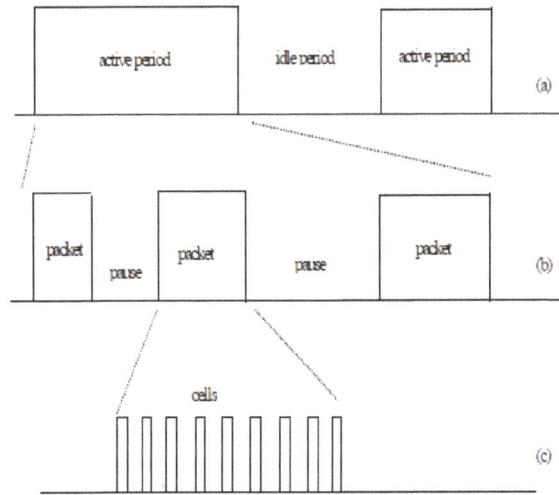

Figure 11. Display of packets and Cell placement for calculation purpose.

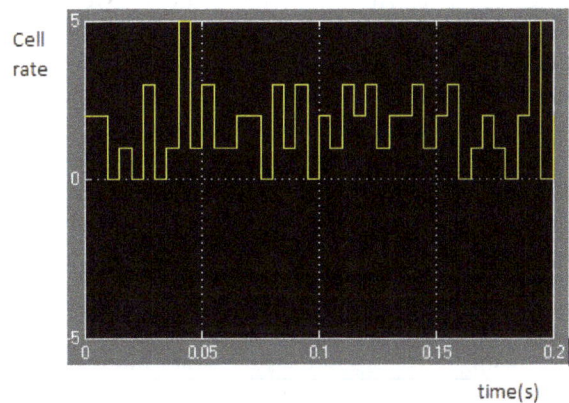

Figure 12. Simulation Result.

In this research work we evaluate the performance properties of the proposed strategy using a 1-node network model, The xmt_ws source is a Poisson distribution source modelled as shown in Figure 4, with a lambda (λ) of 10. The sample "period between packets is 0.0005 seconds. The ATM switch has a finite output buffer (x) of 128 cell places with a reference point(x^{ref}) set to 0.2 cell places. "The controller design variable α is selected equal to 10000, and the ISS Filter period (for calculating new Service Rate) is set at 112040 cell times"[16][18].

Both model in (1); and "MATLAB simulation consider the same input representing a bursty on-off source, shown in Figures 11&12 (MATLAB output). The active and idle periods of the connection are shown in part (a), the packet activity in part (b), and the cell activity in part (c) of Figure 11. The idle period has a geometric distribution with the mean value chosen to adjust the network load. During each active period a number of packets are generated with a geometric distribution and mean number of packets N"[16].

The time evolution of the queue state from MATLAB simulation is presented in Figure 12. From Figure 12, we can observe that there is a reasonable agreement between the proposed model and the required one as described in (1) .

The simulations of the simple model shown in Figure 4. together with the controller in Figure 8 using MATLAB demonstrate the "validity of the model and robustness of the proposed controller with respect to model uncertainties and noise effects"[16].

IDCC (Integrated Dynamic Congestion Controller):

We test a variety of cases in order to allow us to evaluate the responsiveness and robustness of our control design. The results generated when the reference point is set to 0.2, 0.3 and 0.4 are represented below in Table 1, 2 and 3 respectively. For example at t = 0.10 sec the ABR queue is 0.3660 cells. At t=0.50 we want ABR queue to be dropped down to 0.4318 cells. This means that we can either increase the switch service rate available for ABR traffic or regulate at the ABR sources to send less traffic. Also notice that in Table .2 and 3, there was a drop a the same time. Things get more complicated when real-time services get more demanding. This means that the ABR sources must again drop its sending rate. At t=0.5 ABR sources have more bandwidth represented by C (t) in Table 1, 2 and 3. It shows that the controller adapts very quickly to the set reference point. Also a graph of C(t) (which is the bandwidth (capacity, cell service rate) allocated to the ABR traffic at the switch) and X(t)(which is the state of the queue (i.e. ensemble average of number of cells in the queue) at reference point of 0.2,0.3 and 0.4 against the MATLAB simulation time.

Table 5. *Xref = 0.2.*

time	X(t)	C(t)	Xbar = X(t) - Xref
0.1	0.366	154.9	0.266
0.2	0.3729	173.4	0.1729
0.3	0.4035	258.1	0.2035
0.4	0.4474	385.9	0.2474
0.5	0.4318	339.5	0.2318
0.6	0.4309	337	0.2309
0.7	0.3758	181.3	0.1758
0.8	0.4144	289.1	0.2144

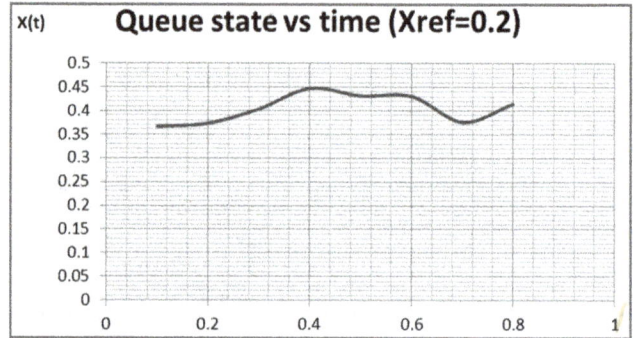

Fig. 13a. *Service Rate vs Time (Xref=0.2).*

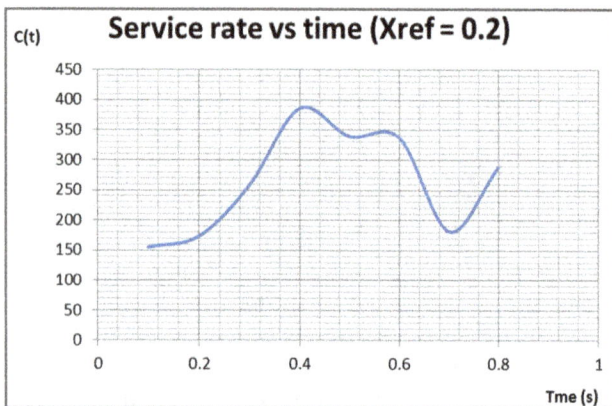

Fig. 13b. *Queue state vs time (Xref=0.2).*

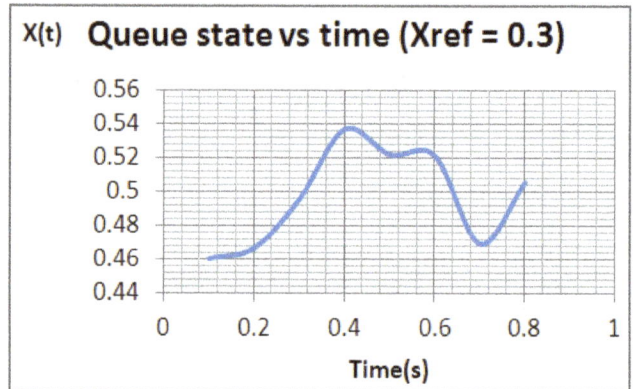

Fig. 14a. *Queue state vs time (Xref=0.3).*

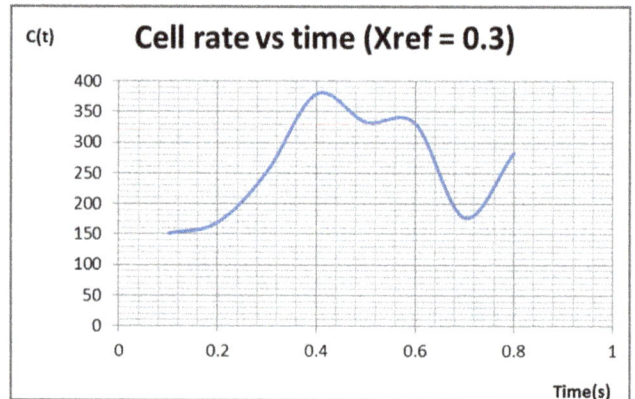

Fig. 14b. *Cell rate vs time (Xref=0.3).*

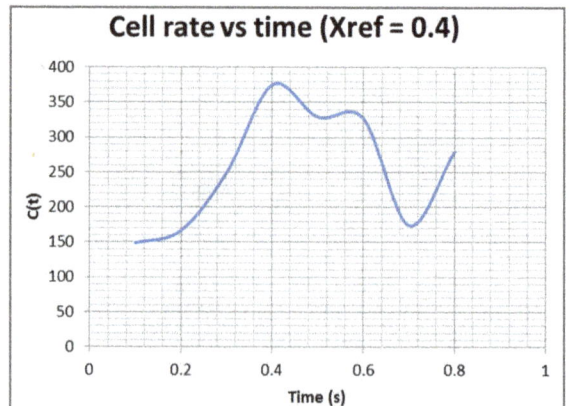

Fig. 15a. *Cell rate vs time (Xref=0.4).*

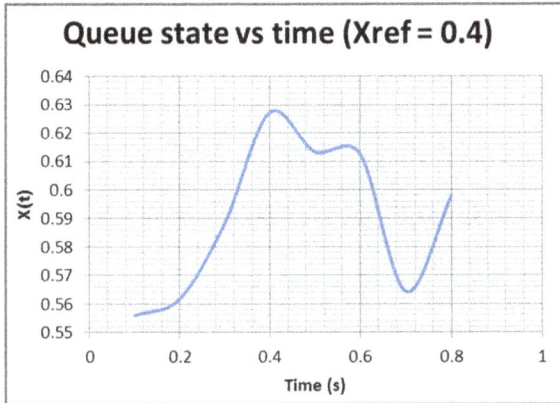

Fig. 15b. *Queue state vs time (Xref=0.4).*

5. Conclusion

A number of simulations were carried out using a computer – based approach and the system was evaluated. The poisson generator model was useful in, reproducing the phenomena in Fig 4.3 of ABR (Available Bit Rate) traffic. A number of network models were developed and results were obtained for the modelled systems.

The proposed scheme for congestion control uses the IDCC approach. A specific challenging formulation for managing traffic, passing through the output port of the switch is represented. IDCC is developed from non-linear control theory using a fluid flow model. This fluid flow model represents the dynamical system behavior, taking advantage of the packet flow conservation considerations and by matching the queue behavior at equi-state. In this way analytical performance limits can be abstracted, for a robust controlled network behaviour. We focused on ABR traffic service (which is a type of real time service). A supplementing study can be carried out to further represent other traffic services and also assess the performance of the system.

References

[1] Dinesh Taukur: What is Congestion,.www.ecomputernote.com

[2] ROBERTS, J 2004, Internet Traffic, Qos, And Pricing. Proceedings Of The IEEE, 92, 1389-1399.

[3] TORSTEN BRAUN & THOMAS STAUB 2008, End-To-End Quality Of Service Over Heterogeneous Networks. Springer. ISBN 978-3-540-79119-5.

[4] WILLIAM STALLINGS, "High Speed Networks TCP/IP And ATM Design Principles", Prentice Hall, 1998.

[5] LEONARD FRANKEN. Quality Of Service Management: A Model-Based Approach. Phd Thesis, Centre For Telematics And Information Technology, 1996.

[6] JACOBSON V 1998, Notes On Using Red For Queue Management And Congestion Avoidance, Viewgraphs From Talk At Nanog 13, Available From Ftp://Ftp.Ee.Lbl.Gov/Talks/Vj-Nanog-Red.Pdf.

[7] NAGLE J 1984, Congestion Control In IP/TCP Internetworks RFC 896.Networking, June 1993.

[8] GERLA M AND KLEINROCK L (1980) Flow Control: A Comparative Survey. IEEE Transactions On Communications 28 (4), 553–574.

[9] CHARNY, A. (1994). "An Algorithm For Rate Allocation In A Cell-Switching Network With Conference Paper At Adekunle Ajasin University, Akungba – Akoko, 2010.

[10] KESHAV, S 1991B, A Control-Theoretic Approach To Flow Control ACM SIGCOMM, Pp. 3–15. ACM Press.

[11] Omijeh. B.O and Biebuma. J.J(2013). Traffic Modelling For Capacity Analysis of CDMA Networks using Lognormal Approximation Method, *IOSR Journal of Electronics and Communication Engineering (IOSR-JECE),Volume 4, Issue 6 (Jan. - Feb. 2013), PP 42-50.*

[12] THOMAS, M & CHEN 2007, Network Traffic Modeling, The Handbook Of Computer Networks, Hossein Bidgoli (Ed.), Wiley, To Appear 2007.Southern Methodist University, Dallas, Texas.

[13] AL-BAHADILI, H. 2012: Simulation In Computer Network Design And Modeling, Use And Analysis. Hershey, PA, IGI Global.

[14] Omijeh, B.O, R. Okinege, R. Ochi (2013): "Traffic Modelling for capacity analysis of CDMA Networks using Gaussian, International Journal of Electronic , Communication and Computer Engineering (IJECCE), Vol 5, N0. 3, 2013.

[15] http://www.eng.auburn.edu

[16] http://www.cs.ucy.ac.cy

[17] Network Congestion :http://en.wikipedia.org/wiki/network_congestion

[18] ARANUWA, F, O 2010, Traffic Control And Bandwidth Allocation In ATM Networks.

[19] Welze (2006): Network congestion Control: Managing Internet Traffic, www.onlinelibrary.wiley.com

13

The Effect of Changing Substrate Material and Thickness on the Performance of Inset Feed Microstrip Patch Antenna

Liton Chandra Paul[1], Md. Sarwar Hosain[2], Sohag Sarker[2], Makhluk Hossain Prio[3], Monir Morshed[4], Ajay Krishno Sarkar[3]

[1]Department of Electronic and Telecommunication Engineering, Pabna University of Science & Technology, Pabna, Bangladesh
[2]Department of Information and Communication Engineering, Pabna University of Science & Technology, Pabna, Bangladesh
[3]Department of Electrical and Electronic Engineering, Rajshahi University of Engineering & Technology, Rajshahi, Bangladesh
[4]Department of Information and Communication Technology, Mawlana Bhashani Science and Technology University,Tangail, Bangladesh

Email address:
litonpaulete@gmail.com (L. C. Paul), sarwar.iceru@gmail.com (M. S.Hosain), sohagsarker5614@gmail.com (S.Sarker), makhlukpr158@gmail.com (M. H.Prio), monirmorshed.mbstu@gmail.com (M.Morshed), sarkarajay139@gmail.com (A. K. Sarkar)

Abstract: In order to design a microstrip patch antenna at first the designer is to select the substrate material and it's thickness. So, if the designer has a clear conception about the effect of changing substrate material and it's thickness on the performance of the antenna, it will be easier to design an antenna. Appropriate selection of dielectric material and it's thickness is an important task for designing a microstrip patch antenna. This paper represents that how antenna performance changes when we vary substrate material and it's thickness. The designed inset feed rectangular microstrip patch antenna operates at 2.4GHz (ISM band).

Keywords: Inset Feed, Dielectric Constant, Substrate Thickness, Bandwidth, Return Loss, Gain, Directivity, Radiation Efficiency

1. Introduction

Microstrip patch antennas consist of a metallic patch on a grounded substrate.The microstrip patch antenna first took form in the early 1970's and interest was renewed in the first microstrip antenna proposed by Deschamps in 1953[1]. Microstrip antennas have found widespread applications for microwave as well as millimeter wave systems [2].Compatible devices are widely used in our daily lives such as mobile phones, laptops with wireless connection, wireless universal serial bus (USB) dongles etc and microstrip patch antenna plays a very significant role for the miniaturization of these devices [3]. The applications in present-day mobile communication systems usually require smaller antenna size in order to meet the miniaturization requirements of mobile units. Thus, size reduction and bandwidth enhancement are becoming major design considerations for practical applications of microstrip antennas. The microstrip patch antennas are well known for their performance and their robust design, fabrication and their extent usage. The inherently narrow impedance bandwidth is the major weakness of a microstrip antenna [4]. Although we used rectangular shaped patch but the radiating patch can be of any geometrical configuration like square, rectangle, circular, elliptical, triangular, E-shaped, H-shaped, L-shaped, U shaped etc. The material which has the dielectric constant in the range of $2.2 \leq \varepsilon_r \leq 12$ can be used as substrate [5]. When we change the substrate material and the thickness of substrate of a microstrip antenna, it changes the system performance. Therefore, in order to introduce appropriate correctness in the design of the antenna, it is important to know the effect of changing dielectric substrate material and substrate thickness. A set of simulation and measurements of inset feed rectangular patch antenna on different substrate material (RT Duroid 5880, GML 1000, RO4003 and FR-4) and on the same substrate material by varying substrate thickness is presented in this research paper. The design, simulation and measurements are performed by advanced design system (ADS) 2009 momentum.

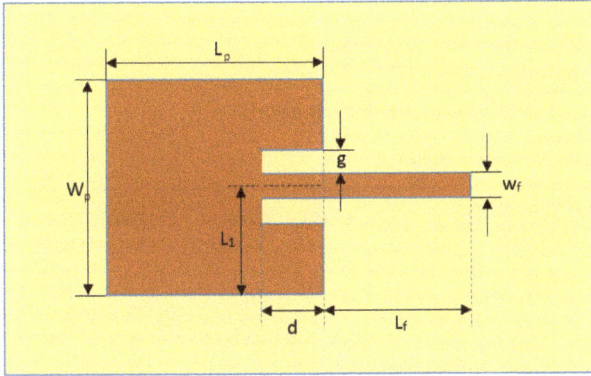

Fig. 1. Inset feed rectangular microstrip patch antenna layout.

2. Feeding Technique

A feedline is used to excite the antenna for making radiation by direct or indirect contact. Microstrip patch antennas can be fed by a variety of methods. Mainly these methods are classified into two groups- contacting and non-contacting. Some popular feeding techniques are microstrip line feed, coaxial probe feed, inset feed/ notch feed/cut feed, aperture coupling, proximity coupling, coupled (indirect) fed etc. The selection of feeding technique for a microstrip patch antenna is an important decision because it directly affects the bandwidth, return loss, VSWR, patch size and smith chart [6]. We chose Inset feed technique because it can be easily fabricated and simplicity in modeling as well as impedance matching [7]. The fig.1 shows the layout of inset feed rectangular microstrip patch antenna with required dimensions. Where, W_p= width of patch, L_p= Length of patch, W_f= Width of feeder, L_f = Length of feeder, d = inset depth, g = notch width / inset width/gap width and L_1 = position of feeder from the left edge of the patch.

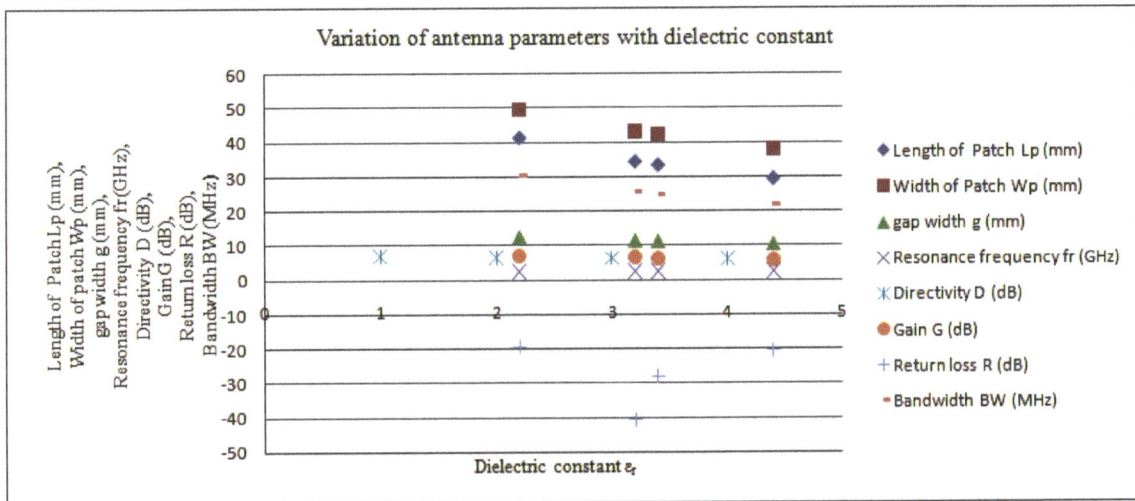

Fig. 2. Variation of antenna parameters with substrate material.

3. Design of Rectangular Microstrip Patch Antenna

In the typical design procedure of rectangular Microstrip patch antenna, three essential parameters are [8]:

1. Resonance frequency f_r
2. Dielectric constant of the substrate, ε_r
3. Thickness of substrate, h

After the proper selection of above three parameters, the next step is to calculate the radiating patch width and length.The designing procedure can be divided following steps:

Step 1: Calculation of width of patch (W_p)

For an efficient radiator, practical width that leads to good radiation efficiencies is [9]

$$W_P = \frac{c}{2f_r}\sqrt{\frac{2}{\varepsilon_r + 1}}$$

Where, c= velocity of light=3×10^8 m/s

f_r= resonance frequency

ε_r=dielectric constant

Step 2: Calculation of effective dielectric constant,

$$\varepsilon_r^{eff} = \frac{\varepsilon_r + 1}{2} + \frac{\varepsilon_r - 1}{2}\left[\frac{1}{\sqrt{1 + \frac{12h}{W_p}}}\right]$$

Step 3: Calculation of effective length of patch,

$$L_{eff} = \frac{c}{2f_r\sqrt{\varepsilon_r^{eff}}}$$

Step 4: Calculation of length extension,

$$\Delta L = 0.412h\frac{\left(\varepsilon_r^{eff} + 0.3\right)\left(\frac{W_p}{h} + 0.264\right)}{\left(\varepsilon_r^{eff} - 0.258\right)\left(\frac{W_p}{h} + 0.8\right)}$$

Step 5: Calculation of actual length of patch,

$$L_p = L_{eff} - 2\Delta L$$

Step 6: calculation of inset depth

$$Z_o = Z_{in} cos^2 (\frac{\pi d}{L_p})$$

Where,

Z_o = Characteristics impedance
Z_{in} = input impedance
d = inset depth/notch depth/gap depth

Table I. Variation of antenna parameters with different substrate material.

Substrate Material Name	Dielectric constant ε_r	Length of Patch L_p(mm)	Width of PatchW_p (mm)	Inset depthd(mm)	Resonance frequencyf_r (GHz)	Directivity D (dB)	GainG(dB)	Return lossR(dB)	Bandwidth BW(MHz)
RT Duroid 5880	2.2	41.408	49.410	12.398	2.406	7.00870	7.00410	-19.402	30.5
GML 1000	3.2	34.483	43.129	11.312	2.411	6.38395	6.37818	-41.363	26
RO4003	3.4	33.472	42.137	11.126	2.408	6.30076	6.29490	-27.994	25
FR-4	4.4	29.479	38.036	10.321	2.408	5.98928	5.68109	-20.516	22

4. Effect of Changing Substrate Material

m3
freq=2.422GHz
dB(insetfeed_mom_a..S(1,1))=-10.065

m2
freq=2.392GHz
dB(insetfeed_mom_a..S(1,1))=-10.405

m1
freq=2.406GHz
dB(insetfeed_mom_a..S(1,1))=-19.402
Min

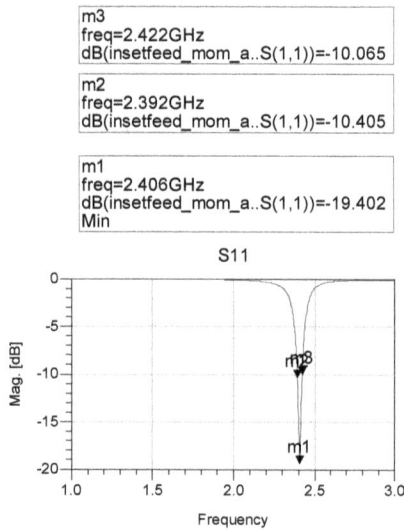

Fig. 3. |S11| for RT Duroid 5880.

m3
freq=2.425GHz
dB(GML1000_mom_a..S(1,1))=-9.187

m2
freq=2.398GHz
dB(GML1000_mom_a..S(1,1))=-9.330

m1
freq=2.411GHz
dB(GML1000_mom_a..S(1,1))=-41.363
Min

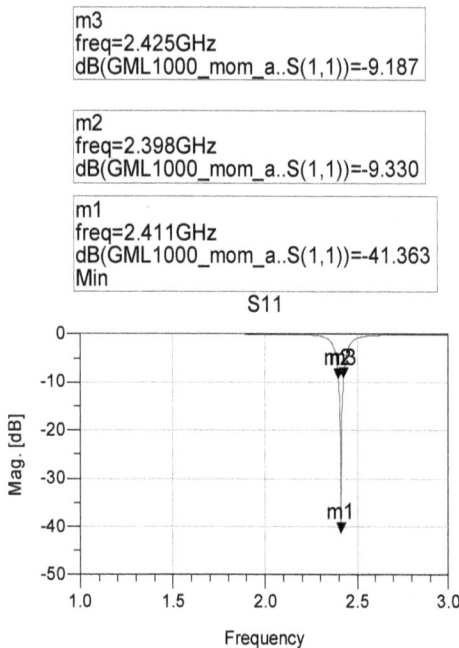

Fig. 4. |S11| for GML 1000.

m3
freq=2.422GHz
dB(insetfeed_mom_a..S(1,1))=-9.476

m2
freq=2.396GHz
dB(insetfeed_mom_a..S(1,1))=-9.795

m1
freq=2.408GHz
dB(insetfeed_mom_a..S(1,1))=-27.994
Min

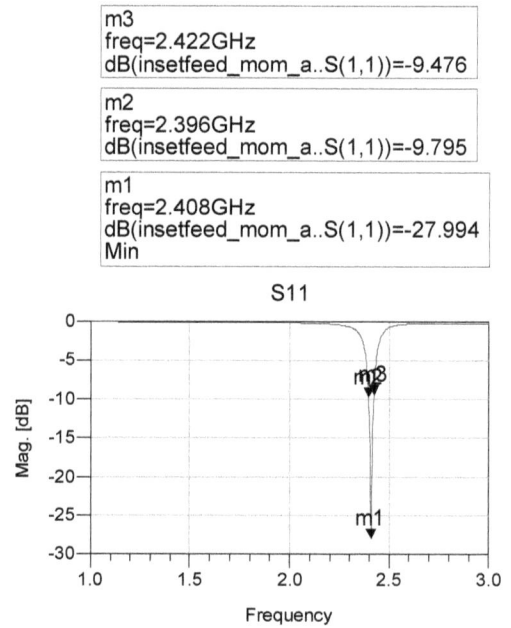

Fig. 5. |S11| for RO4003.

m3
freq=2.420GHz
dB(insetfeed_mom_a..S(1,1))=-9.812

m2
freq=2.398GHz
dB(insetfeed_mom_a..S(1,1))=-10.354

m1
freq=2.408GHz
dB(insetfeed_mom_a..S(1,1))=-20.516
Min

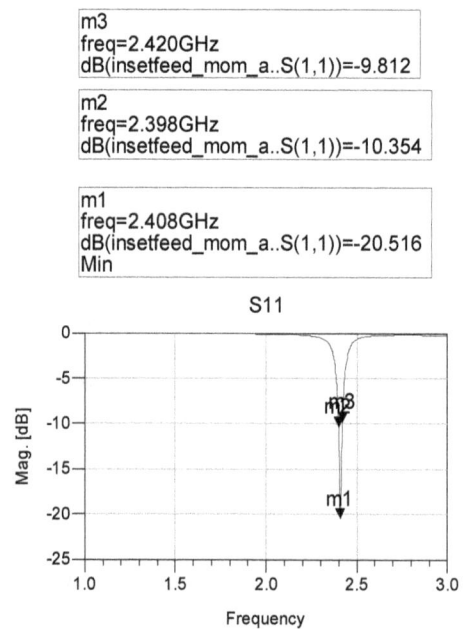

Fig. 6. |S11| for FR-4.

With changing the substrate material, the dielectric constant of the substrate changes i.e. changing the substrate material means the changing the dielectric constant (ε_r). Although, wide variety of substrate materials have been found to exist suitable for microstrip patch antenna design with mechanical, thermal, and electrical properties which are attractive for use in both planar and conformal antenna configurations. However tolerance control of the dielectric constant remains a problem for accurate designs, particularly at higher microwave and millimeter frequencies [10]. Here, we used four different substrate materials – RT Duroid 5880, GML 1000, RO4003 and RF-4 whose dielectric constants are 2.2, 3.2, 3.4 and 4.4 respectively for the same antenna configuration (f_r=2.4GHz, g=1.5mm, h=1.5mm, t=0.1mm, L_f=31.25mm &W_f=4mm). For every different substrate

material we determined the antenna performance parameters like resonance frequency, directivity, gain, return loss, bandwidth as well as the dimension of patch of the antenna (length of patch, width of patch and inset depth). These antennas are designed and simulated by using advanced design system (ADS) 2009 momentum simulator. Table I shows the antenna parameters variation summary with changing substrate material. From the data table I, it is clear that, by using substrate material with higher dielectric constant (ε_r), the length of patch (L_p), width of patch (w_p), inset depth (d), Directivity (D), Gain (G), Bandwidth (BW) decreases. There are also significant changes of return loss (R). The fig.2 shows the graphical representations of the data are listed in table I with respected to dielectric constant of the substrate material.

Table II. *Variation of antenna parameters with substrate thickness.*

h (mm)	L_p (mm)	d (mm)	f_r (GHz)	D (dB)	G (dB)	R (dB)	BW (MHz)
0.5	41.932	12.555	2.421	6.88037	2.62412	-0.836	
1	41.685	12.481	2.406	6.90260	2.69583	-1.316	
1.3	41.522	12.432	2.398	6.91738	2.76126	-1.651	
1.5	41.408	12.398	2.410	7.00735	7.00274	-27.221	31.5
2	41.106	12.308	2.403	7.02175	7.01743	-21.196	44
2.5	40.784	12.212	2.394	7.03605	7.03202	-19.637	56
3	40.448	12.110	2.384	7.04630	7.02344	-20.059	68
3.5	40.099	12.006	2.372	7.06027	7.00487	-22.496	81
4	39.741	11.899	2.362	7.07394	6.98569	-28.280	91
4.5	39.374	11.789	2.354	7.08734	6.96575	-39.232	99

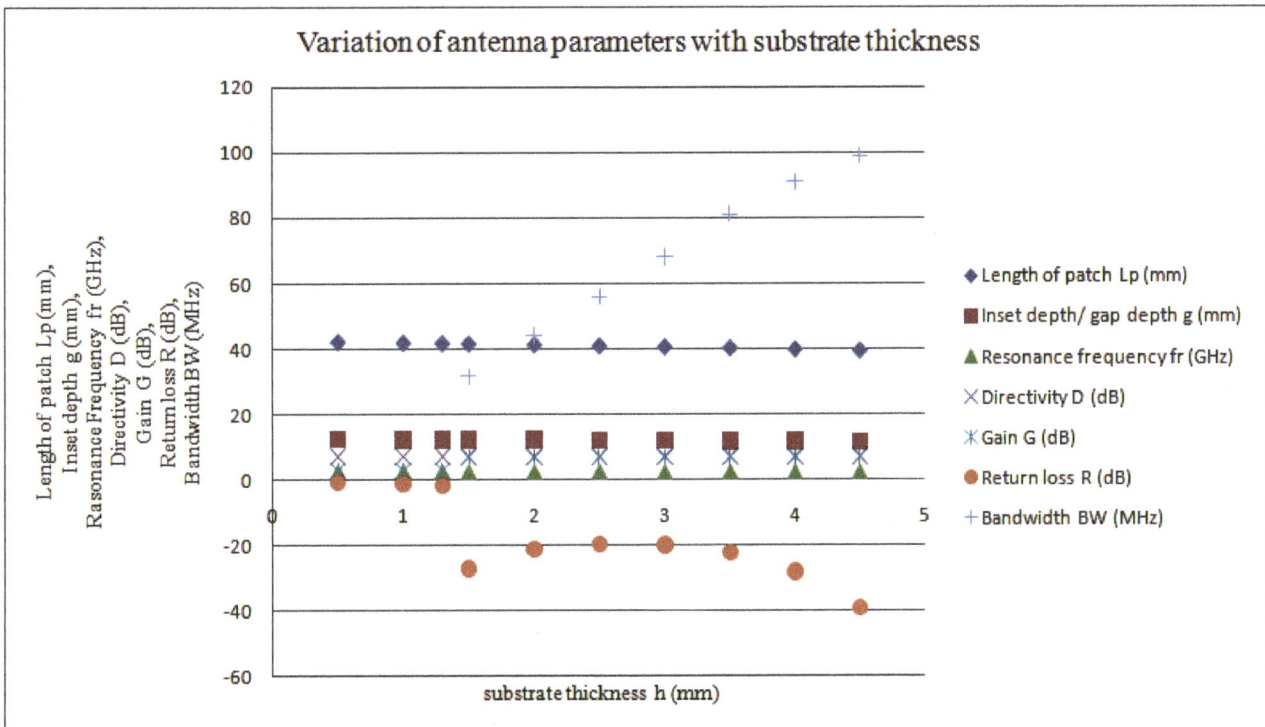

Fig. 7. *Variation of antenna parameters with substrate thickness.*

5.Effect of Substrate Thickness Variation

Selection of proper substrate thickness is another important task in microstrip patch antenna design. To choose appropriate substrate thickness (h), a designer needs to know the effect of changing substrate thickness. Here, we measure the antenna parameters by varying substrate thickness (h) from 0.5 mm to 4.5 mm for an inset feed rectangular microstrip patch antenna. The measured data are listed in table II. RT Duroid 5880 substrate with dielectric constant ε_r=2.2 used for this analysis.

From the data table II, it is seen that with increasing the substrate thickness, the bandwidth increases but the antenna dimension decreases as well as the center operating frequency move away from the desired resonance frequency (for this design 2.4 GHz). Fig.7. shows the graphical representation of the antenna parameters which are given in table II with respect to substrate thickness (h).

6. Conclusion

From the above analysis, we can conclude that the use of substrate material with higher dielectric constant in microstrip patch antenna design, results degradation of antenna performance but size of the antenna reduces. With increasing substrate thickness (h), the resonance frequency decreases but the bandwidth increases. The performance of the antennas was measured for 2.4 GHz operating frequency using inset feeding technique with advanced design system (ADS) 2009 momentum 3D planar electromagnetic simulator.

References

[1] S. S. Holland "Miniaturization of microstrip patch antennas for GPD Applications", M.Sc. thesis, Dept. of Electrical And Computer Engineering,University of Massachusetts Amherst,May 2008.

[2] W. Menzel and W. Grabherr,"A Microstrip Patch Antenna with CoplanarFeed Line",IEEE microwave and guided wave letters, Vol. 1,No. 11,November 1991.

[3] Deepak, D. Parashar, R. S. Pathak and S. K. Bhartiya. "Effect of Change in Feed point on the Micro strip Patch Antenna Performance in Novel H shape Antenna", International Journal of Emerging Trends in Engineering and Development,

Vol.5, Issue 3, September 2013.

[4] C. L. Mak, K. M. Luk , K. F. Lee and Y. L. Chow , "Experimental Study of a Microstrip Patch Antenna with an L-Shaped Probe",IEEE Transactions on antennas and propagation, vol. 48, no. 5, May 2000

[5] S. A. Zaidi and M.R. Tripathy "Design and Simulation Based Study of Microstrip E–Shaped Patch Antenna Using Different Substrate materials ",Advance in Electronic and Electric Engineering.,Volume 4, Number 6 , pp 611, 2014

[6] K. P. Kumar, K. S.Rao, T. Sumanth, N. M.Rao, R. A. Kumar andY.Harish,"Effect of feeding techniques on the radiationcharacteristics of patch antenna design and analysis", International Journal of Advanced Research in Computer and Communication Engineering, Vol. 2, Issue 2, February 2013.

[7] A. I. Salem, A. A. Salama, A. M. Eid, M. Sobhy, and A. watany, "Performance Enhancement of Fabricated and Simulated Inset Fed Microstrip Rectangular Patch Antennas", International Journal of Scientific & Engineering Research, Volume 5, Issue 4, April 2014.

[8] T. k. Raina "Design, fabrication and performance Evaluation of micro-strip patch Antennas for wireless applications", M.Sc thesis, Department of Electronics and Communication Engineering, Thapar University, Patiala, June 2012.

[9] Using aperture coupled feedh. Singh, Y.K.Awasthi and A.K.Verma "Microstrip Patch Antenna with the Defected Ground Structure and Defected Microstrip Structure" proceedings of International Confrence on Microwave, 2008.

[10] Keith R. Carver, and James W. Mink, "Microstrip antenna technology", IEEE transactions on antennas and propagation", Vol AP-29, no.1,pp 21,January 1981.

[11] J. W. Salman, M.M. Ameen and S. O. Hassan "Effects of the Loss Tangent, Dielectric Substrate Permittivity and Thickness on the Performance of Circular Microstrip Antennas"Journal of Engineering and Development, Vol. 10, No.1, March 2006.

[12] K.P. Kumar, K.S. Rao, V. M. Rao, K. Uma, A. Somasekhar and C. M. Mohan "The effect of dielectric permittivity on radiation characteristics of co-axially feed rectangular patch antenna: Design & Analysis",International Journal of Advanced Research in Computer and Communication Engineering, Vol. 2, Issue 2, February 2013.

[13] V. Natarajan and C. Chatterjee, "Effect of substrate permivity and thickness on performance on single layer, wideband, U-slot antennas on microwave substrates", 20th annual review of progress in applied computational Electromagnetics, April 2004.

14

Cryptanalysis of Simplified Data Encryption Standard Using Genetic Algorithm

Purvi Garg, Shivangi Varshney, Manish Bhardwaj

Department of Computer Science and Engineering, SRM University, Modinagar, Utter Pradesh, India

Email address:
purvigarg71@gmail.com (P. Garg), shivangi.varshney100@gmail.com(S. Varshney), aapkaapna13@gmail.com(M. Bhardwaj)

Abstract: Cryptanalysis of cipher text using evolutionary algorithm has gained much interest in the last decade. In this paper, cryptanalysis of SDES has been performed using Genetic Algorithm with Ring Crossover operator. Cryptography has been prone to many attacks but the scope of this paper is limited only to the cipher text attack. Different combinations of keys are generated using the Genetic Algorithm and hence it is concluded that Genetic Algorithm is a better approach than the Brute Force for analyzing SDES.

Keywords: Cryptanalysis, Cipher Text Attack, SDES, Genetic Algorithm, Brute Force, Key Search Space

1. Introduction

Cryptography, a word with Greek origins, means "secret writing." It is basically the science and art of transforming messages to make them secure and immune to attacks. But we cannot completely immune ciphers from different attacks and this attacking or breaking of ciphers is termed as cryptanalysis, in which the intruder converts the cipher text into plaintext with or without knowing the key.

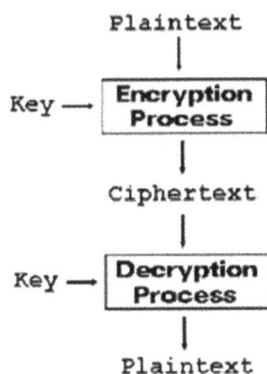

Fig. 1. The process of cryptography [23].

Cryptographic systems have a finite key space so that the intruder can easily search for a key, but still the system remains secure because of the size of the key search space.

And hence, optimization techniques have got significant importance in finding the solution (key), by optimizing key search space. In this paper, we are working on the cryptanalysis of Simplified Data Encryption Standard (S-DES), using Genetic Algorithm and Brute Force.

2. Related Works

Cryptanalysis has got much attention in the last few years. In the year 1993, R.Spillman used Genetic Algorithm to attack the Knapsack cipher [4] and substitution ciphers [5]. The first experimental cryptanalysis of DES using a linear cryptanalysis technique was shown by Matusi in [7]. An important analysis on how different optimization techniques can be used in the field of cryptanalysis is shown in by Clark [6]. In 2006 Nalini [3] used GA, Tabu search and Simulated Annealing techniques to break S-DES. Later in 2008, Garg [1, 2] presented the use of memetic algorithm and genetic algorithm to break a simplified data encryption standard algorithm. Vimalathithan [9] also used GA to attack Simplified-DES. In 2012, Sharma and others [21] showed the breaking of the S-DES using Genetic Algorithms.

3. The S-Des Algorithm

Simplified Data Encryption Standard (S-DES) is developed by Edward Schaefer of Santa Clara University. The S-DES [8, 10] is an encryption algorithm which is basically designed for educational purpose. It is not sufficiently secure. This algorithm takes 8 bit block of plaintext and a 10 bit key as input and gives 8 bit block of cipher text asoutput. This is the encryption process. To get the plaintext back, we again provide the 8 bit block of cipher text and the same 10 bitkey that was given at the time of encryption, as an input to the decryption algorithm and the 8 bit block of plaintext is obtained as the output.

In the process of encryption, five basic functions are used: an initial permutation (IP), a complex function labeled f_k which involves both permutation and substitution operations and depends on a key input, a simple permutation function that switches (SW) the two halves of the data, the function $f_k$again and a permutation function that is the inverse of the initial permutation (IP–1).

Key Generation for f_k

For key generation, a 10-bit key is considered fromwhich two 8-bit sub keys are generated. In this case, the key is first subjected to a permutation P10= [3 5 2 7 4 10 1 98 6], then a shift operation is performed. The numbers in the array represent the value of that bit in the original 10-bit key. The output of the shift operation then passes through a permutation function that produces an 8-bit output P8 = [6 3 74 8 5 10 9] for the first sub key (K1). The output of the shift operation also feeds into another shift operation and another instance of P8 to produce the second sub key

K2. In all bit strings, the leftmost position corresponds to the first bit.

A) Initial and Final Permutations

The input to the algorithm is an 8-bit block of plaintext, which we first permute using the IP function IP= [2 6 3 1 4 8 5 7].This retains all 8-bits of the plaintext but mixes them up. At the end of the algorithm, the inverse permutation is applied; the inverse permutation is done by applying, IP-1 = [4 1 3 5 7 2 8 6] where we have IP-1(IP(X)) =X.

B) The Function f_k

The function f_k, which is the complex component of S-DES, consists of a combination of permutation and substitution functions. The functions are given as follows. Let L, R be the left 4-bits and right 4-bits of the input, then, f_k (L, R) = (L XOR f(R, key), R) Where XOR is the exclusive-OR operation and key is a sub -key. Computation of f(R, key) is done as follows.

1. Apply expansion/permutation E/P= [4 1 2 3 2 3 4 1] to input 4-bits.
2. Add the 8-bit key (XOR).
3. Pass the left 4-bits through S-Box S0 and the right 4-bits through S-Box S1.
4. Apply permutation P4 = [2 4 3 1].

The S-boxes operate as follows:

$$S0= \begin{array}{c} \\ 0 \\ 1 \\ 2 \\ 3 \end{array} \begin{array}{cccc} 0 & 1 & 2 & 3 \\ \left[\begin{array}{cccc} 1 & 0 & 3 & 2 \\ 3 & 2 & 1 & 0 \\ 0 & 2 & 1 & 3 \\ 3 & 1 & 3 & 2 \end{array} \right] \end{array}$$

$$S1= \begin{array}{c} \\ 0 \\ 1 \\ 2 \\ 3 \end{array} \begin{array}{cccc} 0 & 1 & 2 & 3 \\ \left[\begin{array}{cccc} 0 & 1 & 2 & 3 \\ 2 & 0 & 1 & 3 \\ 3 & 0 & 1 & 0 \\ 2 & 1 & 0 & 3 \end{array} \right] \end{array}$$

Fig. 2. *Working of S –box [19].*

The first and fourth input bits are treated as 2-bit numbers that specify a row of the S-box and the second and third input bits specify a column of the S-box. The entry in that row and column in base 2 is the 2-bit output.

C) The Switch Function

The function f_k only alters the leftmost 4 bits of the input. The switch function (SW) interchanges the left and right 4 bits so that the second instance of f_k operates on a different 4 bits. In this second instance, the E/P, S0, S1, and P4 functions are the same. The key input is K2.

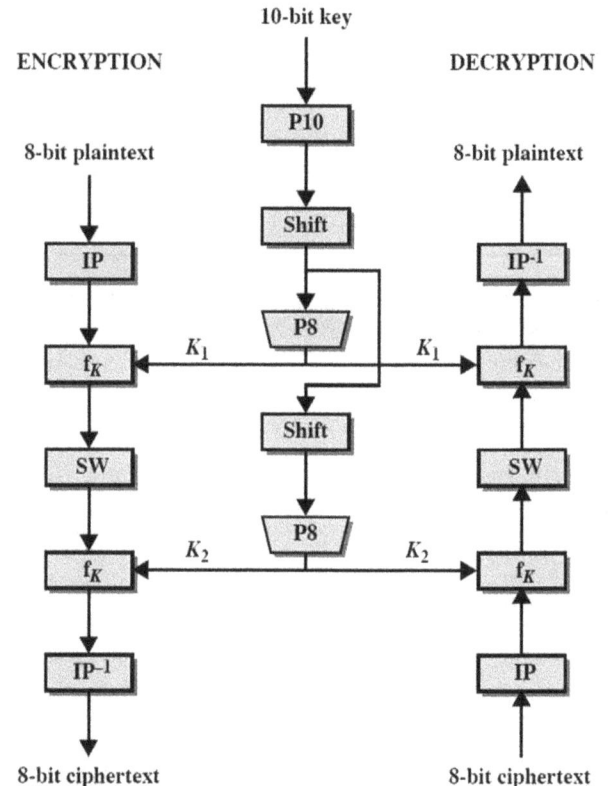

Fig. 3. *The Simplified Data Encryption Algorithm [19].*

4. The Genetic Algorithm

The genetic algorithm is an optimization and search algorithm based on natural selection. GA is a subclass of EA and works on two basic things: Genetic Representation of the solution domain and Fitness function to evaluate solution domain. To some extent, heuristics used in the genetic algorithm is an important reason of its success. The GA involves three basic operations: Selection, Crossover and Mutation.

A) Selection

In this step, we basically decide which chromosomes will take part in evolution process. We use the fitness function to decide the fitness of the chromosome, more the fitness of chromosome, more number of times, it will be selected in the process of evolution.

B) Crossover

In this step, we create a new generation of population by combining the parents and producing off spring. There are different types of crossover operators which can be used to produce new population. In this paper we are using the ring crossover operator. In ring crossover, we combine two parents and form a ring, and then randomly select a crossover point, and with reference to this point, one of the children is created in clock wise direction and the other one is created in anticlockwise direction. In ring crossover operator, swapping and reversing processes are also included.

C) Mutation

Mutation is used to prevent falling all solution in the population into a local optimum of solved problem. In mutation, bits are randomly interchanged or altered to differentiate new population of solution from the existing one.

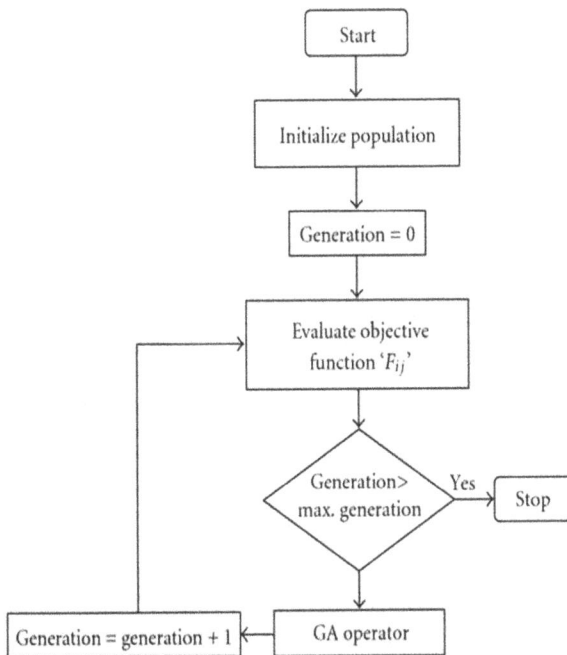

Fig. 4. *Flowchart showing the steps of Genetic Algorithm [22].*

5. Brute Force

Brute Force is a traditional search algorithm. In Brute Force algorithm, all possible keys are checked until the correct one is found. In the worst case, it may be possible that the correct key is the last key of the entire search space. With a key of 56 bits, there are 2^{56} possible keys i.e., 7.2×10^{16} keys approximately. Assuming, on an average half the key space has to be searched, a single machine performing one S-DES encryption per microsecond would take more than a thousand years to break the cipher.

6. Objective

The goal of this paper is to perform a comparison between Brute Force Search Algorithm and Genetic Algorithm and to show the use of Genetic Algorithm in the field of cryptanalysis. The primary goals of this work are to produce aperformance comparison between traditional Brute forcesearch algorithm and genetic algorithm with

improvedparameters based method, and to determine the use oftypical GA-based methods in the field of cryptanalysis.

The procedure to carry out the cryptanalysisusing GA in order to break the key is as follows:
1. Input: cipher text, and the language statistics.
2. Randomly generate an initial pool of solutions (keys).
3. Calculate the fitness value of each of the solutions inthe pool using equation (1).
4. Create a new population by repeating followingsteps until the new population is complete
   a. Select parent (keys) from a current population according to their fitness value (the better fitness, the bigger chance to be selected). Here Tournament selection is used.
   b. With a crossover probability cross over the parents to form new offspring (children). In our genetic algorithm we are using Ring Crossover Operator
   c. For each of the children, perform a mutation operation with some mutation probability to generate new children.
   d. Place new children in the new population
5. Use new generated population for a further run of the algorithm.
6. If the end condition is satisfied, stop, and return the best solution in current population
   A. Cost Function [19]

Equation (1) is a general fitness function usedto determine the suitability of an assumed key (k). Here, Adenotes the language alphabet (i.e., for English, [A. Z, _], where _ represents the space symbol), K and Ddenote known language statistics and decryptedmessage statistics, respectively, and the u, b, and tdenote the unigram, diagram and trigram statisticsrespectively; α, β and γ are the weights assigning different weights to each of the three statistics where α + β + γ = 1. In view of the computational complexity of trigram, only unigram and diagram statistics are used.

$$C^{K} = \alpha \sum (i \, \varepsilon \, \bar{A}) \, |K(i)^{u} - D(i)^{u}| + \beta \sum (i, j \, \varepsilon \, \bar{A}) \, |K(i,j)^{b} - D(i,j)^{b}| + \Upsilon \sum (i, j, k \, \varepsilon \, \bar{A}) \, |K(i,j,k)^{t} - D(i,j,k)^{t}| \quad (1)$$

7. Result & Discussion

There are a variety of cost functions used by other researchers in the past. The most common cost function uses gram statistics. Some use a large amount of grams while others only use a few. Equation 1 is a general formula used to determine the suitability of a proposed key. A number of experiments have been carried out by giving different inputs and applying genetic algorithm and Brute force attacks for breaking Simplified Data Encryption Standard. The results are shown in table1.The table below shows that the key bits matched using GA and Brute Force search algorithm for the given cipher text .the choice of the Genetic Operators play a vital role in GA and are described below:

GA Parameters

The following are the GA parameters used during the experimentation:

Population Size: 100   Crossover: .85
Selection: Tournament Selection operator   Mutation: .02
Crossover Ring Crossover   No. of Generation: 50

Table I. *Comparison of GA and Brute Force Algorithm [19].*

S. No	Amount Of Cipher Text	No. of bits matched using GA	No. of bits matched using Brute Force Search	Time Taken by GA (M)	Time Taken by Brute Force Search (M)
1.	200	5	5	4.7	24.3
2.	400	4	3	2.1	24.7
3.	600	7	6	1.9	23.6
4.	800	8	7	3.1	24.1
5.	1000	9	7	2.6	25.1
6.	1200	9	8	2.1	25.5

From the above table, it is found that both GAworks better than Brute force algorithm in terms of timetaken as well as obtaining number of key bits.

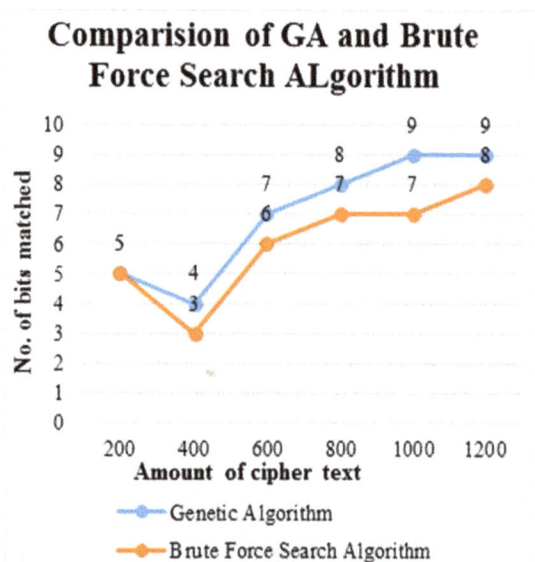

Fig. 5. *Comparison of Genetic algorithm and BruteForce Search algorithm [19].*

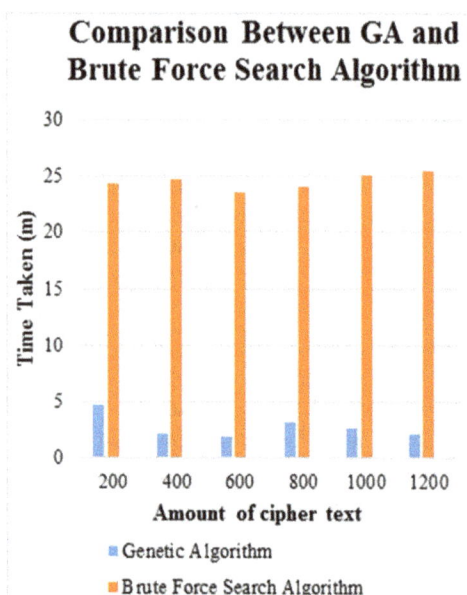

Fig. 6. *The running time comparison of GeneticAlgorithm and Brute Force Search Algorithm [19].*

8. Conclusion

To conclude, in this paper we have discussed the working process of S-DES. We have mentioned the causes for the success of Genetic Algorithm and its 3 basic operations: selection, crossover and mutation. We have pointed out the loopholes of the Brute force attack. Hence proving the success of Genetic Algorithm over brute force in the cryptanalysis of the cipher text.

References

[1] G Poonam, Memetic Algorithm Attack on Simplified Data Encryption Standard algorithm, proceeding of International Conference on Data Management, February 2008, pg 1097-1108.

[2] Garg Poonam, Genetic algorithm Attack on Simplified Data Encryption Standard Algorithm, International journal Research in Computing Science, ISSN1870-4069, 2006.

[3] Nalini, Cryptanalysis of S-DES via Optimization heuristics, International Journal of Computer Sciences and network security, vol 6, No 1B, Jan 2006.

[4] Spillman, R.: Cryptanalysis of Knapsack Ciphers Using Genetic Algorithms.Cryptologia XVII(4), 367– 377 (1993)

[5] Spillman, R., Janssen, M., Nelson, B., Kepner, M.: Use of A Genetic Algorithm in the Cryptanalysis of simple substitution Ciphers. Cryptologia XVII(1), 187–201 (1993)

[6] Clark A and Dawson Ed, "Optimisation Heuristics for the Automated Cryptanalysis of Classical Ciphers", Journal of Combinatorial Mathematics and Combinatorial Computing, Vol.28,pp. 63-86, 1998.

[7] M. Matsui, Linear cryptanalysis method for DES cipher, Lect. Notes Comput. Sci. 765 (1994) 386– 397.

[8] William Stallings, Cryptography and Network Security Principles and Practices, Third Edition, Pearson Education Inc., 2003.

[9] Vimalathithan.R, M.L.Valarmathi, "Cryptanalysis of SDES Using Genetic Algorithm", International Journal of Recent Trends in Engineering, Vol2, No.4, November 2009, pp.76-79.

[10] Schaefer E, "A Simplified Data Encryption Standard Algorithm", Cryptologia, Vol .20, No.1, pp. 77-84, 1996.

[11] Yılmaz Kaya, Murat Uyar, Ramazan Tekdn," A Novel Crossover Operator for Genetic Algorithms: Ring Crossover".

[12] Davis,L. "Handbook of Genetic Algorithm",Van Nostrand Reinhold, New York,1991.

[13] D. E. Goldberg,"Genetic algorithms in search. Optimization and Machine Learning.Reading. M.A. addison -Wesley.1989.

[14] A, Michalewiez and N. Attia." Evolutionary optimization of constrained problems." InProc.3rd annu. Conf. on Evolutionary Programming. 1994.pp 98-108

[15] Z. Michalewicz. "Genetic algorithms+ Data structures = Evolution programs 3rd Ed. New York. Springer,1996.

[16] N.Koblitz, "A Course on number theory and cryptography", Springer-Verlag New York,Inc., 1994.

[17] Alfred J. Menezes. Menezes, Alfred J. Handbook of Applied Cryptography, CRC, 1997.

[18] R. Toemeh, S. Arumugam, Breaking Transposition Cipher with Genetic Algorithhm Electronics and Electrical Engineering,ISSN 1392 – 1215 2007. No. 7(79).

[19] Lavkush Sharma , Bhupendra Kumar Pathak & Ramgopal Sharma Breaking of Simplified Data Encryption Standard Using Genetic Algorithm

[20] Kalyanmoy Deb, Multi-objective Optimization using Evolutionary Algorithms, John Wiley and Sons, 2001.

[21] C.W. Wu and N. F. Rulkov, Studying chaos via 1-Dmaps—atutorial, IEEE Trans. on Circuits and Systems I: Fundamental Theory and Applications, vol. 40, no. 10, pp. 707–721, 1993.

[22] http://www.hindawi.com/journals/mse/2009/540895/fig1/

[23] http://www.decodesystems.com/mt/98oct/crypt.html.

15

Voice over IP End-to-End Delay Measurements

Binyam Shiferaw Heyi

Department of Computer Engineering, Addis Ababa Science and Technology University (AASTU), Addis Ababa, Ethiopia

Email address:

bsh385@gmail.com

Abstract: VoIP refers to a method of transmission of voice data over IP networks rather than using traditional circuit PSTN (Public Switched Telephone Networks).In this paper the researcher discusses about VoIP, about the design issues of RTP/RTCP (real-time transport/real time control protocol), how to measure packet losses at the sender and receiver side and also inter-arrival delay and delay variation (Jitter) , how the receiver and sender sends reports about the sender , how one-way end-to-end delay is measured, and explains the functionalities of static and dynamic jitter buffers and describe how Cisco routers are used to measure packet losses and round trip delays for IP-SLA (Service level agreement).

Keywords: VoIP, End-to-End Delay, RTP, RTCP, IP-SLA

1. Introduction

VoIP is a technology that let us use normal internet to make and receive a call. The most attractive benefits of VoIP cost saving, having single network for voice and data helps for simplification of management and allow you to access your phone system anywhere you have a broadband internet. Although VoIP has different usage, it comes with a challenge; those challenges include speech quality, service reliability, packet loss, delay and delay variations. This paper discusses about the challenges and recommends a better way to use the technology. This chapter discusses the technologies behind VoIP.

Real-time transport protocol (RTP): it is a protocol that is used for application that has real-time characteristics. The protocol provide services that are end-to-end such as payload type identification, sequence numbering, time stamping and delivery monitoring for applications that are interactive videos and audios. [1]

RTP control protocol (RTCP): is a protocol that is used to convey information and monitor the quality of services of about the participants in an on-going session. [1].

As it has been known when there is communicate over an internet, the communication encounters packet losses, delays and out of order delivery of data (this thing is worse for real time communication). So RTP contains header that has fields for sequence number and for timestamps, where their values are going to be inserted by the sender and are read by the receiver used for correct reception of data.

2. Literature Review

In [11], it has been described a method of monitoring VoIP performance, the paper also suggested the measurement point to be taken in the path of the packet to get accurate result. The paper finally recommends more studies should be done in the near future.

In [12], discussing delay and jitter in VoIP traffic. This paper focuses on suggesting proper scheduling scheme especially in low-bandwidth networks that carry VoIP data.

In [13] discusses the end-to-end VoIP quality measurement. The paper focuses on recording voice samples to measure VoIP delay and delay variations and packet loss on the voice signals.

Paper [14], discusses the measurement scheme of one-way delay variation, this paper discusses on removing clock offsets and measuring all variables that is required for one way delay measurements

3. Proposed Methodology

The sender and receiver exchanges RTP Voice packets. Every RTP packet contains 12 byte header which contains 16 bit sequence number which is used to detect packet losses. This sequence number is generated by the sender and incremented by the sender after sending one packet after another. When receiver receives a packet it sees its sequence number and compares it with the expected value if it does not match the receiver concludes that there is some packet lost. If the expected sequence packet matches the receiver calculate   the parameters as in [2]

$$Number\ of\ packet\ losses = Expected\ Number\ of\ Packets - Received\ Number\ of\ Packets \tag{1}$$

$$Number\ of\ packet\ losses = Expected\ Number\ of\ Packets - Received\ Number\ of\ Packets \tag{2}$$

Where

$$Expected\ Number\ of\ Packets = Extended\ Highest\ Sequence\ Number\ Received\ (EHSNR) - Initial$$

$$Sequence\ Number\ of\ Packet\ (ISN) \tag{3}$$

The lost fraction can be calculated

$$Loss\ fraction = \frac{Number\ of\ Packet\ Losses}{Number\ of\ Packet\ Expected}. \tag{4}$$

3.1. Time Stamp Calculations

Time stamp calculations basically depends up on the type of application we use for example for voice and video data it is basically different since the given data is basically for voice I only use the formula for voice data only.

3.1.1. Timestamp Calculation at the Sender Side

For the experimentation purpose, the researcher uses 10 RTP packets .According to [1] since the audio that is used is fixed-rate the sampling period dictates the time stamp clock i.e for each increment by the sample period the time stamp would likely to increment by one. If an output device provide an audio application that has blocks covering 160 sampling period .the sender timestamp would increased by 160 ,regardless of the block has transmitted or dropped.

Table 1 shows the sender timestamp.

Table 1. Sender Timestamps.

Sequence No. i	Si (Timestamp)	Si (ms)
1	0	01:02:02:00
2	160	01:02:02:20
3	320	01:02:02:40
4	480	01:02:02:60
5	640	01:02:02:80
6	800	01:02:02:100
7	960	01:02:02:120
8	1120	01:02:02:140
9	1280	01:02:02:160
10	1440	01:02:02:180

3.1.2. Calculation at the Receiver Side

When a receiver is receives the packet, the timestamp of the received packet according to eq (5)

$$Receiver\ timestamp\ Rec(i) = \frac{Rec(i)in\ terms\ of\ time\ unit \times Sampling\ frequency}{1000} \tag{5}$$

Where, Rec(i) in terms of time unit can be calculated by eq (6) sampling frequency of voice is 8000Samples/Sec

$$Rec(i)in\ terms\ of\ time\ unit = \Big(arrival\ time\ of\ packet\ (I\ )- arrival\ time\ of\ packet\ (i-1)\Big)$$

$$+ Rec(i-1)in\ terms\ of\ time\ unit \tag{6}$$

The receiver calculates the timestamp only when the packet is received. The initial timestamp is zero for the first packet is zero because it is used as a reference. In table 2 shown the packet nine does not arrive to the receiver that's why no calculation has been done for the packet nine

Table 2. Receiver Timestamp.

Seq.	Arrival Time	Rec(i) (ms)	RecTS(i) (timestamp)
1	01:02:01:20	0	0
2	01:02:01:43	23	184
3	01:02:01:63	43	344
4	01:02:01:84	64	512
5	01:02:01:109	89	712
6	01:02:01:130	110	880
7	01:02:01:150	130	1040
8	01:02:01:170	150	1200
9	X	X	X
10	01:02:01:230	210	1680

3.1.3. Inter-Arrival Jitter Calculation Method

In real time communication there is no guarantee that the packet sent, will arrive in order and delivered on time. If the receiver don't receive the packet on time or in order over the network that means delay occur that create a delay gap. The inter-arrival jitter (J) is defined by variation of delay in the network that is perceived by the receiver for each packet. Every packet received contains a timestamp that informs the receiver at which time the received data in the packet should be played back. [1]

The difference in the "transit relative time" D (which is relative time difference between RTP timestamp of the packet and the receiver clock at the time when the packet arrives,) D (x, y) for the two packets x and y can be calculated by equ (7)

$$D\ (x,y) = \big(Rx- Ry\big)- \big(Sx - Sy\big) = \big(Rx - Sy\big)- \big(Rx - Sy\big) \tag{7}$$

Where,

Sx is the RTP timestamp from packet x, and

Rx is the time of arrival in RTP timestamp units for packet x

The receiver can calculate the inter-arrival delay variation or jitter according to eq(8) [1]

$$J(x) = J(x-1) + \frac{|D(x-1,x)| - J(x-1)}{16} \qquad (8)$$

Where

J(x) is the delay variation or Jitter of packet x

Example:

In the given question the first packet is sent at time t=01:02:02:00. The time distance at the sender between each packet is 20 ms and each packet contains 160 octets of voice data. The sampling frequency of the voice signal is 8 kHz (each sampled unit contains one octet).

The calculation of inter-arrival jitter for the packet 1 and 2

are shown below

Sender timestamp for packet 1, S1=: 0

Sender timestamp for packet 2, S2=0+160= 160 timestamp =160/8=20ms

Rec (1) in terms of time unit = 0 ms

Rec (2) in terms of time unit = (43-20) +0=23 ms

Now the receiver timestamp for packet 1,

RecTS(1) = 0

Receiver timestamp for packet 2,

RecTS(2) = 23 * 8000/1000 = 184

Delay: D(1,2) = (184-0)-(160-0) = 24timestamp =24/8Khz=3ms

Jitter: J(2) = 0+(24-0)/16 = 1.5 timestamps =1.5/8Khz =0.188 ms

In similar fashion we can calculate the other inter- arrival jitter values as show in table 3.

Table 3. *Inter-Arrival Jitter.*

Seq. No, i	Si (Timestamp)	Si (ms)	Arrival Time	Rec (i) (ms)	RecTS(i) (Timestamps)	D(i-1,i) timestamp	D(i-1,i) Ms	J(i) Timestamps	Ji (ms)
1	0	0	01:02:01:20	0	0	0	0	0	0
2	160	20	01:02:01:43	23	184	24	3	1.5	0.188
3	320	40	1:02:01:63	43	344	0	0	1.406	0.176
4	480	60	01:02:01:84	64	512	8	1	1.818	0.227
5	640	80	01:02:01:109	89	712	140	17.5	4.204	0.526
6	800	100	01:02:01:130	110	880	8	1	4.442	0.555
7	960	120	01:02:01:150	130	1040	0	0	4.164	0.521
8	1120	140	01:02:01:170	150	1200	0	0	3.904	0.488
9	1280	160	X	X	X	X	X	X	X
10	1440	180	01:02:01:230	210	1680	160	20	13.66	1.708

3.2. *How Senders and Receiver Reports to Each Other in RTCP*

For reporting errors associated to losses and jitter, RTCP packets uses Send Report (SR) and Receive Report (RR) messages. The SR is issued by sender if a sender has sent any data during an interval since issuing the last SR.The sender SR includes absolute timestamp to help the receiver to synchronize with the receiver, if the transmission contains video and audio data transmitted at the same time, both requires independent time stamps.

The participants will issue RR together with SR if they are active senders and receivers; if they are no t active senders they will only issue RR. The RR report contains information about the quality of services of the transmission.

Fig 2 depicts how the report is sent from the two sides

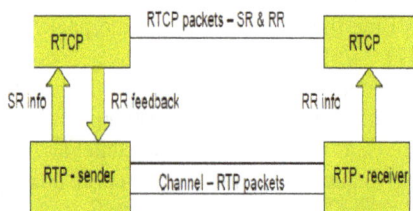

Figure 1. *Details of RTCP.*

4. One-Way Delay Calculation

For calculation of one-way delay in addition to those ones that are mentioned above ,we require the time it takes a packet to travel along the physical links that make up its path through the internet to the destination (what we call propagation delay) and the time it takes to pass through routers between those links (queuing and transmission time).

For calculation one-way delay for active measurement system between two measurement points we must continuously send test traffic into the network that increases either the queuing delay or transmission time this may be accomplished by introducing additional packet to the routers and or by increasing the propagation delay by re-routing the packet to another path and measure the delays and compare it with the normal functioning of the network ,here we must take care of the situation that the traffic we introduce does not going to be above that of the minimal value.[9]

Finally the requirements needed for one way delay are

• Each node in the network sends a packet by setting its own time stamp to other node in the neighbor.

• When the other nodes receive this packet it will mark the arrival time with its own time stamp.

• The difference between this two time stamps is one way delay i.e. one way delay = Receiving timestamp - Sending timestamp

But this measurement is valid only if the two clocks

(receiver and transmitter) are synchronized to the same time base, otherwise the calculation should consider clock drifts.

5. Building up End-to-End Delay Budget

In real-time communication to accurately estimate the performance of the communication, the designer should account and correctly calculate the delay associated with every device inside the network.[5]

The following components should be considered to calculate the end-to-end delay budget.

Coder (processing) delay: it is the delay by the coder or the processor of the signal .This delay varies from one coder to another and also with their speed. Coder delay contains compression and decompression delays since some signals (e.g. voice) may be needed to be compressed or decompressed. [5]

Algorithmic delay: sometimes the system encounters digital signal processing; for correct processing of Sample Block (N) they need to know the behavior of sample block (N+1).So the time spent to overlook for sample block(N+1) is Algorithmic delay.

Coder delay and algorithmic delay are together are called lumped coder delay.

Packetization delay: it's the delay associated with filling up the payload part of the packet with encoded data. This delay is a function of sample block size and the number of block in the frame. [5]

Serialization delay: it is the delay associated with clocking the dat frame on the network interface .It is basically dependant on the speed of the trunk. [5]

Queuing/buffering delay: it is the delay associated with queues inside the routers; it is basically dependant on the trunk speed and the state of the queue. [5]

Network switching delay: it is the most difficult of all delays to quantify is associated with the frame-relay or ATM networks that interconnect the end point locations

De-jitter delay: if the transmitted signal that is constant bit rate signal then, all the bit signal rate delay must be removed before the signal leaves the network; this is done in Cisco routers at the receiver end by transforming the variable delay into a fixed delay. [5]

So, end-to-end dlay comprises the sum of all the above delays.

6. Jitter Buffers

When packet travels inside the network it may be delayed at several points, but the delay associated with this packet is variable so this variation in delay is called jitter. So for giving high quality of service we need to control this jitter. Causes of jitter delay are basically congestion inside the network, or at the access link, route flapping and load sharing between routers.[6]

Jitters are classified into

- Constant jitter,
- Transient jitter, S
- Short term delay variation,

One of the mechanisms to reduce delay jitter is just to introduce jitter buffer so that it buffers each arriving packet for a short interval before playing it out.

There are two types of jitter buffers static (De-Jitter) buffer and dynamic (Adaptive) buffer.

Both fixed and adaptive jitter buffers are capable of automatically adjusting to changes in delay. In many cases the jitter buffer can be considered as a time window with one side (the early side) aligned with the recent minimum delay and the other side (the late side) representing the maximum permissible delay before a packet would be discarded. [6]

De-Jitter buffer: as its name indicates it has a constant size and transforms the variable delay into a fixed delay. It just holds the first sample before playing it out. We must take care the holding time here because if we hold it for longer time it may lead to the buffer to be over-runed and increase the time over all delay to unacceptable level. . The maximum jitter that can be countered by a dè-jitter buffer is equal to the buffering delay introduced before starting the play-out of the media stream. [5]

Dynamic Buffers: this buffer has the ability to adjust its size dynamically which able it to optimization. Dynamic jitter buffers react by either discard events when the measured jitter increase in level. When this event is occurred then the jitter buffer size is increased. If there is no discard event then the size jitter buffer size is reduced. [5]

7. Cisco IP-SLA

Different network equipment inside the network must satisfy service guarantees, validate network performance and the like; Cisco IOS (Internetwork Operating System) IP Service Level Agreements (SLAs) fulfill those requirements that are needed for creating a network that is "performance-aware". [10]

Cisco IP-SLA has unique subset of the following performance metric which includes: Delay, Packet loss, Packet sequencing….

Cisco routers can be used to measure packet loss and round-trip delay. When Cisco router is being used to measure round trip time the routers must be enabled with responder which notes down two timestamps: when the packet arrives and leaves the router interface as can be shown in t fig 2, to calculate the round trip delay four timestamps are needs. At the target router, arrival timestamp (T2) is subtracted from sent timestamp (T3) to produce the time spent processing the test packet. This is represented by delta. This delta value is then subtracted from the overall round-trip time to find the round trip time delay. [10]

Figure 2. Cisco IP SLA Measurements.

$$Round - Trip\ delay = (T4 - T1) - (T3 - T2) \quad (9)$$

8. Conclusion and Recommendation

8.1. Conclusions

This paper presented a method to calculate approximate values of delay in VoIP traffic carried through RTP packets. The suggested procedure different calculation methods to calculate delay and delay variation. These methods give a very good approximation to the real values that is obtained and also the values of VoIP delay which can be helpful in understanding the network behavior when changes in traffic happen.

The suggested methods and formulas can be used in any type of voice traffic independent of network types and independent of signaling protocols.

The calculations were implemented successfully to the 10 sample calls and the results were conclusive with the real measured values.

8.2. Recommendations

The research can further extended to operate on many samples, not only voice but also videos. The improved model can also be extended to include interpolations and regressions.

References

[1] R. Frederick,H. Schulzrinne, V. Jacobson " A Transport Protocol for Real-Time Applications", RFC 3550, July 2003

[2] Kevin Jeffa "RTCP "[online]: available at http://www.cs.unc.edu/~jeffay/courses/comp249f99/Lecture6.pdf Last Access Mar 31,2015

[3] Omer Gurewitz srael Cidon," One-Way Delay Estimation Using Network-Wide Measurements " ,IEEE transactions on information theory" Vol 52 No.6 June 2006

[4] "RTP,RTCP control protocol " [online]: available at http://www.networksorcery.com/enp/protocol/rtcp.htm Last Access Mar 5,2015

[5] "Understanding delay in packet voice network "[online]: available at http://www.cisco.com/en/US/tech/tk652/tk698/technologies_white_paper09186a00800a8993.shtml Last Access Mar 16,2014

[6] Jitter Buffer [online]:available at http://www.voiptroubleshooter.com/indepth /jittersources.html Last Access April 4,2015

[7] Etomic,[online]: available at http://www.etomic.org/index.php Last Access April 17,2015

[8] Sok-hyun jung," One- way delay measurement,"[online]: available at www.apan.net/meetings/fukuoka03/.../measurement/korea-SH J.PPT , 2003 last access April 20,2015

[9] Neville Brownlee,Chris Loosley :"Fundamental of internet measurement" CMG Journal of Computer Resource Management, Issue 102, Spring 2001

[10] Cisco white paper:" Cisco IOS IP service level Agreement",[online] available at http://www.cisco.com Last Access Apr 22,2015

[11] R. G. Cole and J. H. Rosenbluth. 2001. Voice over IP performance monitoring. SIGCOMM Comput. Commun. Rev., Vol. 31, No. 2, April 2001, pp. 9-24.

[12] Karam, M.J.; Tobagi, F.A., "Analysis of the delay and jitter of voice traffic over the Internet," INFOCOM 2001. Twentieth Annual Joint Conference of the IEEE Computer and Communications Societies. Proceedings. IEEE , vol.2, pp.824-833, 2001

[13] ČÁKY et al., End-To-End VOIP Quality Measurement, Acta Electrotechnica et Informatica, Vol. 6, No. 1, 2006, pp.1-5.

[14] Makoto Aoki et al., Measurement Scheme for One-Way Delay Variation with Detection and Removal of Clock Skew, ETRI Journal, Volume 32, Number 6, 2010, pp. 854-862

[15] Ngamwongwattana and Thompson, Measuring One-Way Delay of VoIP Packets Without Clock Synchronization, Proceedings of International Instrumentation and Measurement Technology Conference, Singapore, 5-7 May 2009.

[16] H. Xie and Y. Yang, A Measurement based Study of the Skype Peer-to-Peer VoIP Performance, Prceedings of the 6th International Workshop on Peer-to-Peer Systems, USA, Feb. 2007.

[17] B. Sat and B. Wah. 2007, Playout scheduling and loss-concealments in voip for optimizing conversational voice communication quality. In Proceedings of the 15th international conference on Multimedia (MULTIMEDIA '07),ACM , NY, USA, pp. 137-146.

16

Performance Analysis of Hybrid Web Caching Architecture

Ho Khanh Lam[1], Nguyen Xuan Truong[2]

[1]Faculty of Information Technology, Hung Yen University of Technology and Education, Hung Yen, Vietnam
[2]Training Department, Hung Yen University of Technology and Education, Hung Yen, Vietnam

Email addresses:
lamhokhanh@gmail.com (H. K. Lam), truongutehy@gmail.com (N. X. Truong)

Abstract: The development of next-generation wireless networks combine with the radio network techniques which use technology such as GSM, GPRS, 3G (UMTS, CDMA2000), LTE, WLAN, and WiMAX. It requires the construction and expansion in time of high-speed telecommunication channels for the level of the internet service provider (ISP). In fact, in many developing countries and in Vietnam, the investment rate increased bandwidth capacity can not enough for the demand use the Internet as economic issues, investment procedures. Web caching architecture is one of the effective solutions to save bandwidth while ensuring to satisfy strong demand for internet access. Hybrid web caching architecture (hybrid web caching architecture) is a solution that is used by networks because it takes advantage of the strengths of the web caching architecture stratification and dispersion, reducing connection time and transmission time, helping internet service providers to plan and save network resources at each level in an optimal way. This paper propose a novel produce to hybrid web caching architecture based on the determined time at each level of web-winning network and web time overall winner of the ISP network with n-level network.

Keywords: Hybrid Web Caching Architecture, Performance Analysis

1. Introduction

The architecture of the Internet service provider (ISP) usually organized into four levels: access networks, Institutional networks, regional networks, national backbone, and International backbone. The layer up will have bandwidth communication greater. Institutional networks is access networks which are organized according to the POP point local (poit of presence). it include technologies, telecommunications equipment high speed and allows users to connect to the Internet through the ISP of them. An POP point has some unified address and a set of IP addresses for accessing the Internet from the user (the client). A fact a POP of ISP can stay inside the house telecommunications networks. Where will design and construction of POP and how much bandwidth should be based on the standard of living, population and literacy, the focus of the economic base of schools, etc .. and requires more detailed cost of telecommunication channel capacity, the capacity of access equipment (servers, routers, etc ..).

More than 80 percent of the traffic using the user's web access, via web ISPs provide multimedia services require high-speed, latency and ensure high quality of service (QoS). Thus, the ISP's POP from the outset to ensure the circulation of access for end users from the private LAN, or from the radio access network. Therefore, the solution to improve the performance of web services: web access latency reduction for the client reduce bandwidth costs, it has web caching architecture suitable for ISP network architecture. Based on internet architecture, there are three types of web caching architecture that most ISPs apply, such as: hierarchical web caching architecture, distributed web caching and hybrid web caching. Stratified web caching architecture allows requests from terminal users from network routers low level (level access network) to the higher-level network: Institutional Network, Regional network, national network of the ISP network, if all of the network-level ISP networks are no web content which client requests are required to be transferred to the international Internet. So this is the worst case, it has the largest network latency for a web content which is required by client. Also this architecture for cache hit rate is not high at each network level, but it requires large bandwidth among the network levels. Architecture distributed web caching system is ensure for the peer web cache associated with each other in the network level, so it will ensuring a high rate of secondary web network at each level, and thus saving bandwidth between levels network. However, this architecture requires a large investment costs for the system-level web cache in each network: bandwidth and the web server.

Web caching system is a hybrid solution which usually used by ISP. it combines two types of architectural stratification and distribution. At each level of the network perform distributed caching web architecture, but not all nodes have the web cache system, because the reasons is ensuring cost savings, but only the nodes are high bandwidth requirements by there are large numbers of citizens using the Internet. And all web cache system such links in a peer to peer networks to increase the level of use of web caching system at each level of the network. Web cache architecture is ensure interconnect for these web cache system, so it is hit rate when request of the client moved up on the network layer and also save bandwidth between the network layer. Figure 1 is a diagram of the architecture of the hybrid web-based caching associated with 4 main level. Highest level of entire web caching architecture that combines are web caching system center (level 1) of the national Internet backbone network, ISP CC (Central Cache). At the ISP Regional Network system has the Web cache area (level 2), Regional Cache. Next Level (Level 3) is the web cache system of local networks, Institutional Cache. The client network-level access to the proxy server. We connect with the local access network. The telecommunications provinces buttons POP(Point of Presence Network) is the local node access the Internet. In the POP put the web cache system, IC. The client can be a PC, a mobile phone, directly or through LAN, connected to the Internet via POP in access networks such as ADSL, mobile network. The POP associated with the high-speed transmission with regional networks.

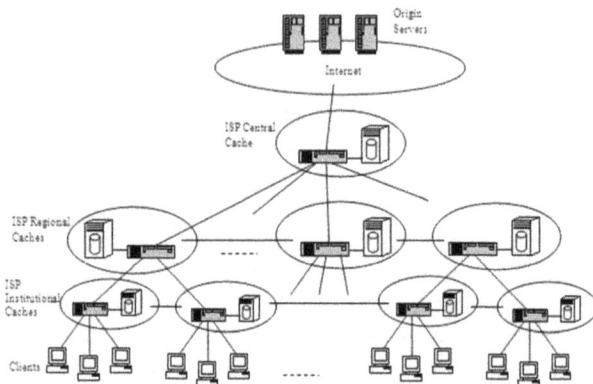

Figure 1. *Architecture of the Internet web caching stratified.*

The web browser client can directly or indirectly through local LAN proxy server sends requests for status online web access to the local network. In the case of a local proxy server does not have the content of the web page request, the proxy server forwards requests from the client to the IC system in the local POP. If the IC system has content client requests (winning IC), the IC transfer will transfer content requirements for the client (also local Proxy server saves the contents of this web page). In case of winning the IC hit time to the requirements of the client (or latency response) is the smallest. When IC miss means that the website content which the client requires not available in the system IC, so the IC system requirements are transferred to the network-level up

network - area network.

At the local network, if the system requirements RC hit, it will transfers the contents to the System IC and IC transition forward to Proxy server and client. If the RC miss request of the client is transferred to the CC system of national networks. If the CC has system contents, CC hit is transferred to the RC system, and IC systems, and to the Proxy server, client. If the CC miss requests of the client is transferred to the international Internet, to web server root.

2. The Study of the Performance of Web Caching Architecture on the Internet

The author Pablo Rodriguez, Christian Spanner, ... [1] gave the model of the web caching architecture stratification and distribution, and analyze the performance of this architecture with connection time, transmission time, delay, and cache hit rate based on the theory of Markov queuing model M/D/1. However, in this study the authors suggest that the average connection time at the network layer is the same and only depend on the network layer. This is incorrect because the actual bandwidth in each different network layer, data transfer speeds are different, different delay systems and web cache different capacity, different level combination (cooperation) and in the lowest layer network - network access layer also depends very much on the access network technology, the concentration of population in the local node and the terminal network.

The author Guangwei Bai and Carey Williamson [2] analyzed the load characteristics of web caching architecture stratification, or those of the authors Balamash Abdullah and Marwan Krunz [3] when analyzing system for caching Web traffic. The study evaluated the performance of web caching system of the authors in the paper [4] [5] [6] [7] [8] [9] showed that at each level of the network must be investigated individual assessment by the difference in traffic. Queues and Markov chains are commonly used to evaluate the analytical performance of web caching architecture, web proxy server.

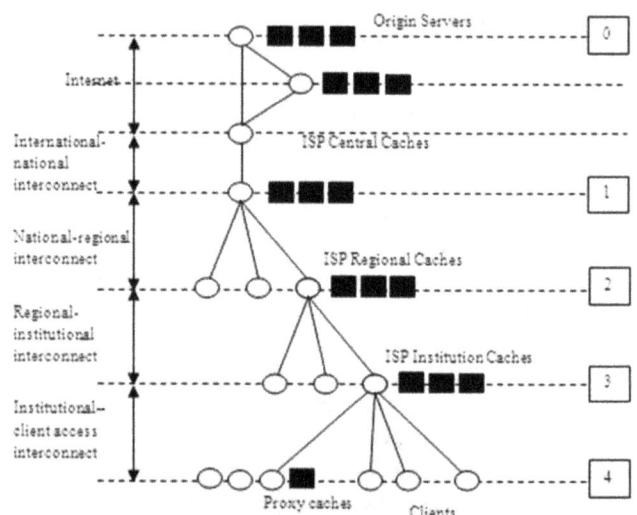

Figure 2. *Model tree of web caching architecture stratified.*

3. Solution Queue Model

Based on the research of the authors in the paper [1], in this proposed model tree diagram of web caching hybrid architecture (Figure 2). Layer 4: the network layer of the user terminal (the proxy server of the LAN, the client separately), Layer 3: local access network (radio access networks, ADSL, the local POP with the Institutional caches), layer 2 network with the regional caches, layer 1 national network with the central caches system , layer 0: international Internet to the origin servers. The highlight rectangle at each network layer represents the peer to peer (P2P) web caching systems.

The hybrid caching architecture makes web access latency of the content of the web client and it was hit ratio decreases if the web hit is high in each network layer. But if the web miss is occurs in multiple network layer the time delay will the greater for access web. To analyze the performance of web caching hybrid system, different methods of analysis of previous studies.

Because each network layer, the web server link peer to peer, so each web caching system at each level network can be considered multi-server connections in parallel and is represented by a model of queuing system type M/M/m/q with both loss and delay. Given the HTTP requests from web caching system independent of the client to each other, and the number of client HTTP requests generated unrestricted. The time between HTTP requests to the system and the service life of the system with exponential distribution. Web caching system has m the same server m = 1, 2, ... We have the capacity cache is limited by the model as the queue to receive HTTP requests with the length q.

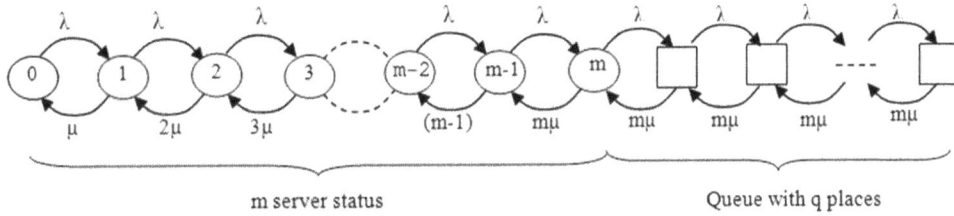

Figure 3. Graph CTMC state's web cache system M/M/m/q.

Because the transmission delay due to network environment at each level different networks so we assume that the average arrival rate of HTTP requests in each level is $\lambda_i; i = 0, 1, 2, ..., n$ where n is the network level, and the average speed service of web caching system(server) at each level of the network is $\mu_i; i = 0, 1, 2, ..., n$. The level of use of each web server is:

$$U = \frac{\lambda_i}{\mu_i}$$

If web caching system in each network layer consists of m parallel server connection and has q positon in queue, we have state of the CTMC shown in figure 3.

For this CTMC chain of flow balance equation is satisfied as follows:

+ where $0 \le k \le m$:

$$p_{ik} = \frac{U_i^k}{k!} p_{i0}; \forall k = 1, 2, ..., m \tag{1}$$

where, p_{i0} is the probability that the client HTTP requests to the web-winning network level i and web cache system is idle, p_{ik} is the probability that the client HTTP requests to the web-winning web caching system and network level i is in state k (k serving HTTP requests).

+ when $m \le k$ but space in the queue from 1 to q:

$$\lambda_i p_{im} = m\mu_i p_{i(m+1)}; \quad p_{i(m+1)} = \frac{U_i^m}{m!}\left(\frac{U_i}{m}\right) p_{i0}$$

$$\lambda_i p_{i(m+1)} = m\mu_i p_{i(m+2)}; p_{i(m+2)} = \frac{U_i^m}{m!}\left(\frac{U_i}{m}\right)^2 p_{i0}$$

$$.\lambda_i p_{i(m+k-1)} = m\mu_i p_{i(m+k)}; p_{i(m+k)} = \frac{U_i^m}{m!}\left(\frac{U_i}{m}\right)^k p_{i0}$$

$$\forall k = 1, 2, ..., q$$

$$\Rightarrow p_{i(m+k)} = \frac{U_i^m}{m!}\left(\frac{U_i}{m}\right)^k p_{i0}$$

+ For model M/M/m/q normalize this condition must be finished, which is the sum of the probabilities must equal 1:

$$1 = \sum_{k=0}^{m+q} p_{ik} = p_{i0}\left(\sum_{k=0}^{m} \frac{U_i^k}{k!} + \frac{U_i^m}{m!}\sum_{k=1}^{q}\left(\frac{U_i}{m}\right)^k\right) = p_{i0}S;$$

$$S = \sum_{k=0}^{m} \frac{U_i^k}{k!} + \frac{U_i^m}{m!}\sum_{k=1}^{q}\left(\frac{U_i}{m}\right)^k$$

Suy ra:

$$p_{i0} = \frac{1}{S} = \left(\sum_{k=0}^{m} \frac{U_i^k}{k!} + \frac{U_i^m}{m!}\sum_{k=1}^{q}\left(\frac{U_i}{m}\right)^k\right)^{-1} \tag{2}$$

+ When the entire m server and q positon in the queue of web caching system of tiered networks are busy, the new HTTP request to the system will be locked (not included in the queue). This is the case of congestion in the network layer i as the entire web caching system with m servers were overloaded. This state is determined by the probability of lock or probability of loss at the network level i and by $B_i = p_{i(m+q)}$:

$$B = p_{m+q} = \frac{U_i^m}{m!} \left(\frac{U_i}{m} \right)^q p_{i0} \qquad (3)$$

Thus the probability of new customers must wait is the probability which the server is busy and the queue is vacancies $p_{i(m+k)}$ với $1 \le k \le q$.

Apply Zipf law and the Internet, we can determine the number of access on a large number of the local. Must have

$$E[N_{iQ}] = \sum_{k=1}^{q} k p_{i(m+k)} = p_{i0} \frac{U_i^m}{m!} \left(\left(\frac{U_i}{m} \right) + 2 \left(\frac{U_i}{m} \right)^2 + 3 \left(\frac{U_i}{m} \right)^3 + ... + q \left(\frac{U_i}{m} \right)^q \right)$$

$$= p_{i0} \frac{U_i^m}{m!} \sum_{k=1}^{q} k \left(\frac{U_i}{m} \right)^k \qquad (4)$$

2) The average waiting time of an HTTP request in the system's web caching for layer network i to be serviced, $E[W_{iQ}]$ is determined by Little law:

$$E[W_{iQ}] = \frac{E[N_{iQ}]}{\lambda_i} = \frac{1}{\lambda_i} \left(p_{i0} \frac{U_i^m}{m!} \sum_{k=1}^{q} k \left(\frac{U_i}{m} \right)^k \right) \qquad (5)$$

3) The average response time $E[C_i]$ web caching system in each layer network i:

This is the average time that a client's HTTP request (web content) is processed in the web cache system (including waiting time in the queue and the time serviced (web content is found):

$$E[C_i] = E[W_{iQ}] + E[S_i] = \frac{E[N_{iQ}]}{\lambda_i} + \frac{1}{\mu_i} \qquad (6)$$

4) In general, if the architecture of a network web caching ISP's network, the client's HTTP request miss web in level network n, the hit web is web caching system at layer network i, where $n > i$, the response of web caching system at the layer network n for HTTP requests from the client by:

$$E[R_{WC}] = E[R_{nH}] + (Miss_n)(E[R_{(n-1)H}] + (Miss_{n-2})(E[R_{(n-2)H}] + ... + (Miss_1)(E[R_{0H}])...) \qquad (7)$$

where, $E[R_i]$ – average response of the system's web caching of the network layer i when the web cache hit layer i; $Miss_n$ – rate cache miss in layer network n.

When cache miss in the local network (proxy server) and client cache hit at the local access network, the average response of web caching system will be:

$$E[R_{3H}] = (D_{4M} + D_{4REQ}) + E[C_3] + D_4 \qquad (8)$$

where, D_{nM} - delays in the web miss layer network n; D_{4M} – delay by the web miss in LAN proxy server; D_{nREQ} - Delay dependent bandwidth transmission channels which require the client HTTP transfer from layer network n to

layer network $n-1$; D_{4REQ} – A delay of a local proxy server to access the network via POP; D_n - delay reply of web content requires the client dependent transmission channel bandwidth from layer network $n-1$ to layer network n, and depending the size of web content; D_4 – Delay reply of web content requests from the client caches intstitutional dependent transmission channel bandwidth from layer network $n-1$ to layer network n;

$$E[C_3] = E[W_{3Q}] + E[S_3] = \frac{E[N_{3Q}]}{\lambda_3} + \frac{1}{\mu_3}.$$

When cache miss in the local network (proxy server) client, cache miss at the local access network, the network-level cache hit, the average response of web caching system will be:

$$E[R_{2H}] = ((D_{4M} + D_{4REQ}) + (D_{3M} + D_{3REQ}) + E[C_2] + D_3 + D_4) \qquad (9)$$

where $E[C_2] = E[W_{2Q}] + E[S_2] = \frac{E[N_{2Q}]}{\lambda_2} + \frac{1}{\mu_2}$

When cache miss in the local network (proxy server) client, cache miss at the local access network, the network-level cache slipped, hit central cache national network level, the average response of web caching system will are:

$$E[R_{1H}] = ((D_{4M} + D_{4REQ}) + (D_{3M} + D_{3REQ}) + (D_{2M} + D_{2REQ}) + E[C_1] + D_2 + D_3 + D_4) \qquad (10)$$

When cache miss in the local network (proxy server) client, cache miss at the local access network, the network-level cache miss, miss central national network-level cache, the cache hit in the international Internet, the response average of web caching system will be

$$E[R_{0H}] = ((D_{4M} + D_{4REQ}) + (D_{3M} + D_{3REQ}) + (D_{2M} + D_{2REQ}) + (D_{1M} + D_{1REQ}) + E[C_0] + D_1 + D_2 + D_3 + D_4) \qquad (11)$$

statistical forecasting residential access in each region, and based on these results build web caching systems to optimize server capacity (CPU and memory capacity) to ensure value for q is contained the maximum number of HTTP requests.

With model M/M/m/q We can be calculated performance parameters for web caching system for each network level i as follows:

1) The number of HTTP requests in the queue of the system's web-based caching i, $E[N_{iq}]$:

Web caching architecture of a hybrid ISP 4 for average

response for Internet access are:

$$E[R_{WC}] = E[R_{3H}] + (Miss_3)(E[R_{2H}] + (Miss_2)(E[R_{1H}] + (Miss_1)(E[R_{0H}]))) \qquad (12)$$

Thus, the worst case is not required web hit at all levels of the ISP network in the national network and only hit the web on an international level at the Internet web server resources. The equation (12) presents mean responses for internet access depending on web caching architecture of each network level, web caching organization, size of web cache, and bandwidth of communication channel (D_n), the protocols and web cache replacement algorithm, the rate cache miss levels ($Miss_n$). Internet network architecture 3 layer is very common: Institutional, regional, and national. However, to response the requirements of small delay for high-speed service and real-time equation(12) can be the basis for design calculations Internet and Web caching architectures suitable for each ISP.

Table 1. *The average response dependency ratios at the network level cache miss calculated (12) and that the value* $E[R_{3H}] = 5ms, E[R_{2H}] = 8ms, E[R_{1H}] = 12ms, E[R_{0H}] = 12ms$.

Only change $Miss_3$				Only change $Miss_2$				Only change $Miss_1$			
$Miss_3$	$Miss_2$	$Miss_1$	$E[R_{WC}]$	$Miss_3$	$Miss_2$	$Miss_1$	$E[R_{WC}]$	$Miss_3$	$Miss_2$	$Miss_1$	$E[R_{WC}]$
0.1	0.7	0.7	7.09	0.7	0.1	0.7	11.89	0.7	0.7	0.1	16.09
0.2	0.7	0.7	9.18	0.7	0.2	0.7	13.18	0.7	0.7	0.2	16.68
0.3	0.7	0.7	11.26	0.7	0.3	0.7	14.46	0.7	0.7	0.3	17.26
0.4	0.7	0.7	13.35	0.7	0.4	0.7	15.75	0.7	0.7	0.4	17.85
0.5	0.7	0.7	15.44	0.7	0.5	0.7	17.04	0.7	0.7	0.5	18.44
0.6	0.7	0.7	17.53	0.7	0.6	0.7	18.33	0.7	0.7	0.6	19.03
0.7	0.7	0.7	19.62	0.7	0.7	0.7	19.62	0.7	0.7	0.7	19.62
0.8	0.7	0.7	21.70	0.7	0.8	0.7	20.90	0.7	0.7	0.8	20.20
0.9	0.7	0.7	23.79	0.7	0.9	0.7	22.19	0.7	0.7	0.9	20.79

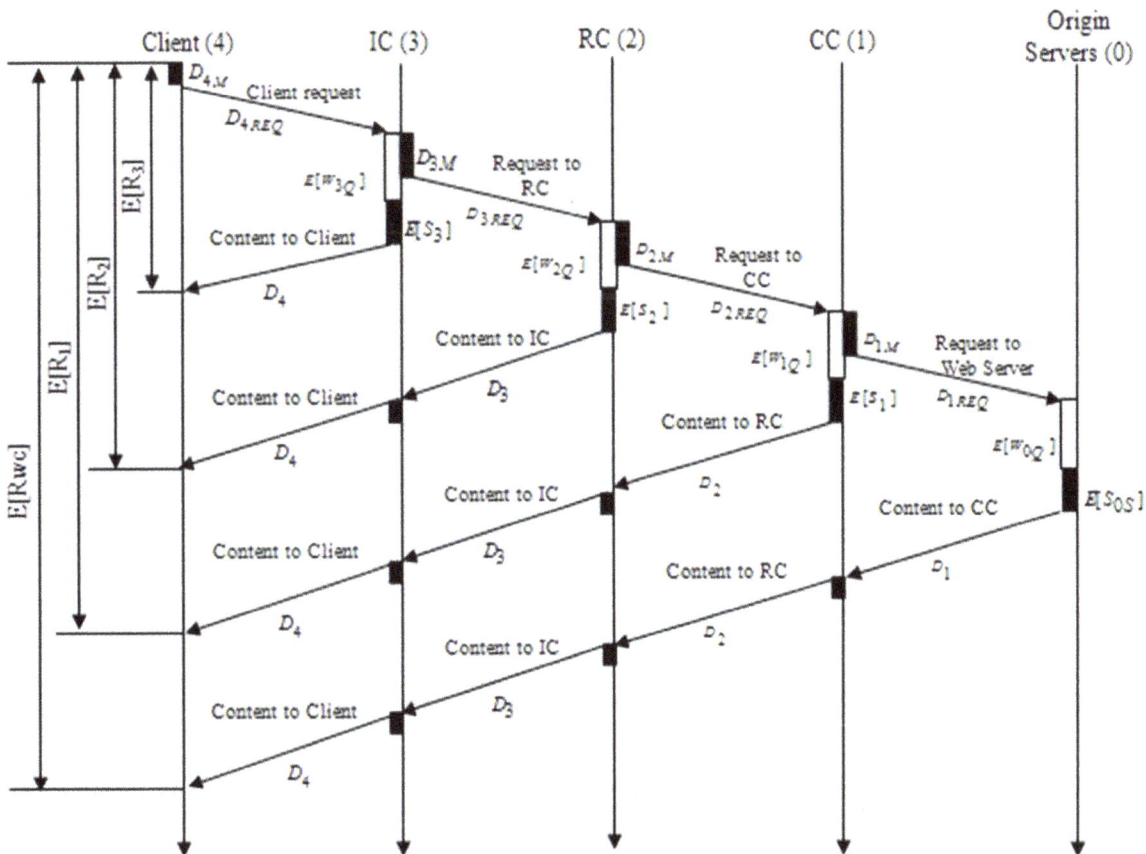

Figure 4. *Delay time chart of HTTP transaction for client on Internet with hybrid Web catching architecture.*

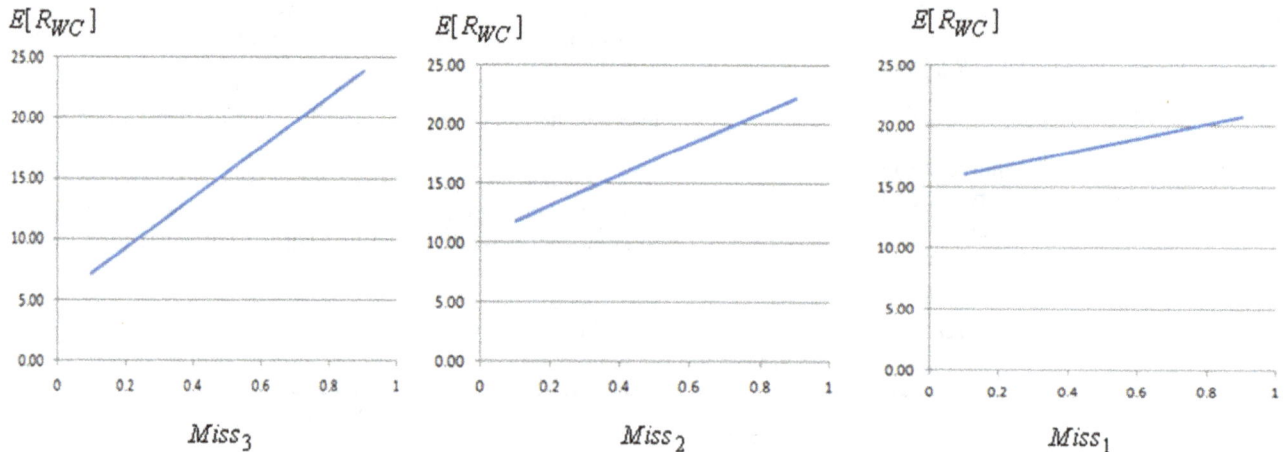

Figure 5. *Comparison of the effects of the ratio cache miss come* $E[R_{WC}]$.

4. Results and Discussion

Table 1 shows the results calculated average response of web caching architecture hybrid equation (12) according to the change of the ratio cache miss in the level Internet. Seen in Figure 5, the ratio cache miss at the local access network greatly affect local average response of web caching architecture. When $Miss_3 = 0.1$, then $E[R_{WC}] = 7.09\,ms$ is smallest, but when $Miss_3 = 0.9$ then $E[R_{WC}] = 23.79\,ms$ is largest with the changes of $Miss_2$, $Miss_1$. Thus, the solution build system Institutional caches in the local access network to ensure the performance of web caching architecture better than a lot of investment and costly network of regional and national level. Organizations in the caching system with POP application cache replacement algorithms and protocols as web caching solutions more economical and more efficient hybrid architecture for web caching. If additional systems of the LAN proxy servers attached to the end user, it also proved that the improved solution for the web proxy cache servers as well as the responsiveness of web caching architecture better.

Based on the formula (12) we can calculate the average response according to the dependence of the size or type of web services, the bandwidth of the access network, the transmission channel at the network level, etc .

5. Conclusions

In the present study, a hybrid web caching architecture for queue model has been suggested to estimate performance based on time at each level of web-winning network and web time overall winner of the ISP network with n-level network. The average response was calculated according to the dependence of the size or type of web services, the bandwidth of the access network, the transmission channel at the network level. Organizations in the caching system with POP application cache replacement algorithms and protocols as web caching solutions more economical and more efficient hybrid architecture for web caching. Moreover, the present solution is improved solution for the web proxy cache servers

as well as the responsiveness of web caching architecture better if we have systems of the LAN proxy servers attached to the end user.

References

[1]   Pablo Rodriguez, Christian Spanner, Ernst W.Biersack: Web Caching Architectures: Hierarchical and Distributed Caching. http://workshop99.ircache.net (4th International WWW Caching Workshop), Institut EUROCOM, france, 1999.

[2]   Guangwei Bai, Carey Williamson: Workload Characterization in Web Caching Hierarchies. 10th IEEE International Symposium on Modeling, analysis, and Simulation of Computer and Telecommunications Systems (MASCOTS'02), 2002.

[3]   Abdullah Balamash, Marwan Krunz and Philippe Nain: Performance Analysis of a Client-Side Caching/Prefetching System for Web traffic. Computer Networks, Volume 51, Issue 13, 12 September 2007, Pages 3673-3692, Copyright @ 2007 Elsevier B.V. All rights reserved.

[4]   Gunter Bolch, Stefan Greiner, Hermann de Meer, and Kishor S. Trivedi,"queueing Networks and Markov Chains Modeling and Performance Evaluation with Computer Science Applications". A Wiley-Interscience Publication Copyright © 1998 by John Wiley & Sons, Inc.

[5]   Carey Williamson, Mudashiru Busari: Simulation Evaluation of Web Caching   Architectures, M.Sc. Thesis, June 2000, Department of Computer science, University of Saskatchewan, http://www.cs.usask.ca/faculty/carey/.

[6]   A. Rousskov, "On Performance of Caching Proxies", In ACM SIGMETRICS, Madison, USA, september 1998.

[7]   C. Maltzahn, J.Richardson, "Performance Issues of Enterprise Level Web Proxies", 1998

[8]   G.N.K.Suresh babu and S.K.Srivatsa "An Analysis of Web Caching Strategies to Improve Web Performance". International Journal of Software and Web Sciences (IJSWS) 13-270; © 2013. ISSN (Print): 2279-0063 ISSN (Online): 2279-0071.

[9] V. Padmapriya and K.Thenmozhi,"Web caching and response time optimization based on eviction method". International Journal of Innovative Research in Science, Engineering and Technology Vol. 2, Issue 4, April 2013. ISSN: 2319-8753.

[10] Dr.K.Ramu1 and Dr.R.Sugumar,"Design and Implementation of Server Side Web Proxy Caching Algorithm" International Journal of Advanced Research in Computer and Communication Engineering VOL. 1, ISSUE 1, MARCH 2012. ISSN 2278 – 1021.

17

Blocking Probability Simulations for FDL Feedback Optical Buffers

Yasuji Murakami

Telecommunications and Computer Networks, Osaka Electro-Communication University, Osaka, Japan

Email address:

mura@isc.osakac.ac.jp

Abstract: Asynchronous optical packet switching seems to a suitable transport technology for the next-generation Internet due to the variable lengths of IP packets. Optical buffers in the output port are an integral part of solving contention by exploiting the time domain. Fiber delay lines (FDLs) are a well-known technique for achieving optical buffers, and various optical buffer architectures using FDLs have been proposed. These are generally classified into two types of structure: feed-forward (FF) or feedback (FB). In the FF buffers, optical packets are delayed at the output ports by passing through step-increasing-length multiple FDLs to avoid contentions, and in the FB buffers, optical packets are delayed by being fed back in re-circulating loop FDLs to avoid contentions. We report the detailed characteristics of optical FB buffers with the Post-Reservation (PostRes) policy and clarify the superiority of the FB buffers through simulations. For comparison, we also show the characteristics of FBSI (FB with step-increasing-length FDLs) and FF buffers. We found that 1) the blocking probabilities in the FB buffer were about 10^{-2} lower than those in the FF buffer and 2) the blocking probabilities for the deterministic case in the FB buffer sharply dropped at $D = 1.0$, where the packet length was equal to the FDL loop length. We carried out 10^8 packet simulations. The results can be applied to the design of WDM optical packet switches and networks with the maximum throughput.

Keywords: Asynchronous Optical Switching, Optical Buffers, Feedback Optical Loop, Blocking Probabilities, Poisson Arrivals, General Packet Length

1. Introduction

The ultimate capacity of the Internet may be constrained by energy density limitations and heat dissipation considerations rather than by the bandwidth of the physical components. Recent studies have shown that optical packet switches do not appear to offer significant throughput improvements or energy savings compared to electronic packet switches [1]. However, the studies also show that optical switch fabrics generally become more energy efficient as the data rate increases [1]. We believe that the development of optical components could lead to a breakthrough in optical packet switches and that the use of all-optical packet switches, in which optical packets are buffered and routed in optical form, will solve the present Internet problems. Asynchronous optical packet switching appears to be suitable as a transport technology for the next-generation Internet due to the variable lengths of IP packets.

In optical packet switches, optical buffers in the output port are an integral part of solving contention by exploiting the time domain. Fiber delay lines (FDLs) are a well-known technique for achieving optical buffers since random access memory (RAM) is not achievable with current optical technologies. However, optical buffers behave differently from electronic RAM. FDLs can only delay the packets for integer multiples of a discrete amount of time, called time granularity; the maximum delay is bound and a packet will be discarded if the maximum delay is not sufficient to avoid contention.

Various optical buffer architectures using DFLs have been proposed. These are generally classified into two types of structure: feed-forward (FF) or feedback (FB) [2-4, 12].

In the FF buffers, optical packets are delayed at the output ports by passing through step-increasing-length multiple FDLs to avoid contentions [3, 5]. The step-increasing-length FDLs cause delays with integer multiples of the time granularity, and output-time differences are added to the packets after passing through the FDLs. There have been many studies on FF buffers [5, 9-11] investigating the blocking probabilities and delays for Poisson arrival packets with generally distributed packet lengths, and highly accurate closed-form expressions for calculating both block probabilities and delays have already been obtained [13, 16].

In contrast, in the FB buffers, optical packets are delayed by being fed back in re-circulating loop FDLs to avoid contentions when another packet already occupies the output, and they arrive again at the input of the switch. On the basis of reservations for re-circulating packets, two types of strategies in the FB buffer have been proposed [4]: Pre-Reservation (PreRes) and Post-Reservation (PostRes). In the PreRes scheme, the output is reserved for re-circulating packets prior to arriving packets, while in the PostRes scheme, there is no priority at the output over newly arriving packets, including re-circulating packets, and all packets are served under first-come-first-serve (FCFS) queueing discipline.

For a cost-efficient packet switch architecture, a share-per-node optical FB butter configuration has been proposed and has been extensively investigated [8, 14-15], where FDLs are shared among the output ports of a node in a feedback configuration. The shared-per-node configuration utilizes the output buffering strategy in the same way as a shared-per-port optical buffer configuration does, where each output port has its own dedicated buffer. Compared with the shared-per-port configuration, naturally, the shared-per-node configuration may achieve a better cost performance. However, it requires more complex switching controls because it has to simultaneously manage and control all packets passing through all output ports.

Generally speaking, for the FB buffers, there has been insufficient theoretical investigation of both blocking probabilities and delays. Simulations [4, 6-8], an approximation for the FB buffer with one FB loop [6], and numerical iterations [12, 14-15] have been performed, but the characteristics of the multi-re-circulation-loop FB buffers with general packet-length distributions have not yet been clarified. Basically, in queueing networks, the existence of feedback loops makes the stochastic processes more complex and difficult to solve [20, 21]. In Jackson-type queueing networks, for example, it is clear that the arrival process for an M/M/1 system with a feedback loop is not Poisson. Then, theoretical investigations for the feedback-M/M/1 system have been carried out by approximated analyses or simulations.

In this work, we report the detailed characteristics of optical FB buffers and clarify the superiority of the FB buffers through simulations. We consider optical buffers with multi-re-circulation-loops at the output ports of an optical packet switch. The re-circulation loops have the same time granularity D and the buffers adopt the PostRes policy. For comparison, we also show the characteristics of the feedback-re-circulation-loop buffers with step-increasing-length FDLs (FBSI) and feedforward buffers with step-increasing-length FDLs (FF). Estimation items for comparing the FB, FBSI, and FF buffers are the blocking probabilities and the delays.

Concerning the optical buffers with feedback re-circulating loops, it is known that the re-circulation numbers should be limited to avoid optical signal degradation [17-19]. Amplified spontaneous emission (ASE) noises in optical amplifiers,

which are inserted to compensate for the transmission losses in the re-circulation loops, and optical crosstalk noises, generated in the optical switches, are accumulated during optical signal re-circulations in the loops and may degrade optical signal-to-noise ratios up to the optical receiver's limits. It has been reported that the maximum re-circulation numbers required mainly depends on each optical device's abilities and optical transmission signal rates, varying largely in the range from 2 to 30 [18]. Our aim in this work is to determine the optimal characteristics of the feedback loop buffers without re-circulation limits, and therefore at first do away with the limits altogether. Next, we discuss an effect of the re-circulation limits on the blocking probabilities in considerations. Up to the present, at least, proper values of the maximum numbers have not been reported and have varied too largely to be considered. We believe that developments of the optical devices could increase the maximum re-circulation numbers, thus enabling us to do away with the limits.

Section 2 presents optical buffer models the FB, FBSI, and FF buffers and explains the packet transfer algorithms used in the models. In Section 3, we present the results of simulations carried out within 10^8 packets, and compare the blocking probabilities and delays of the three optical buffers. Considerations on blocking probability drops in a deterministic distribution case and the maximum re-circulation numbers are presented in Section 4. We conclude in Section 5 with a brief summary.

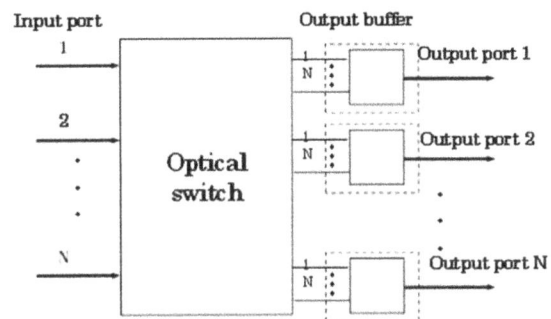

(a) Shared-per-port buffer configuration

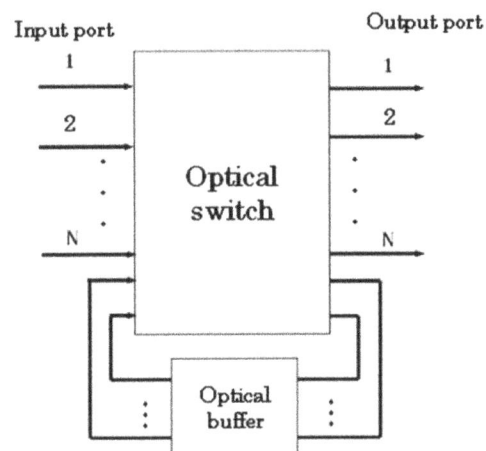

(b) Shared-per-node buffer configuration

Fig. 1. Optical packet switch models.

2. Models and Algorithms

We consider optical buffers at the output ports of an optical packet switch, which is shown in Fig. 1(a). This switch has N input and N output ports, functioning as an $N \times N$ non-blocking switch, and its architecture utilizes the output buffering strategy. Each output port is equipped with a dedicated buffer containing FDLs, which is modeled as the shared-per-port type optical buffer. A switch model with the shared-per-node type optical buffer in a feedback configuration is also shown in Fig. 1(b). In this paper, we concentrate to study at the shared-per-port type optical buffer because we will clarify detailed characteristics of FB buffers compared with those of FF buffers in simpler models.

Figure 2 shows the optical buffer structure models for one output port in the shared-per-port configuration, which is shown in Fig. 1(a). In the figures, (a) and (b) are buffers in the feedback configuration with re-circulation loops, showing (a) re-circulation loops with fixed-length FDLs and (b) those with step-increasing-length FDLs. (c) shows a buffer in the feed-forward configuration with step-increasing-length FDLs. When packets are transferred to the buffer with the step-increasing-length FDLs, FDLs are selected such that packets do not conflict with each other at the output of the buffer. The buffer in Fig. 2(b) thus adopts a new configuration, shown in Fig. 3.

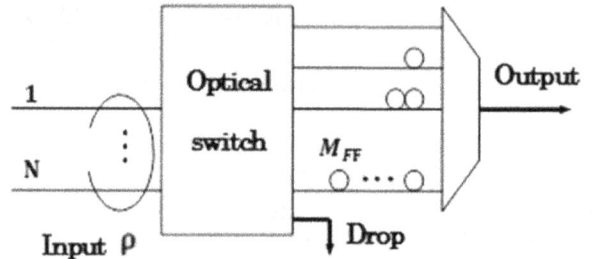

(a) FB buffer

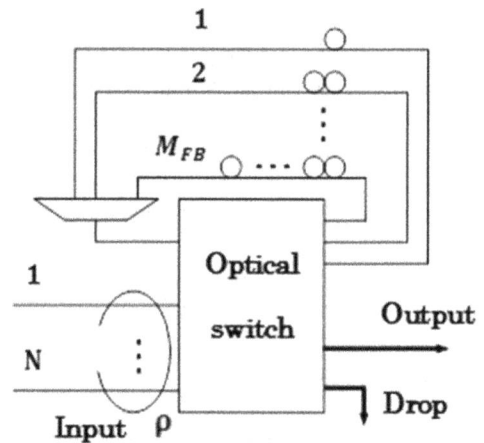

(b) FBSI buffer

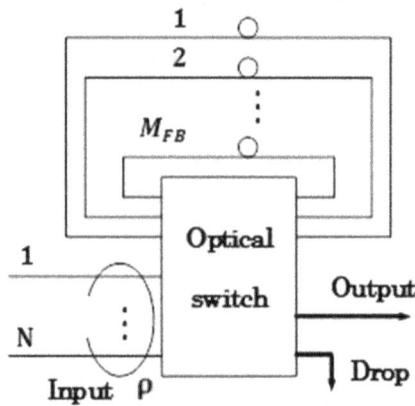

(c) FF buffer

Fig. 2. Optical buffer structure models.

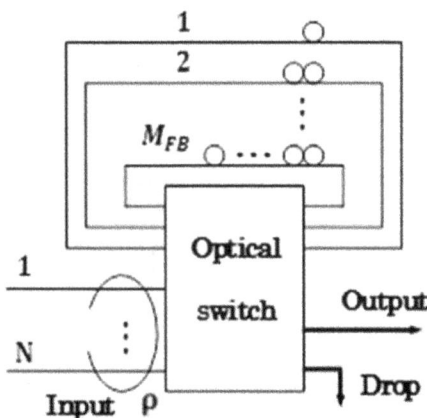

Fig. 3. An FBSI model with a multiplexer.

The detailed structures and packet-transfer algorithms of the three buffer models are as follows.

1) Fixed-length FDL feedback loop buffer (FB buffer)

The FB buffer has an N-port input, a one-port output, M_{FB} re-circulation loops, and a packet-discarding port containing a $(N + M_{FB}) \times (2 + M_{FB})$ switch. The M_{FB} re-circulation loops are all the same fixed-length FDLs with time granularity D; $D = nL/c$, where n is the effective refractive index of the fiber, L FDL length, and c light speed.

In the switch, we adopt a first-come-first-service (FCFS) policy without reservation for all packets, including re-circulated packets. This mechanism is the post-reservation (PostRes) scheme [4]. If the output is free, a packet that arrives at the switch input will be transmitted immediately. Otherwise, it will be transferred to one of the re-circulation loops. If the packet cannot be injected into any loop because all M_{FB} loops are filled with packets at the inputs, the packet will be discarded. All packets are served under the FCFS discipline in the re-circulation loops because FDLs in the re-circulation loops all have the same delay D.

The algorithm in the FB buffer with PostRes is simple, because how to treat a packet arriving at the switch input depends on whether the output is available or not, and if not, depends on whether one of the re-circulation loops is available at the epoch of the arrival. It is also known that the blocking probability of the PostRes buffer is lower than that of the PreRes buffer [4, 6]. At first, we ignore the maximum number of allowable re-circulations for packets, allowing the packets to re-circulate endlessly in the loops. Our aim with these simulations is to present characteristic values for ideal cases of the FB buffer, such as the blocking probabilities, the delay, and re-circulation numbers. The effect of the re-circulation limits on the blocking probabilities will be discussed in considerations.

Figure 4(a) shows an example in which packets are transferred in the FB buffer with $M_{FB} = 2$. Packet 1 is transmitted from loop 1 to the output at t_1. When packet 2 newly arrives at t_2, the output is busy and so it enters loop 1. Packet 3 arrives at t_3 and passes through loop 2 because both the output and loop 1 are busy. Packet 4 arrives at t_4 and is discarded because the output and the loops were all busy.

2) Step-increasing-length FDL feedback loop buffer (FBSI buffer)

The re-circulation loops of the FBSI buffer have M_{FB} step-increasing-length FDLs; the FDL in i-th loop is iD long, where $i = 1, 2, \cdots, M_{FB}$. The policy to transfer packets for the switch is the same FCFS with PostRes as that of the FB buffer. For re-circulation loops, however, by using the step-increasing-length FDLs, re-circulated packets in the loops will be scheduled in time to avoid contentions with each other at the loop output. Therefore, as shown in Fig. 3, we can construct only one port with a multiplexer in the output side of the re-circulation loops.

In a case where a packet that should be transferred to the loops has to wait for at least w units of time, the packet will be inserted into the $(i+1)$-th loop if $iD \leq w < (i+1)D$, and will be discarded if $M_{FB}D < w$. Therefore, a void period τ will be attached to the packet, where

$$\tau = \left\lceil \frac{w}{D} \right\rceil D - w,$$

and $\lceil x \rceil$ means the smallest integer greater than x.

Figure 4(b) shows an example of packets being transferred in the FBSI buffer. Two re-circulation loops are set: loop 1 with a granularity D and loop 2 with $2D$. Packet 1, arriving at t_1, successfully transmits at the output because the output is free at t_1. Packet 2 arrives at t_2 and is inserted into loop 1 because the output is busy at t_2 due to packet 1 transmitting. Packet 3, which arrives at t_3, is inserted into loop 2 because both the output and loop 1 are busy at t_3. Furthermore, void period τ is attached to packet 3 because the waiting time w of packet 2 is in $D < w < 2D$.

Note that the total length of the M_{FB} step-increasing-length FDLs is $M_{FB}(M_{FB} + 1)D/2$, which is $(M_{FB} + 1)/2$ times longer than that of the FB buffer with fixed-length FDLs. We need to take the total length difference into account when comparing packet delays in the buffers.

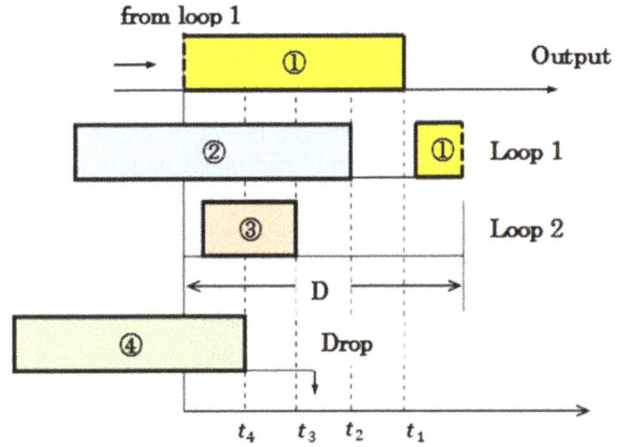

(a) FB buffer

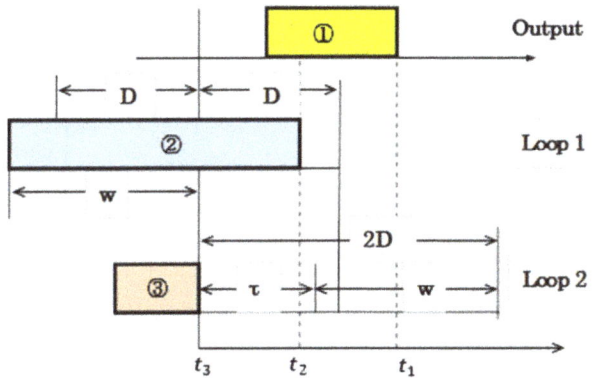

(b) FBSI buffer

Fig. 4. *Packet flows in optical buffers with two re-circulation loops.*

3) Step-increasing-length FDL feed-forwarding buffer (FF buffer)

The FF buffer has M_{FF} step-increasing-length FDLs, where packets in i-th FDL will be delayed with iD $(i = 1, 2, \cdots, M_{FF})$. We adopt the same procedure for selecting the FDLs when the output is busy as that of the FBSI re-circulation loops; i.e., when a packet has to wait for w, it will be inserted into the $(i+1)$-th FDL if $iD \leq w < (i+1)D$, and will be discarded if $M_{FF}D < w$. As of now, extensive studies have been done on this model [9-13], and a highly-accurate approximation method for calculating blocking probabilities and delays has been presented [16].

3. Simulation Results

Simulations for three optical buffer models the FB, FBSI, and FF buffers as shown in Fig. 2 were carried out within 10^8 packets. The countable packet loss limit was therefore 10^{-8}. We assumed Poisson packet arrivals and used three packet-length distributions, namely, exponential, uniform, and deterministic distributions. The time unit was set to be the average packet length. Traffic load ρ was equal to packet arrival rate λ.

3.1. FDL-granularity D Dependency

Figure 5 shows the blocking probabilities of the FB buffer for the case of $\rho = 0.5$ and $M_{FB} = 10$ against D varying from 0 to 2.0, 3.0, 10, and 100. The circles, triangles, and crosses represent the results for the exponential, uniform, and deterministic packet-length distributions, respectively. The blocking probabilities monotonically decreased as D increased for both the exponential and uniform distribution cases, but were saturated around 5.0×10^{-5} with larger-than-3.0 D. It seems the saturation occurred because, by increasing D, the packet delay increases in proportional to D but the void period does not increase and keeps in a fixed value. On the other hand, for the deterministic distribution case, the blocking probabilities sharply dropped at $D = 1.0$ and reached near 1.0×10^{-5}.

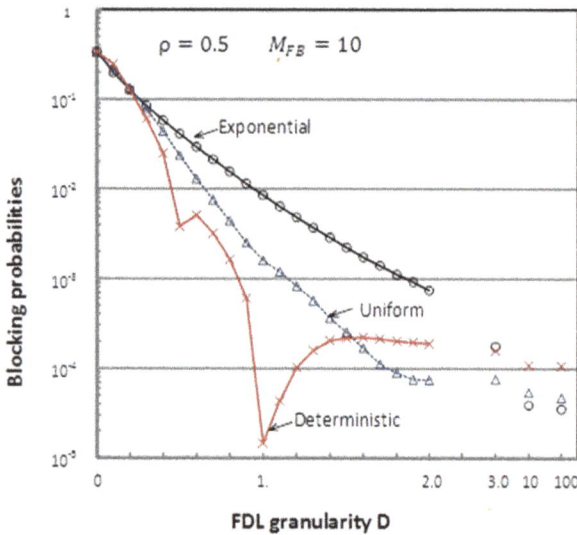

Fig. 5. *Blocking probabilities of FB buffer against D varying from 0 to 2.0, 3.0, 10, and 100, where $\rho = 0.5$ and $M_{FB} = 10$. Circles, triangles, and crosses represent results for exponential, uniform, and deterministic packet-length distributions, respectively.*

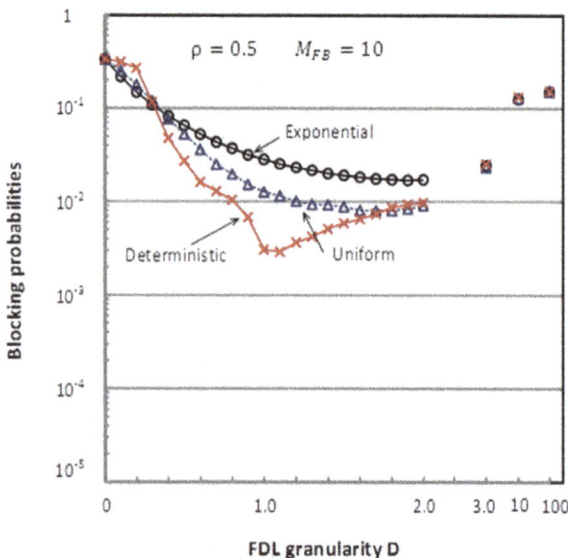

Fig. 6. *Blocking probabilities of FBSI buffer, where $\rho = 0.5$ and $M_{FB} = 10$. Symbols have same meanings as in Fig. 5.*

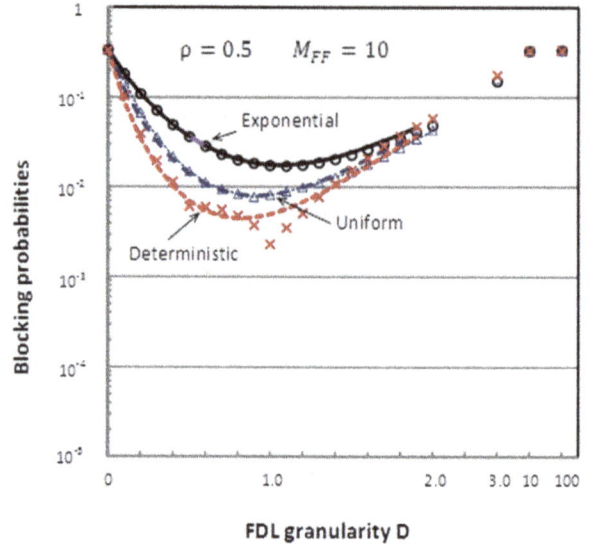

Fig. 7. *Blocking probabilities of FF buffer, where $\rho = 0.5$ and $M_{FF} = 10$. Symbols have same meanings as in Fig. 5.*

Figures 6 and 7 show results for the FBSI and FF buffers, respectively, where $\rho = 0.5$ and $M_{FB} = M_{FF} = 10$. Symbols carry the same meanings as in Fig. 5. In both the FBSI and FF buffers, the blocking probabilities gradually changed with increasing D for both the exponential and uniform distribution cases. For example, in the FBSI buffer, the blocking probabilities were saturated around 2.0×10^{-2} at near $D = 2.0$, and reversely increased with larger-than-3.0 D. In the FF buffer, the blocking probabilities reached a minimum value near $D = 1.0$ and increased with increasing D. For the deterministic distribution case in both the FBSI and FF buffers, however, we found sharp decreases at $D = 1.0$. The thick lines in Fig. 7 show calculations using the forth-order approximation [16], which was established with the assumption that packet virtual waiting times could be expressed by an exponential function. The calculations were in good agreement with the simulations except for near $D = 1.0$ of the deterministic distribution case. We have not yet obtained any exact or sufficient approximation theory for this deterministic distribution case.

The results in Figs. 5-7 suggest that the FB buffer was the best from the viewpoint of blocking probabilities: the probabilities reached a minimum value of about 10^{-2} lower than those of the FBSI and FF buffers, and a sharp decrease of the blocking probabilities appeared at $D = 1.0$ for the deterministic distribution case.

Figure 8 shows the average circulation numbers of re-circulating packets, including discarded packets, with the same parameter values as in Figs. 5 and 6. Results for the uniform distribution case were almost equal to those for the exponential and so have been omitted in the figure. In the FB buffer, the average numbers decreased with D and were saturated at about three circulations. The average numbers in the FBSI buffer were under two circulations in the wide range of D and also were saturated. Observing that the average circulation numbers in the FB buffer were almost twice the averages in the FBSI buffer, we found that the buffer ability of the FB was twice that of the FBSI, making the blocking probabilities in the FB much lower than those in the FBSI buffer.

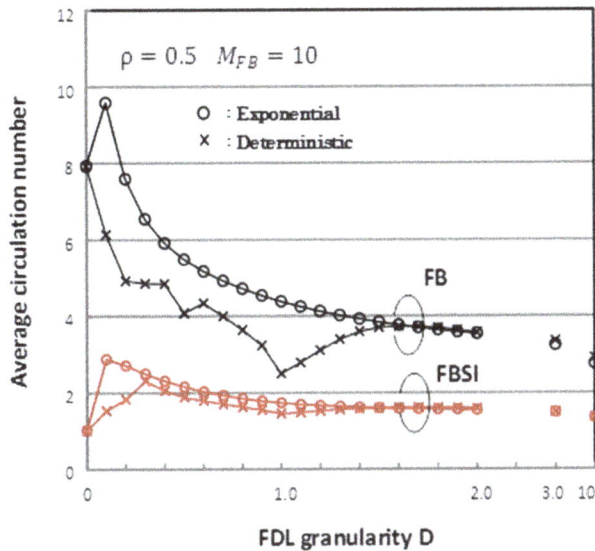

Fig. 8. Average circulation numbers of re-circulating packets, included discarded packets, with same parameter values as those in Figs. 5 and 6.

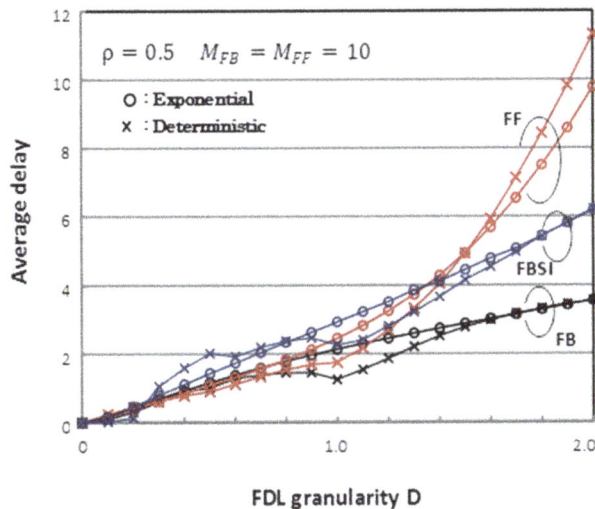

Fig. 9. Average delays of output packets with same parameter values as those in Figs. 5-7.

Figure 9 shows average delays of the output packets, i.e., in which discarded packets were excluded, with the same parameters as in Figs. 5-7. All of the delays were generated in the FDLs. The time unit was set to be the time in which a packet with an average packet length passes through the output. For example, the unit is $0.2\,\mu$sec for 40-Gbps-speed packets with an average packet length of 1000 bytes. If the packets pass through a maximum of 100 switches for end-to-end network transmission, the total delay caused by all switches is $20\,\mu$sec. If we require an average delay, caused by all switches between network ends, to be less than 1 msec, it means that the delay per buffer shall be less than 50. As shown in Fig. 8, the average delays for the less-than-2.0 D range

are less than 12 for all the buffers, meaning that there is sufficient margin against the delay requirement.

We found that the average delays increased with increasing D for all the buffers and that the delays of the FB buffer were the least of the three. The reason for the least delays of the FB buffer in spite of its larger average circulation numbers (Fig. 8) might be that the total FDL lengths of the FBSI and FF buffers were 5.5 times longer than that of the FB buffer.

Blocking probabilities for the case of $\rho = 0.8$ and $M_{FB} = M_{FF} = 40$ are shown in Fig. 10, where only results in the deterministic distribution are illustrated for both the FBSI and FF buffer cases in order to compare them with the FB buffer case.

The same characteristics as in Fig.5 could explicitly be observed: 1) for the exponential and uniform distribution cases in the FB buffer, the blocking probabilities decreased with increasing D and were saturated with larger-than 2.0 D, 2) the probabilities sharply dropped at $D = 1.0$ for the deterministic distribution case, and 3) in both the FBSI and FF buffers, the probabilities reached minimum values near $D = 0.25$ and increased gradually with larger-than-0.25 D. Since the void periods were attached to packets in step-increasing length FDL buffers, such as FBSI and FF, the equivalent load reached 1.0 at $D = 0.5$ when $\rho = 0.8$ [3, 9]. The buffers then failed in excess load conditions over $D = 0.5$.

The difference in the blocking probabilities between the FB, FBSI, and FF buffers is more clear in Fig. 10. The values were 2×10^{-2} at the minimum in the FBSI and FF buffers, whereas the values were under 10^{-3} with larger-than-1.0 D for all the distribution cases in the FB buffer. In particular, the value dropped at $D = 1.0$ and reached 10^{-7} for the deterministic distribution case.

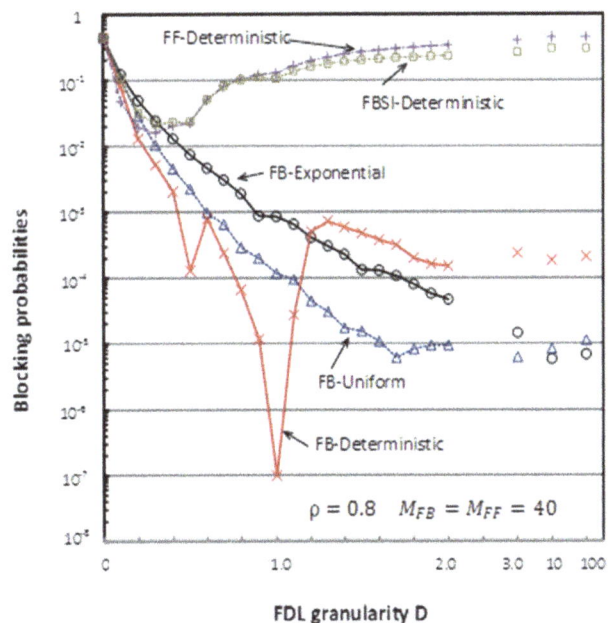

Fig. 10. Blocking probabilities for case of $\rho = 0.8$ and $M_{FB} = M_{FF} = 40$.

Figure 11 shows average circulation numbers with the same

parameters as in Fig. 10. The average number kept at about 1.5 time circulations over the wide range of D in the FBSI buffer, whereas, in the FB buffer, the number largely varied with D and almost became more than 10. Moreover, the number decreased to 5.5 at $D=1.0$ for the deterministic distribution case. Comparing this with the results in Fig. 10, we found that the blocking probabilities decreased when the circulation number decreased for the deterministic distribution case.

Figure 12 shows average delays with the same parameters as in Figs. 10 and 11. The delays rapidly increased with larger-than-0.5 D in both the FF and FBSI buffers, whereas in the FB buffer the delays gradually increased with D and were less than 20. The reason the average delays in the FB buffer became 1/5 of those in the FBSI and FF buffers even though the circulation number in the FB buffer was almost 10 times that in the other two may be that the total FDL length of the FB buffer was 1/20 that of the FBSI and FF buffers.

3.2. Loop Number M_{FB} Dependency

Figure 13 shows the blocking probabilities with $\rho=0.5$ and $D=1.0$ against loop number M_{FB} of the FB buffer and buffer number M_{FF} of the FF buffer. The loop number of the FB buffer became about half of the buffer number of the FF buffer when the blocking probabilities were of an equal value; for example, the probability in the FB deterministic distribution case was 10^{-5} at $M_{FB}=11$, whereas that in the FF case was at $M_{FF}=23$.

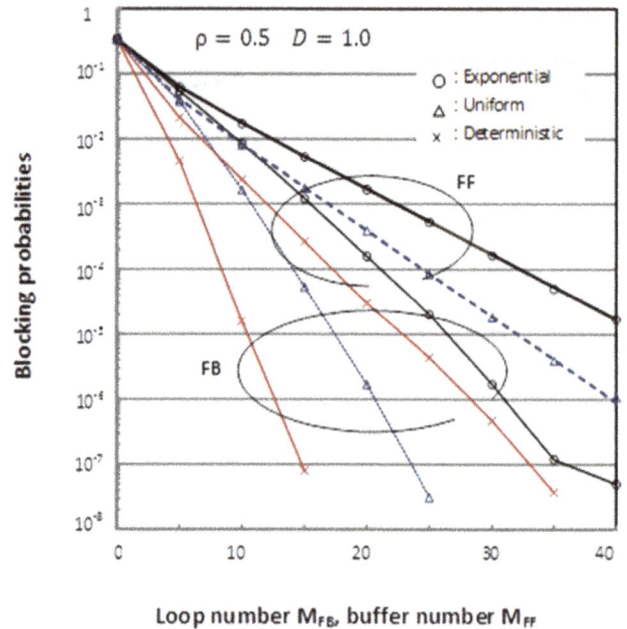

Fig. 11. *Average circulation numbers with same parameters as in Fig. 10.*

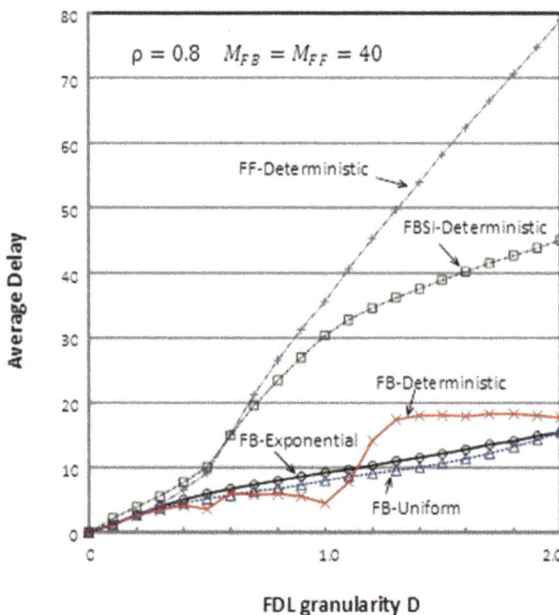

Fig. 13. *Blocking probabilities with $\rho=0.5$ and $D=1.0$ against loop number M_{FB} of the FB buffer and buffer number M_{FF} of the FF buffer.*

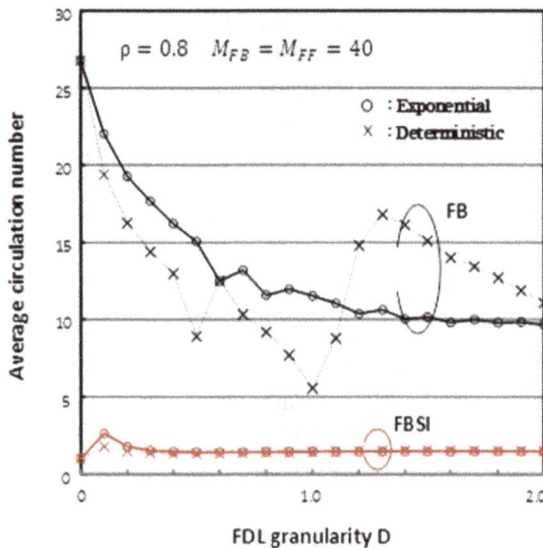

Fig. 12. *Average delays with same parameters as in Figs. 10 and 11.*

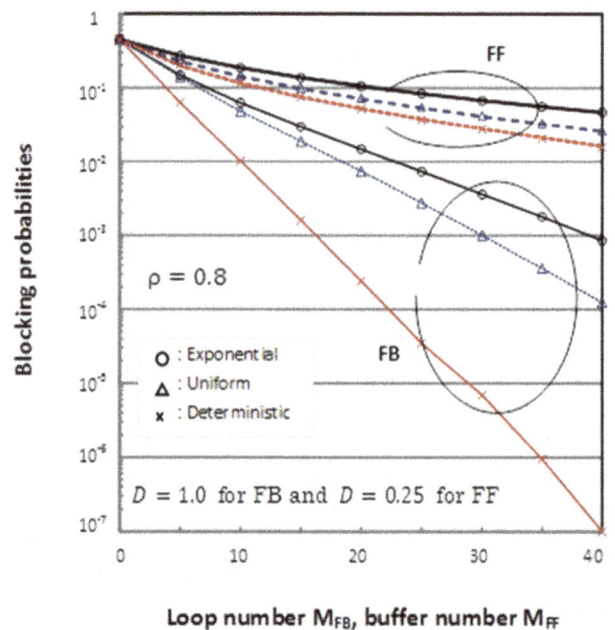

Fig. 14. *Blocking probabilities with $\rho=0.8$. D was set to $D=1.0$ for the FB buffer and $D=0.25$ for the FF buffer.*

Blocking probabilities with $\rho = 0.8$ are shown in Fig. 14, where the granularity D was set to $D = 1.0$ for the FB buffer and $D = 0.25$ for the FF buffer, since the probabilities for the FF buffer became the minimum at $D = 0.25$. Comparing the values in $\rho = 0.5$ and $\rho = 0.8$, the difference between the FB and FF buffers is remarkable: for example, the blocking probability in the FB deterministic-distribution case became 10^{-5} at $M_{FB} = 29$, whereas 10^{-5} probability in the FF case became at $M_{FF} = 200$. A buffer number seven times higher than the FB buffer was required for the FF buffer.

The loop number and the buffer number are added to the switch port number, and the seven times higher buffer number requires a (7 times ports)×(7 times ports) switch. This makes the cost of the FF buffer switch 7^2 times higher than the FB buffer.

4. Considerations

4.1. $D = 1.0$ of the Deterministic Distribution Cases

The simulations reported in Sec. 3 yielded the following.

1) The blocking probabilities in the FB buffer became about 10^{-2} lower than those in the FF buffer, and the buffer number of the FB buffer can be reduced to 1/2 ($\rho = 0.5$) - 1/7 ($\rho = 0.8$) that of the FF buffer when the blocking probabilities require equal values. Then, by using the FB buffer structure, the switch port scale can be drastically reduced.

2) The blocking probabilities for the deterministic case in the FB buffer sharply dropped at $D = 1.0$, where the packet length was equal to the FDL loop length.

Of the above, we are particularly interested in the reason 2).

Figure 15 shows the flow of packets in the FB buffer with three re-circulation loops, when (a) $D < 1.0$, (b) $D = 1.0$, and (c) $D > 1.0$.

For the case shown in Fig. 15(b), each packet in each loop re-circulates without a void period because the packet length coincides with the FDL loop length. Only if the head of the packet reaches the output of the loop when the output is free, i.e., the head lies in between busy periods, that is, from t_1 to t_2 in (b), can the packet escape from the loop and go through the output. The probability that the packet escapes from the loop is $(1 - \rho_{out})/D$, where ρ_{out} is output load, and is expressed by $\rho_{out} = \rho(1 - P_B)$. We conclude for the $D = 1.0$ case that each re-circulated packet occupies each loop and waits to go through the output with the probability of $(1 - \rho_{out})/D$ in the random-service rule.

Figure 15(a) shows the packet flow for the $D < 1.0$ case. The packet has to enter another loop after circulating a first loop because the packet length is longer than the FDL loop length. The head of packet 3 entered loop 2 after circulating loop 1 and was connected by a void period. Therefore, the blocking probabilities become higher than those in the $D = 1.0$ case. With the exception of the $D = 0.5$ case, each re-circulating

packet can occupy two loops and has no void period. This situation is the same as the $D = 1.0$ case in Fig. 15 (b), but effective loop numbers are reduced by half. Then, the blocking probabilities will locally become minimum values.

The $D > 1.0$ case is shown in Fig. 15(c), where each packet occupies one loop. The re-circulating packet numbers are the same three as for the $D = 1.0$ case in Fig. (b), but the probability that the packet escapes from the loop is lower than those for the $D = 1.0$ case because the $(1 - \rho_{out})/D$ value decreases when D increases. At the same time, the circulation numbers increase when D increases and the probabilities decrease. As we can see in Figs. 8 and 11, this is why the average circulation numbers become minimums at $D = 1.0$.

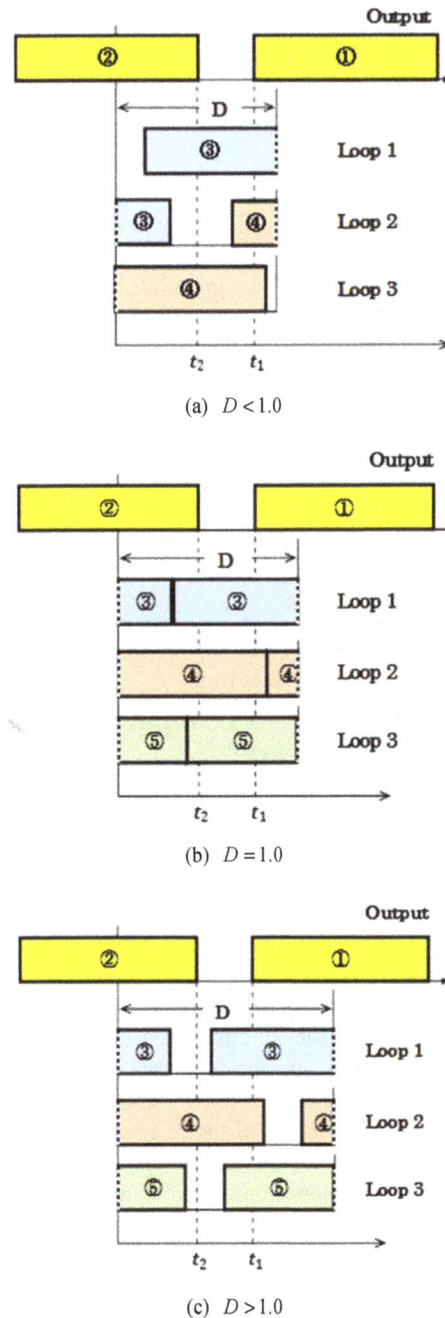

(a) $D < 1.0$

(b) $D = 1.0$

(c) $D > 1.0$

Fig. 15. *Packet flows in the FB buffer with three re-circulation loops.*

4.2. Re-circulation Number Limits

Considering the optical buffer structures (Fig. 2), we know that each packet that circulates the loops suffers from transmission losses and optical noise. These are physical layer impairments caused by losses in FDLs, optical switches, and arrayed waveguide gratings (AWGs), amplified spontaneous emission (ASE) noises in optical amplifiers, and crosstalk noises in AWGs [14-16]. Wavelength selective optical switches constructed by many tunable wavelength converters (TWCs) and AWGs, for example, generate both ASE and crosstalk noises. These impairments degrade signal-to-noise ratios in optical receivers. To avoid degradation, it is known that the re-circulation numbers should be limited. However, such limitations are considered a weakness in the feedback-loop-buffer structures because they increase the probabilities of blocking.

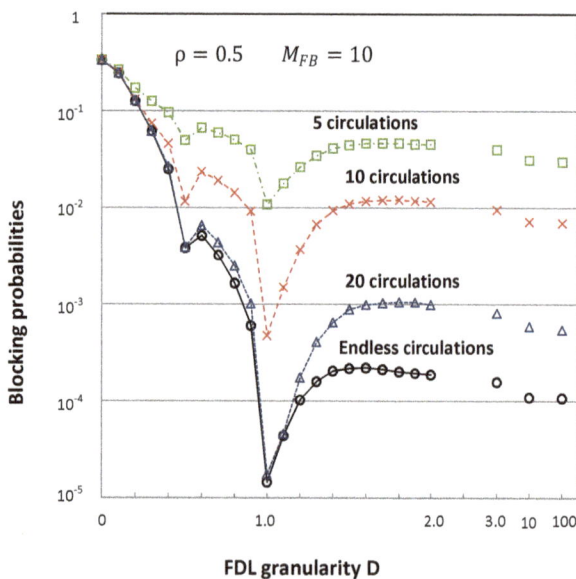

Fig. 16. *Blocking probabilities of the FB buffer with the re-circulation number limits for the case of $\rho = 0.5$ and $M_{FB} = 10$.*

Figure 16 shows the blocking probabilities of the FB buffer with the re-circulation number limits for the case of $\rho = 0.5$ and $M_{FB} = 10$. The results of the deterministic-packet-length distribution case are only shown for simplicity. In the 5-circulation-limit case, the blocking probabilities increase from 10-5 (in the case without circulation number limits; that is endless-circulation case) to 10^{-2} at $D = 1.0$. However, the blocking probabilities in the 20-circulation-limit case show little change from those in the endless-circulation case especially for less-than-$D = 1.0$ range. As shown in Fig. 8, the average circulation numbers of the re-circulating packets are 2.5 at $D = 1.0$. In order that the re-circulation number limits may have little influence on the blocking probabilities, therefore, it is necessary to set the limit numbers more than 7 times the average circulation number.

Considering practical FDL loop lengths, a 40-Gbps optical packet with a 1000-byte length, for example, needs a 0.2-μsec transmission time, and the FDL length with the same time granularity as 0.2 μsec is 40 m. The transmission loss

in a 40-m-long FDL is only 0.008 dB and can be neglected when the fiber loss is 0.2 dB/km. On the other hand, the losses in the optical switches are large, and optical amplifiers are inevitable to compensate for the losses. We therefore conclude that, if the switching losses become lower, the required gains in the optical amplifiers will also become lower and re-circulation number limits can be relaxed.

In order to relax the re-circulation number limits, it is effective to decrease the optical switch's port numbers, since the optical switch losses are inclined to increase in proportion to the port numbers [17]. From Figs. 13 and 14, we found that the loop numbers of the FB buffers, i.e., the switch's port numbers, can be reduced, compared with those of the FF buffers. Moreover, in the deterministic-packet-length case, the average circulation numbers have minimum values at $D = 1.0$, as shown in Figs. 8 and 11, and the limits can be relaxed more. In future, however, we will expect developments of lower loss optical switches to abolish the re-circulation number limits.

5. Conclusions

We reported the detailed characteristics of optical FB buffers with the PostRes policy and clarified the superiority of the FB buffers through simulations. For comparison, we also showed the characteristics of FBSI and FF buffers. Our main findings are as follows.

1) The blocking probabilities in the FB buffer became about 10^{-2} lower than those in the FF buffer, and if blocking probabilities are required in equal values, the buffer number of the FB buffer can be reduced to 1/2 ($\rho = 0.5$) - 1/7 ($\rho = 0.8$) that of the FF buffer.

2) The blocking probabilities for the deterministic case in the FB buffer sharply dropped at $D = 1.0$, where the packet length was equal to the FDL loop length. This sharp dropping is likely because re-circulating packets have no void period in the loops and can occupy their own loops until transmitting through the output.

In this work, we carried out 10^8 packet simulations. The results can be applied to the design of WDM optical packet switches and networks with the maximum throughput. Our future work is to perform theoretical investigations to reinforce the simulation results.

References

[1] R. S Tucker et al, "Evolution of WDM Optical IP networks: A Cost and Energy Perspective," IEEE J. Lightwave Technol., Vol. 27, No. 3, pp. 243-252, 2009.

[2] G. Grasso et al, "Role of Integrated Photonics Technologies in the Realization of Terabit Nodes," J. Opt. Commun. Netw., Vol. 1, No. 3, pp. B111-B119, 2009.

[3] F. Callegati, "Optical Buffers for Variable Length Packets," IEEE Commun. Lett., Vol. 4, No. 9, pp. 292-294, 2000.

[4] C. M. Gauger, "Dimensioning of FDL Buffers for Optical Burst Switching Nodes," in Proc. 6th IFIP Working Conference on Optical Network Design and Modelling, Feb. 2002.

[5] R. C. Almeida, J. U. Pelegrini, and H. Waldman, "A generic-traffic optical buffer modeling for asynchronous optical switching networks," IEEE Commun. Lett., Vol. 3, No. 2, pp. 175-177, 2005.

[6] A. Rostami and S. S. Chakraborty, "On Performance of Optical Buffers with Specific Number of Circulations," IEEE Photo. Tech. Lett., Vol. 17, No. 7, pp. 1570-1572, 2005.

[7] T. Zhang, K. Lu, and J. Jue, "Shared Fiber Delay Line Buffers in Asynchronous Optical Packet Swiches," IEEE J. Sel. Areas Commun. Vol. 24, No. 4, pp. 118-127, 2006.

[8] C. M. Gauger, H. Buchta, and E. Partzak, "Integrated Evaluation of Performance and Technology – Throuput of Optical Burst Switching Nodes Under Dynamic Traffic," IEEE J. Lightwave Technol., Vol. 26, No. 13, pp. 1969-1979, 2008.

[9] Jianming Liu et al., "Blocking and Delay Analysis of Single Wavelength Optical Buffer with General Packet Size Distribution," IEEE J. Lightwave Technol., Vol. 27, No. 8, pp. 955-966, 2009.

[10] H. E. Kankaya and N. Akar, "Exact Analysis of Single-Wavelength Optical Buffers with Feedback Markov Fluid Queues," J. Opt. Commun. Netw., Vol. 1, No. 6, pp. 530-542, 2009.

[11] W. Rogiest, and H. Bruneel, "Exact Optimization Method for an FDL Buffer with Variable Packet Length", IEEE Photon. Technol. Lett., Vol. 22, No. 4, pp. 242-244, 2010.

[12] N. Akar and K. Sohraby, "Retrial Queuing Models of Multi-Wavelength FDL Feedback Optical Buffers," IEEE Trans. Commun., Vol. 59, No. 10, pp. 2832-2840, 2011.

[13] Murakami Y., "An Approximation for Blocking Probabilities and Delays of Optical Buffer With General Packet-Length Distributions," IEEE J. Lightwave Technol., Vol. 30, No. 1, pp. 54-66, 2012.

[14] D. Tafani, C. McArdle, and L. P. Barry, "A Two-Moment Performance Analysis of Optical Burst Switched Networks with Shared Fibre Delay Lines in a Feedback Configuration," Optical Switching and Networking, Elsevier, Vol. 9, No. 4, pp. 323-335, 2012.

[15] N. Akar, and Y. Gunalay, "Dimensioning Shared-per-Node Recirculating Fiber Delay Line Buffers in an Optical Packet Switch," Performance Evaluation, Vo. 7, No. 12, pp. 1059-1071, 2013.

[16] Y. Murakami, "Asymptotic Analysis for Blocking Probabilities of Optical Buffer with General Packet-Length Distributions," American J. Appl. Mathematics, Vol. 2, No. 6-1, pp. 1-10, 2014.

[17] R. Srivastava, R. K. Singh, and Y. N. Singh, "Design Analysis of Optical Loop Memory," IEEE J. Lightwave Technol., Vol. 27, No. 21, pp. 4821-4831, 2009.

[18] Q. Xu et. al "Analysis of Large-Scale Multi-Stage All-Optical Packet Switching Routers," J. Opt. Commun. Netw., Vol. 4, No. 5, pp. 412-425, 2012.

[19] H. Rastegarfar, A. Leon-Garcia, S. LaRochelle, and L. A. Rusch, "Cross-Layer Performance Analysis of Recirculation Buffers for Optical Data Centers," IEEE J. Lightwave Technol., Vol. 31, No. 3, pp. 432-445, 2013.

[20] J. R. Artalejo and A. Gornez-Corral, "Retrial Queueing Systems", Springer, 2008.

[21] E. A. Pekoz and N. Joglekar, "Poisson traffic Flow in a General Feedback Queue," L. Appl. Prob., Vol. 39, pp. 630-636, 2002.

18

Privacy preserving data publishing through slicing

Shivani Rohilla, Megha Sharma, A. Kulothungan, Manish Bhardwaj

Department of Computer science and Engineering, SRM University, NCR Campus, Modinagar, Ghaziabad, India

Email address:

shivani.engineer@gmail.com (S. Rohilla), megha.tech09@gmail.com (M. Sharma), kulosoft@gmail.com (A. Kulothungan), aapkaapna13@gmail.com (M. Bhardwaj)

Abstract: Microdata publishing should be privacy preserved as it may contain some sensitive information about an individual. Various anonymization techniques, generalization and bucketization, have been designed for privacy preserving microdata publishing. Generalization does not work better for high dimensional data. Bucketization failed to prevent membership disclosure and does not show a clear separation between quasi-identifiers and sensitive attributes. There are number of attributes in each record which can be categorized as 1) Identifiers such as Name or Social Security Number are the attributes that can be uniquely identify the individuals. 2)Some attributes may be Sensitive Attributes(SAs) such as disease and salary and 3) Some may be Quasi Identifiers (QI) such as zipcode, age, and sex whose values, when taken together, can potentially identify an individual. Data anonymization enables the transfer of information across a boundary, such as between two departments within an agency or between two agencies, while reducing the risk of unintended disclosure, and in certain environments in a manner that enables evaluation and analytics post-anonymization. Here, we present a novel technique called slicing which partitions the data both horizontally and vertically. It preserves better data utility than generalization and is more effective than bucketization in terms of sensitive attribute.

Keywords: PPDP, AG, CG, PT

1. Introduction

Data Anonymization is a technology that convert clear text into a non-human readable form. Data anonymization technique for privacy-preserving data publishing has received a lot of attention in recent years. Detailed data (also called as micro-data) contains information about a person, a household or an organization.

Data mining is the process of analysing data from different perspectives and summarizing it into useful information. Knowledge discovery from databases, techniques like clustering, association rules, regression, classification, decision trees, genetic algorithm etc. are used nowadays. Data mining is also used in areas of Science and Engineering such as genetics and bioinformatics.

In both generalization and bucketization, one first removes identifiers from the data and then partitions tuples into buckets. The two techniques differ in the next step. Generalization transforms the QI-values in each bucket into "less specific but semantically consistent" values so that tuples in the same bucket cannot be distinguished by their QI values. In bucketization, one separates the SAs from the QIs

by randomly permuting the SA values in each bucket. The anonymized data consists of a set of buckets with permuted sensitive attribute values.

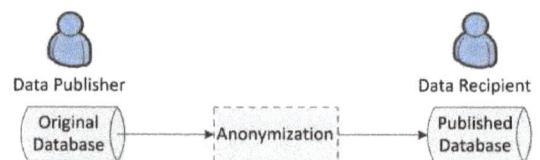

Fig. 1.1. *Anonymization of data*

In this information age, data and knowledge extracted by data mining techniques represent a key asset driving research, innovation, and policy-making activities. Many agencies and organizations have recognized the need of accelerating such trends and are therefore willing to release the data they collected to other parties, for purposes such as research and the formulation of public policies. However the data publication processes are today still very difficult. Data often contains personally identifiable information and therefore releasing such data may result in privacy breaches, this is the case for the examples of micro-data, e.g., census data and medical data. This thesis studies how we can publish and share micro data in a privacy-preserving manner. This present an extensive study of this problem along three dimensions: Designing a simple, intuitive, and robust privacy model, Designing an effective anonymization technique that works on sparse and high-dimensional data and developing a methodology for evaluating privacy and utility tradeoffs.

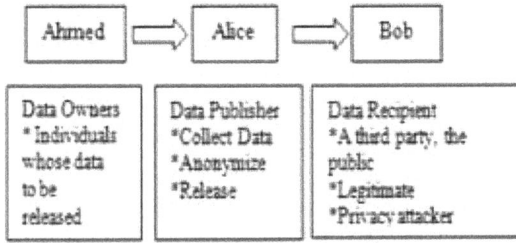

Fig. 1.2. *Privacy preserving model for data*

1.1. Organization

Here, we are studying slicing for privacy-preserving data publishing. Our contributions include the following. First, we introduce slicing as a new technique for privacy preserving data publishing. Slicing has several advantages when compared with generalization and bucketization. It preserves better data utility than generalization. It preserves more attribute correlations with the SAs than bucketization. It can also handle high-dimensional data and data without a clear separation of QIs and SAs.

Second, we show that slicing can be effectively used for preventing attribute disclosure, based on the privacy requirement of ℓ-diversity. We introduce a notion called ℓ-diverse slicing, which ensures that the adversary cannot learn the sensitive value of any individual with a probability greater than 1/ℓ.

Third, we develop an efficient algorithm for computing the sliced table that satisfies ℓ-diversity. Our algorithm partitions attributes into columns, applies column generalization, and partitions tuples into buckets. Attributes that are highly-correlated are in the same column; this preserves the correlations between such attributes. The associations between uncorrelated attributes are broken; the provides better privacy as the associations between such attributes are less-frequent and potentially identifying.

Fourth, we describe the intuition behind membership disclosure and explain how slicing prevents membership disclosure. A bucket of size k can potentially match kc tuples where c is the number of columns. Because only k of the kc tuples are actually in the original data, the existence of the other kc −k tuples hides the membership information of tuples in the original data.

Finally, we conduct extensive workload experiments. Our results confirm that slicing preserves much better data utility than generalization. In workloads involving the sensitive attribute, slicing is also more effective than bucketization. In some classification experiments, slicing shows better performance than using the original data (which may overfit the model). Our experiments also show the limitations of bucketization in membership disclosure protection and slicing remedies these limitations.

2. Existing Technology

There are two existing systems considered in this paper: Generalization and bucketization with which slicing is compared later.

Record linkage model attack and attribute linkage and attribute linkage model attack are different attacks occurred at the time of microdata publishing. There are some principles of privacy preserving as follows:-

2.1. K-Anonymity

Samarati and Sweeney introduced k-anonymity as the property that each record is indistinguishable with at least k-1 other records with respect to the quasi-identifier. In other words, k-anonymity requires that each QI group contains at least k records. k-anonymity is one of the most classic models, which prevents joining attacks by generalizing or suppressing portions of the released micro data so that no individual can be uniquely distinguished from a group of size k. k-Anonymity attributes are suppressed or generalized until each row is identical with at least k-1 other rows.

2.1.1. K-Anonymity using Generalization

The generalization hierarchy transforms the k-anonymity problem into a partitioning problem. Specifically, this approach consists of the following two steps. The first step is to find a partitioning of the dimensional space, where n is the number of attributes in the quasi identifier, such that each partition contains at least k records. Then the records in each partition are generalized so that they all share the same quasi-identifier value. The Generalization method substitutes the values of a given attribute with more general values. Generalization can be applied at the following levels. Attribute Generalization (AG): generalization is performed at the level of column; a generalization step generalizes all the values in the column.

Cell Generalization (CG): generalization is performed on single cells; as a result a generalized table may contain, for a specific column, values at different generalization levels. There are two types of generalization exist namely domain generalization hierarchy and value generalization hierarchy. The domain generalization hierarchy of a domain topic is a lattice, where each vertex represents a generalized table that is obtained by generalizing the involved attributes according to the Corresponding domain tuple and by suppressing a certain number of tuples to fulfill the k-anonymity constraint.

Figure illustrates an example of domain generalization hierarchy obtained by considering marital status and sex its quasi-identifying attributes i.e. by considering the domain tuple (Ma, So). Each path in the hierarchy corresponds to a generalization strategy according to which the original private table PT can be generalized.

(M2,S1)

(M1,S1)   (M2,S1)

(M0,S1)   (M1,S0)

(M0,S0)

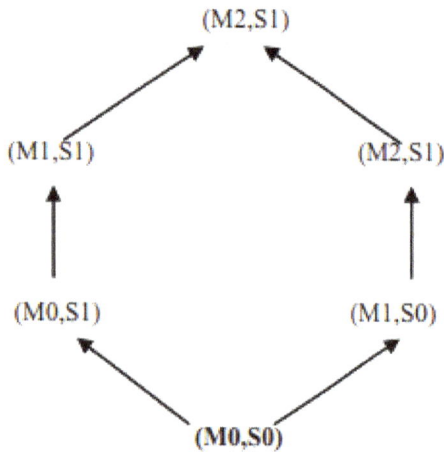

Fig. 1.3. *Domain generalization hierarchy*

K-anonymity model for multiple sensitive attributes mentioned that there are three kinds of information disclosure.

1)Identity Disclosure: When an individual is linked to a particular record in the published data called as identity disclosure.

2)Attribute Disclosure: When sensitive information regarding individual is disclosed called as attribute disclosure.

3)Membership Disclosure: When information regarding individual's information belongs from data set is present or not is disclosed is said to be membership disclosure.

2.1.2. Attacks on k-Anonymity

Here, we studied two attacks on k-anonymity: the homogeneity attack and the background knowledge attack.

1) Homogeneity Attack:

Sensitive information may be revealed based on the known information if the non sensitive information of an individual is known to the attacker. If there is no diversity in the sensitive attributes for a particular block then it occurs. To getting sensitive information this method is also known as positive disclosure.

2) Background Knowledge Attack:

If the user has some extra demographic information which can be linked to the released data which helps in neglecting some of the sensitive attributes, then some sensitive information about an individual might be revealing information. Such a method of revealing information is known as negative disclosure.

2.1.3. Limitations of k-Anonymity

(1) K-anonymity cannot hide whether a given individual is in the database,

(2) K-anonymity reveals individuals' sensitive attributes,

(3) K-anonymity cannot protect against attacks based on background knowledge,

(4) Mere knowledge of the k-anonymization algorithm can be violated by the privacy,

(5) K-anonymity does not applied to high-dimensional data without complete loss of utility.

(6) If a dataset is anonymized and published more than once then special methods are required.

2.2. ℓ-Diverse Slicing

In the above example, tuple t1 has only one matching bucket. In general, a tuple t can have multiple matching buckets. We now extend the above analysis to the general case and introduce the notion of ℓ-diverse slicing. Consider an adversary who knows all the QI values of t and attempts to infer t's sensitive value from the sliced table. He first needs to determine which buckets t may reside in, i.e., the set of matching buckets of t. Tuple t can be in any one of its matching buckets. Let $p(t,B)$ is the probability that t is in bucket B (the procedure for computing $p(t,B)$ will be described later in this section). For example, in the above example, $p(t1,B1) = 1$ and $p(t1,B2) = 0$. In the second step, the adversary computes $p(t, s)$, the probability that t takes a sensitive value s. $p(t, s)$ is calculated using the law of total probability. Specifically, let $p(s|t,B)$ be the probability that t takes sensitive value s given that t is in bucket B, then according to the law of total probability, the probability $p(t, s)$ is:

$$p(t, s) = \sum b \ p(t,B)p(s|t,B) \qquad (1)$$

In the rest of this section, we will show how to compute the two probabilities: $p(t,B)$ and $p(s|t,B)$.Computing $p(t,B)$:

Given a tuple t and a sliced bucket B, the probability that t is in B depends on the fraction of t's column values that match the column values in B. If some column value of t does not appear in the corresponding column of B, it is certain that t is not in B. In general, bucket B can potentially match $|B|c$ tuples, where $|B|$ is the number of tuples in B. Without additional knowledge, one has to assume that the column values are independent; therefore each of the $|B|c$ tuples is equally likely to be an original tuple. The probability that t is in B depends on the fraction of the $|B|c$ tuples that match t.

We formalize the above analysis. We consider the match between t's column values $\{t[C1], t[C2], \cdots, t[Cc]\}$ and B's column values $\{B[C1],B[C2], \cdots ,B[Cc]\}$. Let $f_i(t,B)$ ($1 \leq i \leq c - 1$) be the fraction of occurrences of $t[Ci]$ in $B[Ci]$ and let $f_c(t,B)$ be the fraction of occurrences of $t[Cc - \{S\}]$ in $B[Cc - \{S\}])$. Note that, $Cc - \{S\}$ is the set of QI attributes in the sensitive column. For example, in Table 1(f), $f1(t1,B1) = 1/4 = 0.25$ and $f2(t1,B1) = 2/4 = 0.5$. Similarly, $f1(t1,B2) = 0$ and $f2(t1,B2) = 0$. Intuitively, $f_i(t,B)$ measures the matching degree on column Ci, between tuple t and bucket B. Because each possible candidate tuple is equally likely to be an original tuple, the matching degree between t and B is the product of the matching degree on each column, i.e.,$f(t,B) = Q1\_i\_c \ f_i(t,B)$. Note that Pt $f(t,B) = 1$ and when B is not a matching bucket of t, $f(t,B) = 0$. Tuple t may have multiple matching buckets, t's total matching degree in the whole data is $f(t) = PB \ f(t,B)$. The probability that t is in bucket B is: $p(t,B) = f(t,B)/f(t)$

Computing $p(s|t,B)$. Suppose that t is in bucket B, to determine t's sensitive value, one needs to examine the sensitive column of bucket B. Since the sensitive column contains the QI attributes, not all sensitive values can be t's sensitive value. Only those sensitive values whose QI values match t's QI values are t's candidate sensitive values. Without additional knowledge, all candidate sensitive values (including duplicates) in a bucket are equally possible. Let D(t,B) be the distribution of t's candidate sensitive values in bucket B. Definition 6 (D(t,B)). Any sensitive value that is associated with t[Cc − {S}] in B is a candidate sensitive value for t (there are fc(t,B) candidate sensitive values for t in B, including duplicates). Let D(t,B) be the distribution of the candidate sensitive values in B and D(t,B)[s] be the probability of the sensitive s in the distribution.

For example, in Table 1(f), D(t1,B1) = (dyspepsia :0.5, flu : 0.5) and therefore D(t1,B1)[dyspepsia] = 0.5. The probability p(s|t,B) is exactly D(t,B)[s], i.e., p(s|t,B) =D(t,B)[s]. ℓ-Diverse Slicing. Once we have computed p(t,B) and p(s|t,B), we are able to compute the probability p(t, s) based on the Equation (1). We can show when t is in the data, the probabilities that t takes a sensitive value sum up to 1.

Fact 1. For any tuple t ∈ D, Ps p(t, s) = 1.

ℓ-Diverse slicing is defined based on the probability p(t, s).

Definition(7) for ℓ-diverse slicing: A tuple t satisfies ℓ-diversity iff for any sensitive value s,

p(t, s) ≤ 1/ℓA sliced table satisfies ℓ-diversity iff every tuple in it satisfies ℓ-diversity.

Our analysis directly show that from an ℓ-diverse sliced table, an adversary cannot correctly learn the sensitive value of any individual with a probability greater than 1/ℓ. Note that once we have computed the probability that a tuple takes a sensitive value, we can also use slicing for other privacy measures such as t-closeness.

2.2.1. Attacks on L-Diversity

In this section we studied two attacks on l-diversity: the Skewness attack and the Similarity attack.

1) Skewness Attack :l-diversity cannot prevent attribute disclosure whenever the overall distribution is skewed and satisfied.

2) Similarity Attack :When the sensitive attribute values are distinct but also semantically similar, an adversary can learn important information.

2.2.2. Limitations of L-Diversity

While the l-diversity principle represents an important step with respect to k-anonymity in protecting against attribute disclosure, it has several drawbacks. It is very difficult to achieve l − Diversity and it also may not provide sufficient privacy protection.

3. Slicing

Generally in privacy preserving, there is loss of security due to the presence of the adversary's background knowledge in real life application. Data contains sensitive information about individuals.These data when published violate the privacy. The current practice in data publishing relies mainly on policies and guidelines as to what types of data can be published and on agreements on the use of published data. The approach alone may lead to excessive data distortion or insufficient protection. Privacy-preserving data publishing (PPDP) provides methods and tools for publishing useful information while preserving data privacy. Many algorithms like bucketization, generalization have tried to preserve privacy however they exhibit attribute disclosure. So to overcome this problem an algorithm called slicing is used.

3.1. Architecture of Slicing and Formulization

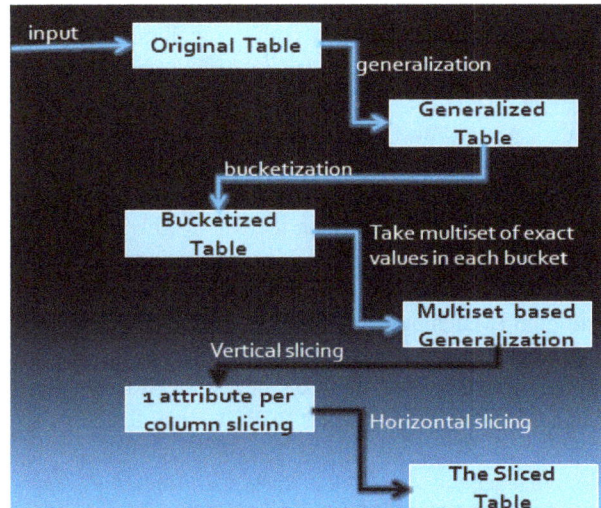

Fig. 1.4. *Slicing Architecture*

Let T be the microdata table to be published. T contains d attributes: A = {A1,A2, . . . ,Ad} and their attribute domains are {D[A1],D[A2], . . . ,D[Ad]}. A tuple t ∈ T can be represented as t = (t[A1], t[A2], ..., t[Ad]) where t[Ai] (1 ≤ i ≤ d) is the Ai value of t.

Definition 1 (Attribute partition and columns). An attribute partition consists of several subsets of A,such that each attribute belongs to exactly one subset. Each subset of attributes is called a column. Specifically, let there be c columns C1,C2, . . . ,Cc, then ∪c i=1Ci = A and for any 1 ≤ i1 6= i2 ≤ c, Ci1 ∩ Ci2 = ∅.

For simplicity of discussion, we consider only one sensitive attribute S. If the data contains multiple sensitive attributes, one can either consider them separately or consider their joint distribution. Exactly one of the c columns contains S. Without loss of generality, let the column that contains S be the last column Cc. This column is also called the sensitive column. All other columns {C1,C2, . . . ,Cc−1} contain only QI attributes.

Definition 2 (Tuple partition and buckets). A tuple partition consists of several subsets of T, such that each tuple belongs to exactly one subset. Each subset of tuples is called a bucket. Specifically, let there be b buckets B1,B2, . . . ,Bb, then ∪b i=1Bi = T and for any 1 ≤ i1 6= i2 ≤ b, Bi1 ∩ Bi2 = ∅.Definition 3 (Slicing). Given a microdata table T, a slicing of T is given by an attribute partition and a tuple partition. For example, Table 1(e) and Table 1(f) are two sliced tables. In Table 1(e), the attribute partition is {{Age}, {Sex},

{Zipcode}, {Disease}} and the tuple partition is {{t1, t2, t3, t4}, {t5, t6, t7, t8}}. In Table 1(f), the attribute partition is {{Age, Sex}, {Zipcode, Disease}} and the tuple partition is {{t1, t2, t3, t4}, {t5, t6, t7, t8}}. Often times, slicing also involves column generalization.

Definition 4 (Column Generalization). Given a microdata table T and a column $C_i = \{A_{i1}, A_{i2}, \ldots, A_{ij}\}$, a column generalization for C_i is defined as a set of non-overlapping j-dimensional regions that completely cover $D[A_{i1}] \times D[A_{i2}] \times \ldots \times D[A_{ij}]$. A column generalization maps each value of C_i to the region in which the value is contained.

Column generalization ensures that one column satisfies the k-anonymity requirement. It is a multidimensional encoding and can be used as an additional step in slicing. Specifically, a general slicing algorithm consists of the following three phases: attribute partition, column generalization, and tuple partition. Because each column contains much fewer attributes than the whole table, attribute partition enables slicing to handle high-dimensional data. A key notion of slicing is that of matching buckets.

Definition 5 (Matching Buckets). Let $\{C_1, C_2, \ldots, C_c\}$ be the c columns of a sliced table. Let t be a tuple, and $t[C_i]$ be the C_i value of t. Let B be a bucket in the sliced table, and $B[C_i]$ be the multiset of C_i values in B. We say that B is a matching bucket of t iff for all $1 \leq i \leq c$, $t[C_i] \in B[C_i]$. For example, consider the sliced table shown in Table 1(f), and consider t1 = (22, M, 47906, dyspepsia). Then, the set of matching buckets for t1 is {B1}.

Table 1: An original microdata table and its anonymized versions using various anonymization techniques

Table 1(a). The original table

Age	Sex	Zipcode	Disease
22	M	47906	Dyspepsia
22	F	47906	Flu
33	F	47905	Flu
52	F	47905	Bronchitis
54	M	47302	Flu
60	M	47302	Dyspepsia
60	M	47304	Dyspepsia
64	F	47304	Gastritis

Table 1(b). The generalized table

Age	Sex	Zipcode	Disease
[20-52]	*	4790*	Dyspepsia
[20-52]	*	4790*	Flu
[20-52]	*	4790*	Flu
[20-52]	*	4790*	Bronchitis
[54-64]	*	4730*	Flu
[54-64]	*	4730*	Dyspepsia
[54-64]	*	4730*	Dyspepsia
[54-64]	*	4730*	Gastritis

Table 1(c). The Bucketized table

Age	Sex	Zipcode	Disease
22	M	47906	Flu
22	F	47906	Dyspepsia
33	F	47905	Bronchitis
52	F	47905	Flu
54	M	47302	Gastritis
60	M	47302	Flu
60	M	47304	Dyspepsia
64	F	47304	dyspepsia

Table 1(d). Multiset based generalization

Age	Sex	Zipcode	Disease
		47905:2, 47906:2	
22:2,33:1, 52:1	M:1,F:3	47905:2,	Dyspepsia
22:2,33:1, 52:1	M:1,F:3	47906:2	Flu
22:2,33:1, 52:1	M:1,F:3	47905:2,	Flu
22:2,33:1, 52:1	M:1,F:3	47906:2	Bronchitis
		47905:2, 47906:2	
		47302:2, 47304:2	
54:1,60:2, 64:1	M:3,F:1	47302:2,	Flu
54:1,60:2, 64:1	M:3,F:1	47304:2	Dyspepsia
54:1,60:2, 64:1	M:3,F:1	47302:2,	Dyspepsia
54:1,60:2, 64:1	M:3,F:1	47304:2	Gastritis
		47302:2, 47304:2	

Table 1(e). One attribute per column slicing

Age	Sex	Zipcode	Disease
22	F	47906	Flu
22	M	47905	Flu
33	F	47906	Dyspepsia
52	F	47905	Bronchitis
54	M	47302	Dyspepsia
60	F	47304	Gastritis
60	M	47302	Dyspepsia
64	M	47304	Flu

Table 1(f). The sliced table

(Age,Sex)	(Zipcode,Disease)
(22,M)	(47905,Flu)
(22,F)	(47906,Dysp.)
(33,F)	(47905,Bronchitis)
(52,F)	(47906,Flu)
(54,M)	(47304,Gast.)
(60,M)	(47302,Flu)
(60,M)	(47302,Dysp.)
(64,F)	(47304,Dysp.)

3.2. Comparison with Generalization

There are several types of recodings for generalization. The recoding that preserves the most information is local recoding. In local recoding, one first groups tuples into buckets and then for each bucket, one replaces all values of one attribute with a generalized value. Such a recoding is local because the same attribute value may be generalized differently when they appear in different buckets. We now show that slicing preserves more information than such a local recoding approach, assuming that the same tuple partition is used. We achieve this by showing that slicing is better than the following enhancement of the local recoding approach. Rather than using a generalized value to replace more specific attribute values, one uses the multiset of exact values in each bucket. For example, Table 1(b) is a generalized table, and Table 1(d) is the result of using multisets of exact values rather than generalized values. For the Age attribute of the first bucket, we use the multiset of exact values {22,22,33,52} rather than the generalized interval. The multiset of exact values provides more information about the distribution of values in each attribute than the generalized interval. Therefore, using multisets of exact values preserves more information than generalization. However, we observe that this multiset-based generalization is equivalent to a trivial slicing scheme where each column contains exactly one attribute, because both approaches preserve the exact values in each attribute but break the association between them within one bucket. For example, Table 1(e) is equivalent to Table 1(d). Now comparing Table 1(e) with the sliced table shown in Table 1(f), we observe that while one-attribute-per-column slicing preserves attribute distributional information, it does not preserve attribute correlation, because each attribute is in its own column. In slicing, one groups correlated attributes together in one column and preserves their correlation. For example, in the sliced table shown in Table 1(f), correlations between Age and Sex and correlations between Zipcode and Disease are preserved. In fact, the sliced table encodes the same amount of information as the original data with regard to correlations between attributes in the same column. Another important advantage of slicing is its ability to handle high dimensional data. By partitioning attributes into columns, slicing reduces the dimensionality of the data.

Each column of the table can be viewed as a sub-table with a lower dimensionality. Slicing is also different from the approach of publishing multiple independent sub-tables in that these sub-tables are linked by the buckets in slicing.

3.3. Comparison with Bucketization

To compare slicing with bucketization, we first note that bucketization can be viewed as a special case of slicing, where there are exactly two columns: one column contains only the SA, and the other contains all the QIs. The advantages of slicing over bucketization can be understood as follows. First, by partitioning attributes into more than two columns, slicing can be used to prevent membership disclosure. Our empirical evaluation on a real dataset shows that bucketization does not prevent membership disclosure. Second, unlike bucketization, which requires a clear separation of QI attributes and the sensitive attribute, slicing can be used without such a separation. For dataset such as the census data, one often cannot clearly separate QIs from SAs because there is no single external public database that one can use to determine which attributes the adversary already knows. Slicing can be useful for such data. Finally, by allowing a column to contain both some QI attributes and the sensitive attribute, attribute correlations between the sensitive attribute and the QI attributes are preserved. For example, in Table 1(f), Zipcode and Disease form one column, enabling inferences about their correlations. Attribute correlations are important utility in data publishing. For workloads that consider attributes in isolation, one can simply publish two tables, one containing all QI attributes and one containing the sensitive attribute.

3.4. Privacy Threats

When publishing microdata, there are three types of privacy disclosure threats. The first type is membership disclosure. When the dataset to be published is selected from a large population and the selection criteria are sensitive (e.g. only diabetes patients are selected), one needs to prevent adversaries from learning whether one's record is included in the published dataset.

The second type is identity disclosure, which occurs when an individual is linked to a particular record in the released table. In some situations, one wants to protect against identity disclosure when the adversary is uncertain of membership. In this case, protection against membership disclosure helps protect against identity disclosure. In other situations, some adversary may already know that an individual's record is in the published dataset, in which case, membership disclosure protection either does not apply or is insufficient.

The third type is attribute disclosure, which occurs when new information about some individuals is revealed, i.e., the released data makes it possible to infer the attributes of an individual more accurately than it would be possible before the release. Similar to the case of identity disclosure, we need to consider adversaries who already know the membership information. Identity disclosure leads to attribute disclosure. Once there is identity disclosure, an individual is re-identified and the corresponding sensitive value is revealed. Attribute disclosure can occur with or without identity disclosure, e.g., when the sensitive values of all matching tuples are the same.

For slicing, we consider protection against membership disclosure and attribute disclosure. It is a little unclear how identity disclosure should be defined for sliced data (or for data anonymized by bucketization), since each tuple resides within a bucket and within the bucket the association across different columns are hidden. In any case, because identity disclosure leads to attribute disclosure, protection against attribute disclosure is also sufficient protection against identity disclosure.

We would like to point out a nice property of slicing that is important for privacy protection. In slicing, a tuple can potentially match multiple buckets, i.e., each tuple can have more than one matching buckets. This is different from previous work on generalization and bucketization, where each tuple can belong to a unique equivalence-class (or bucket). In fact, it has been recognized that restricting a tuple in a unique bucket helps the adversary but does not improve data utility. We will see that allowing a tuple to match multiple buckets is important for both attribute disclosure protection and attribute disclosure protection.

4. Methodology

The key intuition that slicing provides privacy protection is that the slicing process ensures that for any tuple, there are generally multiple matching buckets. Slicing first partitions attributes into columns. Each column contains a subset of attributes. Slicing also partition tuples into buckets. Each bucket contains a subset of tuples. This horizontally partitions the table. Within each bucket, values in each column are randomly permutated to break the linking between different columns. This algorithm consists of three phases: attribute partitioning, column generalization, and tuple partitioning. We now describe the three phases:-

4.1. Attribute Partitioning

Highly correlated attributes are grouped together into one column in this attribute partitioning technique.There are three steps :

4.1.1. Equal Width Partitioning
There are two types of attribute: continuous and categorical. So, in this step, continuous attribute are converted into categorical attribute.

In equal width partitioning, we first divide the range into N intervals of equal size: uniform grid if A and B are the lowest and highest values of the attribute.

Width of intervals will be $W=(B-A)/N$

4.1.2. Measures of Correlation
Here,we calculate relation between two attributes..Let two attributes A_1 and A_2 with domains $\{V_{11}, V_{12}, \ldots \ldots V_1 n_1\}$ and $\{V_{21}, V_{22}, \ldots \ldots V_2 n_2\}$ respectively. Their domain sizes are thus n_1 and n_2. Therefore, Mean square contingency coefficient formula is used.

4.1.3. Attribute Clustering
In this step,k-medoid clustering algorithm is used to partition attribute into columns as follows:-

The most common realisation of *k*-medoid clustering is the Partitioning Around Medoids (PAM) algorithm:

Algorithm 1.1
1. Initialize: randomly select (without replacement) *k* of the *n* data points as the medoids
2. Associate each data point to the closest medoid. ("closest" here is defined using any valid distance metric, most commonly Euclidean distance, Manhattan distance or Minkowski distance)
3. For each medoid *m*
   For each non-medoid data point *o*
   Swap *m* and *o* and compute the total cost of the

configuration
4. Select the configuration with the lowest cost.
5. Repeat steps 2 to 4 until there is no change in the medoid.

There can be a cluster based attribute slicing algorithm also as in existing systems, equal width discretization is used so it cannot handle skew data properly.So,to solve this problem,we proposed a new algorithm in proposed method,we use cluster based attribute algorithm for converting the continuous attribute into categorical attribute.This algorithm shows:

Input: Vector of real valued data $a=(a_1, a_2 \ldots \ldots a_{11})$ and number of clusters to be determined k.

Goal: To find partition of data in k distinct clusters.

Output: The set of cut points $t_0, t_1 \ldots \ldots tk$ with $t_0 < t_1 < \ldots \ldots tn$ that defines discretization of adom(A).

Algorithm 1.2:
1. Compute $amax=max\{a_1, a_2, \ldots \ldots an\}$ and $amin=min\{a_1, a_2, \ldots \ldots an\}$
2. Choose the centres as the first k distinct values of the attribute A.
3. Arrange them in increasing order i.e.$c[1] < c[2] < \ldots \ldots c[k]$.
4. Define boundary points $bo=amin$, $bj = (c[j]+c[j+1])/2$ for $j=1$ to $k-1$, $bk=amax$
5. Find the closest cluster to ai.
6. Recompute the centres of the cluster as the average of the values in each cluster.
7. Find the closest cluster to ai from the possible clusters $\{j-1,j,j+1\}$
8. Determination of cut points:-
   $t_0 = amin$
   for $i= 1$ to $k-1$
   do
   $ti=(c[i]+c[i+1])/2$
9. end for
10. $tk=amax$
11. Apply formula of measures of correlation
12. Apply attribute clustering algorithm
13. Apply attribute partitioning algorithm

4.2. Column Generalization

First, column generalization may be required for identity/membership disclosure protection. If a column value is unique in a column, a tuple with this unique column value can only have one matching bucket. This is not good for privacy protection as in the case of generalization/bucketization where each tuple can belong to one equivalent class.
- Given microdata table T and column $Ci=\{Ai1, Ai2, \ldots Aij\}$
- Column generation for Ci is defined as set of non-overlapping j-dimensional regions that completely cover $D[Ai1] \times D[Ai2] \times \ldots \ldots D[Aij]$
- Column gen. maps each value of Ci to the region in which the value is contained.
- It may be required for membership disclosure

protection and privacy protection.

4.3 Tuple Partitioning

The algorithm maintains two data structures:
1) A queue of buckets Q and
2) A set of sliced buckets SB.

Initially, Q contains only one bucket which includes all tuples and SB is empty. For each iteration, the algorithm removes a bucket from Q and splits the bucket into two buckets . If the sliced table after the split satisfies l-diversity, then the algorithm puts the two buckets at the end of the queue Q Otherwise, we cannot split the bucket anymore and the algorithm puts the bucket into SB.When Q becomes empty, we have computed the sliced table. The set of sliced buckets is SB.

Algorithm 1.3 for Tuple partitioning
1. Q = {T}, SB = φ.
2. While Q is not empty
3. Remove the first bucket B from Q, Q = Q − {B}.
4. Split B into two buckets B1 and B2, as in Mondrian.
5. If diversity-check (T, Q ∪ {B1, B2} ∪ SB, ℓ)
6. Q = Q ∪ {B1, B2}.
7. Else SB = SB ∪ {B}.
8. Return SB.

Algorithm 1.4 for Diversity-Check
1. For each tuple t ∈ T, L[t] = φ.
2. For each b 3. Record f (v) for each column value v in bucket B.
4. for each tuple t ∈ T
5. Calculate p(t,B) and find D(t,B).
6. L[t] = L[t] ∪ {hp (t, B), D (t, B) i}.
7. for each tuple t ∈ T
8. Calculate p(t, s) for each s based on L[t].
9. If p(t, s) ≥ 1/ℓ, return false.
10. Return true…buckets B in T*

5. Experimental Results

5.1. Membership Disclosure Protection

We evaluate the effectiveness of slicing in membership disclosure protection. We first show that bucketization is vulnerable to membership disclosure. In both the OCC-7 dataset and the OCC-15 dataset, each combination of QI values occurs exactly once. This means that the adversary can determine the membership information of any individual by checking if the QI value appears in the bucketized data. If the QI value does not appear in the bucketized data, the individual is not in the original data. Otherwise, with high confidence, the individual is in the original data as no other individual has the same QI value.

We then show that slicing does prevent membership disclosure. We perform the following experiment. First, we partition attributes into c columns based on attribute correlations. We set c ∈ {2, 5}. In other words, we compare 2-column-slicing with 5-column-slicing. For example, when we set c = 5, we obtain 5 columns. In OCC-7,{Age, Marriage,

Gender} is one column and each other attribute is in its own column. In OCC-15, the 5 columns are: {Age, Work class, Education, Education-Num, Cap-Gain, Hours, Salary}, {Marriage, Occupation, Family, Gender}, {Race,Country}, {Final -Weight}, and {Cap-Loss}.

Then, we randomly partition tuples into buckets of size p (the last bucket may have fewer than p tuples). As described above, we collect statistics about the following two measures in our experiments: (1) the number of fake tuples and (2) the number of matching buckets for original v.s. the number of matching buckets for fake tuples. The number of fake tuples. Figure shows the experimental results on the number of fake tuples, with respect to the bucket size p. Our results show that the number of fake tuples is large enough to hide the original tuples. For example, for the OCC-7 dataset, even for a small bucket size of 100 and only 2 columns, slicing introduces as many as 87936 fake tuples, which is nearly twice the number of original tuples (45222). When we increase the bucket size, the number of fake tuples becomes larger. This is consistent with our analysis that a bucket of size k can potentially match kc –k fake tuples. In particular, when we increase the number of columns c, the number of fake tuples becomes exponentially larger. In almost all experiments, the number of fake tuples is larger than the number of original tuples. The existence of such a large number of fake tuples provides protection for membership information of the original tuples. The number of matching buckets. We categorize the tuples (both original tuples and fake tuples) into three categories: (1) ≤ 10: tuples that have at most 10 matching buckets, (2) 10−20: tuples that have more than 10 matching buckets but at most 20 matching buckets, and (3) > 20: tuples that have more than 20 matching buckets. For example, the "original-tuples(≤ 10)" bar gives the number of original tuples that have at most 10 matching buckets and the "fake-tuples(> 20)" bar gives the number of fake tuples that have more than 20 matching buckets. Because the number of fake tuples that have at most 10 matching buckets is very large, we omit the"fake-tuples(≤ 10)"bar from the figures to make the figures more readable.

Our results show that, even when we do random grouping, many fake tuples have a large number of matching buckets.For example, for the OCC-7 dataset, for a small p = 100 and c = 2, there are 5325 fake tuples that have more than 20 matching buckets; the number is 31452 for original tuples. The numbers are even closer for larger p and c values. This means that a larger bucket size and more columns provide better protection against membership disclosure. Although many fake tuples have a large number of matching buckets, in general, original tuples have more matching buckets than fake tuples. As we can see from the figures, a large fraction of original tuples have more than 20 matching buckets while only a small fraction of fake tuples have more than 20 tuples. This is mainly due to the fact that we use random grouping in the experiments. The results of random grouping are that the number of fake tuples is very large but most fake tuples have very few matching buckets. When we aim at protecting membership information, we can design more effective grouping algorithms to ensure better protection against membership disclosure. The design of tuple grouping algorithms is left to future work.

6. Discussions and Future Work

A new approach called slicing is for privacy-preserving microdata publishing. Slicing overcomes the limitations of generalization and bucketization and preserves better utility while protecting against privacy threats. We illustrate how to use slicing to prevent attribute disclosure and membership disclosure. Our experiments show that slicing preserves better data utility than generalization and is more effective than bucketization in workloads involving the sensitive attribute.

The general methodology proposed by this work is that: before anonymizing the data, one can analyze the data characteristics and use these characteristics in data anonymization. The rationale is that one can design better data anonymization techniques when we know the data better. We show that attribute correlations can be used for privacy attacks. We have also shown that cluster based attribute slicing can also be done to achieve attribute partitioning.

This work motivates several directions for future research. First, in this paper, we consider slicing where each attribute is in exactly one column. An extension is the notion of *overlapping slicing*, which duplicates an attribute in more than one columns. This releases more attribute correlations. For example, in Table 1(f), one could choose to include the Disease attribute also in the first column. That is, the two columns are {Age,Sex,Disease} and {Zipcode,Disease}. This could provide better data utility, but the privacy implications need to be carefully studied and understood. It is interesting to study the tradeoff between privacy and utility .

Second, we plan to study membership disclosure protection in more details. Our experiments show that random grouping is not very effective. We plan to design more effective tuple grouping algorithms.

Third, slicing is a promising technique for handling high-dimensional data. By partitioning attributes into columns,we protect privacy by breaking the association of uncorrelated attributes and preserve data utility by preserving the association between highly-correlated attributes. For example, slicing can be used for anonymizing transaction databases,

which has been studied recently in.

Finally, while a number of anonymization techniques have been designed, it remains an open problem on how to use the anonymized data. In our experiments, we randomly generate the associations between column values of a bucket.This may lose data utility. Another direction to design data mining tasks using the anonymized data computed by various anonymization techniques.

References

[1]   Aggarwal.C, "On K-Anonymity and the Curse of Dimensionality," Proc. Int"l Conf.Very Large Data Bases (VLDB), 2005.

[2]   Brickell.J and Shmatikov, "The Cost of Privacy:Destruction of Data Mining Utility in Anonymized Data Publishing", Proc.ACM SIGKDD int"l conf. Knowledge Discovery and Data Mining (KDD), 2008.

[3]   Ghinita.G,Tao.Y, and Kalnis.P, "OnThe Anonymization of Sparse High Dimensional Data," Proc. IEEE 24th Int"l Conf. Data Eng. (ICDE), 2008.

[4]   He.Y and Naughton.J, "Anonymization of Set-Valued Data via Top-Down, local Generalization," Proc.IEEE 25th Int"l Conf.Data Engineering (ICDE), 2009.

[5]   Inan.A,Kantarcioglu.M,and Bertino.e, "Using Anonymized Data for Classification," Proc. IEEE 25th Int"l Conf. Data Eng. (ICDE), pp. 429-440, 2009.

[6]   Li.T and Li.N, "On the Tradeoff between Privacy and Utility in Data Publishing," Proc.ACM SIGKDD Int"l Conf.Knowledge Discovery and Data Mining (KDD), 2009.

[7]   Li.N, Li.T, "Slicing: The new Approach for Privacy Preserving Data publishing", IEEE Transaction on knowledge and data Engineering, vol.24, No, 3, March 2012.

[8]   L. Kaufman and P. Rousueeuw. Finding Groups in Data: an Introduction to Cluster Analysis. John Wiley& Sons, 1990.

[9]   X. Xiao and Y. Tao. Anatomy: simple and effective privacy preservation. In VLDB, pages 139–150, 2006.

[10] X. Xiao and Y. Tao. Output perturbation with query relaxation. In VLDB, pages 857–869, 2008.

[11] Y. Xu, K. Wang, A. W.-C. Fu, and P. S. Yu. Anonymizing transaction databases for publication. In KDD, pages 767–775, 2008

19

Full width at half maximum (FWHM) analysis of solitonic pulse applicable in optical network communication

IS Amiri[1, *], H. Ahmad[1], Hamza M. R. Al-Khafaji[2]

[1]Photonics Research Centre, University of Malaya (UM), 50603 Kuala Lumpur, Malaysia
[2]Wireless Communication Centre, Faculty of Electrical Engineering, Universiti Teknologi Malaysia (UTM), 81310 UTM Skudai, Johor, Malaysia

Email address:
amiri@um.edu.my (I. Amiri), harith@um.edu.my (H. Ahmad), hamza@utm.my (H. M. R. Al-Khafaji)

Abstract: In this paper, we propose a system of microring resonator (MRR). This system uses a laser diode input which can be incorporated with an optical add/drop filter system. When light from the laser diode feedbacks to the fiber ring resonator, the pulses in the form of soliton can be generated by using appropriate fiber ring resonator parameters and also the input power. The filtering process occurs during the propagation of the pulse within the ring resonators. The full width at half maximum (FWHM) or bandwidth characterization of the pulse can be performed using the proposed system. Results obtained have established particular possibilities from the application such as optical network communication. The obtained results show the effects of coupling coefficients and ring radius on the bandwidth of the soliton pulse, where the graph of the FWHM versus the variable parameters such as the radius and coupling coefficient are presented.

Keywords: Optical Network Communication, Soliton, FWHM, Pulse Bandwidth Characterization

1. Introduction

In optical communication, soliton controls within a semiconductor add/drop filter have numerous applications [1, 2]. Microring resonators (MRRs) are the types of Fabry-Pérot resonators, which can be readily integrated in array geometries for useful functions in areas such as optical communication [3, 4], signal processing in the nanoscale regime [5]. Its nonlinear phase response can also be readily incorporated into an interferometer system to produce a specific intensity output function [6, 7]. One interesting result emerges through the use of an add/drop system, which is a good candidate for nanoscale interferometer applications [8, 9]. One new feature of this specific type of ring resonator, which introduces a system of nanoscale-sensing transducers based on the add/drop ring resonator was presented by Amiri et al [10, 11]. They have shown that the multisoliton can be generated and controlled within a modified add/drop ring resonator [12, 13].

Nonlinear behaviors associated with light traveling inside a fiber optic ring resonator can be caused by the effects such as the Kerr effects, four-wave mixing, as well as the external nonlinear pumping electrical power [14, 15]. This sort of nonlinear behaviors usually are called chaos, bistability, in addition to bifurcation. Additional information regarding these kinds of behaviors in a micro ring resonator evidently are defined by Amiri et al [16, 17]. Nonetheless, aside from the penalties of the nonlinear behaviors of light traveling within the fiber ring resonator, there are several benefits that can be employed by the communication methods in order to examine the obtained result. One called chaotic behavior which has been employed to make the benefit within digital or optical communications [18, 19]. In this paper we consider the characterization of soliton pulses using MRRs. The bandwidth and free spectrum range of the soliton pulses can be manipulated using suitable parameters of the system.

2. Theory of The Research

The system of the ring resonator interferometer is shown in Figure 1.

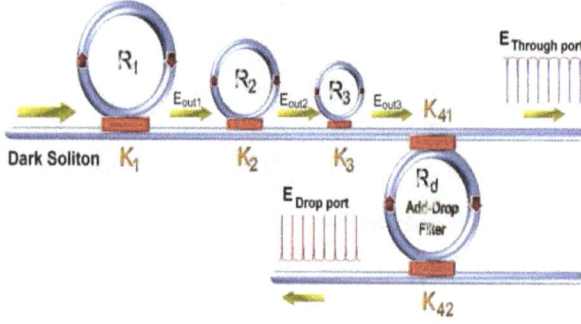

Figure 1. A schematic of the proposed MRR's system, where R_s: ring radii, κ_s: coupling coefficients, R_d: an add/drop ring radius, A_{effs}: effective areas

The input optical fields (E_{in}) in the form of dark soliton can be expressed by [20]

$$E_{in}(t) = A \tanh \left[\frac{T}{T_0} \right] \exp \left[\left(\frac{iz}{2L_D} \right) - i\omega_0 t \right] \tag{1}$$

Here A and z are the optical field amplitude and propagation distance, respectively [21]. T represents the soliton pulse propagation time in a frame moving at the group velocity, $(T = t - \beta_1 \times z)$, where β_1 and β_2 are the coefficients of the linear and second order terms of the Taylor expansion of the propagation constant [22-24]. The dispersion length of the soliton pulse can be defined as $L_D = T_0^2 / |\beta_2|$, where the frequency carrier of the soliton is ω_0 [25]. The intensity of soliton peak is $(|\beta_2 / \Gamma T_0^2|)$, where T_o is representing the initial soliton pulse propagation time [26, 27]. A balance should be achieved between the dispersion length (L_D) and the nonlinear length $(L_{NL} = 1/\Gamma\phi_{NL})$, where $\Gamma = n_2 \times k_0$, is the length scale over which disperse or nonlinear effects causes the beam becomes wider or narrower. Here, $L_D = L_{NL}$ [28]. The total index (n) of the system is given by [29, 30].

$$n = n_0 + n_2 I = n_0 + (\frac{n_2}{A_{eff}})P, \tag{2}$$

where n_0 and n_2 are the linear and nonlinear refractive indices, respectively. I and P are the optical intensity and optical power, respectively [31, 32]. A_{eff} represents the effective mode core area of the device, where in the case of MRRs, the effective mode core areas range from 0.50 to 0.1 μm2. The normalized output of the light field is defined as [33, 34].

$$\left| \frac{E_{out}(t)}{E_{in}(t)} \right|^2 = (1-\gamma) \times \left[1 - \frac{(1-(1-\gamma)x^2)\kappa}{(1-x\sqrt{1-\gamma}\sqrt{1-\kappa})^2 + 4x\sqrt{1-\gamma}\sqrt{1-\kappa}\sin^2(\frac{\phi}{2})} \right] \tag{3}$$

Here, κ is the coupling coefficient, $x = \exp(-\alpha L/2)$ represents a round-trip loss coefficient, $\phi = \phi_0 + \phi_{NL}$, $\phi_0 = kLn_0$ and $\phi_{NL} = kLn_2|E_{in}|^2$ are the linear and nonlinear phase shifts and $k = 2\pi / \lambda$ is the wave propagation number and γ is the

fractional coupler intensity loss [35, 36]. Here L and α are the waveguide length and linear absorption coefficient, respectively [37, 38]. The input power insert into the input port of the add/drop filter system. E_{th} and E_{drop} represent the optical electric fields of the through and drop ports, respectively expressed by equations (4) and (5) [39, 40],

$$|E_{th}|^2 / |E_{out3}|^2 = \frac{(1-\kappa_{41}) - 2\sqrt{1-\kappa_{41}} \cdot \sqrt{1-\kappa_{42}} e^{-\frac{\alpha}{2}L_d} \cos(k_n L_d) + (1-\kappa_{42})e^{-\alpha L_d}}{1 + (1-\kappa_{41})(1-\kappa_{42})e^{-\alpha L_d} - 2\sqrt{1-\kappa_{41}} \cdot \sqrt{1-\kappa_{42}} e^{-\frac{\alpha}{2}L_d} \cos(k_n L_d)}$$

$$|E_{drop}|^2 / |E_{out3}|^2 = \frac{\kappa_{41}\kappa_{42} e^{-\frac{\alpha}{2}L_d}}{1 + (1-\kappa_{41})(1-\kappa_{42})e^{-\alpha L_d} - 2\sqrt{1-\kappa_{41}} \cdot \sqrt{1-\kappa_{42}} e^{-\frac{\alpha}{2}L_d} \cos(k_n L_d)}$$

where $|E_{th}|^2$ and $|E_{drop}|^2$ are the output intensities of the through and drop ports respectively [41, 42].

3. Results and Discussion

Temporal profile of the input dark soliton pulse can be seen from Figure 2, where the input of the pulse is 350 mW with central wavelength of 1.3 μm. The ring radii of the rings are selected to R_1=30μm, R_2=12μm and R_3=5μm, where κ_1=0.7, κ_2=0.9 and κ_3=0.93.

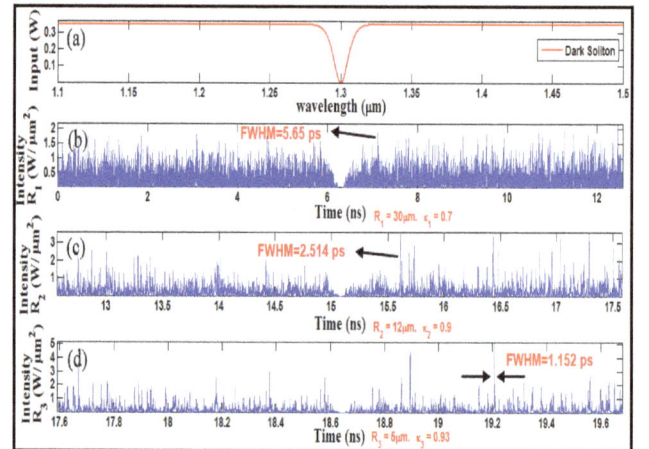

Figure 2. Results of temporal chaotic signals generation within a series of MRRs with dark soliton input

In Figure 2(a), the dark soliton is input and propagates within the ring system. In Figure 2(b), the input pulse is split to many noisy and chaotic signals in the form of temporal signals with FWHM=5.65 ps, where Figures 2(c) and 2(d) show the temporal signals seen within the range of 12.5-17.5 ns and 17.6-19.6 ns with FWHM of 2.514 and 1.152 ps respectively. The shorter bandwidth of the chaotic signals can be obtained by adding more ring resonators. The effects of the ring's radius and coupling coefficients of the ring resonators on the FWHM of the chaotic pulses are shown by Figure 3.

Therefore, the chaotic signals can be generated using the input dark soliton pulse. In order to use the chaotic signals for long distance communication, the use of input bright soliton is recommended, where the security of soliton signals can be performed when the input dark soliton is inserted into the system and split into chaotic signals.

Optical field of the Gaussian pulse can be inserted into the input port of the multi-stage MRR's system shown in Figure 4. Considering the proposed system, the radii of the rings have been selected as $R_1 = 15\mu m$, $R_2 = 9\mu m$, $R_3 = 7\mu m$, and $\kappa_1 = 0.96$, $\kappa_2 = 0.94$, $\kappa_3 = 0.92$, where the add/drop filter has a radius of $R_d = 78\mu m$ and coupling coefficients of $\kappa_4 = \kappa_5 = 0.1$. Some parameters of the system are fixed such as $n_0 = 3.34$ (InGaAsP/InP), $A_{eff} = 0.50$, 0.25 and 0.10μm^2 for the microrings, $\alpha = 0.5$dBmm^{-1}, $\gamma = 0.1$. The nonlinear refractive index of the MRRs is $n_2 = 2.2$ x 10^{-17} m^2/W.

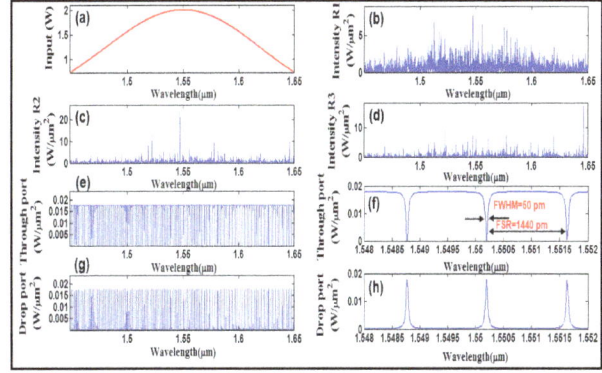

Figure 4. *Results of spatial dark and bright soliton generation, where (a): input Gaussian beam, (b-d): large bandwidth signals, (e-f): dark soliton at the through port with FWHM and FSR of 50 pm and 1440 pm respectively, (g-h): bright soliton at the drop port with FWHM and FSR of 50 pm and 1440 pm respectively*

The advantage of this technique is its ability to operate on the trains of low-power picometer optical pulses. In order to improve the system, narrower soliton pulses are recommended, where the attenuation of such signals during transmission lessens when compared to the conventional peaks of micrometre laser pulses. Therefore, the bandwidth varies with respect to the variation of the coupling coefficients of the add/drop filter system shown in Figure 5.

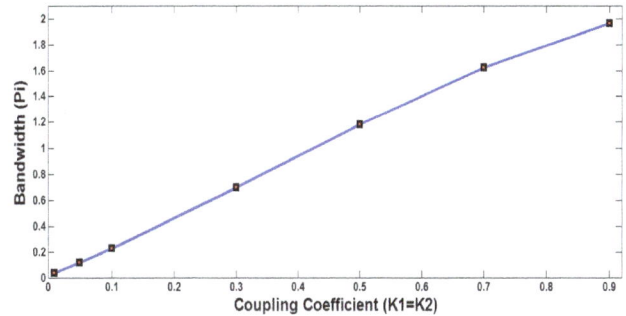

Figure 5. *Result of the single bandwidth manipulation respect to variation of the coupling coefficients of the add/drop filter system*

As it can be seen from the Figure 5, the increase of the coupling coefficients of the add/drop filter system leads to increase of the bandwidth of the soliton pulse. Thus in order to use ultra-short soliton pulses applied in optical communication, lower coupling coefficient is recommended where, the power control can be performed within the system [43-45].

Using this method, the output power of the system can be simulated successfully. This system act as a passive filter system which can be used to split the input power and generate chaotic signals using suitable parameters of the system [46-48]. Therefore input power of Gaussian beam can be sliced to smaller peaks as chaotic signals. The chaotic signals have many applications in optical communications.

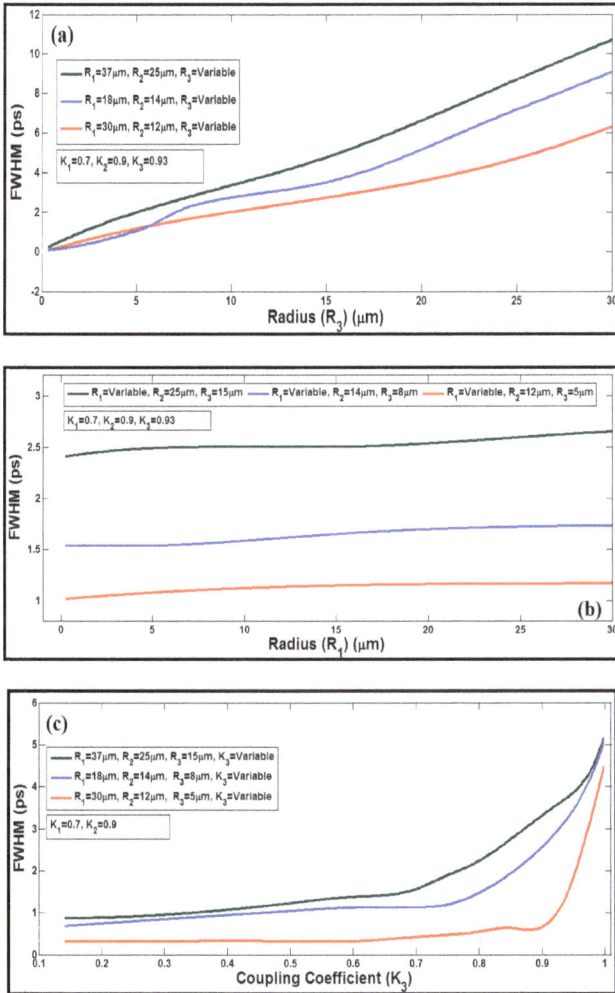

Figure 3. *Results of FWHM, where (a): radius of the third ring varies, (b): radius of the first ring varies, (c): coupling coefficient of the third ring varies*

The input Gaussian laser pulse with power of 2W is introduced into the MRR's system shown in Figure 4(a). The output powers from three ring resonators are shown in Figures 4(b-d), where Figures 4(e-f) show the output power from the throughput port in terms of wavelength. The FWHM and FSR of the spatial soliton pulses are 50 pm and 1440 pm respectively. Figures 4(g-h) show the output power from the drop port of the system with the same FWHM and FSR.

4. Conclusion

Soliton signals can be generated using the input laser

power propagating within a nonlinear ring resonator, where the required signals can be recovered and manipulated by using appropriate parameters of the system such as the ring radius and coupling coefficients. Results obtained have shown that the FWHM of the generated soliton pulses can be affected by vary the parameters. Here, the effects of coupling coefficients and radius on the bandwidth of the soliton pulse have been presented.

References

[1] Iraj Sadegh Amiri, Sayed Ehsan Alavi & Sevia Mahdaliza Idrus, Soliton Coding for Secured Optical Communication Link. USA: Springer, 2014.

[2] IS Amiri & A Afroozeh, Soliton Generation Based Optical Communication, in Ring Resonator Systems to Perform Optical Communication Enhancement Using Soliton, ed USA: Springer, 2014.

[3] I. S. Amiri, S. E. Alavi, H. Ahmad, A.S.M. Supa'at & N. Fisal, (2014) "Numerical Computation of Solitonic Pulse Generation for Terabit/Sec Data Transmission", Optical and Quantum Electronics,

[4] IS Amiri, MZ Zulkifli & H Ahmad, (2014) "Soliton comb generation using add-drop ring resonators", International Research Journal of Telecommunications and Information Technology,

[5] Iraj Sadegh Amiri, Sayed Ehsan Alavi, S. M. Idrus, Abdolkarim Afroozeh & Jalil Ali, Soliton Generation by Ring Resonator for Optical Communication Application. Hauppauge, NY 11788 USA: Nova Science Publishers, 2014.

[6] Amiri, H Ahmad & MZ Zulkifli, (2014) "Integrated ring resonator system analysis to Optimize the soliton transmission", International Research Journal of Nanoscience and Nanotechnology, 1(1), 002-007.

[7] Iraj Sadegh Amiri & Harith Ahmad, Optical Soliton Communication Using Ultra-Short Pulses. USA: Springer, 2014.

[8] I. S. Amiri, S. E. Alavi, Sevia M. Idrus, A. Nikoukar & J. Ali, (2013) "IEEE 802.15.3c WPAN Standard Using Millimeter Optical Soliton Pulse Generated By a Panda Ring Resonator", IEEE Photonics Journal, 5(5), 7901912.

[9] A. Afroozeh, A.Zeinalinezhad, I. S. Amiri & S. E. Pourmand, (2014) "Stop Light Generation using Nano Ring Resonators for ROM", Journal of Computational and Theoretical Nanoscience (CTN), 12(3),

[10] S. E. Alavi, I. S. Amiri, S. M. Idrus, ASM Supa'at, J. Ali & P. P. Yupapin, (2014) "All Optical OFDM Generation for IEEE802.11a Based on Soliton Carriers Using MicroRing Resonators ", IEEE Photonics Journal, 6(1),

[11] IS Amiri, SE Alavi, N Fisal, ASM Supa'at & H Ahmad, (2014) "All-Optical Generation of Two IEEE802.11n Signals for 2×2 MIMO-RoF via MRR System", IEEE Photonics Journal, 6(6),

[12] I. S. Amiri & J. Ali, (2014) "Generating Highly Dark–Bright Solitons by Gaussian Beam Propagation in a PANDA Ring Resonator", Journal of Computational and Theoretical Nanoscience (CTN), 11(4), 1092-1099.

[13] I. S. Amiri, S. E. Alavi & J. Ali, (2013) "High Capacity Soliton Transmission for Indoor and Outdoor Communications Using Integrated Ring Resonators", International Journal of Communication Systems,

[14] IS Amiri, SE Alavi, H Ahmad, ASM Supa'at & N Fisal, (2014) "Generation and Transmission of 3× 3 W-Band MIMO-OFDM-RoF Signals Using Micro-Ring Resonators", Applied Optics, 53(34),

[15] I. S. Amiri, M. Ranjbar, A. Nikoukar, A. Shahidinejad, J. Ali & P. Yupapin, (2012), "Multi optical Soliton generated by PANDA ring resonator for secure network communication", in Computer and Communication Engineering (ICCCE) Conference, Malaysia, 760-764.

[16] I. S. Amiri, S. E. Alavi & H. Ahmad, (2015) "Analytical Treatment of the Ring Resonator Passive Systems and Bandwidth Characterization Using Directional Coupling Coefficients ", Journal of Computational and Theoretical Nanoscience (CTN), 12(3),

[17] Iraj Sadegh Amiri & Abdolkarim Afroozeh, Ring Resonator Systems to Perform the Optical Communication Enhancement Using Soliton. USA: Springer, 2014.

[18] IS Amiri & A Afroozeh, Integrated Ring Resonator Systems, in Ring Resonator Systems to Perform Optical Communication Enhancement Using Soliton, ed USA: Springer, 2014.

[19] I. S. Amiri, S. Soltanmohammadi, A. Shahidinejad & j. Ali, (2013) "Optical quantum transmitter with finesse of 30 at 800-nm central wavelength using microring resonators", Optical and Quantum Electronics, 45(10), 1095-1105.

[20] I. S. Amiri, A. Afroozeh, I. N. Nawi, M. A. Jalil, A. Mohamad, J. Ali & P. P. Yupapin, (2011) "Dark Soliton Array for communication security", Procedia Engineering, 8 417-422.

[21] I. S. Amiri, S. E. Alavi, S. M. Idrus, A. S. M. Supa'at, J. Ali & P. P. Yupapin, (2014) "W-Band OFDM Transmission for Radio-over-Fiber link Using Solitonic Millimeter Wave Generated by MRR", IEEE Journal of Quantum Electronics, 50(8), 622 - 628.

[22] I. S. Amiri & J. Ali, (2013) "Data Signal Processing Via a Manchester Coding-Decoding Method Using Chaotic Signals Generated by a PANDA Ring Resonator", Chinese Optics Letters, 11(4), 041901(4).

[23] S. E. Alavi, I. S. Amiri, M. R. K. Soltanian, A.S.M. Supa'at, N. Fisal & H. Ahmad, (2015) "Generation of Femtosecond Soliton Tweezers Using a Half-Panda System for Modeling the Trapping of a Human Red Blood Cell", Journal of Computational and Theoretical Nanoscience (CTN), 12(1),

[24] I. S. Amiri, S. E. Alavi & S. M. Idrus, Theoretical Background of Microring Resonator Systems and Soliton Communication, in Soliton Coding for Secured Optical Communication Link, ed USA: Springer, 2015, pp. 17-39.

[25] A. Afroozeh, I.S. Amiri, K. Chaudhary, J. Ali & P. P. Yupapin, Analysis of Optical Ring Resonator, in Advances in Laser and Optics Research, ed New York: Nova Science, 2014.

[26] I. S. Amiri, J. Ali & P. P. Yupapin, (2012) "Enhancement of FSR and Finesse Using Add/Drop Filter and PANDA Ring Resonator Systems", International Journal of Modern Physics B, 26(04), 1250034.

[27] I. S. Amiri, B. Barati, P. Sanati, A. Hosseinnia, HR Mansouri Khosravi, S. Pourmehdi, A. Emami & J. Ali, (2014) "Optical Stretcher of Biological Cells Using Sub-Nanometer Optical Tweezers Generated by an Add/Drop Microring Resonator System", Nanoscience and Nanotechnology Letters, 6(2), 111-117.

[28] I. S. Amiri, S. E. Alavi & S. M. Idrus, Introduction of Fiber Waveguide and Soliton Signals Used to Enhance the Communication Security, in Soliton Coding for Secured Optical Communication Link, ed USA: Springer, 2015, pp. 1-16.

[29] I. S. Amiri & J. Ali, (2014) "Femtosecond Optical Quantum Memory generation Using Optical Bright Soliton", Journal of Computational and Theoretical Nanoscience (CTN), 11(6), 1480-1485.

[30] I. S. Amiri & J. Ali, (2013) "Nano Particle Trapping By Ultra-short tweezer and wells Using MRR Interferometer System for Spectroscopy Application", Nanoscience and Nanotechnology Letters, 5(8), 850-856.

[31] P. Sanati, A. Afroozeh, I. S. Amiri, J.Ali & Lee Suan Chua, (2014) "Femtosecond Pulse Generation using Microring Resonators for Eye Nano Surgery", Nanoscience and Nanotechnology Letters, 6(3), 221-226

[32] I. S. Amiri & J. Ali, (2014) "Optical Quantum Generation and Transmission of 57-61 GHz Frequency Band Using an Optical Fiber Optics ", Journal of Computational and Theoretical Nanoscience (CTN), 11(10), 2130-2135.

[33] S. E. Alavi, I. S. Amiri, S. M. Idrus & A. S. M. Supa'at, (2014) "Generation and Wired/Wireless Transmission of IEEE802.16m Signal Using Solitons Generated By Microring Resonator", Optical and Quantum Electronics,

[34] S. E. Alavi, I. S. Amiri, H. Ahmad, N. Fisal & ASM. Supa'at, (2015) "Optical Amplification of Tweezers and Bright Soliton Using an Interferometer Ring Resonator System", Journal of Computational and Theoretical Nanoscience (CTN), 12(4),

[35] I. S. Amiri, A. Nikoukar, A. Shahidinejad, J. Ali & P. Yupapin, (2012), "Generation of discrete frequency and wavelength for secured computer networks system using integrated ring resonators", in Computer and Communication Engineering (ICCCE) Conference, Malaysia, 775-778.

[36] I. S. Amiri & J. Ali, (2014) "Picosecond Soliton pulse Generation Using a PANDA System for Solar Cells Fabrication", Journal of Computational and Theoretical Nanoscience (CTN), 11(3), 693-701.

[37] I. S. Amiri, A. Nikoukar & J. Ali, (2013) "GHz Frequency Band Soliton Generation Using Integrated Ring Resonator for WiMAX Optical Communication", Optical and Quantum Electronics, 46(9), 1165-1177.

[38] I. S. Amiri, M. Ebrahimi, A. H. Yazdavar, S. Gorbani, S. E. Alavi, Sevia M. Idrus & J. Ali, (2014) "Transmission of data with orthogonal frequency division multiplexing technique for communication networks using GHz frequency band soliton carrier", IET Communications, 8(8), 1364 – 1373.

[39] I. S. Amiri, K. Raman, A. Afroozeh, M. A. Jalil, I. N. Nawi, J. Ali & P. P. Yupapin, (2011) "Generation of DSA for security application", Procedia Engineering, 8 360-365.

[40] I. S. Amiri, P. Naraei & J. Ali, (2014) "Review and Theory of Optical Soliton Generation Used to Improve the Security and High Capacity of MRR and NRR Passive Systems", Journal of Computational and Theoretical Nanoscience (CTN), 11(9), 1875-1886.

[41] I. Sadegh Amiri, M. Nikmaram, A. Shahidinejad & J. Ali, (2013) "Generation of potential wells used for quantum codes transmission via a TDMA network communication system", Security and Communication Networks, 6(11), 1301-1309.

[42] I. S. Amiri, A. Afroozeh & M. Bahadoran, (2011) "Simulation and Analysis of Multisoliton Generation Using a PANDA Ring Resonator System", Chinese Physics Letters, 28(10), 104205.

[43] Y. S. Neo, S. M. Idrus, M. F. Rahmat, S. E. Alavi & I. S. Amiri', (2014) "Adaptive Control for Laser Transmitter Feedforward Linearization System", IEEE Photonics Journal 6(4),

[44] I. S. Amiri, S. E. Alavi, M. Bahadoran, A. Afroozeh & H. Ahmad, (2015) "Nanometer Bandwidth Soliton Generation and Experimental Transmission within Nonlinear Fiber Optics Using an Add-Drop Filter System", Journal of Computational and Theoretical Nanoscience (CTN), 12(2),

[45] S. E. Alavi, I. S. Amiri, M. Khalily, A. S. M. Supa' at, N. Fisal, H. Ahmad & S. M. Idrus, (2014) "W-Band OFDM for Radio-over-Fibre Direct-Detection Link Enabled By Frequency Nonupling Optical Up-Conversion", IEEE Photonics Journal 6(6),

[46] I. S. Amiri, R. Ahsan, A. Shahidinejad, J. Ali & P. P. Yupapin, (2012) "Characterisation of bifurcation and chaos in silicon microring resonator", IET Communications, 6(16), 2671-2675.

[47] S. E. Alavi, I.S.Amiri, A. S. M. Supa'at & S. M. Idrus, (2015) "Indoor Data Transmission Over Ubiquitous Infrastructure of Powerline Cables and LED Lighting", Journal of Computational and Theoretical Nanoscience (CTN), 12(4).

[48] I. S. Amiri & A. Afroozeh, Spatial and Temporal Soliton Pulse Generation By Transmission of Chaotic Signals Using Fiber Optic Link in Advances in Laser and Optics Research. vol. 11, ed New York: Nova Science Publisher, 2014.

20

Ontology based fuzzy query execution

Geetanjali tyagi, kumar kaushik, Arnika Jain, Manish Bhardwaj

Department of Computer science and Engineering, SRM University, NCR Campus, Modinagar, Ghaziabad, India

Email address:

git101288@gmail.com (G. Tyagi), kumarkaushik26@gmail.com (K. Kaushik) , jain.arnika2009@gmail.com (A. Jain), aapkaapna13@gmail.com (M. Bhardwaj)

Abstract: Database engineering has been progressed up to the Relational database stage. Fuzzy information administration in databases is a complex process in view of adaptable information nature and heterogeneous database frameworks. Relational Database Management System (RDBMS) can just handle fresh information but cannot handle precise data information. Structured Query Language (SQL) is a very powerful tool but can handle data which is crisp and precise in nature. It is not able to fulfill the requirements for information which is indeterminate, uncertain, inapplicable and imprecise and vague in nature. The goal of this work is to use Fuzzy technique in RDBMS. But, Fuzzy Relational Database Management System (FRDB) requires complex data structures, in most cases, are dependent on the platform in which they are implemented. A solution that involves representing an FRDB using an Ontology as an interface has been defined to overcome this problem. A new Fuzzy Query Ontology is proposed in this dissertation with implementation. The implementation layer, which is responsible for parsing and translating user requests into the corresponding DB implementations in transparent, is required to establish communication between the Ontology and the relational databases management system (RDBMS). This ontology defines a framework for storing fuzzy data by defining those using classes, slots, and instances. An Ontology is an explicit and formal specification of a conceptualization. Ontologies provides a shared understanding of a domain which allows interoperability between semantics.

Keywords: Fuzzy, Ontology, Fuzzy Data, Metadata, DBMS Catalog, Fuzzy Knowledge Representation, Ontology, Databases, Database Modeling

1. Introduction

Relational Database Management System (RDBMS) is used to store crisp data and SQL Language is used to intact with database. A Relational Database Management System (RDBMS) extension for representing fuzzy data is obviously not a new problem. Relational Database can store only crisp, precise data, but can able to store fuzzy values. Complexity normally arises from uncertainty in the form of ambiguity. The computerized system is not capable of addressing uncertain, imprecise data and ambiguous issues whereas, the human have the capacity to reason "approximately". As a result, human, when interacting with the database, want to make complex queries that have a lot of vagueness present in it. So, database which is in use cannot store uncertain data. But in real situations, these are not crisp and deterministic and therefore, cannot be described precisely. The conventional database management system does not handle imprecise, incomplete or vague information such as very high, approximately some values. To triumph over this problem, the

fuzzy database system has been introduced. And SQL Language works with conventional database, so in order to interact with fuzzy database we need to extend SQL into language. This paper is organized as follows: Fuzzy Relational Database is presented in section 2. Section 3 The GEFRED Model. Section 4 ONTOLOGY. Section 5 the implementation detail of criteria that follow.

2. Fuzzy Relational Database

2.1. Fuzzy Relational Database Model

The basic model of fuzzy relational databases is considered the simplest one, and it consists of adding a grade, normally in the [0, 1] interval, to each instance (or tuple). This makes keeping database data homogeneity possible. Nevertheless, the semantic assigned to this grade will determine its usefulness, and this meaning will be utilized in the query processes. This grade may have the meaning of membership degree of each tuple to the relation (Giardina,

1979; Mouaddib, 1994). But it may mean something different, such as the dependence strength level between two attributes, thus representing the relation between them (Baldwin, 1983), the fulfillment Prade, 1997) of each tuple in the relation, among others.The main problem with these fuzzy models is that they do not allow for the representation of imprecise information about a certain attribute of a specific entity (such as the "tall" or "short" values for a height attribute). Besides, the fuzzy character is assigned globally to each instance (tuple), making it impossible to determine the specific fuzzy contribution from each constituting attribute.

2.2. Fuzzy Logic

Fuzzy Logic is a form of many valued Logic. It deals with reasoning that is approximate rather than fixed and exact. Fuzzy Logic has been extended to handle the concept of partial truth where the truth value may range between completely true and completely false. In fuzzy Logic, membership function represents the degree of truth. For any set, membership function lies between [0,1]. The purpose of introducing fuzzy logic in databases is to enhance the classical models such that uncertain and imprecise information can be represented and manipulated. This resulted in numerous contributions, mainly with respect to the popular relational model or to some related form of it.

2.3. Fuzzy Sets

The original interpretation of fuzzy sets arises from a generalization of the classic concept of a subset extended to embrace the description of "vague" and "imprecise" notions.

Definitions 2.3.1: A fuzzy set A over a universe of discourse X (a finite or infinite interval within which the fuzzy set can take a value) is a set of pairs

$$A = \{\mu A(x) / x : x \in X, \mu A(x) \in [0,1] \in \Re\}. \qquad (2.1)$$

Where $\mu A(x)$ is called the membership degree of the element x to the fuzzy set A. This degree ranges between the extremes 0 and 1 of the dominion the real numbers:

• $\mu A(x) = 0$ indicates that x in no way belongs to the fuzzy set A.

• $\mu A(x) = 1$ indicates that x completely belongs to the fuzzy set A.

Sometimes, instead of giving an exhaustive list of all the pairs that make up the set (discreet values), a definition is given for the function $\mu A(x)$, referring to it as characteristic function or membership

The universe X may be called underlying universe

$$\mu A(x): X \rightarrow [0,1] \qquad (2.2)$$

If the membership function produces only values of the set {0,1}, then the set that it generates is not fuzzy, but "crisp" (specific, exact, or precise).As mentioned previously, the universe of discourse X or the set of values being considered can be of two types:

• Finite or discreet universe of discourse X = {x1, x2,..., xn}, where a fuzzy set A can be represented by:

$$A = \mu 1 / x1 + \mu 2 / x2 + ... + \mu n / xn \qquad (2.3)$$

• Infinite universe of discourse, where a fuzzy set A over X can be represented by:

$$A = \int \mu A(x)$$

2.4. Linguistic Labels

If an attribute is able of fuzzy treatment then linguistic labels can be defined on it. These labels will be preceded with the symbol \$ to distinguish them easily. There are two types of labels and they will be used in different fuzzy attribute types:

(1) Labels for attributes with an ordered underlined fuzzy domain (Fuzzy Attributes Type 1 and 2) and (2) Labels for attributes with a non ordered fuzzy domain (Fuzzy Attributes Type 3 and 4).

Example 2.1: If we express the qualitative concept "young" by means of a fuzzy set, where the x-axis represents the universe of discourse "age" (in natural whole numbers) and the y-axis represents the membership degrees in the interval [0,1], then, following Equation 2.3, the fuzzy set that represents that concept could be expressed in the following way (considering a discreet universe):

Young = 1/0 + ... + 1/25 + 0.9/26 + 0.8/27 + 0.7/28 + 0.6/29 + 0.5/30 + ... + 0.1/34

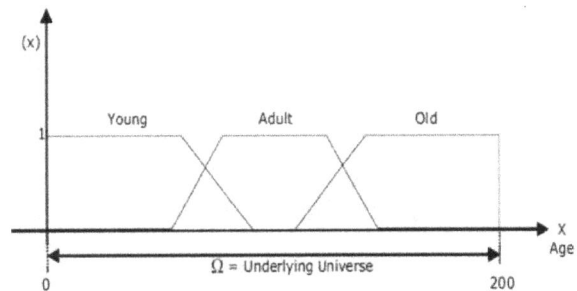

Figure 1. Three linguistic labels of Example 2.1

3. The GEFRED Model

The GEFRED model (GEneralised model Fuzzy heart Relational Database). It is based on the generalized fuzzy domain and generalized fuzzy relation, which include classic domains and classic.It is a possiblistic model, it particularly refers generalized fuzzy domains, but it also includes the case of where the underlying domain is not numeric but scalars of any type. It includes unknown, undefined and null values as well. The GEFRED model is based on the generalized fuzzy domain (D) and generalized fuzzy relation (R), which include classic domains and classic relations, respectively. To model the flexible queries and the concept of the fuzzy attributes, this paper uses an extension of SQL language called SQLf. We present in the following section this language.

3.1.1. Fuzzy Comparators

Besides the typical comparators (=,>, etc.), FSQL includes fuzzy comparators. There are many comparators, the most used are: FEQ (Fuzzy Equal than), (NFEQ: Necessarily FEQ),

FGT (Fuzzy Greater than), (NFGT: Necessarily FGT), FGEQ (Fuzzy Greater or Equal than), (NFGEQ: Necessarily FGEQ), FLT (Fuzzy Less Than), (NFLT: Necessarily FLT), FLEQ (Fuzzy Less or Equal than), (NFLEQ: Necessarily FLEQ), MGT (Much Greater Than), (NMGT: Necessity MGT), MLT (Much Less Than), (NMLT: Necessarily MLT), FINCL (Fuzzy INCLuded in), INCL, FDIF (Fuzzy DIFrent), (NFDIF: Necessary FDIF) [4, 10]. Like in SQL, fuzzy comparators compare one column with one constant or two columns of the same type.

3.1.2. Fuzzy Constants

Besides the typical constants (NULL), FSQL included many constants such as $[a,b,c,d], #n, $label, [n,m], UNKNOWN, UNDEFINED, etc.

3.1.3. Fuzzy Qualifiers

They are of two natures, absolute and relative [10].

3.1.4. Fuzzy Attributes

The classification adopted for the types of attributes is based on the approaches of representation and treatment of the "imprecise" data [9, 10]. These fuzzy attributes may be classified in four data types. This classification is performed taking into account the type of referential or underlying domain. In all of them the values Unknown, Undefined, and Null are included:

These attributes are of four types-

1.) Fuzzy Attributes type I: These attributes are represented as usual attributes because they do not allow fuzzy values. For example, queries of the kind, "Give me employees that earn a lot more than the minimum salary."

2.) Fuzzy Attributes type II: This attribute stores the type of value corresponding to the data that we want to store, indicating its representation. It is an extension of the Type 1 that allows the storage of imprecise information, such as "he is approximately 2 meters tall."

3.) Fuzzy Attributes type 3: In these attributes, some labels are defined that are scalars with a similarity relationship defined over them. For example, the value (1/dark, 0.4/brown), which expresses that a certain person is more likely to be dark than brown-haired but will definitely not be blond or ginger.

*4) Fuzzy Attributes types 4:*They are defined in the same way as Type 3 attributes, without it being necessary for a similarity relationship to exist between the labels (or values) of the domain. A possible example could be the type of role a client plays in a real estate agency, where the degree measures the importance with which a client is seeking or offering a property, without taking into account the similarity between the two roles.

3.1.5. Fuzzy Querying of Fuzzy Databases

Different frameworks for dealing with fuzzy querying of fuzzy databases exist. As mentioned above, Possibillistic Truth Values (PTV's) will be used in this paper. In that case, the evaluation of a criterion will lead to a PTV $\sim t$ (p) which has the advantage over regular satisfaction degrees $s \in [0, 1]$ of also being able to model an unknown satisfaction degree (p) =

$\{(T, 1), (FM),$ or even a partially un-known satisfaction degree (e.g. $i(p) = \{(T,1), (F, 0.5)\}$. So, in case of unknown information in the database (which in a fuzzy databases is a possibility distribution over the domain of the attribute), this can be handled very easily using PTV's. On the other hand, inapplicable information, which in fuzzy databases still requires a special 'null' value, still can't be handled in a natural way, and if nothing else is done, will lead to a satisfaction degree expressed by the PTV $\{(T, 0), (F, 1)\}$ (i.e. False'). Again, this is similar to the two situations above, and is not what is really desired when we want to deliver semantically correct answers to the user so, even when using PTV's, inapplicable information still requires a special approach. Again, it is proposed to use marked 'null' values, but this time only for the handling of inapplicable information because in a fuzzy database there is no need for a `null' value to express unknown information. As in the previous cases, an additional "mark" will be used to indicate to which domain value (including `unknown') the value 'inapplicable' should be treated semantically equivalent in case this value is queried.

4. Ontology

An Ontology is an explicit and formal specification of a conceptualization. Ontologies provides a shared understanding of a domain which allows interoperability between semantics. Components of an ontology:

- Terms
- Relations

The ontology that describes a Fuzzy Database Schema, as defined previously consists of a fuzzy Database schema and the fuzzy data stored in the Database (the tuples). This ontology, however, represent schemas and data simultaneously as ontology class cannot be instantiable twice, therefore two ontologies is defined, one of which describes fuzzy schemas as instances of a Database catalog ontology and the other which describes the same schema as a domain ontology which will allow the data instantiating it to be defined.

4.1. Ontologies vs. Databases

Ontologies allow representing knowledge of any domain in a formal and consensuated way. On the other hand, relational DBs also represent knowledge of any domain but following certain rules specified by ANSI Standard SQL. Nowadays, both technologies coexist together and they can exchange information to take advantage of the information that they represent. Several proposals have been developed for establishing the communication between ontologies and databases. Some of them consist of creating ontologies from a database structures, others populate ontologies using database data, and there are another which represent databases as ontologies.

4.1.1. Fuzzy Query Ontology

An ontology, called from now Fuzzy Query Ontology (FQO), represents all the basic fuzzy relational database query constituents to get a flexible Select clause definition. After the instantiation of this ontology, the query process can start. This Ontology is specially designed for localhost application.

Database queries can be viewed as ontologies from two different perspectives: The first one, a query is a set of descriptions, conditions and rules defined as a set of instances of a Fuzzy Query ontology. In this sense, this query can be viewed as a SQLf statement where any element of the sentence is modeled in the ontology. The second one, a query is described as a a reduced domain ontology where a set of classes, properties and axioms represent a query specification and the ontology instances represent the resulting tuples.

Following fuzzy structures have been added to this ontology to complete the fuzzy RDBMS description:-

* Fuzzy Constraints: These restrictions, which are described in table can only be applied to fuzzy domains and are used either alone or in combination to generate domains such as, for example, not known, undefined, or null values are allowed, or only labels are allowed.
* Fuzzy Domains: These represent a set of values that can be used by one or more attributes. They are defined by a fuzzy data type, one or more fuzzy constraints, and those labels or discrete values that describe this domain.

4.1.2. Why Use Ontology?

The proposed Ontology, whose definition is extended in this paper, provides a frame where fuzzy data are defined in a platform-independent manner.

An implementation layer, which is responsible for parsing and translating user requests into the corresponding DB implementations in transparent is required to establish communication between the Ontology and the relational databases management system(RDBMS).

Fuzzy RDBMS requires complex data structures, in most cases, are dependent on the platform in which they are implemented. FRDB systems are poorly portable and scalable, even when implemented in standard RDBMs. The solution is to use Ontology which makes system independent of platform used. But all the previous work which uses Ontology are used for web application, none of the work is carried out for window application. Here, a new Ontology is proposed for the above problem.

5. Design & Implementation of Fuzzy Query Ontology

5.1. Fuzzy Query Executer Using Ontology

An implementation layer, which is responsible for parsing and translating user requests into the corresponding DB implementations in transparent, is required to establish communication between the Ontology and the relational databases management system (RDBMS). This implementation is designed with respect to corporate.

5.1.1. Parsing Technique Used

A parser generator which generates fully featured ob

ject-oriented frameworks for building compilers, interpreters and other text parsers. In particular, generated frameworks include intuitive strictly-typed abstract syntax trees and tree walkers. SableCC [8] also keeps a clean separation between machine-generated code and user-written code which leads to a shorter development cycle.

An object-oriented framework that generates compilers (and interpreters) in the Java programming language. This framework is based on two fundamental design decisions. Firstly, the framework uses object-oriented techniques to automatically build a strictly-typed abstract syntax tree that matches the grammar of the compiled language and simplifies debugging.

Secondly, the framework generates tree-walker classes using an extended version of the visitor design pattern which enables the implementation of actions on the nodes of the abstract syntax tree using inheritance. These two design decisions lead to a tool that supports a shorter development cycle for constructing compilers

5.1.2. Fuzzy Query Translator

The queries are written in SQLf which is an extention to SQL. Fuzzy Query Executer works as a translator, that translates fuzzy queries to standard, SQL queries, and executes them with RDBMS.

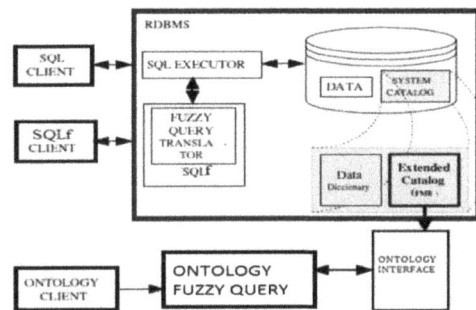

Figure 2. Block Diagram of Fuzzy Query Ontology

SQL Client directly interact with Relational Database Management System through SQL Executer and do not need any translation in between. SQLf is language used in the project tO retrieve data which is fuzzy in nature. SQLf Client need translation because Database can understand only SQL Language. So, Fuzzy Query Translator is used to convert SQLf query into SQL query. Ontology which is used in this project is independent of the platform used. Figure shows ontology is outside of RDBMS. Fuzzy Query Ontology is implemented in this project through various classes and its properties and behavior.A FMB is needed. In relational database,

The FMB will be responsible for organizing all the information related to the inexact nature or context of these attributes. The FMB is contemplated as an extension of the catalogue of the system (Data Dictionary),

FMB is created as shown in Block Diagram figure 2.

FMB contains few tables in database which stores the fuzzy attributes with their parameters to calculate fuzzy degree which handles the fuzzy database. This Block Diagram shows how data is carried in database. An ontology for fuzzy information representation is proposed in this paper. The ontology includes knowledge about how to manage and represent uncertain and imprecise data. Figure 2 shows how the system catalog is related to the ontology modeling it. The Ontology Client module carries out the same operations through the Ontology than the DBMS Clients. The connection between the ontology and the database needs an interface, the Ontology Interface, which establishes the communication and refreshes the data. The fuzzy model representation comprises two well-differentiated parts. Firstly, the ontology must define the necessary classes and slots to represent the metadata. Secondly, the ontology will be able to represent classical or fuzzy information as instances of the relations. In this ontology, metadata establish how the fuzzy information will be stored.

5.2. Verification of Implemented Fuzzy Query Ontology

This project is implemented in context of some workers who are working for a various organization like Academics, government, Industry. The job expertise for organizations are AI, Statistics, Robotics, Expert System. There are various fuzzy attributes like age, salary, expertise and there are various linguistic labels for fuzzy attributes like age there are BABY, YOUNG, MIDDLE, MATURE, OLD and for salary labels are SMALL, MIDDLE, HIGH.

Examples:

1. Select name of worker, location of worker and age of worker from table name workers where fuzzy attribute age is ~young‘;

```
SQL_FQE>select fname,lname, age from workers where ~age is 'young';
+----------+------------+-----+--------------+
| fname    | lname      | age | fuzzy_degree |
+----------+------------+-----+--------------+
| neha     | ghaziabad  |  16 | 0.933333333  |
| mandeep  | meerut     |  23 | 0.466666667  |
| rajesh   | muradnagar |  27 | 0.200000000  |
+----------+------------+-----+--------------+
3 rows in set
OK
SQL_FQE>
```

Figure 3. shows the result of example 1

2. Select name of worker, location of worker and salary of worker from table name workers where fuzzy attribute is salary is ~high‘;

```
SQL_FQE>select fname, lname, salary from workers where ~salary is 'high';
+-----------+------------+--------+--------------+
| fname     | lname      | salary | fuzzy_degree |
+-----------+------------+--------+--------------+
| ravikant  | delhi      |   6500 | 1.000000000  |
| jogi      | delhi      |   6200 | 1.000000000  |
| rajesh    | muradnagar |   3800 | 0.520000000  |
| abhishek  | bihar      |   3000 | 0.200000000  |
+-----------+------------+--------+--------------+
4 rows in set
OK
SQL_FQE>
```

Figure 4. shows the result of example 2

3. Select fuzzy attribute age and function that calculate sum of salary from table called workers group by and order by fuzzy attribute age;

```
SQL_FQE> SELECT ~Age as age, FUZZY_SUM(Salary) FROM workers GROUP BY ~Age ORDER BY ~Age;
+--------+----------------+
| age    | sum            |
+--------+----------------+
| MATURE | 7473.333330100 |
| MIDDLE | 8266.666667800 |
| OLD    | 4333.333335500 |
| YOUNG  | 4026.666666600 |
+--------+----------------+
4 rows in set
OK
SQL_FQE>
```

Figure 5. shows the result of example 3

4. Select similarity label expertise from table name workers in order by Expertise only.

Expertise attribute is of type 3 fuzzy attribute having a similarity relation. This result shows the workers exact expertise and its related expertise with their fuzzy degree. In this

result mandeep and neha are having expertise AI that‘s why fuzzy degree is 1.0000 . rajesh is having expertise in Expert System which is 90% related to AI that is why fuzzy degree is 0.900000. Jogi is having expertise in Robotics which is 60% related to AI that is why 0.60000 is fuzzy degree.

```
Output - Fuzzy_Query_Executor (run)
SQL_FQE>select fname,#expertise as expertise from workers order by #expertise;
+----------+----------------+--------------+
| fname    | expertise      | fuzzy_degree |
+----------+----------------+--------------+
| mandeep  | AI             | 1.000000000  |
| neha     | AI             | 1.000000000  |
| rajesh   | AI             | 0.900000000  |
| jogi     | AI             | 0.600000000  |
| ravikant | AI             | 0.600000000  |
| abhishek | AI             | 0.200000000  |
| rajesh   | Expert Systems | 1.000000000  |
| mandeep  | Expert Systems | 0.900000000  |
| neha     | Expert Systems | 0.900000000  |
| jogi     | Expert Systems | 0.600000000  |
| ravikant | Expert Systems | 0.600000000  |
| abhishek | Expert Systems | 0.200000000  |
| ravikant | Robotics       | 1.000000000  |
| jogi     | Robotics       | 1.000000000  |
| neha     | Robotics       | 0.600000000  |
| rajesh   | Robotics       | 0.600000000  |
| mandeep  | Robotics       | 0.600000000  |
| abhishek | Robotics       | 0.200000000  |
| abhishek | Statistics     | 1.000000000  |
| neha     | Statistics     | 0.200000000  |
| mandeep  | Statistics     | 0.200000000  |
| rajesh   | Statistics     | 0.200000000  |
| ravikant | Statistics     | 0.200000000  |
| jogi     | Statistics     | 0.200000000  |
+----------+----------------+--------------+
24 rows in set
OK
SQL_FQE>
```

Figure 6. shows the result of example 4

5.3. Conclusion and Future Work

The above defined Ontology is specific to data retrieval from database. It is specifically designed for a flexible Select clause which represents all the basic fuzzy relational database query constituents. In this proposal of an ontology which represents a query structure regardless of any RDBMS implementation is defined. This ontology allows generating and executing any query on fuzzy or classical data according to the system where it is executed. The use of ontologies has provided of an independent layer where data can be modeled regardless of any RDBMS implementation particularity. Thus, the client interaction is performed through this ontology and an interface between the ontology and a real RDBMS is in charge of establishing the communication. A new Ontology is defined in this dissertation called Fuzzy Query Ontology.

A Fuzzy Query Ontology design, developed, implemented and verified successfully on given system environment (hardware & software) for data manipulation (data retrieval (select clause)).

References

[1] Rumbaugh, J., Blaha, M., Premerlani, W., Eddy, F. and Lorensen, W. (1991). Object-oriented modeling and design. Englewood Cliffs, New Jersey: Prentice Hall.

[2] Carmen Martinez-Cruz.. "An Ontology to represent Queries in Fuzzy Relational Databases". IEEE 2011

[3] J. Galindo, A. Urrutia, and M. Piattini, "Fuzzy Database Modeling, Design and Implementation". Idea Group Publishing, 2006

[4] "Describing Fuzzy Database DB Schemas as Ontologies: A System Architecture View". Springer 2010. 13 th International Conference, IPMU july 2010.

[5] Sunita M. Mahajan, Vaishali P. Jadhav,"Analysis of Execution Plans in query optimization", International Journal of Scientific & Engineering Research, Volume 3, Issue 2, February-2012 , ISSN 2229-5518.

[6] Protégé, tool for creating and editing ontologies. http://protege.stanford.edu/, 2011.

[7] M. Neunerdt, B. Trevisan, T. C. Teixeira, R. Mathar, and E.- M. Jakobs, "Ontology-based corpus generation for web comment analysis," in ACM conference on Hypertext and hypermedia (HT 2011), (Eindhoven), 05 2011.

[8] Michaela Kreutzová, Jaroslav Porubän, "Automating User Actions on GUI: Defining a GUI Domain-Specific Language". In: CSE 2010: proceedings of International Scientific conference on Computer Science and Engineering: Stará Ľubovňa, Slovakia, 2010 pp. 60-67.

[9] Java Look & feel design guidelines, http://java.sun.com/products/jlf/ed2/book/, 2001

[10] J. Cheng, Z. M. Ma, and L. Yan, "f-SPARQL: a flexible extension of SPARQL," in Proceedings of the 21st international Conference on Database and expert systems applications: Part I, ser. DEXA'10. Berlin, Heidelberg: Springer-Verlag, 2010, pp. 487–494.

21

Security in ad hoc networks

Sanjana Lakkadi, Amit Mishra, Manish Bhardwaj

Department of Computer science and Engineering, SRM University, NCR Campus, Modinagar, Ghaziabad, India

Email address:

sanjanalakkadi@gmail.com (S. Lakkadi), amit_mishra65@outlook.com (A. Mishra), aapkaapna13@gmail.com (M. Bhardwaj)

Abstract: Ad hoc networks are a wireless networking paradigm for mobile hosts. Unlike traditional mobile wireless networks, ad hoc networks do not rely on any fixed infrastructure. Instead, these networks are self-configurable and autonomous systems which are able to support movability and organize themselves arbitrarily. These unique characteristics of ad hoc networks pose a number of challenges for the implementation of security infrastructure in the wireless network system design. In this paper, we study the ad-hoc architecture thus understanding the vulnerabilities and security goals. Further, we discuss the various security attacks and explore approaches to secure the communication.

Keywords: ARPANET, MANET, OSPF, QOS.

1. Introduction

Internet usage has skyrocketed in the last decade, propelled by web and multimedia applications. While the predominant way to access the Internet is still cable or fiber, an increasing number of users now demand mobile, ubiquitous access whether they are at work, at home or on the move. For instance, they want to compare prices on the web while shopping at the local department store, read e-mail while riding a bus or hold a project review while at the local coffee shop or in the airport lounge. The concept of wireless, mobile Internet is not new. When the packet switching technology, the fabric of the Internet, was introduced with the ARPANET in 1969, the Department of Defense immediately understood the potential of a packet switched radio technology to interconnect mobile nodes in the battlefield. The DARPA Packet Radio project helped establish the notion of ad hoc wireless networking. This is a technology that enables untethered, wireless networking in environments where there is no wired or cellular infrastructure (example - battlefield, disaster recovery, etc.); or, if there is an infrastructure, it is not adequate or cost effective. Ad hoc networks may be different from each other, depending on the area of application: For instance, in a computer science classroom an ad hoc network could be formed between students' PDAs and the workstation of the teacher. In another scenario, a group of soldiers are operating in a hostile environment, trying to keep their presence and mission totally unknown from the viewpoint of the enemy. The

soldiers carry wearable communication devices that are able to eavesdrop on the communication between enemy units, shut down hostile devices, divert the hostile traffic arbitrarily or impersonate themselves as the hostile parties. As it can be seen, these two scenarios of ad hoc networking are very different from each other in many ways: In the first scenario the mobile devices need to work only in a safe and friendly environment where the networking conditions are predictable. Thus no special security requirements are needed. On the other hand, in the second and rather extreme scenario the devices operate in an extremely hostile and demanding environment, in which the protection of the communication and the mere availability, access and operation of the network are both very vulnerable without strong protection [2].

The challenge lies exactly in securing the ad hoc network operation, because any malicious or selfish network entity can disrupt, degrade, or even deny communication of other entities. Securing the network operation is paramount for both civilian and tactical applications [1]. Users would have no incentive to embrace new products if, for example, they cannot access their services and get the quality they paid for due to available resources being monopolized by adversarial nodes, or if their privacy is at stake. Similarly, a General or a Police Commissioner would not endorse networking technologies that do not guarantee secure and reliable communications in a battlefield or an emergency situation.

2. Understanding Ad Hoc Network

Ad-hoc network is a collection of nodes that do not rely on a predefined infrastructure. The nodes are often mobile in which case the networks are called as mobile ad hoc networks (MANET).These networks are self-configurable and autonomous systems consisting of routers and hosts, which are able to support mobility and organize themselves arbitrarily. That means the topology of the ad-hoc network changes dynamically and unpredictably. These networks can be formed, merged together or partitioned on the fly with no central administrative server or infrastructure. Thus, it is difficult to distinguish between legal and illegal participants of the network system

The Mobile ad hoc network requires a highly flexible technology for establishing communications in situations which demand a fully decentralized network without any base stations, such as battlefields, military applications, and other emergency and disaster situations.

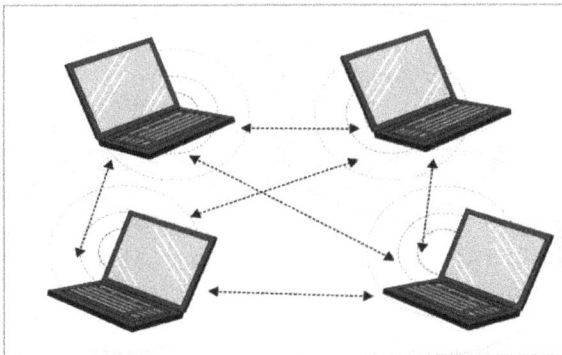

Fig 1. Ad hoc network

Since, all nodes are mobile; the network topology of the MANET is generally dynamic and may vary frequently. Hence, the protocol such as 802.11 to communicate via same frequency require power consumption directly proportional to the distance between hosts and direct single-hop transmissions between two hosts requires significant power that may cause interference. To avoid this problem multi-hop transmissions are used for communication. The router should be able to rank routing information sources from most trustworthy to least trustworthy and accept routing information about any particular destination from the most trustworthy sources first.

A router should provide a mechanism to filter out invalid routes and be careful while distributing routing information provided to them by another party.

Characteristics:
- Distributed Operations: Nodes' functions should be designed in a way so that they can operate efficiently under distributed conditions; supporting security and routing.
- Dynamic network topology: Connectivity of the network must be maintained to allow applications and services to operate undisrupted in a mobile scenario.
- Fluctuating link capacity: Efficient functions for link

layer protection can substantially improve the link quality. Also, the bit-error rates would e high for multi-hop ad hoc networks.
- Low power devices: The nodes may be battery driven which will make the power budget tight for all power-consuming components in a device.

3. Routing Protocols in Ad Hoc

Certain unique combinations of characteristics make routing in ad hoc networks interesting. First, nodes in an ad hoc network are allowed to move in an uncontrolled manner resulting in a highly dynamic network that may cause route failures. A good routing protocol for this environment has to dynamically adapt to the changing network topology. Second, the underlying wireless channel provides much lower and more variable bandwidth than wired networks. The wireless channel working as a shared medium makes available bandwidth per node even lower. So routing protocols should be bandwidth-efficient by expending a minimal overhead for computing routes so that much of the remaining bandwidth is available for the actual data communication. Third, nodes run on batteries which have limited energy supply. In order for nodes to stay and communicate for longer periods, it is desirable that a routing protocol be energy-efficient as well. Thus, routing protocols must meet the conflicting goals of dynamic adaptation and low overhead to deliver good overall performance.

Mobile IP cannot fulfill the requirements for routing in wireless ad hoc networks in which not only the hosts but also the backbone is mobile and multi hop wireless connections composed of many links with varying quality of service (QOS) are allowed. Therefore, more adaptive network layer protocols are required. Proactive or reactive approaches can be followed when designing a routing algorithm for ad hoc networks [5].

A *proactive approach*, often also called a *table-driven approach*, is used by Internet routing algorithms like RIP, OSPF, IS–IS and BGP. In these algorithms, the routers maintain consistent, up-to-date routing information to every other node in the network. Routing tables are updated every time the topology changes. The following are examples of proactive ad hoc routing protocols [8]:
- Destination-sequenced distance vector routing;
- cluster head gateway switch routing;
- Wireless routing.

In *reactive* techniques, also called *on-demand* techniques, topology maintenance, i.e. maintaining up-to-date topology information in every router, is not continuous but is an on-demand effort. When a new packet needs to be delivered and there is not a valid route to carry out this delivery, a new route is discovered. Examples of reactive techniques are:
- flooding;
- Ad hoc on-demand distance vector routing (AODV);
- Dynamic source routing (DSR);
- Temporarily ordered routing;
- Associativity-based routing;

- signal stability routing.

A route may be unnecessarily updated many times before it is used in a proactive approach. On the other hand, the cost of route discovery every time a route is needed may be higher than the cost of maintaining an always up-to-date, consistent view of the network. This depends on the traffic generation and topology change rates. For contemporary wireless ad hoc network applications, reactive techniques such as AODV and DSR are preferred. Let us look at a few protocols in detail:

3.1. Flooding and Gossiping

In *flooding*, each node receiving a packet repeats it by broadcasting unless a maximum number of hops for the packet is reached or the destination of the packet is the node itself. Flooding is a reactive technique and it does not require costly topology maintenance or complex route discovery algorithms. However, it has several deficiencies such as:

- *Implosion* – a situation where duplicated messages are sent to the same node. For example, if node A has *n* neighbors that are also the neighbors of node B, then node B receives *n* copies of the same packet sent by node A.
- The flooding protocol does not take into account the available resources at the nodes or links, i.e. *resource blindness*.

A derivation of flooding is *gossiping*, where nodes do not broadcast but send the incoming packets to a randomly selected neighbor. Once the neighbor node receives the data, it selects another node randomly. Although this approach avoids the implosion problem by just having one copy of a packet at any node, it takes a long time to propagate the message to all nodes.

3.2. Ad Hoc On-Demand Distance Vector Routing (AODV)

AODV is an on-demand ad hoc routing scheme that adapts the distance vector algorithm to run on a network with a mobile backbone. In AODV, every node maintains a routing table where there can be at most one entry for a destination. Each entry has fields like the neighbor node to relay an incoming packet destined to a specific node and the cost of the selected route. AODV differs from the distance vector algorithm by its routing table maintenance mechanism. When a node receives a packet, it first checks its routing table to determine the next hop router for the destination in the packet. If there is an entry for the destination, the packet is forwarded to the next hop router. Otherwise a new route is discovered by broadcasting a route request (RREQ) packet.

A RREQ packet includes the following fields: source address, request id, destination address, source sequence number, destination sequence number and hop count. The source address is the address of the initiator of the route request.

If a node receives a route request that has the same source address and request id fields as those in one of the previous route request packets, it discards the packet. Otherwise it

checks if there is an entry in its routing table for the destination address. If there is, the destination sequence number in the table is compared to the destination sequence number in the route request. If a router has a route for a destination in its routing table, and if it cannot reach the destination through that route, it increments the destination sequence number and sends a route request. Therefore, the destination sequence number indicates the freshness of a route. If a router has an entry for the destination in its table, and the sequence number for the request is smaller than the sequence number for the destination in its table, this means the route known by the router is fresher than the one known by the router that sends the request. In this case the receiver sends a route reply (RREP). The RREP is forwarded back to the source node through the route where the request is received. Again, this routing scheme introduces new security challenges. A malicious node may send RREP messages for every RREQ and make the other nodes forward their packets towards it. It may then sink the incoming packets, forward them to another adversary or gain unauthorized access to their contents.

3.3. Dynamic Source Routing

Another self-forming and self-healing routing protocol for ad hoc networks is dynamic source routing (DSR). It is similar to AODV in that the DSR protocol is also based on 'route discovery' and 'route maintenance' mechanisms and it is a reactive technique. On the other hand, DSR applies source routing instead of relying on the routing tables maintained by the routers. In DSR when a node has a packet and it does not know the route for the destination, it sends out a 'route request' packet. While this packet is being transferred through the network, all the nodes traversed are recorded in the packet header. A node that knows the route to the destination does not forward the packet further, but appends the route to the route information already accumulated in the packet and returns a 'route reply' packet to the source node.

After this, the source node maintains the discovered route in its 'route cache' and delivers the packets to the destination node through the discovered route by using source routing, i.e. the address of each router to visit until reaching the destination is written in the packet header by the source node. If the routing through a previously discovered route fails, a 'route error' message generated by the node that discovers the route failure is sent back to the source node, the failed route is removed from the 'route caches' and a new route discovery procedure is initiated for the destination. DSR also introduces security challenges similar to those in AODV. On the other hand, the source node controls the nodes to be traversed and this can be advantageous for security because unreliable nodes can be avoided by the source node.

4. Security Challenges

Physical security of the network elements forms the basis

for the security architecture. Further, proper key management is crucial for security in networking. The following are the areas are in question:

- Trust models
- Cryptosystems
- Key creation
- Key storage
- Key distribution

Wireless networks are susceptible to several attacks from passive eavesdropping to active impersonation, message replay and message distortion. Active attacks could range from deleting messages to injecting erroneous messages, etc. We need to consider attacks from not only the outside but from within the network as well as these nodes may be in hostile environments with low physical protection.

The following are vulnerabilities due to which security can be breached:

- Vulnerability of channels: In wireless network, messages can be eavesdropped and fake messages can be injected into the network without difficulty of having physical access to network components.
- Absence of Infrastructure: Since ad hoc networks don't work on fixed infrastructure, the classical security solutions based on certification authorities and on-line servers inapplicable.
- Vulnerability of nodes: As nodes do not reside in physically protected places they can be easily captured and fall under the control of the attacker.
- Dynamically changing topology: It is difficult to distinguish whether routing information change is due to topology change or incorrect routing information has been generated by a compromised node.

For high survivability ad hoc networks should have distributed architecture with no central entities as centrality increases vulnerabilities. Dynamic security mechanisms are needed and they should be scalable.

5. Security Goals

- *Availability*: Ensures survivability despite Denial of Service (DOS) attacks. On physical and media access control layer attacker can use jamming techniques to interfere with communication on physical channel. On network layer the attacker can disrupt the routing protocol. On higher layers, the attacker could bring down high level services e.g.: key management service.
- *Confidentiality*: Ensures certain information is never disclosed to unauthorized entities.
- *Integrity*: Message being transmitted is never corrupted.
- *Authentication*: Enables a node to ensure the identity of the peer node it is communicating with. Without which an attacker would impersonate a node, thus gaining unauthorized access to resource and sensitive information and interfering with operation of other nodes.
- *Non-repudiation*: Ensures that the origin of a message cannot deny having sent the message.

- *Non-impersonation*: No one else can pretend to be another authorized member to learn any useful information.
- *Attacks using fabrication*: Generation of false routing messages is termed as fabrication messages. Such attacks are difficult to detect.

6. Security Attacks

There are various types of attacks on ad hoc network which are describing following [10]:

- *Location Disclosure*: Location disclosure is an attack that targets the privacy requirements of an ad hoc network. Through the use of traffic analysis techniques [11], or with simpler probing and monitoring approaches, an attacker is able to discover the location of a node, or even the structure of the entire network.
- *Black Hole*: In a black hole attack a malicious node injects false route replies to the route requests it receives, advertising itself as having the shortest path to a destination [12]. These fake replies can be fabricated to divert network traffic through the malicious node for eavesdropping, or simply to attract all traffic to it in order to perform a denial of service attack by dropping the received packets.
- *Replay*: An attacker that performs a replay attack injects into the network routing traffic that has been captured previously. This attack usually targets the freshness of routes, but can also be used to undermine poorly designed security solutions.
- *Wormhole*: The wormhole attack is one of the most powerful presented here since it involves the cooperation between two malicious nodes that participate in the network [13]. One attacker, e.g. node A, captures routing traffic at one point of the network and tunnels them to another point in the network, to node B, for example, that shares a private communication link with A. Node B then selectively injects tunneled traffic back into the network. The connectivity of the nodes that have established routes over the wormhole link is completely under the control of the two colluding attackers. The solution to the wormhole attack is *packet leashes*.
- *Blackmail*: This attack is relevant against routing protocols that use mechanisms for the identification of malicious nodes and propagate messages that try to blacklist the offender [14]. An attacker may fabricate such reporting messages and try to isolate legitimate nodes from the network. The security property of non-repudiation can prove to be useful in such cases since it binds a node to the messages it generated.
- *Denial of Service*: Denial of service attacks aim at the complete disruption of the routing function and therefore the entire operation of the ad hoc network [15]. Specific instances of denial of service attacks include the *routing table overflow and the sleep deprivation torture*. In a routing table overflow attack the malicious

node floods the network with bogus route creation packets in order to consume the resources of the participating nodes and disrupt the establishment of legitimate routes. The sleep deprivation torture attack aims at the consumption of batteries of a specific node by constantly keeping it engaged in routing decisions.

- *Routing Table Poisoning*: Routing protocols maintain tables that hold information regarding routes of the network. In poisoning attacks the malicious nodes generate and send fabricated signaling traffic, or modify legitimate messages from other nodes, in order to create false entries in the tables of the participating nodes [15]. For example, an attacker can send routing updates that do not correspond to actual changes in the topology of the ad hoc network. Routing table poisoning attacks can result in the selection of non-optimal routes, the creation of routing loops, bottlenecks, and even portioning certain parts of the network.

- *Rushing Attack*: Rushing attack is that results in denial-of-service when used against all previous on-demand ad hoc network routing protocols [16]. For example, DSR, AODV, and secure protocols based on them, such as Ariadne, ARAN, and SAODV, are unable to discover routes longer than two hops when subject to this attack. develop *Rushing Attack Prevention (RAP)*, a generic defense against the rushing attack for on-demand protocols that can be applied to any existing on-demand routing protocol to allow that protocol to resist the rushing attack.

- *Breaking the neighbor relationship*: An intelligent filter is placed by an intruder on a communication link between two ISs(Information system) could modify or change information in the routing updates or even intercept traffic belonging to any data session.

- *Masquerading*: During the neighbor acquisition process, an outside intruder could masquerade an nonexistent or existing IS by attaching itself to communication link and illegally joining in the routing protocol domain by compromising authentication system The threat of masquerading is almost the same as that of a compromised IS.

- *Passive Listening and traffic analysis*: The intruder could passively gather exposed routing information. Such an attack cannot effect the operation of routing protocol, but it is a breach of user trust to routing the protocol. Thus, sensitive routing information should be protected.

- However, the confidentiality of user data is not the responsibility of routing protocol.

7. Exploring the Solutions

Attack *prevention* measures, such as authentication and encryption, can be used as the first line of defense to reduce the possibilities of attacks. Most of the security research efforts in MANET to date, e.g., [33] [27] [29] [30] [31] [32] [28], are on attack prevention techniques. For example,

(session) shared secret key schemes can be used to encrypt messages to ensure the confidentiality, and to some degree the authenticity (group membership), of routing information and data packets; more elaborate public key schemes can be employed to sign and encrypt messages to ensure the authenticity (of individual nodes), confidentiality, and non-repudiation of the communications between mobile nodes. The prevention schemes proposed so far differ in several ways, depending on their assumptions on the intended MANET applications.

7.1. Key and Trust Management: Preventing External Attacks [1]

Encryption, authentication, and key management are widely used to prevent external (outsider) attacks. They however face many challenges in ad-hoc networks. First, we must deal with the dynamic topologies, both in communications and in trust relationship; the assessment of whether to trust a wireless node may change over time. Second, we must deal with the lack of infrastructure support in MANET; any centralized scheme may face difficulties in deployment.

Key management consists of various services, of which each is vital for the security of the networking systems. The services must provide solutions to be able to answer the following questions:

Trust model: It must be determined how much different elements in the network can trust each other. The environment and area of application of the network greatly affects the required trust model. Consequently, the trust relationships between network elements affects the way the key management system is constructed in network.

Cryptosystems: Available for the key management: in some cases only public- or symmetric key mechanisms can be applied, while in other contexts *Elliptic Curve Cryptosystems (ECC)* are available. While public-key cryptography offers more convenience (e.g. by well-known digital signature schemes), public-key cryptosystems are significantly slower than their secret-key counterparts when similar level of security is needed. On the contrary, secret-key systems offer less functionality and suffer more from problems in e.g. key distribution. ECC cryptosystems are a newer field of cryptography in terms of implementations, but they are already in use widely, for instance in smart card systems.

Key creation: it must be determined which parties are allowed to generate keys to themselves or other parties and what kind of keys.

Key storage: In ad-hoc networks there may not be a centralized storage for keys. Neither there may be replicated storage available for fault tolerance. In ad-hoc networks any network element may have to store its own key and possibly keys of other elements as well. Moreover, in some proposals such as in [25], *shared secrets* are applied to distribute the parts of keys to several nodes. In such systems the compromising of a single node does not yet compromise the secret keys.

Key distribution: The key management service must ensure that the generated keys are securely distributed to their owners. Any key that must be kept secret has to be distributed so that confidentiality, authenticity and integrity are not violated. For instance whenever symmetric keys are applied, both or all of the parties involved must receive the key securely. In public-key cryptography the key distribution mechanism must guarantee that private keys are delivered only to authorized parties. The distribution of public keys need not preserve confidentiality, but the integrity and authenticity of the keys must still be ensured.

7.2. Secure Routing Protocols: Preventing Internal Attacks

To create a secure route to transport data, a proper routing protocol in Ad-Hoc networks must create a route accurately and maintain it. It means that it doesn't let the hostile nodes prevent accurate building and maintaining of the route. In general, if, in a protocol, the points such as routing signals don't counterfeit, the manipulated signals can't be injected into the network, routing messages don't change during transporting except protocol routines, routing loops don't create during aggressive activities, the shortest routes don't change by hostile nodes and so on are considered, it can be called a secure protocol [17]. To observe these points, we begin to review several protocols as far as possible.

7.3. DSR (Dynamic Source Routing)

In this protocol, the source node produces a package called RREQ in which it is determined source and target node. It sends these packages through flooding [34]. By receiving a RREQ package of each node, if it doesn't know about target route, then, it add its name to the package list and broadcast it. So, as the package reach to the target, a package includes data of route nodes and its arrangements will be available for the target node. The target node creates RREP and returns it back via available list in RREQ package header. The middle nodes know the target and do it according to the available list. So, the package traverses the route inversely to reach the source node. Although, it is a good method and certainly applicable but increases the network load and uses high band width which resulted in transporting large headers in the network. Increasing rate of header volumes resulted in increasing distance between links this approach may not work properly. OLSR works in a totally distributed manner, e.g. the MPR approach does not require the use of centralized resources. The OLSR protocol specification does not include any actual suggestions for the preferred security architecture to be applied with the protocol. The protocol is, however, adaptable to protocols such as the *Internet MANET Encapsulation Protocol* (*IMEP*), as it has been designed to work totally independently of other protocols. source and target nodes. This volume increase is due to the name of network middle elements name in the package header. Then, data sender can put the target route in the sent data header to inform middle nodes through this route that to whom they send the package. When a node can't deliver data package to

the next one, it produces a package called RERR (Route Error) and returns it back to the route. So, RERR receiving nodes acknowledges about these two nodes disconnection and routing operation will be started again

7.4. AODV (Advanced On-demand Distance Vector)

In contrast to DSR protocol, this protocol doesn't put the route in the package header. But, each node controls it while receiving PREQ according to tables it had before. If the route has the final node it its table, RREP will be sent. Otherwise, it broadcasts RREQ message. Certainly, RREPs can be returned back to RREQ. It is used consecutive number in RREQ messages that a middle node gets inform whether the route is a new one. So, if the number of RREQ consecutive is smaller than route consecutive number, RREP message will be sent bymiddle node.

7.5. SAODV (Secure AODV)

As it is clear from its name, it is provided to create more security in AODV [22]. In this protocol, it is used Hash functions as it is shown in equation (1)

$$h_{n-1} = H(h_n) \qquad (1)$$

In equation (1), H is the function of Hash and h is the related to the hop. In this protocol, it is used hop count to measure the number of hops in which the packages go through. If the hop count becomes more than the amount of Max Count, the package will be ignored. To prevent the changes of hop count amount and make sure about the accuracy of its amount, it is used the noted Hash functions. Due to the equation (1), each node can be sure about its authenticity by receiving a message and controlling equation (1) on it. Number n also indicates the maximum hop that a package can go through

7.6. OLSR

Optimized Link State Routing protocol (*OLSR*) [1, 24], is a proactive and table driven protocol that applies a multi-tiered approach with *multi-point relays* (MPR). MPRs allow the network to apply scoped flooding, instead of full node-to-node flooding, with which the amount of exchanged control data can substantially be minimized. This is achieved by propagating the link state information about only the chosen MPR nodes. Since the MPR approach is most suitable for large and dense ad hoc networks, in which the traffic is random and sporadic, also the OLSR protocol as such works best in these kind of environments. The MPRs are chosen so that only nodes with one-hop symmetric (bi-directional) link to another node can provide the services. Thus in very dynamic networks where there exists constantly a substantial amount of uni-directional.

5. Conclusion

We have shown that the nature of ad hoc networks has intrinsic vulnerabilities which cannot be removed. Evidently, various attacks that exploit these vulnerabilities have been devised and studied. New attacks will no doubt emerge in the future, especially when ad hoc networking becomes widely used. Defense against these attacks can be achieved by key management or secure routing protocols. This is an important and still largely an open research area with many open questions and opportunities for technical advances.

References

[1] Security in Ad Hoc Networks, Vesa Kärpijoki, Helsinki University of Technology, Telecommunications Software and Multimedia Laboratory, Vesa.Karpijoki@hut.fi

[2] Secure Ad Hoc Networking, Panagiotis Papadimitratos, Virginia Polytechnic, Institute and State University, papadp@vt.edu

[3] http://itlaw.wikia.com/wiki/Ad-hoc_mode

[4] Security for Ad Hoc Networks, Hang Zhao.

[5] Data Communication & Networking, Forouzan.

[6] D. M. Blough et al. On the Symmetric Range Assignment Problem in Wireless Ad Hoc Networks. In Proceedings of IFIP Conference on Theoretical Computer Science, pages 71–82, 2002.

[7] Marina and S. R. Das. Routing Performance in the Presence of Unidirectional Links in Multihop Wireless Networks. In Proceedings of ACM MobiHoc, pages 12–23, 2002.

[8] (Haas and Liang, 1999; Royer and Toh, 1999)

[9] Securing Ad Hoc Networks, Lidong Zhou, Department of Computer Science, Zygmunt J. Haas, School of Electrical Engineering

[10] Karan Singh, R. S. Yadav, Ranvijay, International Journal of Computer Science and Security, Volume (1): Issue (1) 52

[11] A REVIEW PAPER ON AD HOC NETWORK SECURITY

[12] K. Balakrishnan, J. Deng, and P.K. Varshney, "TWOACK: Preventing Selfishness in Mobile Ad Hoc Networks" Proc. IEEE Wireless Comm. and Networking Conf. (WCNC '05), Mar. 2005

[13] Mohammad Al-Shurman and Seong-Moo Yoo, Seungjin Park, "Black Hole Attack in Mobile Ad Hoc Networks" ACMSE'04, April 2-3, 2004, Huntsville, AL, USA.

[14] Yih-Chun Hu, Adrian Perrig, and David B. Johnson., "Packet Leashes A Defense against Wormhole Attacks in Wireless Ad Hoc Networks" In Proceedings of the Twenty-Second Annual Joint Conference of the IEEE Computer and Communications Societies (INFOCOM 2003), April 2003.

[15] Patroklos g. Argyroudis and donal o'mahony, "Secure Routing for Mobile Ad hoc Networks", IEEE Communications Surveys & Tutorials Third Quarter 2005

[16] I. Aad, J.-P. Hubaux, and E-W. Knightly, "Denial of Service Resilience in Ad Hoc Networks," Proc. MobiCom, 2004

[17] Yih-Chun Hu, Adrian Perrig, David B. Johnson, "Rushing Attacks and Defense in Wireless Ad Hoc Network Routing

Protocols" WiSe 2003, September 19, 2003, San Diego, California, USA.

[18] S. Prakash, J.P. Saini, S.C. Gupta, "Methodologies and Applications of Wireless Mobile Ad-hoc Networks Routing Protocols", International Journal of Applied Information Systems, Vol. 1, No. 6, pp. 5-15, February 2012.

[19] D. Johnson, D. Maltz, Y. Hu, "The Dynamic Source Routing Protocol for Mobile Ad Hoc Networks (DSR)", IETF Internet-Draft, 2011.

[20] N.S.M. Usop, A. Abdullah, "Performance Evaluation of AODV, DSDV & DSR Routing Protocol in Grid Environment", IJCSNS International Journal of Computer Science and Network Security, Vol. 9, No.7, pp.261-268, July 2009.

[21] A. Akbari, M. Soruri, A. Khosrozadeh, "A New AODV Routing Protocol in Mobile Adhoc Networks", World Applied Sciences Journal, Vol. 19, No. 4, pp. 478-485, 2012.

[22] D. Benetti, M. Merro, L.Viganò, "Model Checking Ad Hoc Network Routing Protocols: ARAN vs. endairA", IEEE 8th International Conference on Software Engineering and Formal Methods (SEFM), Pisa, pp. 191-202, Sep 2010.

[23] Smith, S. Murthy, J.J. Garcia-Luna-Aceves, "Securing Distance Vector Routing Protocols", in Internet Society Symposium on Network and Distributed System Security, the 7th International Workshop on Security Protocols, San Diego, CA, USA, pp. 85-92, Feb 1997.

[24] Y.C. Hu, D.B. Johnson, A. Perrig, "Secure efficient distance vector routing in mobile wireless ad hoc networks", in Proceedings of the 4th IEEE Workshop on Mobile Computing Systems and

[25] Applications(WMCSA'02) , pp. 3-13, June 2002.

[26] Jacquet, P. et al. Optimized Link-State Routing Protocol (OLSR). IETF draft, 18 July 2000. [referred 25.9.2000] <http://www.ietf.org/internet-drafts/draft-ietf-manet-olsr-02.txt> [in ASCII format]

[27] Zhou, L. and Haas, Z. Securing Ad Hoc Networks. 1999. [referred 25.9.2000]<http://www.ee.cornell.edu/~haas/Publications/netw ork99.ps > [in PostScript format]

[28] S. Basagni, K. Herrin, D. Bruschi, and E. Rosti. Secure pebblenets. In Proceedings of the 2001 ACM International Symposium on Mobile Ad Hoc Networking and Computing (MobiHoc 2001), Long Beach, CA, October 2001.

[29] J. Binkley and W. Trost. Authenticated ad hoc routing at the link layer for mobile systems. Wireless Networks, 7(2): 139-145, 2001.

[30] Y. Hu, A. Perrig, and D. Johnson. Ariadne: A secure on-demand routing protocol for ad hoc networks. In Proceedings of ACM MOBICOM'02, 2002.

[31] S. Jacobs and M. S. Corson. MANET authentication architecture. Internet draftdraft-jacobs-imep-auth-arch-01.txt, expired 2000, February 1999.

[32] P. Papadimitratos and Z. J. Hass. Secure routing for mobile ad hoc networks. In Proceedings of SCS Communication Networks and Distributed Systems Modeling and Simulation Conference (CNDS), San Antonio, TX, January 2002.

[33] A. Perrig, R. Canetti, J.D. Tygar, and D. Song. Spins: Security protocols for sensor networks. In Proceedings of the Seventh Annual ACM International Conference on Mobile Computing and Networks (MobiCom 2001), Rome, Italy, July 2001.

[34] B. R. Smith, S. Murthy, and J.J. Garcia-Luna-Aceves. Securing distancevector routing protocols. In Proceedings of Internet Society Symposium on Network and Distributed System Security, pages 85–92, San Diego, California, February 1997

[35] M. Zapata and N. Asokan. Securing ad hoc routing protocols. In Proceedings of the ACM Workshop on Wireless Security (WiSe 2002), Atlanta, GA, September 2002.

22

Forward-Secure Identity-Based Shorter Blind Signature from Lattices

Yanhua Zhang[*], **Yupu Hu**

State Key Laboratory of Integrated Service Networks, Xidian University, Xi'an, China

Email address:

yhzhangxidian@163.com (Yanhua Zhang), yphu@mail.xidian.edu.cn (Yupu Hu)

[*]Corresponding author

Abstract: Blind signature (BS) plays one of key ingredients in electronic cash or electronic voting system. However, the key exposures bring out very serious problems in insecure mobile devices. Forward-secure blind signatures preserve the validity of past signatures and prevent a forger from forging past signatures even if current secret key has been compromised. In this paper, we propose the first forward-secure identity-based shorter blind signature scheme from lattices which can resist quantum attack, and prove that our scheme satisfies the security requirements of blindness, unforgeability, and forward secrecy in the random oracle model. Furthermore, we also extend our construction to a forward-secure identity-based shorter blind signature in the standard model.

Keywords: Forward-Secure, Blind Signature, Unforgeability, Lattice, Random Oracle Model

1. Introduction

Blind signature (BS) was first introduced by Chaum [1] to protect the privacy of an individual. In a BS scheme, a signer is requested to sign on a blinded message from a requester. It means the signer knowing nothing about the original message, the requester can unblind the signature to obtain a signature on the original message from the signer. Even the signature on the original message is opened later, the signer cannot link it with the actual signing process. Thus due to the property of strong blindness and untraceability, BS has several significant applications in areas such as electronic cash systems [2] and electronic voting systems [3], etc.

Identity-based signature (IBS) was first introduced by Shamir [4] to reduce the complexity for managing the public key infrastructure. In an IBS scheme, the identity of a signer is regarded as the public key. Then the key generation center (PKC) generates the secret key corresponding to that identity information.

As far as we know, the security of most BS depends on the assumption that the signing secret keys are absolutely secure. However, the key exposures seem more likely to occur in the real life such as the explosive use of mobile, the unprotected devices. Once the signing secret keys are exposed, no matter the past or the future blind signatures will be compromised. Moreover the key exposures bring out more serious problems in electronic systems in which money is directly involved.

One of the most promising solutions to resolve the key exposure problems is forward-secure. And the concept of non interactive forward-secure was first introduced by Anderson [5] and further formalized by Bellare and Miner [6]. In the model of [6], the whole lifetime of the system is divided into N time periods labeled $0, 1, \cdots, N-1$, and a different secret key is used in each time period while the public key remains the same. The initial signing secret key is SK_0 and at the end of time period i, a new signing secret key SK_{i+1} is computed for the next time period $i+1$ using the signing secret key SK_i and then SK_i is deleted.

Forward-secure blind signature (FSBS) was introduced by Duc et al. [7] to preserve the validity of past blind signatures and prevent a forger from forging past blind signatures even if current signing secret key has been compromised. A large number of FSBS schemes [8–12] have been proposed so far based on large integer factoring or discrete logarithms.

However, recent studies show that cryptographic schemes based on large integer factoring and discrete logarithms have been unable to resist quantum attacks. And as one of the most promising candidates for post-quantum cryptography, lattice cryptography has attracted significant interest, due to several potential benefits: asymptotic efficiency, worst-case hardness assumptions, security against quantum computers. To design secure and efficient lattice-based cryptographic constructions are interesting and challenging. With the first signature and identity-based encryption schemes over lattice proved to be secure proposed by Gentry et al. [13], lattice cryptography enters into a rapid development period and large numbers of schemes are constructed, such as public key encryption (PKE) schemes [14–20], identity-based encryption (IBE) schemes [13, 21–27], fully homomorphic encryption schemes [28–32], signature schemes [33–36] and signature schemes with some particular characteristics [37–44].

In this paper, we propose the first forward-secure identity based shorter blind signature (FSIBBS) over lattice. Then, we prove that our construction satisfies security requirements of blindness, forward secrecy, and unforgeability in the random oracle model. Furthermore, we extend the above construction to be a FSIBBS scheme in the standard model

2. Preliminaries

2.1. Notation

In this paper, the set of real numbers (integers) is denoted by $\mathbb{R}(\mathbb{Z})$. The function log denotes natural logarithm. Vectors are in column form and denoted by the bold lower-case letter (e.g., x). The i-th component of x will be denoted by x_i. We view a matrix as the set of its column vectors and denoted by the bold capital letter (e.g., X). The Euclidean norm of x is denoted as $\|x\|$, and define the norm of X as the norm of its longest column (i.e., $\|X\| = \max_i \|x_i\|$). The security parameter throughout this paper is n, and all other quantities are implicit function of n. Let $poly(n)$ denote an unspecified function $f(n) = O(n^c)$ for any $c>0$. We use standard notation O, ω to classify the growth of functions. If $f(n) = O(g(n) \cdot \log^c n)$, we denote it as $f(n) = \tilde{O}(g(n))$. And we use $negl(n)$ to denote a negligible function $f(n) = O(n^{-c})$ for $c>0$, and a probability is overwhelming if it is $1-negl(n)$.

2.2. Lattices

Let $B = \{b_1, b_2, \cdots, b_m\} \in \mathbb{R}^{m \times m}$ be an $m \times m$ matrix with a list of linearly independent vectors $b_1, b_2, \cdots, b_m \in \mathbb{R}^m$. The lattice Λ generated by B is as follows:

$$\Lambda = \mathcal{L}(B) = \{y \in \mathbb{R}^m, s.t. \exists s \in \mathbb{Z}^m, y = \sum_{i=1}^{m} s_i b_i\} \quad (1)$$

Here, we focus on integer lattices, i.e., $\mathcal{L}$ is contained in $\mathbb{Z}^m$.

Definition 1 For a prime q, matrix $A \in \mathbb{Z}_q^{n \times m}$ and a vector u in $\mathbb{Z}_q^n$, define:

$$\Lambda_q(A) = \{e \in \mathbb{Z}^m \ s.t. \ \exists s \in \mathbb{Z}_q^n, A^T s = e \bmod q\} \quad (2)$$

$$\Lambda_q^u(A) = \{e \in \mathbb{Z}^m \ s.t. \ Ae = u \bmod q\} \quad (3)$$

$$\Lambda_q^\perp(A) = \{e \in \mathbb{Z}^m \ s.t. \ Ae = 0 \bmod q\} \quad (4)$$

Lemma 1 [45] Let a prime $q \geq 3$ and $m \geq 6n \log q$. There is a probabilistic polynomial-time (PPT) algorithm TrapGen(q,n) that outputs two matrices $A \in \mathbb{Z}_q^{n \times m}$ and $T \in \mathbb{Z}^{m \times m}$ such that A is statistically close to a uniform matrix in $\mathbb{Z}_q^{n \times m}$ and T is a basis for $\Lambda_q^\perp(A)$ satisfying $\|\tilde{T}\| \leq O(\sqrt{n \log q})$ and $\|T\| \leq O(n \log q)$ with all but a negligible probability in n.

2.3. Discrete Gaussian Distributions

For any $s > 0$, define the gaussian function on $\mathbb{R}^m$, centered at c with parameter s:

$$\forall x \in \mathbb{R}^m, \rho_{s,c}(x) = exp(-\pi \|x - c\|^2 / s^2) \quad (5)$$

For any $c \in \mathbb{R}^m$, real $s > 0$, m-dimensional lattice Λ, define the discrete gaussian distribution over Λ as:

$$\forall x \in \mathbb{R}^m, D_{\Lambda,s,c}(x) = \frac{\rho_{s,c}(x)}{\rho_{s,c}(\Lambda)} = \frac{\rho_{s,c}(x)}{\sum_{x \in \Lambda} \rho_{s,c}(x)} \quad (6)$$

The subscripts s and c are taken to be 1 and 0 (respectively) when omitted.

Lemma 2 [13] Assume that the columns of matrix $A \in \mathbb{Z}_q^{n \times m}$ generate $\mathbb{Z}_q^n$, let $\epsilon \in (0, 1/2)$, $s \geq \eta_\epsilon(\Lambda^\perp(A))$. For $e \sim D_{\mathbb{Z}^m, s}$, the distribution of syndrome $u = A \cdot e \bmod q$ is within statistical distance 2ϵ of uniform over $\mathbb{Z}_q^n$. Furthermore, fix $u \in \mathbb{Z}_q^n$ and let vector $t \in \mathbb{Z}^m$ be an arbitrary solution to $At = u \bmod q$. The conditional distribution of $e \sim D_{\mathbb{Z}^m, s}$ given $A \cdot e = u \bmod q$ is $t + D_{\Lambda^\perp, s, -t}$.

Definition 2 [46] For any m-dimensional lattice Λ and real $\epsilon > 0$, the smoothing parameter η_ϵ is the smallest real $s > 0$ such that $\rho_{1/s}(\Lambda^* \setminus \{0\}) \leq \epsilon$.

Lemma 3 [46] Let $q > 2$, $A \in \mathbb{Z}_q^{n \times m}$ and $0 < \epsilon < 1$. Let T be a basis for $\Lambda_q^\perp(A)$, $s \geq \|\tilde{T}\| \cdot \omega(\sqrt{\log m})$. For $c \in \mathbb{R}^m, u \in \mathbb{Z}_q^n$:

1). $\Pr_{x \sim D_{\Lambda,s,c}}[\|x - c\| > s\sqrt{m}] \leq \frac{1+\epsilon}{1-\epsilon} \cdot 2^{-m}$.

2). There exits a PPT algorithm SampleGau(A, T, s, c) that returns $x \in \Lambda_q^\perp(A)$ drawn from a distribution statistically close to $D_{\Lambda_q^\perp(A),s,c}$.

3). There exits a PPT algorithm $\mathrm{Sample\,Pre}(A,T,u,s)$ that returns $x \in \Lambda_q^u(A)$ sampled from a distribution statistically close to $D_{\Lambda_q^u(A),s}$.

2.4. Useful Facts

In this subsection, we recall several useful facts on lattices in literatures.

Lemma 4 [21] On input a matrix $A \in \mathbb{Z}_q^{n \times m}$ and a gaussian parameter $r > \|T\| \cdot \omega(\sqrt{\log n})$. There exits a PPT algorithm $\mathrm{RandBasis}(A,T,r)$ that given a basis T of $\Lambda_q^\perp(A)$, outputs a short basis $T' \in \mathbb{Z}_q^{m \times m}$ for $\Lambda_q^\perp(A)$ such that $\|T'\| \le r\sqrt{m}$, and no any information specific to T is leaked by T'.

Next, we describe the basis delegation technique proposed by Agrawal et al. [23].

Lemma 5 Let $q \ge 3, m > 2n \log q$, $\sigma_R = \sqrt{n \log q}\,\omega(\sqrt{\log m})$. $D_{m \times m}$ denotes distribution on matrices in $\mathbb{Z}^{m \times m}$ and is defined as $(D_{\mathbb{Z}^m,\sigma_R})^m$ conditioned on the matrix being invertible. On input $A \in \mathbb{Z}_q^{n \times m}$, a invertible matrix R sampled from $D_{m \times m}$ or a product of such, a parameter $\sigma > \|\tilde{T}_A\| \cdot \sigma_R \sqrt{m} \cdot \omega(\log^{3/2} m)$. There is a PPT algorithm $\mathrm{BasisDel}(A,R,T_A,\sigma)$ that given a short basis T_A of $\Lambda_q^\perp(A)$, outputs a short basis T_B for $\Lambda_q^\perp(B)$, satisfying $\|\tilde{T}_B\| \le \sigma_R / \omega(\sqrt{\log m})$, where $B = AR^{-1} \in \mathbb{Z}_q^{n \times m}$.

2.5. Hardness Assumption

The small integer solution (SIS) problem was suggested to be hard on average by Ajtai [47] and formally defined by Micciancio and Regev [46].

Definition 3 The SIS problem in Euclidean norm is that given a prime q, a matrix $A \in \mathbb{Z}_q^{n \times m}$ and real β, find a non-zero m-dimensional vector $e \in \mathbb{Z}^m$ such that $Ae = 0 \bmod q, \|e\| \le \beta$. The average-case $\mathrm{SIS}_{q,m,\beta}$ problem is defined similarly, where A is uniformly random.

The problem was shown to be as hard as certain worst-case lattice problems, first by Ajtai [47], then by Micciancio and Regev [46] and Gentry et al. [13].

Lemma 4 [13] For poly-bounded m, $\beta = poly(n)$ and prime $q \ge \beta\omega(\sqrt{n \log n})$, the average-case $\mathrm{SIS}_{q,m,\beta}$ problem is as hard as approximating the shortest independent vector problem, among others, in worst-case to within $\gamma = \beta \cdot \tilde{O}(\sqrt{n})$ factors.

3. Forward-Secure Identity-Based Blind Signature

We now formalize the definition and security requirements of forward-secure identity-based blind signature (FSIBBS) in Refs. [7, 48]. Here, the whole lifetime of the system is divided into N time periods labeled $0, 1, \cdots, N-1$.

3.1. Syntax of FSIBBS

A FSIBBS scheme consists of five algorithms, namely, Setup, Extract, Update, Sign and Verify. All are described as follows:

FSIBBS-Setup: A PPT algorithm takes security parameter n as input, and outputs a master secret key msk and a master public key mpk by the private key generator (PKG).

FSIBBS-Extract: A PPT algorithm takes msk, mpk and an identity $id \in \{0,1\}^*$ as inputs, and outputs the initial secret key $sk_{id,0}$, which is sent to the signer in a secure way.

FSIBBS-Update: A PPT algorithm takes mpk, msk and $sk_{id,i}$ of the signer with identity id at the i-th time period as inputs and outputs a new signing secret key $sk_{id,i+1}$ for the next time period $i+1$, then deletes $sk_{id,i}$.

FSIBBS-Sign: A PPT algorithm takes a secret key $sk_{id,i}$ of a signer with identity id at the i-th time period and a message m as inputs,

a. Blind: Taking a message m which to be signed together with a blinding factor r as inputs, it outputs a blinded message M of m.

b. Sign: Taking the blinded message M as input, it outputs a signature sig' on M.

c. Unblind: Taking the signature sig' on a blinded message M, a blinding factor r as inputs, it outputs the final unblinded signature sig on m.

FSIBBS-Verify: A deterministic algorithm takes mpk, and a signer with identity id, a message m and a signature (i, sig) as inputs and outputs "accept" if (i, sig) is a valid signature or "reject", otherwise.

The correctness is that if sig is a valid signature generated by $\mathrm{FSIBBS\text{-}Sign}(id, m, i)$, then we have

$$\mathrm{FSIBBS\text{-}Verify}(id, m, i, sig) = \text{"accept"}.$$

3.2. Security Requirements for FSIBBS

We give the security requirements for FSIBBS in [7, 49].

Blindness: Let S be a signer or any adversary that controls the signer and U_0, U_1 be two honest users. A FSIBBS scheme is blind if a PPT dishonest S wins the following game with a negligible advantage:

a. The PKG generates the master secret key msk, the master public key mpk, and a initial signing secret key for S, whose identity is id and then S chooses two messages m_0 and m_1.

b. The referee chooses a random bit $b \in \{0,1\}$, and then (m_b, id, mpk), (m_{1-b}, id, mpk) are given to U_0, U_1 respectively.

c. U_0 and U_1 engage with S to get signature on m_b and m_{1-b}, respectively (Note: not necessary in two different time periods since blindness property must be satisfied for all signatures, not just for signatures issued in one time period).

d. U_0 and U_1 output two valid blind signatures (m_b, id, sig_b) (m_{1-b}, id, sig_{1-b}), then these signatures are given to S.

e. Finally, S outputs a guess $\tilde{b}$ for b, and S wins the game if $\tilde{b} = b$.

If the probability that S wins the above game is no better than the probability of guessing a random bit b (probability of $1/2$), S cannot link a signature to its owner. We say that the blindness property is satisfied.

Unforgeability: Let $\mathcal{A}$ be any PPT adversary and $\mathcal{C}$ be a challenger. A FSIBBS scheme is unforgeable if $\mathcal{A}$ wins the game with a negligible advantage:

Setup phase: $\mathcal{C}$ runs algorithm FSIBBS-Setup to generate the master secret key msk, the master public key mpk and then sends mpk to $\mathcal{A}$.

Queries phase: $\mathcal{A}$ is allowed to make poly-bounded queries as follows:

a. Key Exit query: On receiving a query for the initial signing secret key of a signer with identity id. $\mathcal{C}$ returns $sk_{id,0}$ to $\mathcal{A}$.

b. Signing secret key query: On receiving a query of (id, j), where $1 \le j \le N-1$. $\mathcal{C}$ returns $sk_{id,j}$ to $\mathcal{A}$.

c. Signing query: On receiving a query of (id, i, M), where $1 \le i \le N-1$ and M is a blinded message of m. Using $sk_{id,i}$ for the time period i, $\mathcal{C}$ returns sig' to $\mathcal{A}$.

Forgery phase: Finally, the adversary $\mathcal{A}$ outputs a tuple of (id^*, i^*, m^*, sig^*). Adversary $\mathcal{A}$ is considered to be succeed if the following conditions hold:

a. FSIBBS-Verify(id^*, i^*, m^*, sig^*) = "accept".

b. id^* has not been issued as a KeyExt query.

c. (id^*, i^*, M^*) has not been issued as a signing query, here M^* is a blinded message of m^*.

Forward secrecy: In the different cryptographic schemes, forward secrecy owns different meanings depending on the security goals for the schemes. In a blind signature context, forward secrecy means that unforgeability of signatures is valid in previous time periods even if current signing secret key of the signer is compromised.

4. A FSIBBS Scheme from Lattices

4.1. Description of the Scheme

Inspired by the basis delegation technique proposed by Agrawal et al. [23], we construct the first FSIBBS scheme

from lattices. The main steps of our construction are provided as follows:

FSIBBS-Setup: On inputting the security parameter n, a prime $q \ge 3$, $m \ge 6n \log q$, and two collision-resistance hash functions $H_1 : \{0,1\}^* \to \mathbb{Z}_q^{m \times m}$, and $H_2 : \{0,1\}^* \to \mathbb{Z}_q^n$. For each time period, the PKG sets two series of gaussian parameters $\sigma = (\sigma_0, \sigma_1, \cdots, \sigma_{N-1})$, $\delta = (\delta_0, \delta_1, \cdots, \delta_{N-1})$. Then, the PKG does as follows:

1). Using algorithm TrapGen(q,n), the PKG gets a matrix $A \in \mathbb{Z}_q^{n \times m}$ together with a short basis $T \in \mathbb{Z}_q^{m \times m}$ for $\Lambda_q^{\perp}(A)$.

2). The PKG outputs the master public key and the master secret key: $mpk = A \in \mathbb{Z}_q^{n \times m}$, $msk = T \in \mathbb{Z}_q^{m \times m}$.

FSIBBS-Extract: On receiving the identity id of a signer, the PKG generates the initial signing secret key as follows:

1). Let $R_{id,0} = H_1(id,0)$, the PKG gets $A_{id,0} = A \cdot R_{id,0}^{-1}$.

2). Using algorithm BasisDel$(A, R_{id,0}, T, \sigma_0)$, the PKG can generate the initial signing secret key $sk_{id,0} \in \mathbb{Z}_q^{m \times m}$, and then sends it to the signer in a secure way.

FSIBBS-Update: On inputting $(id, i, sk_{id,i-1})$, where i is the current time period, $sk_{id,i-1}$ is the signing secret key associated with previous time period i-1. A signer with the identity id does as follows:

1). Let $R_{id,i-1} = H_1(id, i-1) \cdot H_1(id, i-2) \cdots H_1(id, 0) \in \mathbb{Z}_q^{m \times m}$ and $A_{id,i-1} = A \cdot R_{id,i-1}^{-1} \in \mathbb{Z}_q^{n \times m}$.

2). Let $R_i = H_1(id, i) \in \mathbb{Z}_q^{m \times m}$.

3). Using algorithm BasisDel$(A_{id,i-1}, R_i, sk_{id,i-1}, \sigma_i)$, a singer with an identity id generates a secret key $sk_{id,i} \in \mathbb{Z}_q^{m \times m}$ for the time period i, and then deletes $sk_{id,i-1}$.

FSIBBS-Sign: On inputting an original message m, a user requests a signer with an identity id to make a blind signature. The user interacts with the signer as follows:

1). Once receiving a blind signature request, a signer with an identity id sends current time period i to the requester.

2). Once obtaining current time period i from a signer with identity id, the user randomly chooses $v \leftarrow D_{\mathbb{Z}^m, \sigma_R}, V \leftarrow D_{n \times n}$, where, $\sigma_R = \sqrt{n \log q} \cdot \omega(\sqrt{\log m})$.

3). The user computes $A_{id,i} = A \cdot R_{id,i}^{-1} \in \mathbb{Z}_q^{n \times m}$, where $R_{id,i} = H_1(id,i) \cdot H_1(id, i-1) \cdots H_1(id, 0) \in \mathbb{Z}_q^{m \times m}$.

4). The user computes $u_i = H_2(id, i, m)^{\mathrm{T}} \cdot V + A_{id,i} \cdot v \in \mathbb{Z}_q^n$ as a blinded message of m and sends it to the signer.

5). Using SamplePre$(A_{id,i}, sk_{id,i}, u_i, \delta_i)$, the signer generates a vector $sig'_i \in \mathbb{Z}_q^m$, and sends (id, i, u_i, sig'_i) to the user.

6). Once getting (id, i, u_i, sig'_i), the user can compute $sig_i = (sig'_i - v)V^{-1} \in \mathbb{Z}_q^m$ and outputs the final blind signature $(id, i, m, \boldsymbol{sig}_i)$.

FSIBBS-Verify: On inputting $(id, i, m, \boldsymbol{sig}_i)$, where id is a signer identity, i is an index of time period, m is an original message, and $\boldsymbol{sig}_i$ is the corresponding blind signature. Then, the verifier does as follows:

1). To verify that $\boldsymbol{sig}_i \in \mathbb{Z}_q^m$ is a small but non-zero vector, which means $0 < \|\boldsymbol{sig}_i\| \leq \delta_i \sqrt{m}$.

2). To verify that $A_{id,i} \cdot \boldsymbol{sig}_i = H_2(id, i, m)^{\mathrm{T}}$, where $A_{id,i} = A \cdot R_{id,i}^{-1}$, and $R_{id,i} = H_1(id, i) \cdot H_1(id, i-1) \cdots H_1(id, 0)$.

3). If both the above conditions are satisfied, the verifier outputs "accept"; otherwise "reject".

4.2. Security Analysis of Our Construction

We now analysis the security requirements of correctness, blindness, unforgeability, forward secrecy for the proposed FSIBBS construction in the random oracle model.

Correctness: If both the user and the signer with a identity id interacted with each other honestly, then once getting a signature $(id, i, m, \boldsymbol{sig})$, we have the following equations:

$$A_{id,i} \cdot \boldsymbol{sig} = A_{id,i} \cdot (\boldsymbol{sig}' - \boldsymbol{v}) \cdot V^{-1} = A_{id,i} \cdot \boldsymbol{sig}' \cdot V^{-1} - A_{id,i} \cdot \boldsymbol{v} \cdot V^{-1}$$
$$= \boldsymbol{u}_i \cdot V^{-1} - A_{id,i} \boldsymbol{v} \cdot V^{-1} = H_2(id, i, m)^{\mathrm{T}} + A_{id,i} \boldsymbol{v} \cdot V^{-1} - A_{id,i} \boldsymbol{v} \cdot V^{-1}$$
$$= H_2(id, i, m)^{\mathrm{T}}.$$

So it is clear that the proposed construction is correct.

Theorem 1 The proposed FSIBBS construction satisfies the blindness property.

Proof: We show that the view of an adversarial signer S with an identity id is perfectly from the value of b. From the above game, the PKG generates the master secret key msk, the master public key mpk and then sends mpk to S. S chooses two messages m_0 and m_1.

The referee chooses a random bit $b \in \{0,1\}$. (m_b, id, mpk), (m_{1-b}, id, mpk) are given to honest users U_0, U_1 respectively. U_0 and U_1 interact with S to get signatures on m_b and m_{1-b}. U_0 and U_1 blind m_b and m_{1-b} by

$$\boldsymbol{u}_b = H_2(id, i, m_b)^{\mathrm{T}} \cdot V_b + A_{id,i} \cdot \boldsymbol{v}_b,$$

$$\boldsymbol{u}_{1-b} = H_2(id, j, m_{1-b})^{\mathrm{T}} \cdot V_{1-b} + A_{id,j} \cdot \boldsymbol{v}_{1-b}.$$

Here H_2 is a collision resistance hash function, V_b and V_{1-b} are two random matrices sampled from $D_{n \times n}$, $\boldsymbol{v}_b$ and $\boldsymbol{v}_{1-b}$ are random vectors sampled from $D_{\mathbb{Z}^m, \sigma_R}$. Due to the property of collision resistance hash function H_2, and the random matrix V_b or V_{1-b}, $H_2(\cdot)^{\mathrm{T}} \cdot V_b$ or $H_2(\cdot)^{\mathrm{T}} \cdot V_{1-b}$ is perfect independent from the value of b. From lemma 2, the distribution of $A_{id,j} \boldsymbol{v}_b, A_{id,j} \boldsymbol{v}_{1-b}$ are uniform over $\mathbb{Z}_q^n$. So $\boldsymbol{u}_b$ and $\boldsymbol{u}_{1-b}$ are uniform over $\mathbb{Z}_q^n$ and indistinguishable. Therefore, we have that the distributions of $(id, i, \boldsymbol{u}_b, \boldsymbol{sig}_b)$ and

$(id, i, \boldsymbol{u}_{1-b}, \boldsymbol{sig}_{1-b})$ are perfect independent from the view of S.

From all the above, the distributions of the view of the signer S with the identity id for U_0 and U_1 are equivalent and independent from the value of b. Therefore the proposed FSIBBS construction satisfies the blindness property.

Theorem 2 Under the $\mathrm{SIS}_{q,m,\beta}$ assumption, the proposed FSIBBS construction is unforgeability.

Proof: Assume that there is an adversary $\mathcal{A}$ that can forge a valid signature in the proposed construction with a non negligible advantage ε. Now we construct a challenger $\mathcal{C}$ that simulates an attack environment and uses the adversary $\mathcal{A}$ to create a solution for the $\mathrm{SIS}_{q,m,\beta}$ problem with non-negligible advantage ε'.

$\mathcal{C}$ gets a random instance of the $\mathrm{SIS}_{q,m,\beta}$ problem and is asked to return an admissible solution.

1). $\mathcal{C}$ is supplied with a matrix $B \in \mathbb{Z}_q^{n \times m}$ from the uniform distribution.

2). $\mathcal{C}$ wants to get a vector $e \in \mathbb{Z}^m$ such that: $Be = 0 \bmod q$, $0 < \|e\| \leq \beta$.

First of all, we assume that:

3). For each time period $i = 0, 1, \cdots, N-1$, the adversary $\mathcal{A}$ makes poly-bounded H_1 queries on any identity adaptively.

4). Assume that $\mathcal{A}$ has queried H_1 at time period $j < i$, when it makes an H_1 query on any identity at time period i.

5). Assume that adversary $\mathcal{A}$ has made all relevant H_1 query beforehand, when it makes a signing secret key query for any signer.

The operations performed between $\mathcal{A}$ and $\mathcal{C}$ are as follows:

Setup phase: $\mathcal{C}$ runs algorithm TrapGen(q, n) to generate a matrix $A \in \mathbb{Z}_q^{n \times m}$ together with a basis $T \in \mathbb{Z}_q^{m \times m}$ for $\Lambda_q^{\perp}(A)$. For each time period, $\mathcal{C}$ sets two series of gaussian parameters $\sigma = (\sigma_0, \sigma_1, \cdots, \sigma_{N-1})$, $\delta = (\delta_0, \delta_1, \cdots, \delta_{N-1})$. Then $\mathcal{C}$ publishes the master public key $mpk = A$, and keeps the master secret key $msk = T$ secret.

Queries phase: Firstly, $\mathcal{C}$ randomly guesses $1 \leq i^* \leq N-1$ as the time period when $\mathcal{A}$ forges a valid blind signature, then $\mathcal{C}$ simulates the random oracles H_1, H_2 as follows. Without loss of generality, assume that $\mathcal{A}$ has queried H_2 on every message m for id and i before making a blind signature query. $\mathcal{C}$ maintains five lists in its local storage, namely, l_1, l_2, l_3, l_4 and l_5 list, which are all set to be empty initially.

H_1 queries: For time period $i = 0, 1, \cdots, N-1$, $\mathcal{A}$ can query H_1 on any identity adaptively. For any (id, i) query, $\mathcal{C}$ looks up l_1 list to check if the value of H_1 was previously defined. If it was, the value is returned. Otherwise, $\mathcal{C}$ randomly samples a low norm matrix $R_{id,i} \in \mathbb{Z}_q^{m \times m}$ from $D_{m \times m}$, stores $(id, i, R_{id,i})$ in l_1 list and returns it to $\mathcal{A}$.

KeyExt queries: $\mathcal{A}$ randomly chooses a scalar

$l \in \{1, \cdots, Q\}$, where Q denotes the maximum number of KeyExt queries, $\mathcal{C}$ does as follows:

If id is not the l-th query, $\mathcal{A}$ queries on identity id at the initial time period, $\mathcal{C}$ looks up l_1 list to find its hash value. If it was previously defined, and the value will be returned. $\mathcal{C}$ gets $A_{id,0} = A \cdot H_1(id,0)^{-1}$ and runs $\text{BasisDel}(A, H_1(id,0), T, \sigma_0)$ to generate the initial signing secret key $sk_{id,0}$ and then sends it to $\mathcal{A}$. If it was not defined before, $\mathcal{C}$ randomly samples a low norm matrix $R_{id,0} \in \mathbb{Z}_q^{m \times m}$ from $D_{m \times m}$, and stores $(id,0,R_{id,0})$ in l_1 list and runs algorithm $\text{BasisDel}(A, R_{id,0}, T, \sigma_0)$ to generate $sk_{id,0} \in \mathbb{Z}_q^{m \times m}$, the $\mathcal{C}$ returns it to $\mathcal{A}$ and stores $(id,0,sk_{id,0})$ in l_2 list.

2). If id is the l-th query, $\mathcal{C}$ aborts the simulations

Signing secret key queries: $\mathcal{A}$ queries the secret key on the signer with identity id at time period $i+1$, $\mathcal{C}$ does as follows:

1). If id is not the l-th query, for each $1 \le i \le N-1$, since we have assumed that $\mathcal{A}$ has made H_1 query on (id, j) for $j < i$. For H_1 query on (id, i), $\mathcal{C}$ looks up l_1 to find a corresponding low norm matrix $R_{id,i}$. $\mathcal{C}$ runs $\text{BasisDel}(A_{id,i-1}, R_{id,i}, sk_{id,i-1}, \sigma_i)$ to generate $A_{id,i} = A \cdot H_1(id,0)^{-1} \cdots H_1(id,i-1)^{-1} \cdot R_{id,i}^{-1} \in \mathbb{Z}_q^{m \times m}$, and a signing secret key $sk_{id,i} \in \mathbb{Z}_q^{m \times m}$. Then, $\mathcal{C}$ returns it to $\mathcal{A}$ and stores $(id,i,A_{id,i},sk_{id,i})$ in l_3 list.

1). If id is the l-th query, $\mathcal{C}$ does as follows:

a. If $i \le i^*$, $\mathcal{C}$ randomly samples a low norm matrix $R \in \mathbb{Z}_q^{m \times m}$ from $D_{m \times m}$, and returns (id,i,A,R) to $\mathcal{A}$ and stores it in l_3 list.

b. If $i = i^*+1$, $\mathcal{C}$ runs algorithm $\text{TrapGen}(q,n)$ to generate a matrix $A_{id,i^*+1} \in \mathbb{Z}_q^{m \times m}$ together with a basis $sk_{id,i^*+1} \in \mathbb{Z}_q^{m \times m}$. $\mathcal{C}$ returns it to $\mathcal{A}$ and stores $(id,i^*+1,A_{id,i^*+1},sk_{id,i^*+1})$ in l_3 list.

c. If $i^*+1 \le i \le N-1$, $\mathcal{C}$ does as before in case that id is not the l-th query.

H_2 queries: For any (id,i,m) query, $\mathcal{C}$ looks up l_4 list to check if the value of H_2 was previously defined. If it was, the value is returned. Otherwise, $\mathcal{C}$ looks up l_1 list and l_3 list to get $(id,i,R_{id,i}),(id,i,A_{id,i},sk_{id,i})$. If they are found, $\mathcal{C}$ randomly chooses $sig_i \leftarrow D_{\mathbb{Z}^m,\sigma_R}$, here $\sigma_R = \sqrt{n \log q} \omega(\sqrt{\log m})$, returns $A_{id,i} sig_i$ to $\mathcal{A}$ and stores $(id,i,m,sig_i,A_{id,i} sig_i)$ in l_4 list. If they are not found, $\mathcal{C}$ regenerates and stores them in l_1 and l_3 list respectively as before and carries on the operation mentioned above. According to lemma 2, this is identity to the uniformly random value of $H_2(id,i,m)$ in the real system.

Signing queries: Once receiving a signing query on u, a blinded message of m for a signer with the identity id at the time period i, the corresponding blinded factor is $v_i \in \mathbb{Z}_q^m$

and $V_i \in \mathbb{Z}_q^{m \times n}$. $\mathcal{A}$ and $\mathcal{C}$ do as follows:

1). If id is not the l-th query, $\mathcal{C}$ will look up l_5 list to find its signing value. If it was previously defined, the value will be returned. And if it was not defined before, $\mathcal{C}$ runs algorithm $\text{Sample} \Pr e(A_{id,i}, sk_{id,i}, u, \delta_i)$ to get the signature $sig_i' \in \mathbb{Z}_q^m$. $\mathcal{C}$ returns it to $\mathcal{A}$ and stores (id,i,u,sig_i') in l_5 list. $\mathcal{A}$ unblinds sig_i' to get the finial signature $sig_i = (sig_i' - v_i) \cdot V_i^{-1} \in \mathbb{Z}_q^m$.

2). If id is the l-th query, when $i^* < i \le N-1$, $\mathcal{C}$ generates the signature as above. Otherwise, $\mathcal{C}$ aborts the simulations.

Forgery phase: Finally, the adversary $\mathcal{A}$ outputs a tuple of (id^*, t^*, m^*, sig^*). Adversary $\mathcal{A}$ is considered to be succeed if the following conditions hold:

1). $1 \le t^* < j$.

2). id^* has not been issued as a KeyExt query.

3). (id^*, i^*, u^*) has never been issued as a Signing query, u^* is a blinded message of m^*, and

$$\text{FSIBBS-Verify}(id^*, t^*, m^*, sig^*) = \text{"accept"}.$$

Once $\mathcal{A}$ outputs a valid forgery (id^*, t^*, m^*, sig^*), $\mathcal{C}$ does as follows:

1). To check if id^* is the l-th query and $i^* = t^*$. If any of them does not hold, $\mathcal{C}$ aborts its run, otherwise, the view of $\mathcal{C}$ is perfectly simulated. As we know, sig^* is a forgery signature such that $id^*,(id^*,t^*),(id^*,t^*,u^*)$ are all not equal to the queries to KeyExt query, Signing secret key query, and Signing query, here u^* is a blinded message of m^*.

Before forging a signature, for query to H_2 on (id^*,t^*,m^*), $\mathcal{C}$ stores $(id^*,t^*,m^*,sig_t^*,A_{id^*,i^*} \cdot sig_t^*)$ in l_4 list, here $A_{id^*,i^*} = B$. By the preimage min-entropy property of the hash family, the min-entropy of sig_t^* given $A_{id^*,i^*} sig_t^*$ (the rest of the view of $\mathcal{A}$, which is independent of sig_t^*) is $\omega(\log n)$, so the signature $sig_t^* \ne sig^*$ except with a negligible probability $2^{-\omega(\log n)}$. We also know that $\mathcal{A}$ wins the above game only if sig^* is a valid blind signature on (id^*,i^*,u^*), u^* is a blinded message of m^*. Thus, we have: $0 < \|sig^*\| \le \delta_i \sqrt{m}$ and $A_{id^*,i^*} \cdot sig^* = B \cdot sig^* = $

$H_2(id^*,t^*,m^*) = A_{id^*,i^*} \cdot sig_t^* = B \cdot sig_t^*$. Therefore, we obtain a vector $e = sig^* - sig_t^*$ as a solution to $B \cdot e = 0 \mod q$.

1). If id^* is not the l-th query or $i^* \ne t^*$, $\mathcal{C}$ will abort its run. From both the above, the success probability of $\mathcal{C}$ in solving the $\text{SIS}_{q,m,\beta}$ problem is the same as that of $\mathcal{A}$ in forging a valid blind signature except for a factor of $1/QN$

due to the aborting events. Since the advantage of solving the $SIS_{q,m,\beta}$ problem is negligible, so that the advantage of $\mathcal{A}$ in forging a valid blind signature is negligible which means that the proposed scheme satisfies the unforgeability property.

Theorem 3 The proposed construction is forward-secure.

Proof: Let the time period $j < i$, the user cannot obtain a signature $(id, j, \boldsymbol{u}_j, \boldsymbol{sig}'_j)$ in the time period i. In time period i, the user sends blinded message $\boldsymbol{u}_j = H_2(id, j, m)^{\mathrm{T}} \cdot V + A_{id,j}\boldsymbol{v}$ to the signer. Using $\mathrm{SamplePre}(A_{id,i}, \boldsymbol{sk}_{id,i}, \boldsymbol{u}_j, \delta_i)$, the signer generates a vector $\boldsymbol{sig}'_i \in \mathbb{Z}_q^m$ satisfying $A_{id,i} \cdot \boldsymbol{sig}'_i = \boldsymbol{u}_j$. In the unblind phase, the user only obtains $\boldsymbol{sig}_i = (\boldsymbol{sig}'_i - \boldsymbol{v})V^{-1} \in \mathbb{Z}_q^m$, which cannot satisfies the equation $A_{id,j} \cdot \boldsymbol{sig}' = H_2(id, j, m)^{\mathrm{T}}$. Therefore, the proposed scheme satisfies the forward secrecy.

4.3. Efficiency Comparison

In this subsection, we will compare the sizes of the signing public key, the signing secret key and the signature with three classical IBS schemes from lattices in Refs. [39–40, 48]. The details of the efficiency comparison are described in Table 1. We denote *spk* and *ssk* as the signing public key and signing secret key. Assume that the random number is k, and for consistency, we also assume that the size of time period in the proposed construction is k. Table 1 shows that the sizes of *spk*, *ssk* and the signature in the proposed FSIBBS scheme is the same as that in Ref. [48], and much shorter than that in Refs. [39–40]. Meanwhile, the proposed construction enjoys the blindness property, which can provide perfect anonymity and the forward secrecy property, which can resolve the key exposure problems.

Table 1. *Efficiency comparison.*

Schemes	$\|spk\|$	$\|ssk\|$	Signature size	Blindness	Forward-secure
[39]	$2nm\log q$	$4m^2\log q$	$2m\log q + k$	No	No
[40]	$3nm\log q$	$4m^2\log q$	$(3m+1)\log q$	No	No
[48]	$nm\log q$	$m^2\log q$	$m\log q + k$	No	Yes
Our	$nm\log q$	$m^2\log q$	$m\log q + k$	Yes	Yes

5. A FSIBBS from Lattices in the Standard Model

In this section, we extend our construction to a FSIBBS scheme from lattices in the standard model. Here, the whole lifetime of the system is also divided into N time periods. The main steps of our construction are provided as follows:

FSIBBS-Setup: On inputting the security parameter n, a prime $q \geq 3$, a integer $m \geq 6n\log q$, and a collision-resistance hash function $H : \{0,1\}^* \to \{0,1\}^l$, where $l = poly(n)$. For each time period, the PKG sets two series of gaussian parameters $\sigma = (\sigma_0, \sigma_1, \cdots, \sigma_{N-1})$, and $\delta = (\delta_0, \delta_1, \cdots, \delta_{N-1})$. The PKG does as follows:

1). $2lN$ matrices $\left\{ \boldsymbol{R}_{i,k}^0, \boldsymbol{R}_{i,k}^1 \right\}_{0 \leq i \leq N-1, 1 \leq k \leq l} \in \mathbb{Z}_q^{m \times m}$ are sampled from $D_{m \times m}$ randomly and a nonzero vector $\boldsymbol{u}$ is sampled from $D_{\mathbb{Z}^n, \sigma_R}$, $\sigma_R = \sqrt{n\log q} \cdot \omega(\sqrt{\log m})$ by the PKG. The PKG also draws $l+1$ independent vectors $\boldsymbol{c}_0, \boldsymbol{c}_1, \cdots, \boldsymbol{c}_l \in \mathbb{Z}_q^n$.

2). Using algorithm $\mathrm{TrapGen}(q, n)$, the PKG gets a matrix $A \in \mathbb{Z}_q^{n \times m}$ together with a short basis $\boldsymbol{T} \in \mathbb{Z}_q^{m \times m}$ for $\Lambda_q^\perp(A)$.

3). The PKG outputs the master public key and the master secret key:

$$mpk = \left\{ A, \left\{ \boldsymbol{R}_{i,k}^0, \boldsymbol{R}_{i,k}^1 \right\}_{0 \leq i \leq N-1, 1 \leq k \leq l}, \boldsymbol{u}, \left\{ \boldsymbol{c}_i \right\}_{0 \leq i \leq l} \right\}, msk = \boldsymbol{T}.$$

FSIBBS-Extract: On receiving the identity *id* of a signer, the PKG generates the initial secret key $\boldsymbol{sk}_{id,0}$ as follows:

1). Let $\boldsymbol{h}_0 = H(id, 0) \in \{0,1\}^l$, the PKG computes

$$\boldsymbol{R}_{id,0} = \boldsymbol{R}_{0,l}^{h_{0,l}} \cdot \boldsymbol{R}_{0,l-1}^{h_{0,l-1}} \cdots \boldsymbol{R}_{0,1}^{h_{0,1}} \in \mathbb{Z}_q^{m \times m}, A_{id,0} = A \cdot \boldsymbol{R}_{id,0}^{-1} \in \mathbb{Z}_q^{n \times m}.$$

2). Using algorithm $\mathrm{BasisDel}(A, \boldsymbol{R}_{id,0}, \boldsymbol{T}, \sigma_0)$, the PKG can generate the initial signing secret key $\boldsymbol{sk}_{id,0} \in \mathbb{Z}_q^{m \times m}$ and sends it to the signer in a secure way.

FSIBBS-Update: On inputting $(id, i, \boldsymbol{sk}_{id,i-1})$, where i is the current time period, $\boldsymbol{sk}_{id,i-1}$ is the signing secret key associated with the time period i-1. A signer with the identity *id* does as follows:

1) Let $\boldsymbol{h}_{i-1} = H(id, i-1) \in \{0,1\}^l$, the signer computes

$$\boldsymbol{R}_{id,i-1} = \boldsymbol{R}_{i-1,l}^{h_{i-1,l}} \cdots \boldsymbol{R}_{i-1,1}^{h_{i-1,1}}, A_{id,i-1} = A \cdot (\boldsymbol{R}_{id,i-1} \cdots \boldsymbol{R}_{id,0})^{-1} \in \mathbb{Z}_q^{n \times m}.$$

2) Let $\boldsymbol{h}_i = H(id, i)$, $\boldsymbol{R}_{id,i} = \boldsymbol{R}_{i,l}^{h_{i,l}} \cdots \boldsymbol{R}_{i,1}^{h_{i,1}} \in \mathbb{Z}_q^{m \times m}$.

3) Using algorithm $\mathrm{BasisDel}(A_{id,i-1}, \boldsymbol{R}_{id,i}, \boldsymbol{sk}_{id,i-1}, \sigma_i)$, the signer generates the signing secret key $\boldsymbol{sk}_{id,i} \in \mathbb{Z}_q^{m \times m}$ for time period i, and then deletes $\boldsymbol{sk}_{id,i-1}$.

FSIBBS-Sign: On inputting the message $\boldsymbol{m} \in \{0\} \times \{0,1\}^l$, a user requests the signer with the identity *id* to make a blind signature. The user then interacts with the signer as follows:

1). Once receiving a blind signature request, a signer with identity *id* sends the current time period i to the requester.

2). Once obtaining current time period i from a signer with identity *id*, the user randomly chooses

$v \leftarrow D_{\mathbb{Z}^m, \sigma_R}, V \leftarrow D_{n \times n}$, where, $\sigma_R = \sqrt{n \log q} \cdot \omega(\sqrt{\log m})$.

3). Let $h_i = H(id, i) \in \{0,1\}^l$, the user computes matrices

$$R_{id,i} = R_{i,l}^{h_{i,l}} \cdot R_{i,l-1}^{h_{i,l-1}} \cdots R_{i,1}^{h_{i,1}}, \quad A_{id,i} = A \cdot (R_{id,i} \cdots R_{id,0})^{-1} \in \mathbb{Z}_q^{n \times m}.$$

4). The user computes $u_i = \sum_{i=0}^{l} (-1)^{m_i} \cdot c_i^{\mathrm{T}} \cdot V + A_{id,i} \cdot v$ as a blinded message of m and sends it to the signer.

5). Using $\mathrm{Sample\,Pre}(A_{id,i}, sk_{id,i}, u_i, \delta_i)$, the signer gets a vector $sig_i' \in \mathbb{Z}_q^m$, and sends (id, i, u_i, sig_i') to the user.

6). Once getting (id, i, u_i, sig_i'), the user computes a vector $sig_i = (sig_i' - v)V^{-1} \in \mathbb{Z}_q^m$, and outputs the final blind signature (id, i, m, sig_i).

FSIBBS-Verify: On inputting (id, i, m, sig_i), id is a signer identity, i is an index of time period, m is an original message and sig_i is the corresponding bind signature. Then the verifier does as follows:

1). To verify that sig_i is a small but non-zero vector, which means $0 < \|sig_i\| \le \delta_i \sqrt{m}$.

2). To verify that $A_{id,i} \cdot sig_i = \sum_{i=0}^{l} (-1)^{m_i} \cdot c_i^{\mathrm{T}} \in \mathbb{Z}_q^n$, where $A_{id,i} = A \cdot (R_{id,i} \cdots R_{id,0})^{-1} \in \mathbb{Z}_q^{n \times m}$, $R_{id,i} = R_{i,l}^{h_{i,l}} \cdots R_{i,1}^{h_{i,1}} \in \mathbb{Z}_q^{m \times m}$.

3). If both the above conditions are satisfied, the verifier outputs "accept"; otherwise "reject".

6. Conclusion

The key exposures bring out very serious problem in some insecure mobile devices. Forward-secure blind signatures preserve the validity of past blind signatures and prevent a forger from forging past blind signatures even if the current signing secret key has been compromised. In this paper, we used basis delegation technique to construct the first FSIBBS scheme over lattice and proved its security requirements of blindness, unforgeability and forward secrecy in the random oracle model. Compared with three IBS schemes from lattice, our construction enjoys a shorter public key, secret key and signature. What is more, our scheme can deal with the key exposure problems even in the post-quantum cryptographic era. Furthermore, we also extended our scheme to a FSIBBS scheme in the standard model. In the future work, we attempt to give a security analysis of our second construction.

Acknowledgements

This paper is supported by the Nature Science Foundation of China (61472309).

References

[1] D. Chaum, "Blind signatures for untraceable payments," Proceedings of the Cryptology Conference (CRYPTO'82): Santa Barbara, CA, USA, pp. 199–203, August 23–25, 1982

[2] D. Chaum, "Untraceable electronic cash," Proceedings of the Cryptology Conference (CRYPTO'88). Santa Barbara, CA, USA, vol. 403, pp. 319–327, August 21–25, 1988

[3] S. B. Wang, H. Fan, and G. H. Cui, "A proxy blind signature schemes based DLP and applying in e-voting," Proceedings of the International Conference on Electronic commerce (ICEC'05). Xi'an, China, pp. 641–645, August 15–17, 2005

[4] A. Shamir, "Identity-based cryptosystem and signature schemes," Proceedings of the Cryptology Conference (CRYPTO'84). Santa Barbara, CA, USA, vol. 196, pp. 47–53, August 19–22, 1984

[5] R. Anderson, "Two remarks on public key cryptology (invited lecture)," Proceedings of the ACM conference on Computer and Communications Security (CCS'97). Zurich, Switzerland, pp. 135–147, May 21–24, 1997

[6] M. Bellare and S. K. Miner, "A forward-secure digital signature scheme," Proceedings of the Cryptology Conference (CRYPTO'99). Santa Barbara, CA, USA, vol. 1666, pp. 431–438, August 15–19, 1999

[7] D. N. Duc, J. H. Cheon, and K. Kim, "A forward-secure blind signature scheme based on the strong RSA assumption," Proceedings of the International Conference on Information and Communications Security (ICICS'03). Huhehaote, China, vol. 2836, pp. 11–21, October 10–13, 2003

[8] Y. P. Lai and C. C. Chang, "A simple forward secure blind signature scheme based on master keys and blind signatures," Proceedings of the International Conference on Advanced Information Networking and Applications (AINA'05). Taipei, Taiwan, vol. 2, pp. 139–144, March 28–30, 2005

[9] H. F. Huang and C. C. Chang, "A new forward-secure blind signature scheme," Journal of Engineering and Applied Sciences. 1rd ed., vol. 2, 2007, pp. 230–235

[10] J. Yu, F. Y. Kong, and G. W. Li, "Forward-secure multi-signature, threshold signature and blind signature schemes," Journal of Networks. 6rd ed., vol. 5, 2010, pp. 634–641

[11] X. Zhang and H. H. Hang, "A new forward-secure blind signature scheme," Journal of Wuhan University: Natural Science Edition. 5rd ed., vol. 57, 2010, pp. 434–438. (in Chinese)

[12] J. J. He, F. Sun, and C. D. Qi, "A forward-secure blind signature scheme based on quadratic residue," Computer Applications & Software. 7rd ed., vol. 30, 2013, pp. 54–56. (in Chinese)

[13] C. Gentry, C. Peikert, and V. Vaikuntanathan, "How to use a short basis: trapdoors for hard lattices and new cryptographic constructions," Proceedings of the ACM Symposium on Theory of Computing (STOC'08). Victoria, BC, Canada, pp. 197–206, May 17–20, 2008

[14] O. Regev, "On lattices, learning with errors, random linear codes and cryptography," Proceedings of the ACM Symposium on Theory of Computing (STOC'05). Maryland, USA, pp. 84–93, May 21–24, 2005

[15] V. Lyubashevsky, C. Peikert, and O. Regev, "On ideal lattices and learning with errors over rings," Proceedings of the International Conference on the Theory and Applications of Cryptographic Techniques (EUROCRYPT'10). French Riviera, vol. 6110, pp. 1–23, May 30–June 3, 2010

[16] D. Stehlé and R. Steinfeld, "Making NTRU as secure as worst-case problems over ideal lattices," Proceedings of the International Conference on the Theory and Applications of Cryptographic Techniques (EUROCRYPT'11).Tallinn, Estonia, vol. 6632, pp. 27–47, May 15–19, 2011

[17] D. Micciancio and C. Peikert, "Trapdoors for lattices: simpler, tighter, faster, smaller," Proceedings of the International Conference on the Theory and Applications of Cryptographic Techniques (EUROCRYPT'12). Cambridge, UK, vol. 7237, pp. 700–718, April 15–19, 2012

[18] S. Garg, C. Gentry, and S. Halevi, "Candidate multilinear maps from ideal lattice," Proceedings of the International Conference on the Theory and Applications of Cryptographic Techniques (EUROCRYPT'13). Athens, Greece, vol. 7881, pp. 1–17, May 26–30, 2013

[19] C. Peikert, "Lattice cryptography for the Internet," Proceedings of the International Conference on Post-Quantum Cryptography (PQCRYPTO'14). Waterloo, ON, Canada, vol. 8772, pp. 197–219, October 1–3, 2014

[20] S. Gorbunov, V. Vaikuntanathan, and H. Wee, "Predicate encryption for circuits from LWE," Proceedings of the Cryptology Conference (CRYPTO'10). Santa Barbara, CA, USA, vol. 9216, pp. 503–523, August 16–20, 2015

[21] D. Cash, D. Hofheinz, E. Kiltz, et al, "Bonsai trees, or how to delegate a lattice basis," Proceedings of the International Conference on the Theory and Applications of Cryptographic Techniques (EUROCRYPT'10). French Riviera. vol. 6110, pp. 523–552, May 30–June 3, 2010

[22] S. Agrawal, D. Boneh, and X. Boyen, "Efficient lattice (H)IBE in the standard model," Proceedings of the International Conference on the Theory and Applications of Cryptographic Techniques (EUROCRYPT'10). French Riviera. vol. 6110, pp. 553–572, May 30–June 3, 2010

[23] S. Agrawal, D. Boneh, and X. Boyen, "Lattice basis delegation in fixed dimension and shorter-ciphertext hierarchical IBE," Proceedings of the Cryptology Conference (CRYPTO'10). Santa Barbara, CA, USA, vol. 6223, pp. 98–115, August 15–19, 2010

[24] S. Agrawal S, D. M. Freeman, V. Vaikuntanathan, "Functional encryption for inner product predicates from learning with errors," Proceedings of the International Conference on the Theory and Application of Cryptology and information security (ASIACRYPT'11). Seoul, South Korea, vol. 7073, pp. 22–41, December 4–8, 2011

[25] S. Agrawal, X. Boyen, V. Vaikuntanathan, et al, "Functional encryption for threshold functions (or fuzzy IBE) from lattices," Proceedings of the International Conference on Practice and Theory in Public Key Cryptography (PKC'12). Darmstadt, Germany, vol. 7279, pp. 280–297, May 21–23, 2012

[26] R. Bendlin, S. Krehbiel, and C. Peikert, "How to share a lattice trapdoor: threshold protocols for signatures and (H)IBE," Proceedings of the International Conference on Applied Cryptography and Network Security (ACNS'13). Banff, AB, Canada, pp. 218–236, vol. 7954, June 25–28, 2013

[27] L. Ducas, V. Lyubashevsky, and T. Prest, "Efficient identity based encryption over NTRU lattices," Proceedings of the International Conference on the Theory and Application of Cryptology and information security (ASIACRYPT'14). Kaoshiung, Taiwan, vol.8874, pp. 22–41, December 7–11, 2014

[28] C. Gentry, "Fully homomorphic encryption using ideal lattices," Proceedings of the ACM Symposium on Theory of Computing (STOC'09). Bethesda, USA. pp. 169–178, May 31–June 2, 2009

[29] C. Gentry, "Toward basing fully homomorphic encryption on worst-case hardness," Proceedings of the Cryptology Conference (CRYPTO'10). Santa Barbara, CA, USA, vol. 6223, pp. 116–137, August 15–19, 2010

[30] J. H. Cheon, J. S. Coron, J. Kim, et al, "Batch fully homomorphic encryption over integer," Proceedings of the International Conference on the Theory and Applications of Cryptographic Techniques (EUROCRYPT'13). Athens, Greece, vol. 7881, pp. 315–335, May 26–30, 2013

[31] J. S. Coron, T. Lepoint, and M Tibouchi, "Scale-invariant fully homomorphic encryption over the integers," Proceedings of the International Conference on Practice and Theory in Public Key Cryptography (PKC'14). Buenos Aires, Argentina, vol. 6056, pp. 311–328, March 26–28, 2014

[32] K. Nuida and K. Kurosawa, "(Batch) fully homomorphic encryption over integers for non-binary message spaces," Proceedings of the International Conference on the Theory and Applications of Cryptographic Techniques (EUROCRYPT'15). Sofia, Bulgaria, vol. 9056, pp. 537–555, April 26–30, 2015

[33] X. Boyen, "Lattice mixing and vanishing trapdoors: a framework for fully secure short signature and more," Proceedings of the International Conference on Practice and Theory in Public Key Cryptography (PKC'10). Paris, France, vol. 6056, pp. 499–517, May 26–28, 2010

[34] V. Lyubashevsky, "Lattice signatures without trapdoors," Proceedings of the International Conference on the Theory and Applications of Cryptographic Techniques (EUROCRYPT'12). Cambridge, UK, vol. 7237, pp. 738–755, April 15–19, 2012

[35] L. Ducas, A. Durmus, T. Lepoint, et al, "Lattice signatures and bimodal gaussians," Proceedings of the Cryptology Conference (CRYPTO'13). Santa Barbara, CA, USA, vol. 8042, pp. 40–56, August 18–22, 2013

[36] L. Ducas and D. Micciancio, "Improved short lattice signatures in the standard model," Proceedings of the Cryptology Conference (CRYPTO'14). Santa Barbara, CA, USA, vol. 8616, pp. 335–352, August 17–21, 2014

[37] S. D. Gordon, J. Katz, and V. Vaikuntanathan, "A group signature scheme from lattice assumptions," Proceedings of the International Conference on the Theory and Application of Cryptology and information security (ASIACRYPT'10). Singapore, vol. 6477, pp. 395–412, December 5–9, 2010

[38] M. Rückert, "Lattice-based blind signatures," Proceedings of the International Conference on the Theory and Application of Cryptology and information security (ASIACRYPT'10). Singapore, vol. 6477, pp. 413–430, December 5–9, 2010

[39] M. Rückert, "Strongly unforgeable signatures and hierarchical identity-based signatures from lattices without random oracles," Proceedings of the International Conference on Post-Quantum Cryptography (PQCRYPTO'10). Darmstadt, Germany, vol. 6061, pp. 182–200, May 25–28, 2010

[40] Z. H. Liu, Y. P. Hu, X. S. Zhang, et al, "Efficient and strongly unforgeable identity-based signature scheme from lattice in the standard model," Security & Communication Networks, 1rd ed., vol. 6, 2013, pp.69–77

[41] R. E. Bansarkhani and J. Buchmann, "Towards lattice based aggregate signatures," Proceedings of the International Conference on Cryptology in Africa, Marrakesh, Morocco. 2014, vol. 8469, pp. 336–355, May 28–30, 2014

[42] A. Langlois, S. Ling, K. Nguyen, et al, "Lattice-based group signature scheme with verifier-local revocation," Proceedings of the International Conference on Practice and Theory in Public-Key Cryptography (PKC'14). Buenos Aires, Argentina, vol. 8383, pp. 345–361, March 26–28, 2014

[43] S. Ling, K. Nguyen, and H. X. Wang, "Group signature from lattices: simpler, tighter, shorter, ring-based," Proceedings of the International Conference on Practice and Theory in Public-Key Cryptography (PKC'15). Gaithersburg, MD, USA, vol. 9020, pp. 427–449, March 30–April 1, 2015

[44] P. Q. Nguyen, J. Zhang, and Z. F. Zhang, "Simpler efficient group signature from lattices," Proceedings of the International Conference on Practice and Theory in Public-Key Cryptography (PKC'15). Gaithersburg, MD, USA, vol. 9020, pp. 401–426 March 30–April 1, 2015

[45] J. Alwen and C. Peiker, "Generating shorter bases for hard random lattices," Journal of Theory of Computing Systems. 3rd ed., vol. 48, 2011, pp. 535–553

[46] D. Micciancio and O. Regev, "Worst-case to average-case reductions based on Gaussian measures," SIAM Journal on Computing Archive, 1rd ed., vol. 37, 2007, pp. 267–302

[47] M. Ajtai, "Generating hard instances of lattice problems (extended abstract)," Proceedings of the ACM Symposium on Theory of Computing (STOC'96). Philadelphia, Pa, USA, pp. 99–108, May 22-24, 1996

[48] X. J. Zhang, C. X. Xu, C. H. Jin, et al, "Efficient forward secure identity-based shorter signature from lattice," Computers and Electrical Engineering, 6rd ed., vol. 40, 2014, pp. 1963–1971

[49] N. A. Ebri, J. Baek, A. Shoufan, et al, "Efficient generic construction of forward-secure identity-based signature," Proceedings of the International Conference on Availability, Reliability and Security (ARES'12). Washington, DC, USA, vol. 329, pp. 55–64, August 20–24, 2012

23

Aloha Based Resource Allocation Scheme to Efficient Call Holding Times in Public Safety Network

Tuyatsetseg Badarch[1], Otgonbayar Bataa[2], Bat-Enkh Oyunbileg[2]

[1]Department of Information Technology, Mongolian National University, Ulaanbaatar, Mongolia
[2]Department of Information Technology, Mongolian University of Science and Technology, Ulaanbaatar, Mongolia

Email address:

ba.tuyatsetseg@mnu.edu.mn (T. Badarch), b_otgonbayar2002@yahoo.com (O. Bataa), o_bat_enkh@yahoo.com (Bat-Enkh Oyunbileg)

Abstract: The allocation of limited resources such as links, servers, and agents to support as many as emergency call responses as possible while maintaining reasonable quality of service is a principal requirement in public safety networks. It is difficult to handle the highest priority emergency call responses for a congested period with a limited number of dedicated incoming links from commercial networks. Therefore, in this paper, we present the adaptive scheme based on a different call conversation time for an achieving efficient resource scheduling during a congested period without investing extra resources. The proposed adaptive scheduling approach improves accuracy of the network performance with the dynamic call holding time approach.

Keywords: Aloha Protocol, Resource Allocation, Public Safety Network

1. Introduction

The ability to access emergency services by dialing a nationally designed fixed numbers from the Public Switched Telephone Networks (PSTNs) is a vital component of public safety and emergency preparedness [1]. The "real" emergency call handling is a specific service, since all calls have the same highest priority service. Therefore, it is difficult to handle all these highest priority call responses for a congested period, with a limited number of dedicated incoming links from commercial networks.

Therefore, one main criteria of the emergency call handling network performance to enhance SLA (Service Level Agreement) requirements is the limited call holding time, which a customer may use as the common resource within a time after it has been allocated to a system resource. There are standards, for instance, the document serves as a standard for answering 9-1-1 calls operating procedure for the call taking function within Public Safety Answering Points (PSAPs) presents the standard that 90% of all 9-1-1 calls arriving at the PSAP shall be answered within 10 seconds during the busy hour and 95% of all 9-1-1 calls should be answered within 20 seconds [2], [3]. This shows that limited conversation time is critical function for the emergency system SLA.

In general, the recent study shows that it is common in the SLA for workloads whose intensity varies widely and unpredictably that the dynamic resource allocation maximizes performance of a system [4]-[5]. It is a well known that the dynamic allocation is the more efficient solution than the static allocation scheme. In traditional static resource allocation approaches, when network resources are depleted, more resources (links, trunks, operators, agents, servers) are added to the network. These additions may resolve congestion conditions temporarily, but it is not a long term cost effective solution.

In the literature, a resource allocation discipline, such as time dependent priority, selfish scheduling, shortest to longest remaining time scheduling, and processor sharing scheduling, may prove to be optimal for certain application requirements [6] - [7]. A dynamic holding time allocation scheme that uses adaptive quota ensures an efficiency approaching that of First Come First Serve (FCFS), especially under heavy load conditions [8]. Moreover, the priority resource allocation is proposed to support ongoing activities in the ITU and IETF for developing the International Emergency Preparedness Scheme (IEPS) [9]-[11].

Emergency voice incoming calls for specific emergency services such as ambulance, police, fire, and hazard which are directed to emergency – specific PSAPs transferred through the bulk of the existing E1 trunks across PSTNs are the main part of the network used for the peak period performance evaluation. The reason is that the dedicated limited number of inbound trunks is built under the government regulation.

If the number of calls addressed to the system is too large during the peak period, the incoming trunks will be overloaded and operators cannot handle the volume of emergency calls. Hence, our aspect of the peak period performance is to characterize the system capacity using the model of the dynamic call holding time scheduling approach handling the overload of incoming calls.

The rest of the paper is organized as follows: Section 2 presents the dynamic call holding time scheduling scheme based on the Bernoulli theory, Section 3 describes the analytical model formulation of this dynamic scheme. Section 4 and Section 5 present the results based on the proposed method.

2. Proposed Model Formulation

The ALOHA protocol is an interesting example of a MAC protocol of the contention-type. It is the precursor of Ethernet and its subsequent standardization as IEEE 802.3[12].

Based on the protocol, this model considers the vulnerable (non retrial) and non vulnerable (retrial) period of call holding time of the PSN [13]. Under this scheme, a dynamic quota with a vulnerability coefficient k is used to determine a threshold level to allow a number of customers to occupy a resource (agent) within a call holding time limitation, especially under heavy load conditions. Applying the Bernoulli theory under the assumption of multiple Poisson arrival process is the fundamental analytics used.

We consider a customer who is granted permission to use a resource without waiting, is allowed to use a resource for the total CHT with the following condition:

$$T = kT_v \qquad (1)$$

where T is the proposed total Call Holding Time (CHT), T_v is the threshold value of the CHT, and k is the vulnerability coefficient.

When CHT is exceeded the threshold T_v, the system may initiate retrial calls or drop-off calls. The threshold can be measured by k. When $k < 1$, it describes a call success in the threshold value and the condition $k > 1$.

May describe retrials or drop off CHT. When $k = 1$, the CHT equals to threshold value of the time. For the idle system, we assume $k = 0$.

In this scheme, we consider the condition that the T_v which may be equal to the proposed CHT (1), it may be computed within the limited time: $T_{min} < T_v < T_{max}$.

In general, to compute the theoretical throughput in the proposed model, first we assume the arrival process is Poisson, which means the probability P, of getting no call at T time interval $P = [0, t]$ is:

$$P(T) = e^{-\lambda T} \qquad (2)$$

where $\lambda = 1/T$ is the mean call arrival rate. Then, the probability P_T, of getting some calls at T time interval:

$$P_T = P(T) = 1 - e^{-\lambda T} \qquad (3)$$

We assume the condition that the mean service time equals to the mean call holding time plus a constant small amount of the processing time. This small processing time is negligible in this study. Then,

$$T_v = T_{max} - T_{min} = T / k \qquad (4)$$

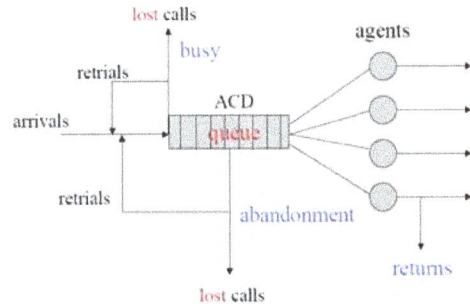

Fig. 1. Feature of emergency call operational scheme.

Offered load λ^* is described by the total call arrival rate divided by the total service rate (inverse of call holding time in the case of the negligible small amount processing time) as follows:

$$\lambda^* = \lambda \times T_v \qquad (5)$$

Under the condition, using the Bernoulli theorem, we can get the probability of the holding time of retrials, P_{T_v}, of the network getting at least one call within the range of kT_v. The call departs from the system with probability $1 - P_{T_v}$. If calls passed within the threshold time, P_{T_v} is zero at an agent tier level.

We have the following formula of the carried load considering the probability of retrials P_T under the model in Fig.1.

$$L_c = \lambda + P_T L_c \qquad (6)$$

where L_c is the carried load, the arrival rate λ describes the offered load, it shows that the carried load may be depend on strongly the probability of retrials time P_T. Hence, the probability P_T, of getting some calls within at kT_v time interval (Fig. 1):

$$P_T = P(T \le T_v) = 1 - e^{-L_c T} \qquad (7)$$

Using (7) substitute P_T in (6) and rewriting, the call arrival rate λ has the following relationship within the call holding time T considering k coefficient.

$$\lambda = L_c e^{-kL_c T_v} \qquad (8)$$

From (5), we easily update the offered load :

$$\lambda^* = \lambda \times kT_v = (kL_c T_v)e^{-kL_c T_v} \qquad (9)$$

Where the normalized total carried load is described as follows:

$$L_{c.n} = kL_c T_v \qquad (10)$$

We often consider that the call holding time T equals to the number of calls N in unit time divided by the call arrival rate capacity μ. If all offered traffic can be serviced, so that the offered load λ^* may be described by the throughput. If the offered load is greater than this value, the system becomes unstable.

We sample throughput data points by feeding calls. We get the throughput vs. total carried load including retrials. It proves that CHT has the best performance in terms of the maximum throughput. The maximum achievable throughput is presented by the differentiation of the right side of equation (8).

3. Results of the Proposed Approach

We monitored the incoming trunks at edge port of IP-PBX of the EIN network [14]. The experimental period was one week including week days, weekends, and the biggest national holiday Naadam, therefore, the workload analysis enables us to account for the existing system's peak period performance.

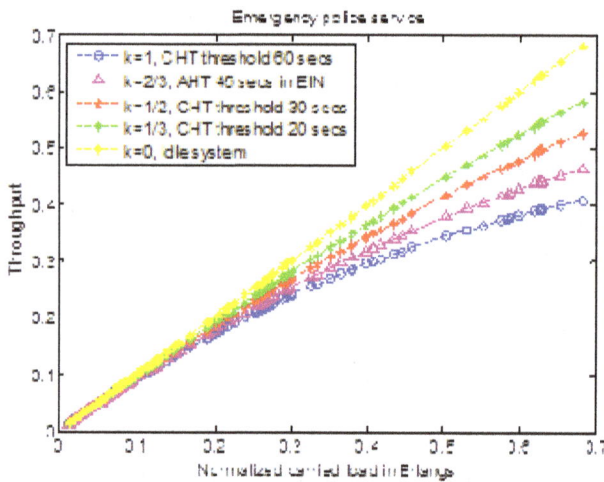

(a) Load vs. Throughput of "102".

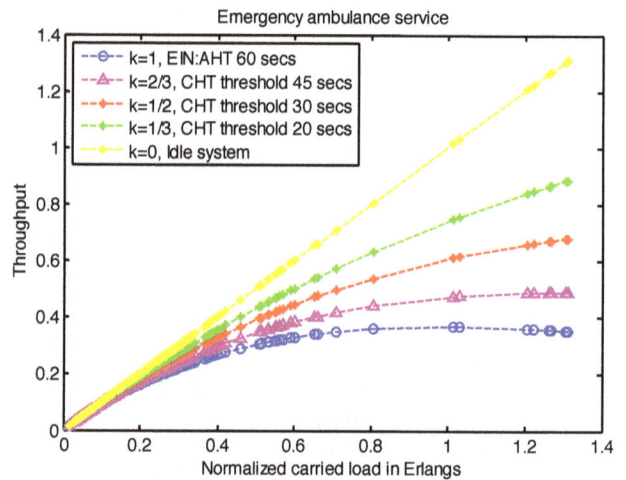

(b). Load vs. Throughput of "103".

Fig 2. Throughput performance for a high load.

In Fig.2a, there are maximum emergency users who generate as high as 1.2 Erlangs of the ambulance class as well as light users generating less than 0.03 Erlangs of the hazards class.

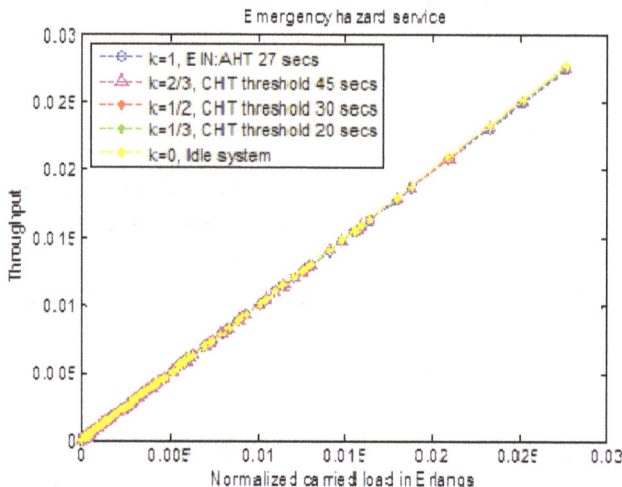

(a). Load vs. Throughput of "105".

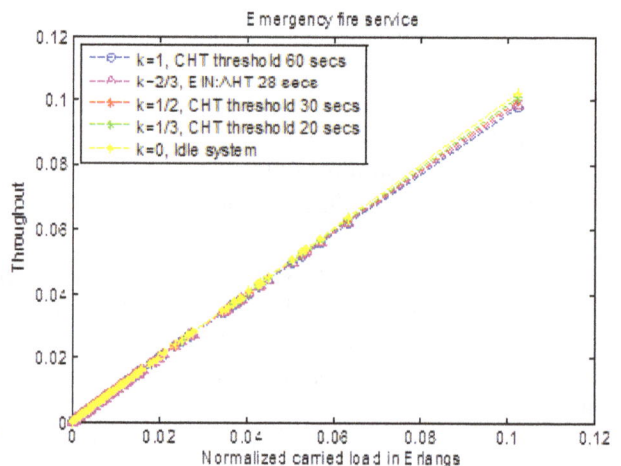

(b). Load vs. Throughput of "101".

Fig 3. Throughput performance for a light load.

During the period (Fig.3a). The figure presents the CHT value of all "103-ambulance" calls arriving at the PSAP is 60 seconds in average during the peak week and 45 seconds show the "102-police" CHT in average. As a result of the proposed approach, the phenomena in the case of "103" is clear the throughput is increased by approximately two times more than the existing throughput when the CHT is reduced to 20 seconds.

It is clear to see there is high demand to use the proposed scheme under the priority scheduling for the heavy load situation. Therefore, even though the arrival rate is increased up to a maximum load level, the calls of each class may be served by Long Holding Time (LHT), Medium Holding Time (MHT), or Short Holding Time (SHT) scheduling approaches. However, the idle system presents linear relationship between the throughput and traffic load, the existing EIN system has shown the exponential relationship between the throughput and traffic load.

Fig. 3 shows the idle case performance of the scheme.

Reducing the CHT, we can keep the system more idle compared to the congested system.

According to the results, in the light load of the emergency fire ("101") and hazard call ("105") classes, there is no difference between the long and short call holding time scheduling advantage. Then, FIFO scheme may be the best approach with a light load, then a dynamic holding policy may not be a main metric in the light load (See Fig. 3).

4. Simulation Verification

We also used Matlab Simulink to support the proposed model. Using the discrete event simulation SimEvents library, we built the model diagram. The diagram is omitted, due to the space limitations. The simulation was the time-based 24 hours period - simulation for both LHT, MHT, and SHT respectively. The block generates Poisson arrival events packed with the random exponential CHT attribute that we extract the data as the experimental data set for the Matlab Simulink simulator from statistics of the EIN by averaging the week into the day.

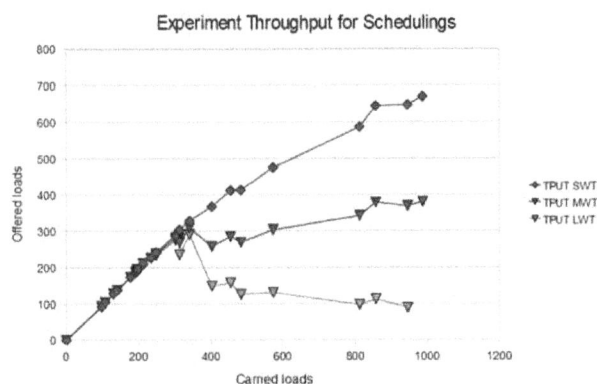

Fig 4. *Carried vs. Offered number of incoming calls depends on a call conversation time.*

The simulator provided output on queue length, throughput, and utilization. Our analysis focuses on the relationship between workload and throughput. The result (Fig. 4) proved that the calls into the network depend on a call conversation time for the long-LWT, medium-MWT, and short-SWT call conversation times, it clearly shows the same pattern as our analytical result this supports (8).

The upper limit thresholds have advantages over other resource allocation policies and is efficient to implement, it still must be considered against traditional resource management approaches, e.g., using excess capacity in the network to control a peak load [15].

5. Conclusion

The paper showed the benefits of a limited call holding time through throughput vs. carried load analysis for the dynamic resource scheduling policy in a public safety network. The new method uses the adaptive time quota ensureing an efficiency approaching the dynamic call holding time. We evaluated the performance of the heterogeneous EIN in Mongolia using the proposed model. Under this priority scheduling with the Bernoulli - loop, the theoretical, experimental and simulation results show the larger the throuhgput, the smaller the call holding time that the existing system produces during the peak period. In the light load, there is no difference of the throughput between the long and short call holding time scheme. Therefore, the paper brings the contribution which results in the smaller CHTs for emergency calls such as ambulance, police, fire and hazard calls of the existing network to influence an efficient dynamic resource allocation. The scheme can be a simple and generic approach, it can be applied to similar time - conscious systems such as time dependent call centers.

References

[1] B. Rosen, H. Schulzrinne, U. Columbia, J. Polk, and A. Newton, "Framework for Emergency Calling Using Internet Multimedia", RFC 6443, December 2011.

[2] National Emergency Number Association (NENA) Standard Operating Procedures Committee, Call taking Working Group, "NENA Call Answering Standard/Model Recommendation", Document 56-005, June 10, 2006.

[3] NENA Guidelines for Minimum Response to Wireless 9-1-1 Calls, Document 56-001 (revised November 18, 2004).

[4] Bennani, N. Mohamed, A. Daniel Menasce, "Resource Allocation for Autonomic Data Centers Using Analytic Performance Models." *In Proc. Sec.Int.Conference on. IEEE,* 2005.

[5] D. Ardagna, B. Panicucci, M. Trubian, and L. Zhang, "Energy-Aware Autonomic Resource Allocation in Multitier Virtualized Environments", *IEEE Transactions on Services Computing*, pp. 2 – 19, vol. 5, 2012.

[6] L. Kleinrock, P. Roy Finkelstein "Time Dependent Priority Queues Source", Journal of Operations Research, vol. 15, no. 1, pp. 104-116, 1967.

[7] E. Angel, E. Bampis, and F. Pascual, "Truthful Algorithms for Scheduling Selfish Tasks on Parallel Machines", *In Theoretical Computer Science,* 2006.

[8] A. N. Tantawy, A. N. Tantawi, D. N. Serpanos, "An Adaptive Scheduling Scheme for Dynamic Service Time Allocation on a Shared Resource", *Proc. of the 12th International Conference on Distributed Computing Systems*, pp. 294-300, 1992.

[9] H. Schulzrinne, "Requirements for Resource Priority Mechanisms for the Session Initiation Protocol (SIP)", *RFC 3487,* Feb 2003.

[10] K.Carlberg, I. Brown, C. Beard, "General Requirements for Emergency Telecommunications Services (ETS)", *RFC 3689,* Feb 2004.

[11] ITU, Telecommunications Standardization Sector, "International Emergency Preparedness Scheme", E-106 Recommendation, March, 2000.

[12] Norman M. Abramson, "Development of the ALOHANET", IEEE Transactions on Information Theory 31(2), 119-123 (1985).

[13] G. Koole, A. Mandelbaum, "Queueing Models of Call Centers: An Introduction", Annals of Operations Research, vol. 113, pp 41-59, 2002.

[14] B. Tuyatsetseg, "Parametric Modeling Approach for Call Holding Times of IP based Public Safety Networks," WASET journal. issue 73, pp. 568-575. Available: http://www.waset.org/journals/waset/v73.php.

[15] C.Beard, V.Frost, "Prioritized Resource Allocation for Stressed Networks", *IEEE/ACM Trans, Network,* vol. 6, no. 5, pp. 618-633, Oct. 2001.

24

Efficient routing with reduced routing overhead and Retransmission of Manet

Megha Sharma, Shivani Rohilla, Manish Bhardwaj

Department of Computer science and Engineering, SRM University, NCR Campus, Modinagar, Ghaziabad, India.

Email address:

Megha.tech09@gmail.com (M. Sharma), shivani.engineer@gmail.com (S. Rohilla), aapkaapna13@gmail.com (M. Bhardwaj)

Abstract: Adhoc network is a collection of mobile nodes that dynamically form a temporary network and are capable of communicating with each other without the use of a network infrastructure. In manet nodes can change location and configure itself, the mobility of nodes causes continuous link breakage due to which frequent path failure occur and route discovery is required. The fundamental construction for a route discoveries is broadcasting in which the receiver node blindly rebroadcast the first received route request packet unless it has route to the destination. The routing overhead associated with route discovery is very high which leads to poor packet delivery ratio and a high delay to be victorious this type of routing overhead we are proposing the new technique using NCPR. To intended NCPR method is used to determine the rebroadcast order and obtain the more precise additional coverage ratio by sensing neighbor coverage knowledge. We can also define connectivity factor to provide node density adaptation. By combining the additional coverage ratio and connectivity factor, rebroadcast probability is determined. The approach can signify improvement in routing performance and decrease the routing overhead by decreasing the number of retransmission.

Keywords: Mobile Adhoc Networks, Neighbor Coverage Network Connectivity, Probabilistic Rebroadcast, Routing Overhead

1. Introduction

A manet is a type of network where topology changes randomly and no infrastructure is available for the network to establish an active connection .Mobile adhoc network consist of a collection of mobile nodes which can move freely. These nodes can be dynamically self –organized into arbitrary topology networks without a fixed infrastructure .one of the fundamental challenge of manet is the design of dynamic routing protocols with good performance and less overhead. Many routing protocols, such as adhoc on demand distance vector routing (AODV) [1] and dynamic source routing (DSR)[2] have been proposed, these protocol performance not constant, it is more ever variable due high mobility, network load and network size Whereas AODV creates no extra traffic for communication along existing links. The adhoc on demand distance vector routing algorithm is a routing protocol designed for adhoc mobile networks. AODV is capable of both unicast and multicast routing. It is an on demand algorithm, meaning that it builds routes between

nodes only as desired by source nodes. It maintains these routes as long as they are needed by the sources [19]. the dynamic source routing(DSR),it is simple and efficient routing protocol designed specifically for use in multi hop wireless adhoc networks of mobile nodes.DSR allows the network to be completely self-organizing and self-configuring, without the need for any existing network infrastructure or administration. The above two protocols are on-demand routing protocols and they could improve the scalability of manet by limiting the routing overhead of routing protocols and reduce the packet delivery ratio and increasing the end to end delay [4].

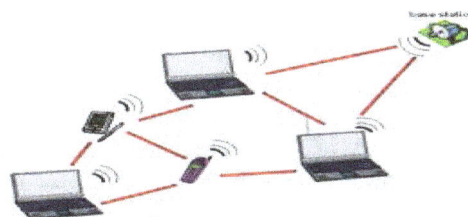

Figure 1. Multi hop ad hoc network

The main announcement of this paper as:

- We propose a novel scheme to calculate the rebroadcast delay. The rebroadcast delay is to determine the forwarding order. The node which has more common neighbors with the previous node has the lower delay. If this node rebroadcast delay enables the information that the nodes have transmitted the packet spread to more neighbors.
- We also propose a novel scheme to calculate the rebroadcast probability .the scheme considers the information about the uncovered neighbors (UCN), connectivity metric and local node density to calculate the rebroadcast probability is composed of two parts: a) additional coverage ratio, which is the ratio of the number of nodes that should be covered by a single broadcast to the total number of neighbors. b) Connectivity factor which reflects the relationship of network connectivity and the number of neighbor of a given node.

1.1. Study the Paper Scope

Since limiting the number of rebroadcast can effectively optimize the broadcasting and the neighbor knowledge methods perform better than the area-based ones and the probability based ones then we propose a neighbor coverage based probability (NCPR).

1.2. Additional Coverage Ratio

In order to effectively exploit the neighbor coverage knowledge, we need a novel rebroadcast delay to determine the rebroadcast order and then we can obtain a more precise additional coverage ratio.

1.3. Connectivity Factor

In order to keep the network connectivity and reduce the redundant retransmissions, we need a metric named connectivity factor to determine how many neighbors should receive the RREQ packet.

2. Literature Review

Broadcasting is an effective mechanism for route discovery, But the routing overhead associated with the broadcasting Can be quite large, especially in high dynamic networks [9]. Ni et al. [5] studied the broadcasting protocol analytically And experimentally, and showed that the rebroadcast is Very costly and consumes too much network resource. The Broadcasting incurs large routing overhead and causes many problems such as redundant retransmissions, contentions, and collisions [5]. Thus, optimizing the broadcasting.

In route discovery is an effective solution to improve the routing performance. Haas et al. [10] proposed a gossip based approach, where each node forwards a packet with a Probability. They showed that gossip-based approach can save up to 35 percent overhead compared to the flooding. However, when the network density is high or the traffic load is heavy,

the improvement of the gossip-based approach is limited [9]. Kim et al. [8] proposed a probabilistic broadcasting scheme based on coverage area and neighbor confirmation. This scheme uses the coverage. Area to set the rebroadcast probability, and uses the neighbor confirmation to guarantee reach ability. Peng and Lu [11] proposed a neighbor knowledge scheme named Scalable Broadcast Algorithm (SBA). This scheme determines the rebroadcast of a packet according to the fact whether this rebroadcast would reach additional nodes. Abdulai et al. [12] proposed a Dynamic Probabilistic Route Discovery (DPR) scheme based on neighbor coverage. In This approach, each node determines the forwarding Probability according to the number of its neighbors and the set of neighbors which are covered by the previous Broadcast. This scheme only considers the coverage ratio by the previous node, and it does not consider the neighbours receiving the duplicate RREQ packet. Thus, there is a room of further optimization and extension for the DPR protocol.

Several robust protocols have been proposed in recent years besides the above optimization issues for broadcasting.

Chen et al. [13] proposed an AODV protocol with Directional Forward Routing (AODV-DFR) which takes the Directional forwarding used in geographic routing into AODV protocol. While a `route breaks, this protocol can automati cally find the next-hop node for packet forwarding.

Keshavarz-Haddad et al. [14] proposed two deterministic Timer-based broadcast schemes: Dynamic Reflector Broadcast (DRB) and Dynamic Connector-Connector Broadcast (DCCB). They pointed out that their schemes can achieve full reach ability over an idealistic lossless MAC layer, and for the situation of node failure and mobility, their schemes are robustness. Stann et al. [15] proposed a Robust Broadcast Propagation (RBP) protocol to provide near-perfect reliability for flooding in wireless networks, and this protocol also has a good efficiency. They presented a new perspective for broadcasting: not to make a single broadcast more efficient but to make a single broadcast more reliable, which means by reducing the frequency of upper layer invoking flooding to improve the overall performance of flooding. In our protocol, we also set a deterministic rebroadcast delay, but the goal is to make the dissemination of neighbor knowledge much quicker.

3. Analysis of Problem

In mobile adhoc network nodes are moving continuously due to node mobility in manet frequent link breakages may lead to frequent path failures and route. They broadcast a route request (RREQ) packet to the network and the broadcasting induces excessive redundant retransmission of (RREQ) packet and causes the broadcast storm problem [5], which leads to a considerable number of packet collisions, especially in dense networks.

The broadcast storm problem:

A forwarding order approach to perform broadcast is by flooding .a host, on receiving a broadcast packet for the first time, has the obligation to rebroadcast the packet. Since, this costs n transmission in a Manet of n hosts.

- Redundancy: when a mobile host broadcasts a packet if many of its neighbors decide to rebroadcast a broadcast packet to its neighbor, all of its neighbor might already have heard the packet.
- Contention: After a mobile host broadcasts a packet if many of its neighbor decide to rebroadcast the packet, these transmission may severely contend with each other.

In flooding, a broadcast packet is forwarded by every node in the network exactly once. The broadcast packet is absolutely to be received by every node in the network providing there is no packet loss caused be collision in the MAC layer and there is no high speed movement of nodes during the broadcast process. Figure show a network w th six nodes .when node v broadcast a packet all neighboring nodes .u, w, x and y receive the packet due to the broadcast nature of wireless communication media. All neighbors will then forward the packet to each other. Apparently, the two transmissions from nodes u and x are unnecessary.

3.1. Proposed Work

To calculate the rebroadcast delay and rebroadcast probability of the proposed protocol . using the upstream coverage ratio of an RREQ packet received from the previous node to calculate the rebroadcast delay and use the additional coverage ratio of the RREQ packet and the connectivity factor to calculate rebroadcast probability in our protocol, which requires that each node needs its 1- hop neighborhood information.

3.2. Uncovered Neighbor Set And Rebroadcast Delay

The node receives the RREQ packet from its earlier   node s, to use the neighbor list in the RREQ packet to estimate how many its neighbors have been not covered by the RREQ packet from s. the node ni has more neighbor not covered by the RREQ packet from the source and the RREQ packet form can reach more additional neighbor nodes when node ni rebroadcast the RREQ packet . To quantify of the uncovered neighbor (UCN) set u (ni) of node.

$$U (ni) = N (ni) - [N (ni) \cap N(s)] - \{s\} \qquad (1)$$

The rebroadcast delay Td (Ni) of node Ni.

$$Tp (ni) = 1 - |N(s) \cap N(ni)| / |N(s)|$$

$$Td(Ni) = \text{max delay} * Tp(ni) \qquad (2)$$

3.3. Neighbor Knowledge and Rebroadcast Probability

The node which has a larger rebroadcast delay may listen to RREQ packets from the nodes which have lower one. For example if node ni receives a duplicate RREQ packet from its neighbor nj , it knows that how many its neighbors have been covered by the RREQ packet from nj.thus node ni could further adjust its UCN set according to the neighbor list in the RREQ packet from nj.

$$U(Ni) = u(ni) - [u(ni) \cap n(nj)] \qquad (3)$$

Now we study how to use the final UCN set to set the rebroadcast probability.

$$Ra (ni) = |U(ni)| / |N(ni)| \qquad (4)$$

$$Fc (ni) = Nc / |N(ni)| \qquad (5)$$

Nc=5.1774 log n , the n is the number of nodes in the network $|N(ni)| > Nc, fc(ni) < 1$ in the dense area of the network. $|N(ni)| < Nc, fc(ni) > 1$ in the sparse area of the network, then node ni should forward the RREQ packet in order to approach network connectivity .combining the additional coverage ratio and connectivity factor, to obtain the rebroadcast probability pre(ni) of node ni :

$$Pre(ni) = f c(ni) . ra(ni) \qquad (6)$$

Where , if the pre(ni) is >1 , to set the pre (ni) to 1.
Algorithm description:
The formal description of the Neighbor Coverage-based Probabilistic Rebroadcast for reducing routing overhead in route discovery is shown in Algorithm 1.
Algorithm 1. NCPR
Definitions:
RREQ v: RREQ packet received from node v.
Rv.id: the unique identifier (id) of RREQ v.
N(u): Neighbor set of node u.
U(u, x): Uncovered neighbors set of node u for RREQ whose
id is x.
Timer(u, x): Timer of node u for RREQ packet whose id is
x.
{Note that, in the actual implementation of NCPR protocol, every different RREQ needs a UCN set and a Timer.}
1: if ni receives a new RREQs from s then
2: {Compute initial uncovered neighbors set U(ni,Rs,id) for RREQ s:}
3: U(ni,R,:id) = N(ni) –[N(ni) \ N(s)]- (s)
4: {Compute the rebroadcast delay Td(ni):}
5: Tp(ni) = 1-|N(s)∩N(ni)|/|N(s)|
6: Td(Ni) = max delay *Tp(ni)
7: Set a Timer(ni,Rs,id) according to Td(ni)
8: end if
9:
10: while ni receives a duplicate RREQ j from nj before Timer (Ni, Rs, id) expires do
11: {Adjust U (ni, Rs, id) :}
12: U(ni,Rs,id)= (ni,Rs,id) – [U(ni,R,:id) \ N(nj)]
13: discard(RREQ j)
14: end while
15:
16: if Timer (ni,Rs,id) expires then
17: {Compute the rebroadcast probability Pre(ni):}
18: Ra(ni) = |U(ni,Rs,id)|/|N(ni)|
19: Fc (ni) =Nc/|N(ni)|
20: Pre(ni)=fc(ni).ra(ni)
21: if Random (0,1) <= Pre(ni) then
22: broadcast (RREQs)
23: else

24: discard (RREQs)
25: end if
26: end if

4. Result and Experiment

This result will show the comparison of AODV and NCPR using the term throughput, packet delivery, end to end delay .

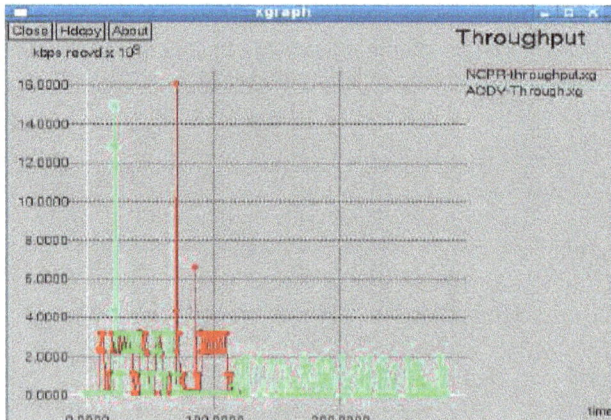

Fig (3). show the throughput

Throughput between the time and packet sent kbps. This graph show comparison of throughput between the AODV and NCPR protocol .the AODV protocol gives the efficient throughput at the end of the packet delivery. But the NCPR protocol gives the most efficient throughput in starting itself.

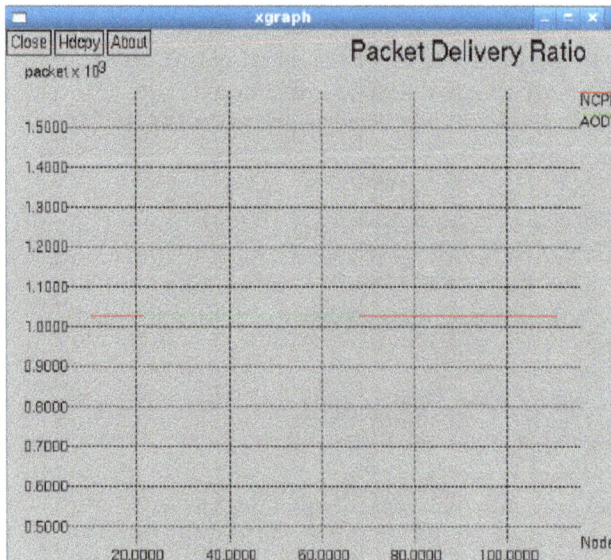

Fig (4). show the packet delivery ratio

Packet delivery ratio comparison between AODV and NCPR protocol. The packet delivery ratio means plot the graph between node and average packet delivery .it shows that PDR of AODV is efficient.

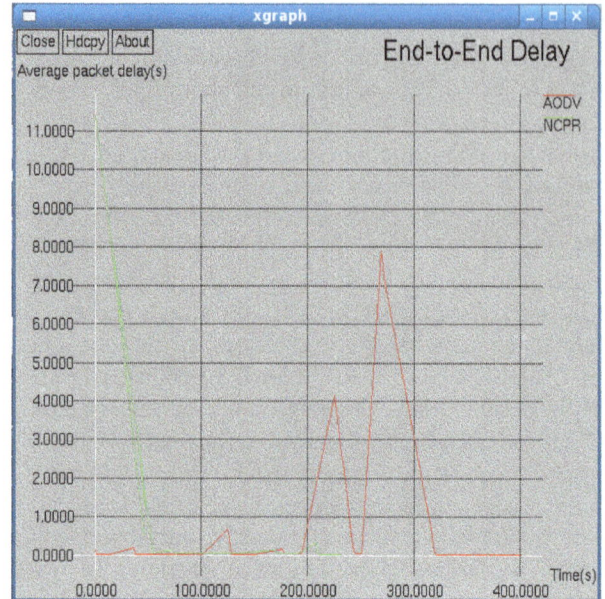

Fig (5). shows the end to and delay

The end to end delay of NCPR is high at initial time interval but when time increases as compared to AODV. In average packet delay of NCPR protocol is more efficient than AODV protocol packet delay

5. Conclusion

In this paper introduced probabilistic rebroadcast mechanism based on neighbor coverage to reduce the routing overhead in manet.the Paper to be constructed on the basis of good performance of the network is in high density or the traffic load is high. We proposed a new scheme to dynamically calculate the rebroadcast delay, which is used to determine the forwarding order and more exploit the neighbor coverage knowledge. Because of less redundant rebroadcast the proposed protocol mitigates the network collision and contention, so as to increase the packet delivery ratio and throughput and decrease the average end to end delay.

Reference

[1]   C. Perkins, E. Belding-Royer, and S. Das, Ad Hoc On-Demand Distance Vector (AODV) Routing, IETF RFC 3561, 2003.

[2]   D. Johnson, Y. Hu, and D. Maltz, The Dynamic Source Routing Protocol for Mobile Ad Hoc Networks (DSR) for IPv4, IETF RFC 4728, vol. 15, pp. 153-181, 2007.

[3]   H. AlAamri, M. Abolhasan, and T. Wysocki, "On Optimising Route Discovery in Absence of Previous Route Information in MANETs," Proc. IEEE Vehicular Technology Conf. (VTC), pp. 1-5, 2009.

[4]   X. Wu, H.R. Sadjadpour, and J.J. Garcia-Luna-Aceves, "Routing Overhead as a Function of Node Mobility: Modeling Framework and Implications on Proactive Routing," Proc. IEEE Int'l Conf. Mobile Ad Hoc and Sensor Systems (MASS '07), pp. 1- 9, 2007.

[5]   S.Y. Ni, Y.C. Tseng, Y.S. Chen, and J.P. Sheu, "The Broadcast Storm Problem in a Mobile Ad Hoc Network," Proc. ACM/IEEE Mob iCom, pp. 151-162, 1999.

[6]   A. Mohammed, M. Ould-Khaoua, L.M. Mackenzie, C. Perkins, and J.D. Abdulai, "Probabilistic Counter-Based Route Discovery for Mobile Ad Hoc Networks," Proc. Int'l Conf. Wireless Comm. And Mobile Computing: Connecting the World Wirelessly (IWCMC '09), pp. 1335-1339, 2009.

[7]   B. Williams and T. Camp, "Comparison of Broadcasting Techniques for Mobile Ad Hoc Networks," Proc. ACM MobiHoc, pp. 194- 205, 2002.

[8]   J. Kim, Q. Zhang, and D.P. Agrawal, "Probabilistic Broadcasting Based on Coverage Area and Neighbor Confirmation in Mobile Ad Hoc Networks," Proc. IEEE GlobeCom, 2004.

[9]   J.D. Abdulai, M. Ould-Khaoua, and L.M. Mackenzie, "Improving Probabilistic Route Discovery in Mobile Ad Hoc Networks," Proc. IEEE Conf. Local Computer Networks, pp. 739-746, 2007.

[10] Z. Haas, J.Y. Halpern, and L. Li, "Gossip-Based Ad Hoc Routing," Proc. IEEE INFOCOM, vol. 21, pp. 1707-1716, 2002.

[11] W. Peng and X. Lu, "On the Reduction of Broadcast Redundancy in Mobile Ad Hoc Networks," Proc. ACM MobiHoc, pp. 129-130, 2000.

[12] J.D. Abdulai, M. Ould-Khaoua, L.M. Mackenzie, and A. Mohammed, "Neighbour Coverage: A Dynamic Probabilistic Route Discovery for Mobile Ad Hoc Networks," Proc. Int'l Symp. Performance Evaluation of Computer and Telecomm. Systems (SPECTS'08), pp. 165-172, 2008.

[13] J. Chen, Y.Z. Lee, H. Zhou, M. Gerla, and Y. Shu, "Robust Ad Hoc Routing for Lossy Wireless Environment," Proc. IEEE Conf. Military Comm. (MILCOM '06), pp. 1-7, 2006.

[14] A. Keshavarz-Haddady, V. Ribeirox, and R. Riedi, "DRB and DCCB: Efficient and Robust Dynamic Broadcast for Ad Hoc and Sensor Networks," Proc. IEEE Comm. Soc. Conf. Sensor, Mesh, and Ad Hoc Comm. and networks (SECON '07), pp. 253-262, 2007.

[15] F. Stann, J. Heidemann, R. Shroff, and M.Z. Murtaza, "RBP: Robust Broadcast Propagation in Wireless Networks," Proc. Int'l Conf. Embedded Networked Sensor Systems (SenSys '06), pp. 85-98, 2006.

[16] F. Xue and P.R. Kumar, "The Number of Neighbors Needed for Connectivity of Wireless Networks," Wireless Networks, vol. 10, no. 2, pp. 169-181, 2004.

[17] X.M. Zhang, E.B. Wang, J.J. Xia, and D.K. Sung, "An Estimated Distance Based Routing Protocol for Mobile Ad Hoc Networks," IEEE Trans. Vehicular Technology, vol. 60, no. 7, pp. 3473-3484, Sept. 2011.

[18] Y.-C. Tseng, S.-Y. Ni, Y.-S.Chen, and J.-P. Sheu, "The Broadcast Storm Problem in a Mobile Ad Hoc Network," Wireless Networks, vol. 8, nos. 2/3, pp. 153-167, Mar.-May 2002.

[19] http://moment.cs.ucsb.edu/AODV/aodv.html

[20] http://www.cs.cmu.edu/~dmaltz/dsr.html Ambarish R. Bhuyar et al, / (IJCSIT) International Journal of Computer Science and Information Technologies, Vol. 5 (1) , 2014,390-393

25

On Basic Frequency and Harmony Parameters with Different Frequencies of Mongolian Traditional Musical Instruments

Bat-Enkh Oyunbileg[1], Baatarkhuu Tsagaan[1], Chuluuntsetseg Jamyaan[2], Tuyatsetseg Badarch[3], Battugs Oyunbileg[2]

[1]Department of Information Technology, School of Information and Telecommunication, MUST, Ulaanbaatar, Mongolia
[2]Mongolian University of Art and Culture, Ulaanbaatar, Mongolia
[3]Mongolian National University, Department of Information Technology, Ulaanbaatar, Mongolia

Email address:
o_bat_enkh@yahoo.com (Bat-Enkh O.), baatarkhuuc@yahoo.com (Baatarkhuu T.), ba.tuyatsetseg@mnu.edu.mn (Tuyatsetseg B.)

Abstract: An instrument is a device that measures and/or regulates physical quantity/process variables as flow, temperature, level, or pressure. Instruments include many varied contrivance that can be as simple as valves and transmitters, and as complex as analyzers. We are determining Mongolian basic five instruments' electrical parameters, such as frequency and Hertz.

Keywords: Basic harmony, electrical parameters, Hertz and Db, harmonic series

1. Introduction

Sound is air in motion – pushed, pulled, blown, plucked, talked, or sung into motion. Music is sound's highest achievement, a wonderfully varied mixture of patterned vibrations sent into the air by all kinds of instruments, from a cricket's hind legs to a massive pipe organ. Most people first visualize the frequency range from 20 Hz to 20,000 Hz.

Pitched musical instruments are often based on an approximate harmonic oscillator such as a string or a column of air, which oscillates at numerous frequencies simultaneously. At these resonant frequencies, waves travel in both directions along the string or air column, reinforcing and canceling each other to form standing waves. Interaction with the surrounding air causes audible sound waves, which travel away from the instrument.

The harmonic series is an arithmetic series ($1 \times f$, $2 \times f$, $3 \times f$, $4 \times f$, $5 \times f$, …). In terms of frequency (measured in cycles per second, or Hertz (Hz) where f is the fundamental frequency), the difference between consecutive harmonic is therefore constant and equal to the fundamental. But because our ears respond to sound nonlinearly, we perceive higher harmonics as "closer together" than lower ones. On the other hand, the octave series is a geometric progression ($2 \times f$, $4 \times f$, $8 \times f$, $16 \times f$, …), and we hear these distances as "the same" in the sense of musical interval. In terms of what we hear, each octave in the harmonic series is divided into increasingly "smaller" and more numerous intervals [2, 3].

The second harmonic (or first overtone), twice the frequency of the fundamental, sounds an octave higher; the third harmonic, three times the frequency of the fundamental, sounds a perfect fifth above the second. The fourth harmonic vibrates at four times the frequency of the fundamental and sounds a perfect fourth above the third (two octaves above the fundamental). Double the harmonic number means double the frequency (which sounds an octave higher).

2. Physical Characteristics of Mongolian Traditional Musical Instrument

There are Mongolian traditional basic five instruments, such as morin khuur, shudarga, khuuchir, limbe and yochin. The traditional musical instruments of Mongolia are most close to their life and customs and have constantly been used in their life. The Mongolian life and thought are peculiar and the folk music is distinctive in its melodious feature and harmonious expression of the truth of the life.

Limbe: It is a wind instrument. The instrument is frequently used in accompaniment, occasionally also as a solo instrument. The sound reflects what is heard in the nature or the sounds of the natural and social environment.

Morin Khuur: (string instrument – horse-head-violin).

The Morin khuur is a typical Mongolian two-stringed instrument. The body and the neck are carved from wood. The end of the neck has the form of a horse-head and the sound is similar to that of a violin or a cello. The strings are made of dried deer or mountain sheep sinews.

Although, these traditional musical instruments are different in their mode, sound, structure and form they are compatible and correlated [6–8].

First, most of the above-mentioned musical instruments are stringed and bowed instruments, but only the limbe is a wind instrument and it plays a connecting role in softening and harmony of the sounds of the instruments.

Second, for their timbres they all are based on the five basic types of the pentatonic timbre of the folk music. For example, it is noted that the notes C, D, F, G and B, a popular pentatonic scale are similar to the composition of the pitches of the instruments including the shudarga, morinkhuur, and khuuchir. According to it the sound of each musical instrument has its peculiar timbre. The sounds of different musical instruments are proper high and low and they are different, depending on their timbre. The sound of a musical instrument is the physical phenomenon and feeling. The difference of the timbres depends on the proper vibrations of the waves of their double sounds. In other words, if the vibration of generating of sound is slow and wide in scope its continuation is slow. The vibration is an exact concept and first of all it is a combination of the vibrations of many parts of one body.

For example, according to the strings they vibrate only in its length and the vibration on the parts 2, 3, 4, 5, and 6 produces their own sound. Therefore, when the string fully vibrates its main tone is specially heard and it produces overtone. In other words, it is a sound refraction and one sound is the refraction of many micro sounds.

When a sound system in the bourdon system possesses a huge sound space in the stringed instrument, using the inner resource of a note it uses ultrahigh and infra low frequency. It is explained that although the sounds which are heard below the basic sound that can define the ascending overtone coincidence or the pitch, high sounds, accompanying sounds or double pitches sound above the basic sound, they belong to the ultrahigh sounds of the music [1–5, 7, 8].

According to the acoustics there are the additional sounds or resonance sounds except the basic sounds. The sounds of a musical instrument consist of the basic frequency and harmonies with different frequencies. For example the note G (Sol) of the lower octave of the shudarga unites many harmonies except the basic harmony (Table 1).

It in fact produces the harmony or timbre. The method to find the structure of the harmony of the sounds is disintegration of the frequency.

Table 1. G (Sol) Note of Lower Octave.

Harmonic Series	Hertz	Db
1	387.7	31.9
2	578.8	40.6
3	775.3	46.3
Basic	966.4	47.4
4	1161	42.6
5	1363	40.4
6	1552	37.4
7	1748	36.9
8	1939	31.6
9	2134	34.7
10	2327	34.5

3. Parameters of "morin khuur" Instrument

First, we defined some electrical parameters of Mongolian traditional instrument "Morin khuur."

"Fig. 1" shows waveform and spectrogram of instrument "Morin khuur," which illustrate Fa note of lower octave.

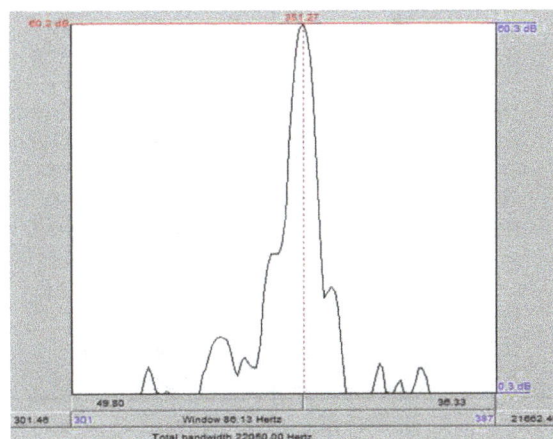

Figure 1. Electrical signals of Fa note of lower octave.

"Fig. 2" shows spectrum slice of instrument "Morin khuur", which illustrates Fa note of lower octave. In this case, total duration is equal to 2.5 seconds and total bandwidth is equal to 22050 Hertz [3,10].

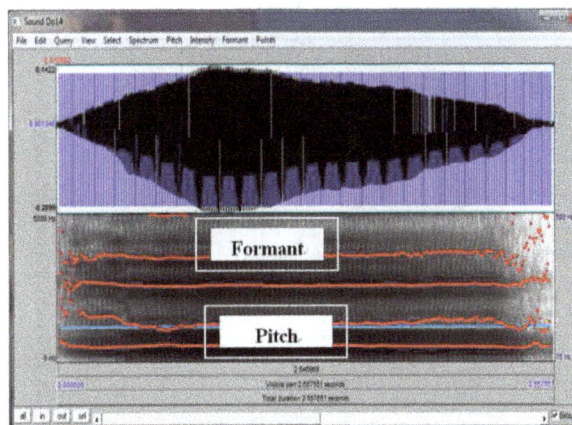

Figure 2. Electrical signals of "Morin khuur" instrument.

Fa note of lower octave for "Morin khuur" instrument include many harmonics. Its harmonic series, its Hertz and Db are presented in following Table 2. In this research we determined all electrical parameters of instrument "Morin khuur." In following table (Table 3), there are basic, F1, F2, F3 and F4 harmonic series, its hertz and Db.

Table 2. Fa Note of Lower Octave.

Harmonic Series	Hertz	Db
1	175.6	56.6
2	351.2	60.2
3	526.7	52.1
4	702.3	49.3
5	878	46.1
6	1053	40.6
7	1229	37.1
8	1404	36.5
9	1580	29.4

Table 3. Electrical Signal of "Morin khuur" Instrument.

		Basic	F1	F2	F3	F4
Do	Hz	173.98	348	523.4	697.9	872.3
C	Db	91.5	66	75.1	72.7	56
Re	Hz	231.2	464.28	698.7	931.8	1164.8
D	Db	84.5	83.7	78.9	54.4	62.4
Mi	Hz	174.7	349.6	523.6	697.5	872.6
E	Db	91.5	71.2	76.2	75.1	58.1
Fa	Hz	697.5	1046	1396	1744	2094
F	Db	84	61.8	54.9	51.8	68.6
So	Hz	522.5	1045	1570	2090	2615
G	Db	80	57.9	49.3	66.4	64.5
La	Hz	3491	4187	4886	5585	6284
A	Db	74.1	57.6	49.5	55.6	53.1
Si	Hz	2637	3516	4397	5279	6154
B	Db	77.7	74.8	67.7	64.9	59.9

4. Parameters of "Limbe" Instrument

We also defined basic frequency and harmony parameters parameters of Mongolian traditional instrument "Limbe". For example, we determined that basic parameters of Do note for

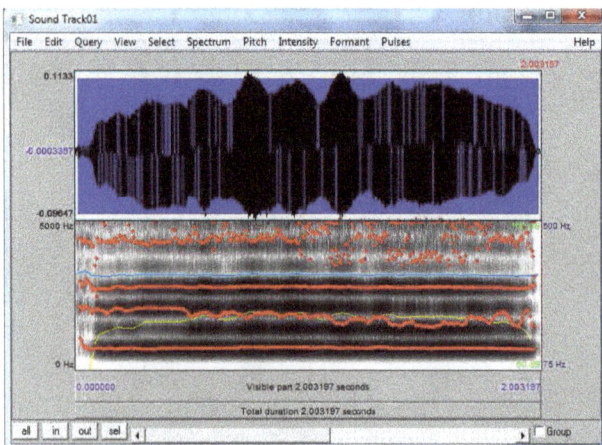

Figure 3. Electrical signals of Do note of lower octave.

"Limbe" are 63.0 Db, 689.7 Hertz.

"Fig. 3" and "Fig. 4" show waveform, spectrogram and spectrum slice of instrument "Limber", which illustrate do note of lower octave.

Also results of W (Bandwidth) and Harmonics (F1, F2, F3, F4) are presented in following Table 4 [3,10].

In this research we determined basic electrical parameters of instrument "Limbe".

Table 4. Parameters of Do note for "Limbe" Instrument.

F1 (Hz)	W	F2 (Hz)	W	F3 (Hz)	W	F4 (Hz)	W
716.3	92.6	1770.2	312.3	2741.9	106.1	4228.3	1133

In following Table 5, there are basic, F1, F2, F3 and F4 harmonic series, its hertz and Db.

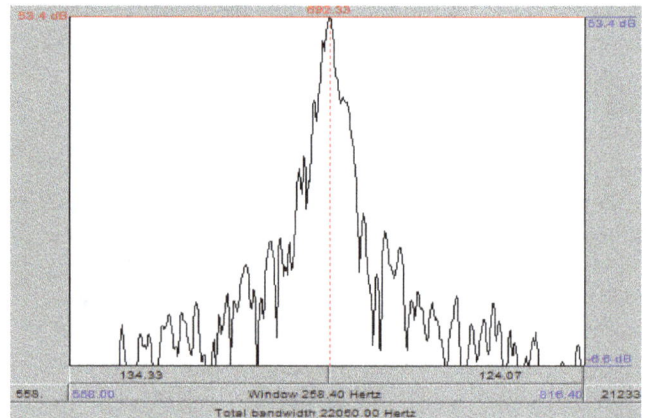

Figure 4. Electrical signals of "Limbe" instrument .

We used Praat 4 version. Praat is a software that is used to acoustic analysis. In other words, Praat is a free scientific computers software package for the analysis of speech in phonetics.

Table 5. Electrical Signal of "Limbe" Instrument.

		Basic	F1	F2	F3	F4
Do	Hz	691	1381	2074	2765	3453
C	Db	84.2	67.3	64.5	54	40.9
Re	Hz	459	920.79	1383	1849	2304
D	Db	78.9	66	62.9	45.1	51.4
Mi	Hz	458.6	921	1385	1850	2308
E	Db	77.4	60.6	60.6	48.8	51.4
Fa	Hz	923	1861	2788	3718	4645
F	Db	80.8	55.6	65.1	37.5	54.8
So	Hz	1390	2074	2770	4165	5560
G	Db	83.2	36.2	50.7	56.6	49.8
La	Hz	1860	2798	3736	5612	7467
A	Db	72.6	36.8	58	62.9	42.1
Si	Hz	2366	2783	3014	4745	7079
B	Db	82.7	28.9	29.1	41.3	40.6

5. Conclusion

Using Praat 4 software, we define basic frequency and harmony parameters of Mongolian traditional instrument. Mongolian basic five instruments that were enriched to become orchestra are kept until our century. Further we need to research a consistency between other countries and Mongolian traditional instruments.

References

[1] J.G. Roederer. The Physics and Psychophysics of Music, p. 106, ISBN 0-387-94366-8, 1995.

[2] William Forde Thompson, Music, Thought, and Feeling: Understanding the Psychology of Music. p. 46. ISBN 978-0-19-537707-1, 2008.

[3] Chuluuntsetseg. J, Bat-Enkh. O. Physical characteristics of Mongolian traditional musical instruments. Mongolian people culture in a globalizing space: Proceedings of the International Scientific conference, Elista. Russia, 2012.

[4] J.R. Pierce, Consonance & Scales. In: Perry R. Cook. Music, Cognition, and Computerized Sound. MIT Press. ISBN 978-0-262-53190-0, 2001.

[5] Martha Goodway, Jay Scott Odell, The Historical Harpsichord Volume Two: The Metallurgy of 17th- and 18th-Century Music Wire. Pendragon Press. ISBN 978-0-918728-54-8, 1987.

[6] Jantsannorov.N Theory of 5 tone of Mongolian musical instrument. UB, Mongolia, pp. 42–43, 2006.

[7] Enkhtsetseg .D. Khuuriin tatlaga, that is phenomenon of musical cultural in Mongolia. UB, 2011.

[8] Ganbaatar.A Foreign terminology and vocabulary words used in art field, UB, 2006.

[9] G. Eason, B. Noble, and I. N. Sneddon, "On certain integrals of Lipschitz-Hankel type involving products of Bessel functions," Phil. Trans. Roy. Soc. London, vol. A247, pp. 529–551, April 1955. (references).

[10] Bat-Enkh.O, Baatarkhuu Ts, Chuluuntsetseg J, Battugs O,. Some electrical parameters of Mongolian traditional musical instrument. WIT Transactions on Information and Communication Technologies,. Vol. 72, 2015, pp. 660-666. doi: 10.2495/AETIE140921.

26

A review of ultra-short soliton pulse generation using InGaAsP/InP microring resonator (MRR) system

I. S. Amiri[1, *], H. Ahmad[1], Hamza M. R. Al-Khafaji[2]

[1]Photonics Research Centre, University of Malaya (UM), 50603 Kuala Lumpur, Malaysia
[2]Wireless Communication Centre, Faculty of Electrical Engineering, Universiti Teknologi Malaysia (UTM), 81310 UTM Skudai, Johor, Malaysia

Email address:
amiri@um.edu.my (I. S. Amiri), harith@um.edu.my (H. Ahmad), hamza@utm.my (H. M. R. Al-Khafaji)

Abstract: System of microring resonators (MRRs) incorporating with an add/drop MRR system are presented to generate single and multi-temporal and spatial ultra-short soliton pulses applicable in optical soliton communications. The chaotic signals caused by the nonlinear condition could be generated and propagated within the system. The Kerr effect in the MRR system induces the nonlinearity condition. The proposed MRR systems are used to generate ultra-short soliton pulse within the system. Using the appropriate MRR parameters, ultra-short spatial and temporal signals are generated spreading over the spectrum. In this work, narrow soliton pulses could be localized within the proposed systems. Here soliton pulses of 0.7 ps, 0.83 fs and 19 pm are generated using series of MRRs connected to an add/drop MRR system. The nonlinear refractive index is $n_2 = 2.2 \times 10^{-17}$ m^2/W. Using the panda ring resonator system, the ultra-short soliton pulses with full width at half maximum (FWHM) and free spectral range (FSR) of 5 MHz and 2 GHz, were generated at the throughput port. The output signals pulses with FWHM of 10 MHz and FSR of 2 GHz could be obtained at the drop port of the system. As second results using this system, multi-carrier soliton pulses with FWHM=20 MHz are localized within this system with respect to 20,000 roundtrips of the input pulse. Localized optical tweezers could be generated using the half-Panda MRR system, where the peaks have FWHM and FSR of 8.9 nm and 50 nm respectively. The nonlinear refractive index was selected to $n_2 = 2.5 \times 10^{-17}$ m^2/W.

Keywords: Microring Resonator (MRR), Ultra-Short Soliton, Spatial and Temporal Soliton, Kerr Effect

1. Introduction

Ultra-short soliton generation is very attractive especially when it uses quantum cryptography in an optical communication network, which is reported by Amiri [1-4]. A spectrum of light over a broad range can be generated, thus an optical soliton pulse is advised as a powerful laser pulse which can be utilized to generate chaotic filter features when spreading within microring resonators (MRRs) [5-8]. Therefore, the capacity of the transmission data can be secured and increased when the chaotic packet switching is utilized provided using this technique, [9-11]. In this study, we propose a systems of MRRs that uses ultra-short localized spatial and temporal soliton pulses to form the high capacity communication [12-16]. The device parameters are simulated associating to the practical device parameters [17-21]. Orthogonal frequency division multiplexing (OFDM) is a combination of modulation and multiplexing [22-25]. Modulation refers to the process of changing the carrier phase, frequency, amplitude, or their combination with a modulated signal that typically contains information to be transmitted [26-29]. However, the aim of multiplexing is to share a bandwidth. Single-carrier modulation is a technology that modulates information onto only one carrier [30-33]. The main problem of this technology is satisfying the need for high bandwidth in fixed spectrum limits of one single-carrier [34-38]. High data rate in one carrier causes a high symbol rate. As the duration of one symbol or bit becomes smaller, the system becomes more susceptible to loss of information from impulse noise, signal reflections, and other impairments [39-43]. These impairments can impede the ability to recover information sent. In addition, as the bandwidth used by a single-carrier system increases, the susceptibility to interference from other continuous signal sources becomes greater. This type of interference is commonly labelled as a carrier wave (CW) or frequency interference [44-47]. Dark-Gaussian soliton controls within a semiconductor add/drop multiplexer has numerous applications in optical communication [48-53]. Nano optical tweezers technique has become a powerful tool for manipulation of micrometer-sized particles/photons in three spatial dimensions [54-62]. It has the unique ability to trap and manipulate molecules/photons at mesoscopic scales with widespread applications in biology and physical sciences [63-69]. The output is achieved when the high optical field is set up as an optical tweezers [70-73]. For communication's application purposes, the optical tweezers can be used to generate entangled photon within the proposed network system [74-78].

Internet security becomes an important function in the modern internet service. However, the security technique known as quantum cryptography has been widely used and investigated in many applications, using ultra-short optical solitons [79-83]. Amiri et al, have proposed the new design of optical switching used in optical communication systems [84-89]. This method uses nonlinear behaviour of light in MRR which can be applicable for high-capacity transmission and switching [90-94]. Transmission of all-optical OFDM can be implemented by generating multiple optical subcarriers, separating these subcarriers via optical devices, and finally modulating each subcarrier separately [95-97]. Optical carrier generation thus constitutes the basic building block to implement OFDM transmission fully in the optical domain [98-101]. The MRR system provides a viable means to generate this building block that represents the generation of the necessary multi-carriers [102-104]. A soliton solution of the nonlinear wave equation is always stable over a long distance link, and this stability of soliton signals is even more remarkable than the possibility of balancing dispersion and non-linearity [105-110]. One important aspect of the MRR system is that suitable tuning of the system parameters allows for desired soliton carriers with specific key characteristics, such as full width at half maximum (FWHM) and stability, to be obtained at the drop/through ports of the system [111-116].

2. Picosecond Soliton Pulses Generation

The schematic diagram of the proposed system is shown in Fig. 1. A soliton pulse with 20 ns pulse width, peak power at 500 mW is input into the system. The suitable ring parameters are used, for instance, ring radii R_1=10 μm, R_2=5μm, and R_3=4μm. In order to make the system associate with the practical device, the selected parameters of the system are fixed to λ_0=1.55μm, n_0=3.34 (InGaAsP/InP) [117-121], A_{eff}=0.50, 0.25μm^2 and 0.12μm^2 for the different radii of MRRs respectively [122-124], α=0.5dBmm^{-1}, γ=0.1 [125-127]. The coupling coefficient (κ) of the MRR ranged from 0.50 to 0.975 [128-130].

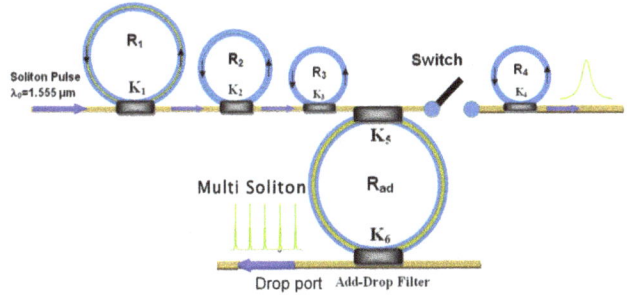

Fig. 1. *Schematic diagram of soliton pulse generation, Rs, ring radii and κs, coupling coefficients*

The soliton pulse is introduced into the proposed system. The input optical field (E_{in}) can be in the form of bright soliton (equation 1), dark soliton (equation 2) and Gaussian laser beam (equation 3) [131-136]. Here, the bright soliton pulse is input into the system. The input optical field (E_{in}) of the bright soliton, dark soliton and Gaussian laser beam are given by:

$$E_{in} = A \sec h\left[\frac{T}{T_0}\right]\exp\left[\left(\frac{z}{2L_D}\right) - i\omega_0 t\right] \qquad (1)$$

$$E_{in} = A \tanh\left[\frac{T}{T_0}\right]\exp\left[\left(\frac{z}{2L_D}\right) - i\omega_0 t\right], \qquad (2)$$

$$E_{in}(t) = A \exp\left[\left(\frac{z}{2L_D}\right) - i\omega_0 t\right] \qquad (3)$$

A and z are the optical field amplitude and propagation distance, respectively [137, 138]. T is a soliton pulse propagation time in a frame moving at the group velocity [139-141], $T = t - \beta_1 \times z$, where β_1 and β_2 are the coefficients of the linear and second order terms of Taylor expansion of the propagation constant [142-144]. $L_D = T_0^2/|\beta_2|$ is the dispersion length of the soliton pulse [145, 146]. The frequency shift of the soliton is ω_0.

This solution reports a pulse that maintains its temporal width invariance as it spreads, and thus is called a temporal soliton [147-149]. When a soliton peak intensity $\left(|\beta_2/\Gamma T_0^2|\right)$ is established, then T_o is recognized [150-152]. For the soliton pulse in the MRR system, a balance should be accomplished between the dispersion length (L_D) and the nonlinear length ($L_{NL} = 1/\Gamma\varphi_{NL}$) [153-155], where $\Gamma = n_2 \times k_0$, is the length scale over which dispersive or nonlinear effects makes the beam gets wider or narrower [156, 157].

Since for a soliton pulse, there is a balance between dispersion and nonlinear lengths, therefore $L_D = L_{NL}$ [158, 159]. Whenever light spreads within the nonlinear medium, the refractive index (n) of light within the medium is contributed by [160-162].

$$n = n_0 + n_2 I = n_0 + \left(\frac{n_2}{A_{eff}}\right)P, \qquad (4)$$

where n_0 and n_2 represent the linear and nonlinear refractive indices, respectively. I and P are the optical intensity and powers, respectively [163, 164]. The effective mode core area of the device is introduced by A_{eff}. For the MRR and NRR, the effective mode core areas range 0.50 to 0.10 µm². When a soliton pulse is input and propagated within a MRR as shown in Fig. 1, the resonant output can be performed, hence, the normalized output of the light field is defined by the ratio between the output and input fields presented by $E_{out}(t)$ and $E_{in}(t)$ in each roundtrip, and can be expressed as [165-168]

$$\left|\frac{E_{out}(t)}{E_{in}(t)}\right|^2 = (1-\gamma)\left[1-\frac{(1-(1-\gamma)x^2)\kappa}{(1-x\sqrt{1-\gamma}\sqrt{1-\kappa})^2+4x\sqrt{1-\gamma}\sqrt{1-\kappa}\sin^2(\frac{\phi}{2})}\right] \quad (5)$$

The equation 5 suggests that a MRR in the special case is very similar to a Fabry-Perot cavity, having an input and an

output mirrors with a field reflectivity, $(1-\kappa)$, and a fully reflecting mirror [169, 170]. κ is the coupling coefficient, and $x=\exp(-\alpha L/2)$ symbolizes a roundtrip loss coefficient, $\Phi_0=kLn_0$ and $\Phi_{NL}=kLn_2|E_{in}|^2$ are the linear and nonlinear phase shifts [171, 172], $k=2\pi/\lambda$ is the wave propagation number in a vacuum. Where L and α are the waveguide length and linear absorption coefficient, respectively [173, 174]. In this work, the iterative method is brought in to obtain the results as shown in equation 3, likewise, when the output field is associated and input into the other ring resonators. The nonlinear refractive index is $n_2=2.2 \times 10^{-17}$ m²/W. In this case, the waveguide loss utilized is 0.5dBmm⁻¹. As shown in Figure 2, the signal is chopped into a smaller signals spreading over the spectrum. The attenuation of the optical power within a MRR is necessitated in order to maintain the constant output gain.

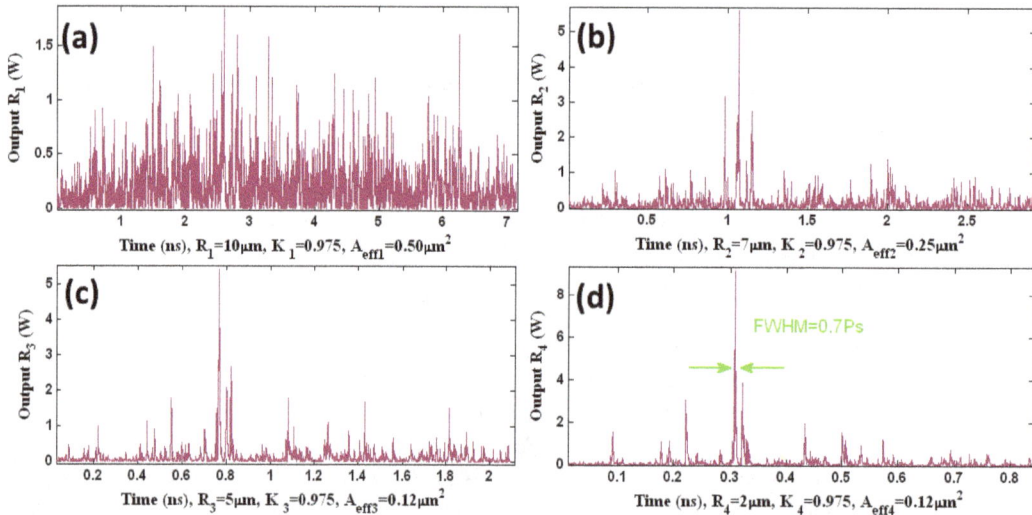

Fig. 2. Results obtained when temporal soliton is localized within a MRR, where (a): Chaotic signals from R_1, (b): Chaotic signals from R_2, (c): temporal soliton, (d): temporal soliton with FWHM of 0.7 ps.

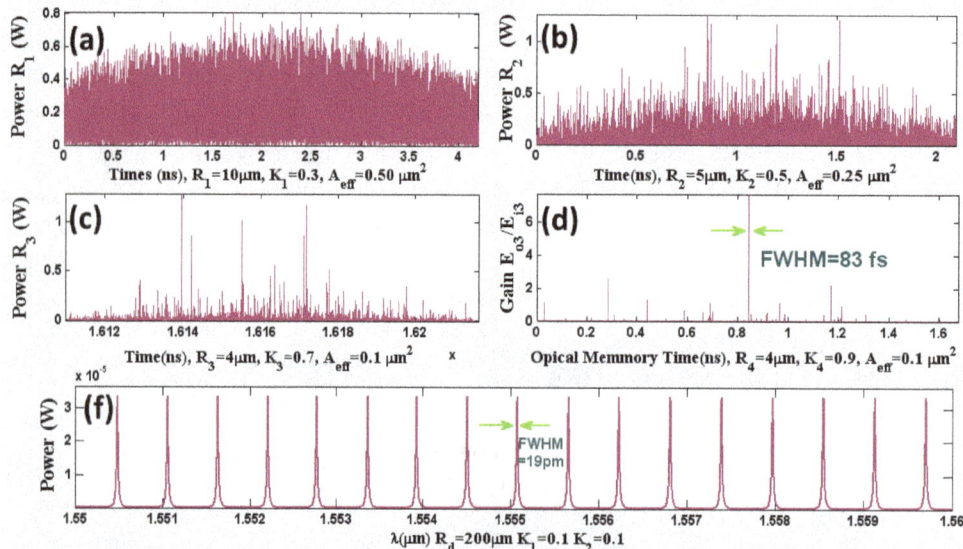

Fig. 3. Results of temporal and spatial soliton generation, where (a): Chaotic signals from R_1, (b): Chaotic signals from R_2, (c): filtering signals, (d): Temporal soliton with FWHM of 83 fs, (e): Spatial soliton with FWHM=19 pm

Figure 3 demonstrates the results while temporal and spatial optical soliton pulses are localized within a MRR and add/drop MRR systems with 20,000 roundtrips, thus the optical pulse of 83 fs can be generated. Here, the ring radii are $R_1= 10$ μm, $R_2= 5$ μm, $R_3= 4$ μm, $R_4=4$ μm and $R_{ad}= 200$ μm with coupling coefficient of $\kappa_1=0.3$, $\kappa_2=0.5$, $\kappa_3=0.7$, $\kappa_4=0.9$, $\kappa_5=0.1$ and $\kappa_6=0.1$.

3. Soliton Generation in Frequency domain for Communication Application

The system of soliton frequency band generation is shown in Figure 4. Here, series of ring resonators are connected to a panda ring resonator system. The filtering process of the input soliton pulses is performed via the ring resonators, where frequency band ranges 40–60 GHz can be obtained via the output signals of the panda ring resonator system. The soliton pulse shown by equation 1 is introduced into the nonlinear system.

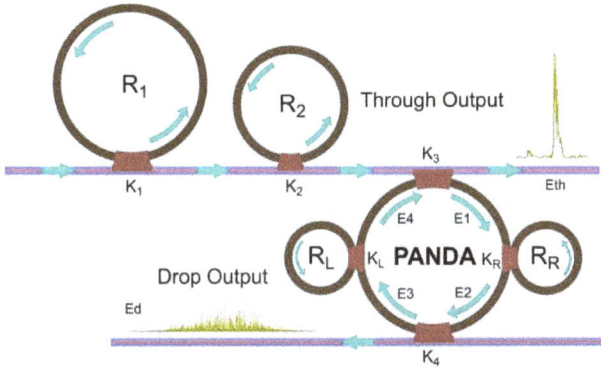

Fig. 4. *Optical frequency band generation system using a panda ring resonator connected to a series of MRRs*

A frequency soliton pulse can be formed and trapped within the panda ring resonator system with suitable ring parameters. The centred ring of the panda ring resonator system has a radius of 100 μm and coupling coefficients of $\kappa_3 = 0.35$ and $\kappa_4 = 0.30$, where the right and left rings have radii of 18 μm and 8 μm, respectively. The effective core areas of the right and left rings are $A_{eff1} =A_{eff2} =0.25\mu m^2$, where the coupling coefficients have been selected to $\kappa_R = 0.22$ and $\kappa_L =0.10$. Filtering of the interior soliton signals can be performed when the pulses pass through the couplers, κ_3 and κ_4. The output signals from the throughput and drop ports of the system can be seen in Figure 5, where soliton range of 40–60 GHz are generated and used in many optical communication applications, such as wireless personal area networks (WPANs) and wireless local area networks (WLANs). The throughput output (E_{th}) shows localized ultra-short soliton pulses with an FWHM and free spectral range (FSR) of 5 MHz and 2 GHz, respectively, where soliton pulses at frequencies of 50 GHz and 52 GHz are generated. The drop output signals expressed by E_d are shown in Figure

5(b-c), where, pulses with FWHM of 10 MHz and FSR of 2 GHz are obtained.

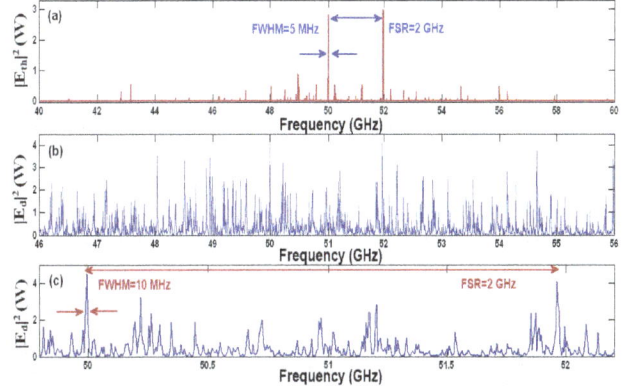

Fig. 5. *Results of localized solitons: (a): throughput output signal with FWHM=5 MHz and FSR=2 GHz, (b): chaotic soliton pulses, (c): expansion of the soliton pulses ranges 50–52 GHz with FWHM=10 MHz and FSR=2 GHz*

A bright soliton with a central frequency of 92.5 GHz and power of 2 W is introduced into the first MRR. The results of the chaotic signal generation are shown in Figure 6. The rings' radii and coupling coefficients are $R_1 = 6\mu m$, $\kappa_1 = 0.3$, $R_2 = 4\mu m$ and $\kappa_2 = 0.3$. The ring of the Panda ring resonator system has a radius of 120 μm and coupling coefficients of $\kappa_3 = 0.3$ and $\kappa_4 =0.5$, where the right and left rings have the same radii of $R_R =3\mu m$ and $R_L =3\mu m$. The effective core areas of the rings are $A_{effR}=A_{effL}=0.10\mu m^2$, where the coupling coefficients have been selected to $\kappa_R = 0.2$ and $\kappa_L =0.1$. The wave-guided and fractional coupler intensity losses are $\alpha=0.5$ dBmm-1, and $\gamma=0.1$ respectively. The through put output shows localized narrow bandwidth soliton pulses with FWHM and FSR of 3 MHz and 200 MHz, respectively, where soliton pulses at frequencies of 92.6, 92.8 and 93 GHz are generated. The drop port output signals are shown in Figure 6(c-d), where multi-soliton pulses with FWHM of 11 MHz and FSR of 188 MHz could be generated.

As it can be concluded from the results, the efficiency of this system can be evaluated by the output solitonic carrier signal generated by the MRRs. Long distance fiber communication link is implemented using this method and can be integrated to short distance indoor wireless link for ultra-high data rate transmission.

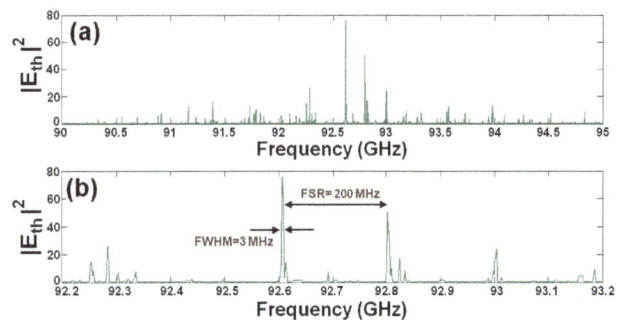

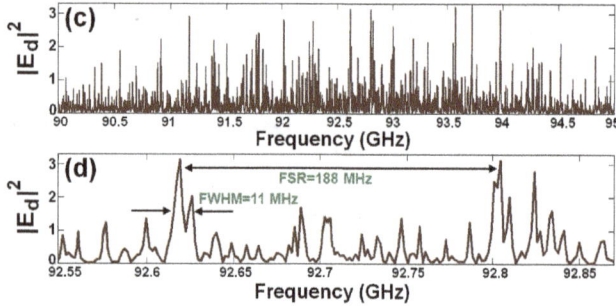

Fig. 6. (a-b) throughput output signal (W/μm^2) with FWHM=3 MHz and FSR=200MHz, (c-d) drop port multi-solitons output signal (W/μm^2) with FWHM=11 MHz and FSR=188 MHz

4. Ultra-Short Soliton Tweezers Generation

The array of dark soliton pulses expressed by equation 2 are introduced into the input port half-Panda MRR system shown in Figure (7). This system consists of an add/drop MRR system connected to a smaller ring resonator on the right side.

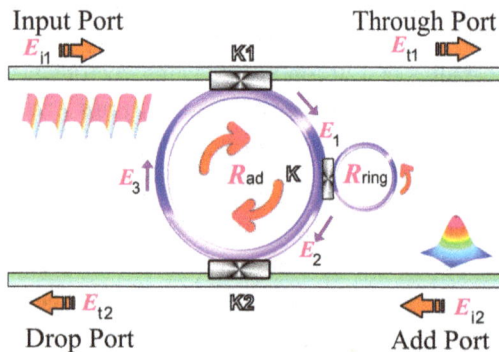

Fig. 7. A schematic diagram of Half-Panda MRR system

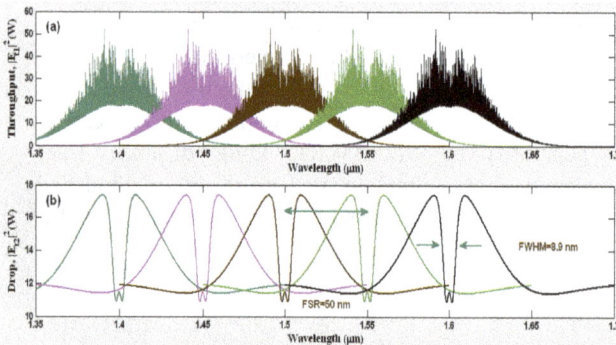

Fig. 8. (a): Through port chaotic output signals (b): drop port output with FWHM=8.9 and FSR=50 nm

Input optical dark solitons and Gaussian laser beam (input into the add port) with powers 2W and 1W respectively are inserted into the half-Panda MRR system. The add/drop MRR system has radius of R_{ad}=15 μm where the coupling coefficients are $\kappa_1 = 0.35$ and $\kappa_2 = 0.25$. The dark solitons are propagating inside the half-Panda MRR system with central wavelengths of $\lambda_0 = 1.4$ μm, 1.45 μm, 1.5 μm, 1.55

μm, 1.6 μm. In order to make the system associate with the practical device (InGaAsP/InP), the selected parameters of the system are fixed to $n_0 = 3.34$ and $n_2 = 2.5 \times 10^{-17}$ m^2/W. Filtered and clear optical tweezers are seen in figure 8 where the peaks have FWHM and FSR of 8.9 nm and 50 nm respectively. In the case of communication networks, generation of narrower signals is recommended to compensate the fiber loss and system's attenuation.

By using suitable dark-Gaussian soliton input powers, tunable optical tweezers can be controlled. This provides the entangled photon as the dynamic optical tweezers probe. The required data can be retrieved via the through and drop ports of the add/drop MRR. High capacity data transmission can be obtained by using more wavelength carriers. The advantage of this study is that ultra-short nano optical tweezers can be generated and transmitted via a network system thus improving the sensitivity and capacity.

5. Multi-Carries Soliton Generation for Wired/Wireless Optical Communication Systems

Using the system shown by Figure 1, the optical soliton carrier can be generated. Figure 9 shows the results when soliton pulses are localized within the MRR and add/drop MRR systems with 20,000 roundtrips, where the soliton pulses with FWHM=20 MHz could be generated as result of the fourth ring resonator shown in Figure 9(d). The output results from the drop port of the system can be seen in the Figure 9(e-f). Here, the ring radii are R_1=15 μm, R_2=9 μm, R_3=6 μm, R_4=6 μm and R_{ad}=100 μm having coupling coefficients of $\kappa_1 = 0.98$, $\kappa_2 = 0.98$, $\kappa_3 = 0.96$, $\kappa_4 = 0.92$, $\kappa_5 = 0.1$ and $\kappa_6 = 0.1$. The nonlinear refractive index is 2.4×10^{-17}. A_{eff1} (μm^2)=0.50, A_{eff2} (μm^2)=0.25 and A_{eff3-4} (μm^2)=0.10.

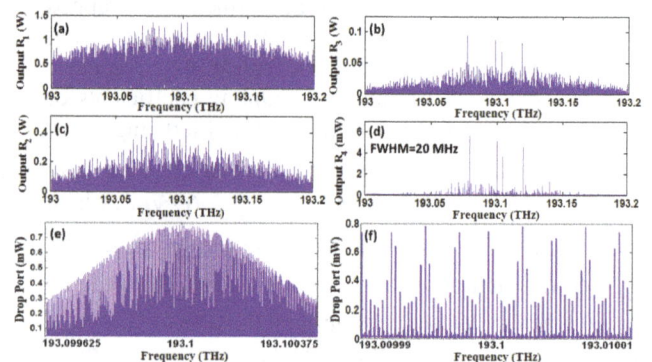

Fig. 9. (a): output from first ring resonator, (b): output from second ring resonator, (c): output from third ring resonator, (d): multi-solitons output from the drop port, (e): multi-carriers generation

Thus, filtering of the input pulse within the system allowed for generation of single- and multi-soliton pulses to be used in a multiple-input and multiple-output (MIMO)-OFDM communication systems.

6. Conclusion

We have proposed an interesting concept of the ultra-short soliton pulse generation using microring resonators (MRRs), in which the single and multiple temporal and spatial soliton pulses could be achieved. The balance established between the dispersion and nonlinear lengths of the soliton pulse presents the soliton behavior known as self-phase modulation, which introduces the optical output constant, meaning that the light pulse can be localized coherently within the nano-waveguide. We have demonstrated that a large bandwidth of the arbitrary soliton pulses can be generated and compressed within a microring waveguide. The chaotic signal generation by means of a soliton pulse in the nonlinear MRRs has been presented. Selected light pulse can be localized and used to perform the high capacity of optical communication due to generate ultra-short bandwidth of the pulses. Localized spatial and temporal soliton pulse are useful to generate optical communication signals applicable for wired/wireless networks. As an application, the classical information and security codes can be formed by using the temporal and spatial soliton pulses, respectively.

Acknowledgements

Amiri and Ahmad would like to acknowledge the financial support from University Malaya/MOHE under grant number UM.C/625/1/HIR/MOHE/SCI/29 and RU002/2013.

References

[1] S. E. Alavi, I. S. Amiri, S. M. Idrus, ASM Supa'at, J. Ali & P. P. Yupapin, (2014) "All Optical OFDM Generation for IEEE802.11a Based on Soliton Carriers Using MicroRing Resonators ", *IEEE Photonics Journal*, 6(1),

[2] Iraj Sadegh Amiri & Harith Ahmad, Optical Soliton Communication Using Ultra-Short Pulses. USA: Springer, 2014.

[3] J. Ali, I. S. Amiri, A. Jalil, A. Kouhnavard, B. Mitatha & P. Yupapin, (2010) "Quantum internet via a quantum processor", presented at the *International Conference on Photonics (ICP 2010)*, Langkawi, Malaysia

[4] I. S. Amiri, A. Nikoukar, J. Ali & P. P. Yupapin, (2012) "Ultra-Short of Pico and Femtosecond Soliton Laser Pulse Using Microring Resonator for Cancer Cells Treatment", *Quantum Matter*, 1(2), 159-165.

[5] J. Ali, K. Raman, A. Afroozeh, I. S. Amiri, M. A. Jalil, I. N. Nawi & P. P. Yupapin, (2010) "Generation of DSA for security application", presented at the *2nd International Science, Social Science, Engineering Energy Conference (I-SEEC 2010)*, Nakhonphanom, Thailand.

[6] Amiri, H Ahmad & MZ Zulkifli, (2014) "Integrated ring resonator system analysis to Optimize the soliton transmission", *International Research Journal of Nanoscience and Nanotechnology*, 1(1), 002-007.

[7] J. Ali, I. S. Amiri, M. A. Jalil, F. K. Mohamad & P. P. Yupapin, (2010) "Optical dark and bright soliton generation and amplification", presented at the *Nanotech Malaysia, International Conference on Enabling Science & Technology*, KLCC, Kuala Lumpur, Malaysia.

[8] J. Ali, M. Kouhnavard, M. A. Jalil & I. S. Amiri, (2010) "Quantum signal processing via an optical potential well", presented at the *Nanotech Malaysia, International Conference on Enabling Science & Technology*, Kuala Lumpur, Malaysia

[9] I. S. Amiri, S. E. Alavi & J. Ali, (2013) "High Capacity Soliton Transmission for Indoor and Outdoor Communications Using Integrated Ring Resonators", *International Journal of Communication Systems*, 28(1), 147–160.

[10] IS Amiri & A Afroozeh, Integrated Ring Resonator Systems, in *Ring Resonator Systems to Perform Optical Communication Enhancement Using Soliton*, ed USA: Springer, 2014.

[11] D. Gifany, I. S. Amiri, M. Ranjbar & J. Ali, (2013) "LOGIC CODES GENERATION AND TRANSMISSION USING AN ENCODING-DECODING SYSTEM", *International Journal of Advances in Engineering & Technology (IJAET)*, 5(2), 37-45

[12] A. Afroozeh, I.S. Amiri, K. Chaudhary, J. Ali & P. P. Yupapin, Analysis of Optical Ring Resonator, in *Advances in Laser and Optics Research*, ed New York: Nova Science, 2015.

[13] A. Afroozeh, I. S. Amiri, M. Kouhnavard, M. Bahadoran, M. A. Jalil, J. Ali & P. P. Yupapin, (2010) "Dark and Bright Soliton trapping using NMRR", presented at the *International Conference on Experimental Mechanics (ICEM)*, Kuala Lumpur, Malaysia.

[14] J. Ali, M. Roslan, M. Jalil, I. S. Amiri, A. Afroozeh, I. Nawi & P. Yupapin, (2010) "DWDM enhancement in micro and nano waveguide", presented at the *AMN-APLOC International Conference*, Wuhan, China.

[15] I. S. Amiri, J. Ali & P. P. Yupapin, (2012) "Enhancement of FSR and Finesse Using Add/Drop Filter and PANDA Ring Resonator Systems", *International Journal of Modern Physics B*, 26(04), 1250034.

[16] I. S. Amiri, (2011), "FWHM Measurement of Localized Optical Soliton", in *The International Conference for Nano materials Synthesis and Characterization*, Malaysia.

[17] I. S. Amiri, S. E. Alavi & H. Ahmad, (2015) "Analytical Treatment of the Ring Resonator Passive Systems and Bandwidth Characterization Using Directional Coupling Coefficients ", *Journal of Computational and Theoretical Nanoscience (CTN)*, 12(3),

[18] M. Bahadoran, I. S. Amiri, A. Afroozeh, J. Ali & P. P. Yupapin, (2011) "Analytical Vernier Effect for Silicon Panda Ring Resonator", presented at the *National Science Postgraduate Conference, NSPC* Universiti Teknologi Malaysia.

[19] I. S. Amiri, R. Ahsan, A. Shahidinejad, J. Ali & P. P. Yupapin, (2012) "Characterisation of bifurcation and chaos in silicon microring resonator", *IET Communications*, 6(16), 2671-2675.

[20] I. S. Amiri & J. Ali, (2014) "Characterization of Optical Bistability In a Fiber Optic Ring Resonator", *Quantum Matter*, 3(1), 47-51.

[21] J. Ali, M. A. Jalil, I. S. Amiri & P. P. Yupapin, (2010) "Fast and slow lights via an add/drop device", presented at the *ICEM*, Legend Hotel, Kuala Lumpur, Malaysia.

[22] IS Amiri, SE Alavi, N Fisal, ASM Supa'at & H Ahmad, (2014) "All-Optical Generation of Two IEEE802.11n Signals for 2×2 MIMO-RoF via MRR System", *IEEE Photonics Journal*, 6(6),

[23] SE Alavi, IS Amiri, H Ahmad, ASM Supa'at & N Fisal, (2014) "Generation and Transmission of 3× 3 W-Band MIMO-OFDM-RoF Signals Using Micro-Ring Resonators", *Applied Optics*, 53(34), 8049-8054.

[24] I. S. Amiri, S. E. Alavi, Sevia M. Idrus, A. Nikoukar & J. Ali, (2013) "IEEE 802.15.3c WPAN Standard Using Millimeter Optical Soliton Pulse Generated By a Panda Ring Resonator", *IEEE Photonics Journal*, 5(5), 7901912.

[25] Iraj Sadegh Amiri, Sayed Esan Alavi, Sevia Mahdaliza Idrus & Mojgan Kouhnavard, MICRORING RESONATOR FOR SECURED OPTICAL COMMUNICATION. USA: Amazon, 2014.

[26] C. Tanaram, C. Teeka, R. Jomtarak, P. P. Yupapin, M. A. Jalil, I. S. Amiri & J. Ali, (2011) "ASK-to-PSK generation based on nonlinear microring resonators coupled to one MZI arm", *Procedia Engineering*, 8 432-435.

[27] IS Amiri, MRK Soltanian & H Ahmad, bright soliton array (BSA) for optical communication using the wired/wireless link, in *Optical Communication Systems: Fundamentals, Techniques and Applications*, ed New York: Novascience Publisher, 2015.

[28] I. S. Amiri, S. Babakhani, G. Vahedi, J. Ali & P. Yupapin, (2012) "Dark-Bright Solitons Conversion System for Secured and Long Distance Optical Communication", *IOSR Journal of Applied Physics (IOSR-JAP)*, 2(1), 43-48.

[29] S. E. Alavi, I. S. Amiri, S. M. Idrus & A. S. M. Supa'at, (2014) "Generation and Wired/Wireless Transmission of IEEE802.16m Signal Using Solitons Generated By Microring Resonator", *Optical and Quantum Electronics*,

[30] I. S. Amiri, A. Nikoukar, G. Vahedi, A. Shojaei, J. Ali & P. Yupapin, (2012) "Frequency-Wavelength Trapping by Integrated Ring Resonators For Secured Network and Communication Systems", *International Journal of Engineering Research and Technology (IJERT)*, 1(5),

[31] I. S. Amiri, A. Nikoukar, A. Shahidinejad, J. Ali & P. Yupapin, (2012), "Generation of discrete frequency and wavelength for secured computer networks system using integrated ring resonators", in *Computer and Communication Engineering (ICCCE) Conference*, Malaysia, 775-778.

[32] J. Ali, M. Kouhnavard, I. S. Amiri, A. Afroozeh, M. A. Jalil, I. Naim & P. P. Yupapin, (2010) "Localization of soliton pulse using nano-waveguide", presented at the *ICAMN, International Conference*, Prince Hotel, Kuala Lumpur, Malaysia.

[33] Abdolkarim Afroozeh, Iraj Sadegh Amiri & Alireza Zeinalinezhad, Micro Ring Resonators and Applications. Saarbrücken, Germany: LAP LAMBERT Academic Publishing, 2014.

[34] I. S. Amiri, S. Ghorbani, P. Naraei & H. Ahmad, (2015) "Chaotic Carrier Signal Generation and Quantum Transmission Along Fiber Optics Communication Using Integrated Ring Resonators", *Quantum Matter*, 4(2),

[35] I. S. Amiri & J. Ali, (2013) "Data Signal Processing Via a Manchester Coding-Decoding Method Using Chaotic Signals Generated by a PANDA Ring Resonator", *Chinese Optics Letters*, 11(4), 041901(4).

[36] I. S. Amiri, K. Raman, A. Afroozeh, M. A. Jalil, I. N. Nawi, J. Ali & P. P. Yupapin, (2011) "Generation of DSA for security application", *Procedia Engineering*, 8 360-365.

[37] I. S. Amiri, S. E. Alavi, M. Bahadoran, A. Afroozeh & H. Ahmad, (2015) "Nanometer Bandwidth Soliton Generation and Experimental Transmission within Nonlinear Fiber Optics Using an Add-Drop Filter System", *Journal of Computational and Theoretical Nanoscience (CTN)*, 12(2),

[38] J. Ali, I. S. Amiri, M. Jalil, M. Kouhnavard, A. Afroozeh, I. Naim & P. Yupapin, (2010) "Narrow UV pulse generation using MRR and NRR system", presented at the *ICAMN, International Conference*, Prince Hotel, Kuala Lumpur, Malaysia

[39] I. S. Amiri, S. Soltanmohammadi, A. Shahidinejad & j. Ali, (2013) "Optical quantum transmitter with finesse of 30 at 800-nm central wavelength using microring resonators", *Optical and Quantum Electronics*, 45(10), 1095-1105.

[40] M. Kouhnavard, I. S. Amiri, M. Jalil, A. Afroozeh, J. Ali & P. P. Yupapin, (2010) "QKD via a quantum wavelength router using spatial soliton", *AIP Conference Proceedings*, 1347 210-216.

[41] I. S. Amiri & A. Nikoukar, (2010-2011) "Quantum Information Generation Using Optical Potential Well", presented at the *Network Technologies & Communications (NTC) Conference*, Singapore.

[42] I. S. Amiri, S. E. Alavi & S. M. Idrus, Results of Digital Soliton Pulse Generation and Transmission Using Microring Resonators, in *Soliton Coding for Secured Optical Communication Link*, ed USA: Springer, 2015, pp. 41-56.

[43] I. S. Amiri & A. Nikoukar, (2010-2011) "Secured Binary Codes Generation for Computer Network Communication", presented at the *Network Technologies & Communications (NTC) Conference*, Singapore.

[44] I. S. Amiri, H. Ahmad & P. Naraei, (2015) "Optical Transmission Characteristics of an Optical Add-Drop Interferometer System", *Quantum Matter*, 4(5),

[45] A. Afroozeh, I. S. Amiri, M. Bahadoran, J. Ali & P. P. Yupapin, (2012) "Simulation of Soliton Amplification in Micro Ring Resonator for Optical Communication", *Jurnal Teknologi (Sciences and Engineering)*, 55 271-277.

[46] Iraj Sadegh Amiri, Ali Nikoukar & Sayed Ehsan Alavi, Soliton and Radio over Fiber (RoF) Applications. Saarbrücken, Germany: LAP LAMBERT Academic Publishing, 2014.

[47] I. S. Amiri, S. E. Alavi & S. M. Idrus', (2014) "Solitonic Pulse Generation and Characterization by Integrated Ring Resonators", presented at the *5th International Conference on Photonics 2014 (ICP2014)*, Kuala Lumpur.

[48] J. Ali, K. Raman, M. Kouhnavard, I. S. Amiri, M. A. Jalil, A. Afroozeh & P. P. Yupapin, (2011) "Dark soliton array for communication security", presented at the *AMN-APLOC International Conference*, Wuhan, China.

[49] J. Ali, M. A. Jalil, I. S. Amiri & P. P. Yupapin, (2010) "Dark-bright solitons conversion system via an add/drop filter for signal security application", presented at the *ICEM*, Legend Hotel, Kuala Lumpur, Malaysia.

[50] J. Ali, M. A. Jalil, I. S. Amiri, A. Afroozeh, M. Kouhnavard & P. P. Yupapin, (2010) "Generation of tunable dynamic tweezers using dark-bright collision", presented at the *ICAMN, International Conference*, Prince Hotel, Kuala Lumpur, Malaysia

[51] IS Amiri & A Afroozeh, Introduction of Soliton Generation, in *Ring Resonator Systems to Perform Optical Communication Enhancement Using Soliton*, ed USA: Springer, 2014.

[52] A. Nikoukar, I. S. Amiri, A. Shahidinejad, A. Shojaei, J. Ali & P. Yupapin, (2012), "MRR quantum dense coding for optical wireless communication system using decimal convertor", in *Computer and Communication Engineering (ICCCE) Conference*, Malaysia, 770-774.

[53] J. Ali, M. Kouhnavard, I. S. Amiri, M. A. Jalil, A. Afroozeh & P. P. Yupapin (2010) "Security confirmation using temporal dark and bright soliton via nonlinear system", presented at the *ICAMN, International Conference*, Prince Hotel, Kuala Lumpur, Malaysia

[54] I. S. Amiri & J. Ali, (2014) "Deform of Biological Human Tissue Using Inserted Force Applied by Optical Tweezers Generated By PANDA Ring Resonator", *Quantum Matter*, 3(1), 24-28.

[55] S. E. Alavi, I. S. Amiri, M. R. K. Soltanian, A.S.M. Supa'at, N. Fisal & H. Ahmad, (2015) "Generation of Femtosecond Soliton Tweezers Using a Half-Panda System for Modeling the Trapping of a Human Red Blood Cell", *Journal of Computational and Theoretical Nanoscience (CTN)*, 12(1),

[56] I. S. Amiri & J. Ali, (2012) "Generation of Nano Optical Tweezers Using an Add/drop Interferometer System", presented at the *2nd Postgraduate Student Conference (PGSC)*, Singapore.

[57] A. Nikoukar, I. S. Amiri & J. Ali, (2013) "Generation of Nanometer Optical Tweezers Used for Optical Communication Networks", *International Journal of Innovative Research in Computer and Communication Engineering*, 1(1), 77-85.

[58] I. Sadegh Amiri, M. Nikmaram, A. Shahidinejad & J. Ali, (2013) "Generation of potential wells used for quantum codes transmission via a TDMA network communication system", *Security and Communication Networks*, 6(11), 1301-1309.

[59] I. S. Amiri, A. Nikoukar, A. Shahidinejad, M. Ranjbar, J. Ali & P. P. Yupapin, (2012) "Generation of Quantum Photon Information Using Extremely Narrow Optical Tweezers for Computer Network Communication", *GSTF Journal on Computing (joc)*, 2(1), 140.

[60] I. S. Amiri & J. Ali, (2013) "Nano Particle Trapping By Ultra-short tweezer and wells Using MRR Interferometer System for Spectroscopy Application", *Nanoscience and Nanotechnology Letters*, 5(8), 850-856.

[61] I. S. Amiri & J. Ali, (2013) "Nano Optical Tweezers Generation Used for Heat Surgery of a Human Tissue Cancer Cells Using Add/Drop Interferometer System", *Quantum Matter*, 2(6), 489-493.

[62] I. S. Amiri & J. Ali, (2013) "Optical Buffer Application Used for Tissue Surgery Using Direct Interaction of Nano Optical Tweezers with Nano Cells", *Quantum Matter*, 2(6), 484-488.

[63] S. E. Alavi, I. S. Amiri, H. Ahmad, N. Fisal & ASM. Supa'at, (2015) "Optical Amplification of Tweezers and Bright Soliton Using an Interferometer Ring Resonator System", *Journal of Computational and Theoretical Nanoscience (CTN)*, 12(4),

[64] I. S. Amiri, B. Barati, P. Sanati, A. Hosseinnia, HR Mansouri Khosravi, S. Pourmehdi, A. Emami & J. Ali, (2014) "Optical Stretcher of Biological Cells Using Sub-Nanometer Optical Tweezers Generated by an Add/Drop Microring Resonator System", *Nanoscience and Nanotechnology Letters*, 6(2), 111-117.

[65] I. S. Amiri, A. Nikoukar, A Shahidinejad, T. Anwar & J. Ali, (2014) "Quantum Transmission of Optical Tweezers Via Fiber Optic Using Half-Panda System", *Life Science Journal*, 10(12s), 391-400.

[66] J. Ali, I. S. Amiri, A. Afroozeh, M. Kouhnavard, M. Jalil & P. Yupapin, (2010) "Simultaneous dark and bright soliton trapping using nonlinear MRR and NRR", presented at the *ICAMN, International Conference*, Prince Hotel, Kuala Lumpur, Malaysia

[67] I. S. Amiri, M. A. Jalil, F. K. Mohamad, N. J. Ridha, J. Ali & P. P. Yupapin, (2010) "Storage of Atom/Molecules/Photon using Optical Potential Wells", presented at the *International Conference on Experimental Mechanics (ICEM)*, Kuala Lumpur, Malaysia.

[68] I. S. Amiri, A. Shahidinejad, A. Nikoukar, J. Ali & P. Yupapin, (2012) "A Study oF Dynamic Optical Tweezers Generation For Communication Networks", *International Journal of Advances in Engineering & Technology (IJAET)*, 4(2), 38-45

[69] N. Suwanpayak, S. Songmuang, M. A. Jalil, I. S. Amiri, I. Naim, J. Ali & P. P. Yupapin, (2010) "Tunable and storage potential wells using microring resonator system for bio-cell trapping and delivery", *AIP Conference Proceedings*, 1341 289-291.

[70] Iraj Sadegh Amiri & Abdolkarim Afroozeh, Ring Resonator Systems to Perform the Optical Communication Enhancement Using Soliton. USA: Springer, 2014.

[71] I. S. Amiri & J. Ali, (2013) "Single and Multi Optical Soliton Light Trapping and Switching Using Microring Resonator", *Quantum Matter*, 2(2), 116-121.

[72] A. A. Shojaei & I. S. Amiri, (2011) "Soliton for Radio wave generation", presented at the *International Conference for Nanomaterials Synthesis and Characterization (INSC)*, Kuala Lumpur, Malaysia.

[73] J. Ali, A. Mohamad, I. Nawi, I. S. Amiri, M. Jalil, A. Afroozeh & P. Yupapin, (2010) "Stopping a dark soliton pulse within an NNRR", presented at the *AMN-APLOC International Conference*, Wuhan, China

[74] I. S. Amiri, G. Vahedi, A. Shojaei, A. Nikoukar, J. Ali & P. P. Yupapin, (2012) "Secured Transportation of Quantum Codes Using Integrated PANDA-Add/drop and TDMA Systems", *International Journal of Engineering Research and Technology (IJERT)*, 1(5),

[75] M. Kouhnavard, A. Afroozeh, M. A. Jalil, I. S. Amiri, J. Ali & P. P. Yupapin, (2010), "Soliton Signals and the Effect of Coupling Coefficient in MRR Systems", in *Faculty of Science Postgraduate Conference (FSPGC)*, Universiti Teknologi Malaysia.

[76] I. S. Amiri, D. Gifany & J. Ali, (2013) "Ultra-short Multi Soliton Generation for Application in Long Distance Communication", *Journal of Basic and Applied Scientific Research (JBASR)*, 3(3), 442-451.

[77] I. S. Amiri, A. Afroozeh, M. Bahadoran, J. Ali & P. P. Yupapin, (2011) "Up and Down Link of Soliton for Network Communication", presented at the *National Science Postgraduate Conference, NSPC*, Universiti Teknologi Malaysia.

[78] S. E. Alavi, I. S. Amiri, M. Khalily, A. S. M. Supa' at, N. Fisal, H. Ahmad & S. M. Idrus, (2014) "W-Band OFDM for Radio-over-Fibre Direct-Detection Link Enabled By Frequency Nonupling Optical Up-Conversion", *IEEE Photonics Journal* 6(6),

[79] A. Afroozeh, I. S. Amiri, M. A. Jalil, M. Kouhnavard, J. Ali & P. P. Yupapin, (2011) "Multi Soliton Generation for Enhance Optical Communication", *Applied Mechanics and Materials*, 83 136-140.

[80] I. S. Amiri, (2011), "Optical Soliton Trapping for Quantum Key Generation", in *The International Conference for Nano materials Synthesis and Characterization*, Malaysia.

[81] IS Amiri & A Afroozeh, Soliton Generation Based Optical Communication, in *Ring Resonator Systems to Perform Optical Communication Enhancement Using Soliton*, ed USA: Springer, 2014.

[82] I. S. Amiri, S. E. Alavi, S. M. Idrus, A. S. M. Supa'at, J. Ali & P. P. Yupapin, (2014) "W-Band OFDM Transmission for Radio-over-Fiber link Using Solitonic Millimeter Wave Generated by MRR", *IEEE Journal of Quantum Electronics*, 50(8), 622 - 628.

[83] J. Ali, A. Afroozeh, I. S. Amiri, M. Jalil & P. Yupapin, (2010) "Wide and narrow signal generation using chaotic wave", presented at the *Nanotech Malaysia, International Conference on Enabling Science & Technology*, Kuala Lumpur, Malaysia.

[84] Y. S. Neo, S. M. Idrus, M. F. Rahmat, S. E. Alavi & I. S. Amiri', (2014) "Adaptive Control for Laser Transmitter Feedforward Linearization System", *IEEE Photonics Journal* 6(4),

[85] I. S. Amiri, A. Nikoukar & J. Ali, (2013) "GHz Frequency Band Soliton Generation Using Integrated Ring Resonator for WiMAX Optical Communication", *Optical and Quantum Electronics*, 46(9), 1165-1177.

[86] I. S. Amiri, S. E. Alavi, H. Ahmad, A.S.M. Supa'at & N. Fisal, (2014) "Numerical Computation of Solitonic Pulse Generation for Terabit/Sec Data Transmission", *Optical and Quantum Electronics*,

[87] I. S. Amiri & J. Ali, (2014) "Optical Quantum Generation and Transmission of 57-61 GHz Frequency Band Using an Optical Fiber Optics ", *Journal of Computational and Theoretical Nanoscience (CTN)*, 11(10), 2130-2135.

[88] I. S. Amiri, S. E. Alavi & S. M. Idrus, Theoretical Background of Microring Resonator Systems and Soliton Communication, in *Soliton Coding for Secured Optical Communication Link*, ed USA: Springer, 2015, pp. 17-39.

[89] I. S. Amiri, M. Ebrahimi, A. H. Yazdavar, S. Gorbani, S. E. Alavi, Sevia M. Idrus & J. Ali, (2014) "Transmission of data with orthogonal frequency division multiplexing technique for communication networks using GHz frequency band soliton carrier", *IET Communications*, 8(8), 1364 – 1373.

[90] I. S. Amiri, A. Shahidinejad, A. Nikoukar, M. Ranjbar, J. Ali & P. P. Yupapin, (2012) "Digital Binary Codes Transmission via TDMA Networks Communication System Using Dark and Bright Optical Soliton", *GSTF Journal on Computing (joc)*, 2(1), 12.

[91] A. A. Shojaei & I. S. Amiri, (2011) "DSA for Secured Optical Communication", presented at the *International Conference for Nanomaterials Synthesis and Characterization (INSC)*, Kuala Lumpur, Malaysia.

[92] I. S. Amiri, D. Gifany & J. Ali, (2013) "Long Distance Communication Using Localized Optical Soliton Via Entangled Photon", *IOSR Journal of Applied Physics (IOSR-JAP)*, 3(1), 32-39.

[93] I. S. Amiri, P. Naraei & J. Ali, (2014) "Review and Theory of Optical Soliton Generation Used to Improve the Security and High Capacity of MRR and NRR Passive Systems", *Journal of Computational and Theoretical Nanoscience (CTN)*, 11(9) 1875-1886.

[94] IS Amiri, MZ Zulkifli & H Ahmad, (2014) "Soliton comb generation using add-drop ring resonators", *International Research Journal of Telecommunications and Information Technology*, 1(1), 002-008.

[95] S. E. Alavi, I.S.Amiri, A. S. M. Supa'at & S. M. Idrus, (2015) "Indoor Data Transmission Over Ubiquitous Infrastructure of Powerline Cables and LED Lighting", *Journal of Computational and Theoretical Nanoscience (CTN)*, 12(4),

[96] A. Shahidinejad, A. Nikoukar, I. S. Amiri, M. Ranjbar, A. Shojaei, J. Ali & P. Yupapin, (2012), "Network system engineering by controlling the chaotic signals using silicon micro ring resonator", in *Computer and Communication Engineering (ICCCE) Conference*, Malaysia, 765-769.

[97] I. S. Amiri, S. E. Alavi & H. Ahmad, (2015) "RF signal generation and wireless transmission using PANDA and Add/drop systems", *Journal of Computational and Theoretical Nanoscience (CTN)*,

[98] M. A. Jalil, I. S. Amiri, C. Teeka, J. Ali & P. P. Yupapin, (2011) "All-optical Logic XOR/XNOR Gate Operation using Microring and Nanoring Resonators", *Global Journal of Physics Express*, 1(1), 15-22.

[99] C. Teeka, S. Songmuang, R. Jomtarak, P. Yupapin, M. Jalil, I. S. Amiri & J. Ali, (2011) "ASK-to-PSK Generation based on Nonlinear Microring Resonators Coupled to One MZI Arm", *AIP Conference Proceedings*, 1341(1), 221-223.

[100] I. S. Amiri, S. E. Alavi & S. M. Idrus, Introduction of Fiber Waveguide and Soliton Signals Used to Enhance the Communication Security, in *Soliton Coding for Secured Optical Communication Link*, ed USA: Springer, 2015, pp. 1-16.

[101] I. S. Amiri & H. Ahmad, (2015) "Multiplex and De-multiplex of Generated Multi Optical Soliton By MRRs Using Fiber Optics Transmission Link", *Quantum Matter*, 4(4),

[102] I. S. Amiri & J. Ali, (2014) "Generating Highly Dark–Bright Solitons by Gaussian Beam Propagation in a PANDA Ring Resonator", *Journal of Computational and Theoretical Nanoscience (CTN)*, 11(4), 1092-1099.

[103] J. Ali, A. Afroozeh, I. S. Amiri, M. A. Jalil, M. Kouhnavard & P. P. Yupapin, (2010) "Generation of continuous optical spectrum by soliton into a nano-waveguide", presented at the *ICAMN, International Conference*, Prince Hotel, Kuala Lumpur, Malaysia

[104] IS Amiri & A Afroozeh, Mathematics of Soliton Transmission in Optical Fiber, in *Ring Resonator Systems to Perform Optical Communication Enhancement Using Soliton*, ed USA: Springer, 2014.

[105] A. Afroozeh, I. S. Amiri, A. Zeinalinezhad, S. E. Pourmand & H. Ahmad, (2015) "Comparison of Control Light using Kramers-Kronig Method by Three Waveguides", *Journal of Computational and Theoretical Nanoscience (CTN)*,

[106] I. S. Amiri, M. A. Jalil, A. Afroozeh, M. Kouhnavard , J. Ali & P. P. Yupapin, (2010), "Controlling Center Wavelength and Free Spectrum Range by MRR Radii", in *Faculty of Science Postgraduate Conference (FSPGC)*, Universiti Teknologi Malaysia.

[107] J. Ali, A. Afroozeh, I. S. Amiri, M. A. Jalil & P. P. Yupapin, (2010) "Dark and Bright Soliton trapping using NMRR", presented at the *ICEM*, Legend Hotel, Kuala Lumpur, Malaysia.

[108] M. A. Jalil, I. S. Amiri, M. Kouhnavard, A. Afroozeh, J. Ali & P. P. Yupapin, (2010), "Finesse Improvements of Light Pulses within MRR System", in *Faculty of Science Postgraduate Conference (FSPGC)*, Universiti Teknologi Malaysia.

[109] I. S. Amiri, A. Afroozeh, J. Ali & P. P. Yupapin, (2012) "Generation Of Quantum Codes Using Up And Down Link Optical Solition", *Jurnal Teknologi (Sciences and Engineering)*, 55 97-106.

[110] J. Ali, I. S. Amiri, M. A. Jalil, A. Afroozeh, M. Kouhnavard & P. Yupapin, (2010) "Novel system of fast and slow light generation using micro and nano ring resonators", presented at the *ICAMN, International Conference*, Prince Hotel, Kuala Lumpur, Malaysia

[111] IS Amiri, SE Alavi, A Shahidinejad, A Nikoukar, T Anwar, ASM Supa'at, SM Idrus & N. K. Yen, (2014) "Characterization of Ultra-Short Soliton Generation Using MRRs", presented at the *The 2014 Third ICT International Student Project Conference (ICT-ISPC2014)*, Thailand

[112] N. J. Ridha, F. K. Mohamad, I. S. Amiri, Saktioto, J. Ali & P. P. Yupapin, (2010) "Controlling Center Wavelength and Free Spectrum Range by MRR Radii", presented at the *International Conference on Experimental Mechanics (ICEM)*, Kuala Lumpur, Malaysia.

[113] I. S. Amiri & J. Ali, (2013) "Controlling Nonlinear Behavior of a SMRR for Network System Engineering", *International Journal of Engineering Research and Technology (IJERT)*, 2(2),

[114] A. Afroozeh, I. S. Amiri, J. Ali & P. P. Yupapin, (2012) "Determination Of Fwhm For Solition Trapping", *Jurnal Teknologi (Sciences and Engineering)*, 55 77-83.

[115] IS Amiri, H. Ahmad & Hamza M. R. Al-Khafaji, (2015) "Full width at half maximum (FWHM) analysis of solitonic pulse applicable in optical network communication", *American Journal of Networks and Communications*, 4(2-1), 1-5.

[116] J. Ali, M. Jalil, I. S. Amiri, A. Afroozeh, M. Kouhnavard, I. Naim & P. Yupapin, (2010) "Multi-wavelength narrow pulse generation using MRR", presented at the *ICAMN, International Conference*, Prince Hotel, Kuala Lumpur, Malaysia

[117] I. S. Amiri, M. Nikmaram, A. Shahidinejad & J. Ali, (2012) "Cryptography Scheme of an Optical Switching System Using Pico/Femto Second Soliton Pulse", *International Journal of Advances in Engineering & Technology (IJAET)*, 5(1), 176-184.

[118] A. Afroozeh, M. Kouhnavard, I. S. Amiri, M. A. Jalil, J. Ali & P. P. Yupapin, (2010), "Effect of Center Wavelength on MRR Performance", in *Faculty of Science Postgraduate Conference (FSPGC)*, Universiti Teknologi Malaysia.

[119] S. Saktioto, S. Daud, M. A. Jalil, I. S. Amiri & P. P. Yupapin, (2010) "FBG sensing system for outdoor temperature measurement", presented at the *ICEM*, Legend Hotel, Kuala Lumpur, Malaysia.

[120] M. Kouhnavard, A. Afroozeh, I. S. Amiri, M. A. Jalil, J. Ali & P. P. Yupapin, (2010) "New system of Chaotic Signal Generation Using MRR", presented at the *International Conference on Experimental Mechanics (ICEM)*, Kuala Lumpur, Malaysia.

[121] Iraj Sadegh Amiri, Falah Jabar Rahim, Ari Sabir Arif, Sogand Ghorbani, Parisa Naraei, David Forsyth & Jalil Ali, (2014) "Single Soliton Bandwidth Generation and Manipulation by Microring Resonator", *Life Science Journal*, 10(12s), 904-910.

[122] I. S. Amiri, A. Afroozeh, I. N. Nawi, M. A. Jalil, A. Mohamad, J. Ali & P. P. Yupapin, (2011) "Dark Soliton Array for communication security", *Procedia Engineering*, 8 417-422.

[123] I. S. Amiri, D. Gifany & J. Ali, (2013) "Entangled Photon Encoding Using Trapping of Picoseconds Soliton pulse", *IOSR Journal of Applied Physics (IOSR-JAP)*, 3(1), 25-31.

[124] S. E. Alavi, I. S. Amiri, S. M. Idrus & J. Ali, (2013) "Optical Wired/Wireless Communication Using Soliton Optical Tweezers", *Life Science Journal*, 10(12s), 179-187.

[125] I. S. Amiri, G. Vahedi, A. Nikoukar, A. Shojaei, J. Ali & P. Yupapin, (2012) "Decimal Convertor Application for Optical Wireless Communication by Generating of Dark and Bright Signals of soliton", *International Journal of Engineering Research and Technology (IJERT)*, 1(5),

[126] I. S. Amiri, A. Afroozeh, M. Bahadoran, J. Ali & P. P. Yupapin, (2012) "Molecular Transporter System for Qubits Generation", *Jurnal Teknologi (Sciences and Engineering)*, 55 155-165.

[127] A. Afroozeh, I. S. Amiri, M. Kouhnavard, M. Jalil, J. Ali & P. Yupapin, (2010) "Optical dark and bright soliton generation and amplification", *AIP Conference Proceedings*, 1341 259-263.

[128] A. Afroozeh, A. Zeinalinezhad, SE. Pourmand & IS. Amiri, (2014) "Determination of Suitable Material to Control of Light", *International Journal of Biology, Pharmacy and Allied Sciences (IJBPAS)*, 3(11), 2410-2421.

[129] S. Saktioto, S. Daud, J. Ali, M. A. Jalil, I. S. Amiri & P. Yupapin, (2010) "FBG simulation and experimental temperature measurement", presented at the *ICEM*, Legend Hotel, Kuala Lumpur, Malaysia.

[130] I. S. Amiri & J. Ali, (2014) "Picosecond Soliton pulse Generation Using a PANDA System for Solar Cells Fabrication", *Journal of Computational and Theoretical Nanoscience (CTN)*, 11(3), 693-701.

[131] J. Ali, M. A. Jalil, I. S. Amiri & P. P. Yupapin, (2010) "Effects of MRR parameter on the bifurcation behavior", presented at the *Nanotech Malaysia, International Conference on Enabling Science & Technology* KLCC, Kuala Lumpur, Malaysia

[132] A. Afroozeh, M. Bahadoran, I. S. Amiri, A. R. Samavati, J. Ali & P. P. Yupapin, (2011) "Fast Light Generation Using Microring Resonators for Optical Communication", presented at the *National Science Postgraduate Conference, NSPC*, Universiti Teknologi Malaysia.

[133] P. P. Yupapin, M. A. Jalil, I. S. Amiri, I. Naim & J. Ali, (2010) "New Communication Bands Generated by Using a Soliton Pulse within a Resonator System", *Circuits and Systems*, 1(2), 71-75.

[134] I. S. Amiri, A. Nikoukar, A. Shahidinejad & Toni Anwar, (2014) "The Proposal of High Capacity GHz Soliton Carrier Signals Applied for Wireless Commutation", *Reviews in Theoretical Science*, 2(4), 320-333.

[135] A. Afroozeh, A.Zeinalinezhad, I. S. Amiri & S. E. Pourmand, (2014) "Stop Light Generation using Nano Ring Resonators for ROM", *Journal of Computational and Theoretical Nanoscience (CTN)*, 12(3),

[136] J. Ali, I. S. Amiri, M. A. Jalil, M. Hamdi, F. K. Mohamad, N. J. Ridha & P. P. Yupapin, (2010) "Trapping spatial and temporal soliton system for entangled photon encoding", presented at the *Nanotech Malaysia, International Conference on Enabling Science & Technology*, Kuala Lumpur, Malaysia.

[137] Ali Shahidinejad, Iraj Sadegh Amiri & Toni Anwar, (2014) "Enhancement of Indoor Wavelength Division Multiplexing-Based Optical Wireless Communication Using Microring Resonator", *Reviews in Theoretical Science*, 2(3), 201-210.

[138] I. S. Amiri, A. Nikoukar & J. Ali, (2013) "Nonlinear Chaotic Signals Generation and Transmission Within an Optical Fiber Communication Link", *IOSR Journal of Applied Physics (IOSR-JAP)*, 3(1), 52-57.

[139] J. Ali, A. Afroozeh, I. S. Amiri, M. Hamdi, M. Jalil, M. Kouhnavard & P. Yupapin, (2010) "Entangled photon generation and recovery via MRR", presented at the *ICAMN, International Conference*, Prince Hotel, Kuala Lumpur, Malaysia.

[140] J. Ali, I. S. Amiri, M. A. Jalil, A. Afroozeh, M. Kouhnavard & P. P. Yupapin, (2010) "Multi-soliton generation and storage for nano optical network using nano ring resonators", presented at the *ICAMN, International Conference*, Prince Hotel, Kuala Lumpur, Malaysia

[141] A. Afroozeh, I. S. Amiri, M. Kouhnavard, M. Bahadoran, M. A. Jalil, J. Ali & P. P. Yupapin, (2010) "Optical Memory Time using Multi Bright Soliton", presented at the *International Conference on Experimental Mechanics (ICEM)*, Kuala Lumpur, Malaysia.

[142] M. Imran, R. A. Rahman & I. S. Amiri, (2010), "Fabrication of Diffractive Optical Element using Direct Writing CO2 Laser Irradiation", in *Faculty of Science Postgraduate Conference (FSPGC)*, Universiti Teknologi Malaysia.

[143] J. Ali, M. A. Jalil, I. S. Amiri & P. P. Yupapin, (2010) "MRR quantum dense coding", presented at the *Nanotech Malaysia, International Conference on Enabling Science & Technology*, KLCC, Kuala Lumpur, Malaysia

[144] I. S. Amiri, A. Nikoukar & J. Ali, (2013) "New System of chaotic signal generation based on coupling coefficients applied to an Add/Drop System", *International Journal of Advances in Engineering & Technology (IJAET)*, 6(1), 78-87.

[145] A. Afroozeh, M. Bahadoran, I. S. Amiri, A. R. Samavati, J. Ali & P. P. Yupapin, (2012) "Fast Light Generation Using GaAlAs/GaAs Waveguide", *Jurnal Teknologi (Sciences and Engineering)*, 57 17-23.

[146] J. Ali, H. Nur, S. Lee, A. Afroozeh, I. S. Amiri, M. Jalil, A. Mohamad & P. Yupapin, (2010) "Short and millimeter optical soliton generation using dark and bright soliton", presented at the *AMN-APLOC International Conference*, Wuhan, China.

[147] I. S. Amiri & J. Ali, (2014) "Femtosecond Optical Quantum Memory generation Using Optical Bright Soliton", *Journal of Computational and Theoretical Nanoscience (CTN)*, 11(6), 1480-1485.

[148] I. S. Amiri, A. Afroozeh & M. Bahadoran, (2011) "Simulation and Analysis of Multisoliton Generation Using a PANDA Ring Resonator System", *Chinese Physics Letters*, 28(10), 104205.

[149] I. S. Amiri & A. Afroozeh, Spatial and Temporal Soliton Pulse Generation By Transmission of Chaotic Signals Using Fiber Optic Link in *Advances in Laser and Optics Research*. vol. 11, ed New York: Nova Science Publisher, 2015.

[150] I. S. Amiri, H. Ahmad & A. Shahidinejad, (2015) "Generating of 57-61 GHz Frequency Band Using a Panda Ring Resonator", *Quantum Matter*, 4(4),

[151] J. Ali, M. Kouhnavard, A. Afroozeh, I. S. Amiri, M. A. Jalil & P. P. Yupapin, (2010) "Optical bistability in a FORR", presented at the *ICEM*, Legend Hotel, Kuala Lumpur, Malaysia.

[152] IS Amiri, SE Alavi, MRK Soltanian & H Ahmad, (2015) "Tunable Channel Spacing of Soliton Comb Generation Using Add-drop Microring Resonators (MRRs)", *Journal of Computational and Theoretical Nanoscience (CTN)*,

[153] A. Afrozeh, A. Zeinalinezhad, S. E. Pourmand & I. S. Amiri, (2014) "Attosecond Pulse Generation Using Nano Ring Waveguides", *INTERNATIONAL JOURNAL OF CURRENT LIFE SCIENCES*, 4(9), 7573-7575.

[154] I. S. Amiri, Light Detection and Ranging Using NIR (810 nm) Laser Source. Germany: LAP LAMBERT Academic Publishing, 2014.

[155] J. Ali, A. Afroozeh, M. Hamdi, I. S. Amiri, M. A. Jalil, M. Kouhnavard & P. Yupapin, (2010) "Optical bistability behaviour in a double-coupler ring resonator", presented at the *ICAMN, International Conference*, Prince Hotel, Kuala Lumpur, Malaysia

[156] S. E. Alavi, I. S. Amiri, S. M. Idrus, ASM. Supa'at & J. Ali, (2013) "Chaotic Signal Generation and Trapping Using an Optical Transmission Link", *Life Science Journal*, 10(9s), 186-192.

[157] Iraj Sadegh Amiri, Sayed Ehsan Alavi & Sevia Mahdaliza Idrus, Soliton Coding for Secured Optical Communication Link. USA: Springer, 2014.

[158] J. Ali, S. Saktioto, M. Hamdi & I. S. Amiri, (2010) "Dynamic silicon dioxide fiber coupling polarized by voltage breakdown", presented at the *Nanotech Malaysia, International Conference on Enabling Science & Technology*, KLCC, Kuala Lumpur, Malaysia.

[159] I. S. Amiri & J. Ali, (2014) "Simulation of the Single Ring Resonator Based on the Z-transform Method Theory", *Quantum Matter*, 3(6), 519-522.

[160] F. K. Mohamad, N. J. Ridha, I. S. Amiri, J. A. Saktioto & P. P. Yupapin, (2010) "Effect of Center Wavelength on MRR Performance", presented at the *International Conference on Experimental Mechanics (ICEM)*, Kuala Lumpur, Malaysia.

[161] N. J. Ridha, F. K. Mohamad, I. S. Amiri, Saktioto, J. Ali & P. P. Yupapin, (2010) "Soliton Signals and The Effect of Coupling Coefficient in MRR Systems", presented at the *International Conference on Experimental Mechanics (ICEM)*, Kuala Lumpur, Malaysia.

[162] J. Ali, K. Kulsirirat, W. Techithdeera, M. A. Jalil, I. S. Amiri, I. Naim & P. P. Yupapin, (2010) "Temporal dark soliton behavior within multi-ring resonators", presented at the *Nanotech Malaysia, International Conference on Enabling Science & Technology*, Malaysia

[163] P. Sanati, A. Afroozeh, I. S. Amiri, J.Ali & Lee Suan Chua, (2014) "Femtosecond Pulse Generation using Microring Resonators for Eye Nano Surgery", *Nanoscience and Nanotechnology Letters*, 6(3), 221-226

[164] I. S. Amiri, M. A. Jalil, F. K. Mohamad, N. J. Ridha, J. Ali & P. P. Yupapin, (2010) "Storage of Optical Soliton Wavelengths Using NMRR", presented at the *International Conference on Experimental Mechanics (ICEM)*, Kuala Lumpur, Malaysia.

[165] F. K. Mohamad, N. J. Ridha, I. S. Amiri, J. A. Saktioto & P. P. Yupapin, (2010) "Finesse Improvements of Light Pulses within MRR System", presented at the *International Conference on Experimental Mechanics (ICEM)*, Kuala Lumpur, Malaysia.

[166] Iraj Sadegh Amiri, Sayed Ehsan Alavi, S. M. Idrus, Abdolkarim Afroozeh & Jalil Ali, Soliton Generation by Ring Resonator for Optical Communication Application. New York: Novascience Publishers, 2014.

[167] A. Shahidinejad, S. Soltanmohammadi, I. S. Amiri & T. Anwar, (2014) "Solitonic Pulse Generation for Inter-Satellite Optical Wireless Communication", *Quantum Matter*, 3(2), 150-154.

[168] S. Saktioto, M. Hamdi, I. S. Amiri & J. Ali, (2010) "Transition of diatomic molecular oscillator process in THz region", presented at the *International Conference on Experimental Mechanics (ICEM)*, Legend Hotel, Kuala Lumpur, Malaysia.

[169] I. S. Amiri, M. Ranjbar, A. Nikoukar, A. Shahidinejad, J. Ali & P. Yupapin, (2012), "Multi optical Soliton generated by PANDA ring resonator for secure network communication", in *Computer and Communication Engineering (ICCCE) Conference*, Malaysia, 760-764.

[170] J. Ali, M. Aziz, I. S. Amiri, M. Jalil, A. Afroozeh, I. Nawi & P. Yupapin, (2010) "Soliton wavelength division in MRR and NRR Systems", presented at the *AMN-APLOC International Conference*, Wuhan, China

[171] J. Ali, I. S. Amiri, M. A. Jalil, M. Hamdi, F. K. Mohamad, N. J. Ridha & P. P. Yupapin, (2010) "Proposed molecule transporter system for qubits generation", presented at the *Nanotech Malaysia, International Conference on Enabling Science & Technology*, Malaysia

[172] A Nikoukar, IS Amiri, SE Alavi, A Shahidinejad, T Anwar, ASM Supa'at, SM Idrus & L. Y. Teng', (2014) "Theoretical and Simulation Analysis of The Add/Drop Filter Ring Resonator Based on the Z-transform Method Theory", presented at the *The 2014 Third ICT International Student Project Conference (ICT-ISPC2014)*, Thailand.

[173] I. S. Amiri, M. H. Khanmirzaei, M. Kouhnavard, P. P. Yupapin & J. Ali, Quantum Entanglement using Multi Dark Soliton Correlation for Multivariable Quantum Router, in *Quantum Entanglement* A. M. Moran, Ed., ed New York: Nova Science Publisher, 2012, pp. 111-122.

[174] A. Afroozeh, I. S. Amiri, A. Samavati, J. Ali & P. Yupapin, (2012), "THz frequency generation using MRRs for THz imaging", in *International Conference on Enabling Science and Nanotechnology (EsciNano)*, Kuala Lumpur, Malaysia, 1-2.

27

Tone Quality Recognition of Instruments Based on Multi-feature Fusion of Music Signal

Zhe Lei[1], Mengying Ding[1], Xiaohong Guan[1], Youtian Du[1, *], Jicheng Feng[2], Qinping Gao[2], Zheng Liu[2]

[1]The School of Electronic and Information Engineering, Xi'an Jiaotong University, Xi'an, China
[2]Xi'an Conservatory of Music, Xi'an, China

Email address:
lzleizhe@163.com (Zhe Lei), corianderherb@foxmail.com (Mengying Ding), xhguan@mail.xjtu.edu.cn (Xiaohong Guan),
duyt@mail.xjtu.edu.cn (Youtian Du), fjc-fagott@hotmail.com (Jicheng Feng), gaoqinping@yahoo.com.cn (Qinping Gao),
composerliu@hotmail.com (Zheng Liu)
[*]Corresponding author

Abstract: The traditional expert-based instrumental music evaluation strategy can't meet the requirements of the rapidly accumulated audio data. The traditional strategy not only takes a high cost of human's energy and time but also may have some problems on consistency and fairness of judgment. This paper aims at designing a complete recognition and evaluation strategy to automatically identify the timber of wind instruments. We take the clarinet as example and propose a strategy based on multi-feature fusion and random forest. First, we use the identification of fundamental frequency algorithm to automatically distinguish the notes performed by the instruments. Second, we extract 3 types of features including MFCC, brightness and roughness to describe the instrumental signals. Then, considering two kinds of variants: note and tone quality, we design 5 strategies to remove the influence of different notes in the evaluation of tone quality. By analyzing these strategies, we explore the optimal strategy for the recognition. The final evaluation results over 840 music slices demonstrate the effectiveness of this method.

Keywords: Tone Quality, Timbre Analysis, Audio Signal Processing, Random Forest

1. Introduction

With the rapid development of digital music processing and storage technologies, the audio data accumulate continuously. Many art colleges have amassed vast amount of instrumental music data. At present, the evaluation of instrumental music is generally based on experts. However, the expert-based evaluation strategy not only takes a high cost of human's energy and time but also may have some problem on consistency and fairness of judgment. Hence, using information processing method to help evaluate the tone quality of instrumental music has great theoretical significance and practical values.

Now, the research on instrumental analysis and application mainly focuses on the fields including instrument category recognition, musical information retrieval, computer-assisted music understanding and instrumental tone quality analysis [1]. It is hard to retrieve the common digital music contents because they contain huge amounts of time-series data which have no well-defined semantic and structural style [2]. Some recent researches introduced the indexing and retrieval of instrumental tone quality based on distance matching technology [3]. Some researchers studied more common chords music in the real music environment to extend the previous research on the evaluation of single channel tone quality. They extracted the tone quality features and employed new data mining algorithms to identify and retrieve the interesting objects in massive audio files [4]. Computer-aided music understanding mainly refers to using computer technology and other relevant information technologies to analyze music structures and contents [5]. The work can make computers automatically analyze the music structures, and thereby reduces the burden on related experts. Computer-aided music understanding is mainly applied in inferring musical form, music genres, music styles and music types [6]. Guo et al. presented a new approach to the instrumental tone quality analysis that can evaluate the level of the performer by instrumental tone quality recognition [7].

This paper focuses on the analysis of tone quality of wind instruments, and evaluates the quality level of wind instruments through the recognition results of tone quality. We take clarinets as example and propose an identifying and evaluating method of tone quality based on multi-feature fusion. The proposed method includes 3 steps: 1) Note recognition. We first distinguish the notes performed by instruments based on the identification of fundamental frequency. 2) Multi-feature extraction. We extract 3 typical features including MFCC, roughness and brightness with good identifying and evaluating performance. 3) We use random forest algorithm to construct the basic classifier and then design 5 strategies, at last we analyze and evaluate these strategies. In summary, the main contribution of this paper is to propose an identifying and evaluating method of tone quality of wind instruments, analyze and compare 5 strategies, explore the influence on tone quality analysis raising by note difference from 5 types of strategies and give the optimal recognition strategy. At last, we do the experiment in the complete audio signals by using the optimal strategy.

2. Methodology

2.1. Scheme Description

In this paper we choose clarinet as example and propose a scheme of instrument tone quality recognition shown in Fig. 1.

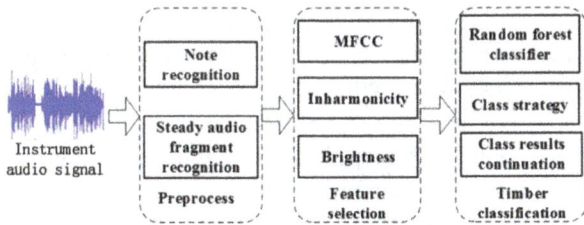

Figure 1. Recognition strategy of instrumental tone quality.

2.2. Note Recognition

Given an audio fragment $X = \{x_m \mid m = 1, 2, \cdots, M\}$, where x_m represents sampled point of the signal and M is the length. Different notes can show different characteristics which may influence the analysis results of tone quality, hence the note should be recognized before tone quality evaluation. In general, a note is denoted by note name and octave. In this paper, we take 28 different notes covering 4 octaves. Each octaves have 7 neighboring notes denoted by $\{c_i, c\#_i, d\#_i, f_i, g_i, a_i, a\#_i\}$ which correspond to $\{do, re, mi, fa, so, la, si\}$ respectively, where the letters and

character ' # ' represent note name, and the subscript $i \in \{1, 2, 3, 4\}$ represents the index of octaves.

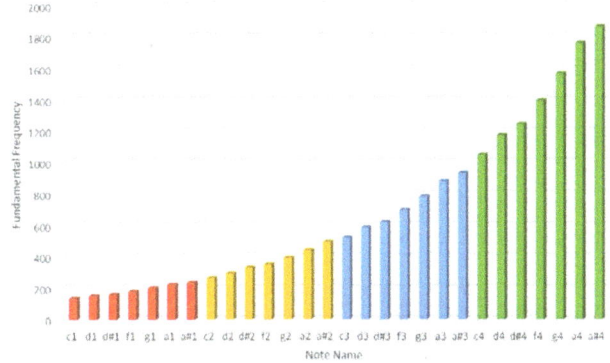

Figure 2. Fundamental frequency of each note. The horizontal axis represents 28 different notes, the vertical axis represents the fundamental frequency, and the different colors indicate the different octaves.

As shown in Fig. 2, different notes correspond to different fundamental frequency. We first use the method presented by Cheveigne et al. [8] to identify the fundamental frequency of notes, which is briefly described below:

1) Computing cumulative mean normalized difference function

$$d_t'(\tau) = \begin{cases} 1 & , if\ \tau = 0 \\ \dfrac{d_t(\tau)}{\sum_{j=1}^{\tau} d_t(j)/\tau} & , otherwise \end{cases} \quad (1)$$

where

$$d_t(\tau) = \sum_{j=1}^{W} (x_j - x_{j+\tau})^2 \quad (2)$$

and searching for the values of τ to make the value of Eq. (2) zero.

2) Setting an absolute threshold and choosing the smallest value of τ that gives a minimum of d' deeper than that threshold.

3) Making parabolic interpolation. In the previous steps, if the period is not a multiple of the sampling period, the estimation may be incorrect by up to half the sampling period. Hence, this step is to reduce the estimation error by parabolic interpolation.

4) Best local estimation. For each time index t, we search for a minimum of $d_\theta'(T_\theta)$ for θ within a small interval $[t - T_{max}/2, t + T_{max}/2]$, where T_θ is the estimation at time θ and is the largest expected period. Generally $T_{max} = 25ms$. The fundamental frequency of the audio signal is the inverse of its estimated period T.

We split the audio signal into multiple frames with the length of 882 points and the overlap of 50%. Every segment's fundamental frequency is estimated. When the ratio of the estimated pitches of two adjacent segments is between $2^{-1/24}$ and $2^{1/24}$, the two segments can be regarded as belonging to one note.

Generally, each note has three states: transient state, quasi-steady state and decay state. In our work, we only focus on the quasi-steady part because it is the most important component for tone quality evaluation and it excludes disturbances of transient state and decay state on feature extracting. To avoid the inaccuracy caused by transitional state between adjacent notes, we extract 80% in the middle of each note as the quasi-steady segment. As shown in Fig. 3, the segment between the two green dashed lines shows a complete note, and the segment between two red dashed lines denotes the quasi-steady state of this note.

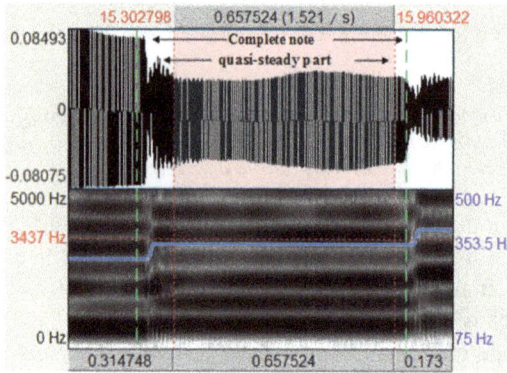

Figure 3. The quasi-steady state of a note.

2.3. Feature Extraction

The information in frequency domain of instrumental signal plays an important role in instrument tone quality analysis. This paper adopts 3 kinds of features to describe the instrumental signal, they are Mel-Frequency Cepstral Coefficients (MFCC), roughness and brightness.

2.3.1. MFCC Feature

In previous research of instrument type recognition, MFCC is a very typical feature, and it integrates the instrumental pronouncing mechanism and human auditory perception effectively [9]. Some research suggests that MFCC is one of the most outstanding property in the single-feature classification scheme [10]. Based on this intuition, we choose MCFF as features for the instrumental tone quality recognition. In this paper, we take 12 dimensional MFCC feature vector, which is denoted by F_{MFCC}.

2.3.2. Roughness Feature

Roughness is mainly used to describe the harmonious degree of a sound segment [11]. When two acoustic sources are with unequal frequency, the harmonic sound will make people feel turbid and harsh, and this feeling calls inharmony. Roughness is one of the features that can evaluate the quality of musical instruments. It is calculated through the dissonance of every two harmonic components of sound.

We apply Fourier transform to each note segment and extract the frequency and amplitude of each harmonic component and represent them by f_i and A_i respectively. Then, we calculate the dissonance of every two harmonic components and sum them over all the component pairs follows:

$$D_F = \frac{1}{2}\sum_{i=1}^{n}\sum_{j=1}^{n}d\left(f_i, f_j, A_i, A_j\right) \qquad (3)$$

where $d(f_1, f_2, A_1, A_2) = \min(A_1, A_2)[e^{-b_1 s(f_2-f_1)} - e^{-b_2 s(f_2-f_1)}]$, $s = \dfrac{x^*}{s_1 f_1 + s_2}$, and b_1, b_2, s_1, s_2 are coefficients that can be determined by experience, x^* is the point in which the roughness achieve the maximum value.

2.3.3. Brightness Feature

Brightness is another important audio feature that denotes the percentage of energy of spectrum segment in which the frequency is higher than the cut-off frequency f_k [12].

$$B = \frac{\sum_{n=k}^{N} x_n e^{-\frac{2\pi i}{N}kn}}{\sum_{n=0}^{N} x_n e^{-\frac{2\pi i}{N}kn}} \qquad (4)$$

where k is the index of cut-off frequency.

2.4. Recognition Scheme Based on Random Forest

2.4.1. Random Forest

Each audio sample can be represented by a feature vector X based on the aforementioned feature extraction. Our work uses random forest as the basic classifier. Random forest is a kind of combined classifier, which combines Bootstrap method and Classification And Regression Tree (CART) algorithm together and constructs an ensemble of decision tree classifier model [13]. The classifier can be denoted by $\{h_1(X), h_2(X), \ldots h_N(X)\}$, and each component denotes a tree. The final classification result is determined by all of the decision trees in the ensemble.

In the stage of constructing decision trees, we first create a new data set by extracting t samples from training set using BootStrap method and selecting features f from the feature vector randomly, and then use CART algorithm to build a decision tree. Finally, by repeating these processes N times, we can obtain an ensemble of N decision trees $\{h_1(X), h_2(X), \ldots h_N(X)\}$. Fig. 4 shows an example of the achieved trees.

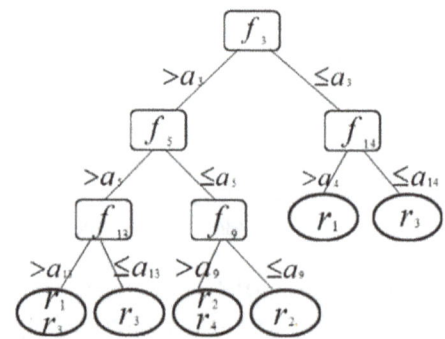

(a) h_1

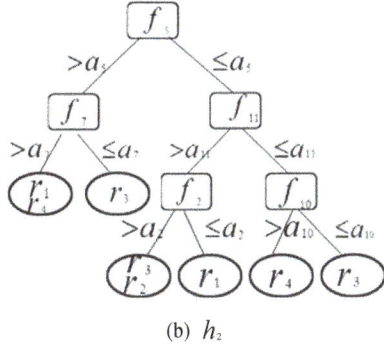

(b) h_2

Figure 4. *An example of random forest.*

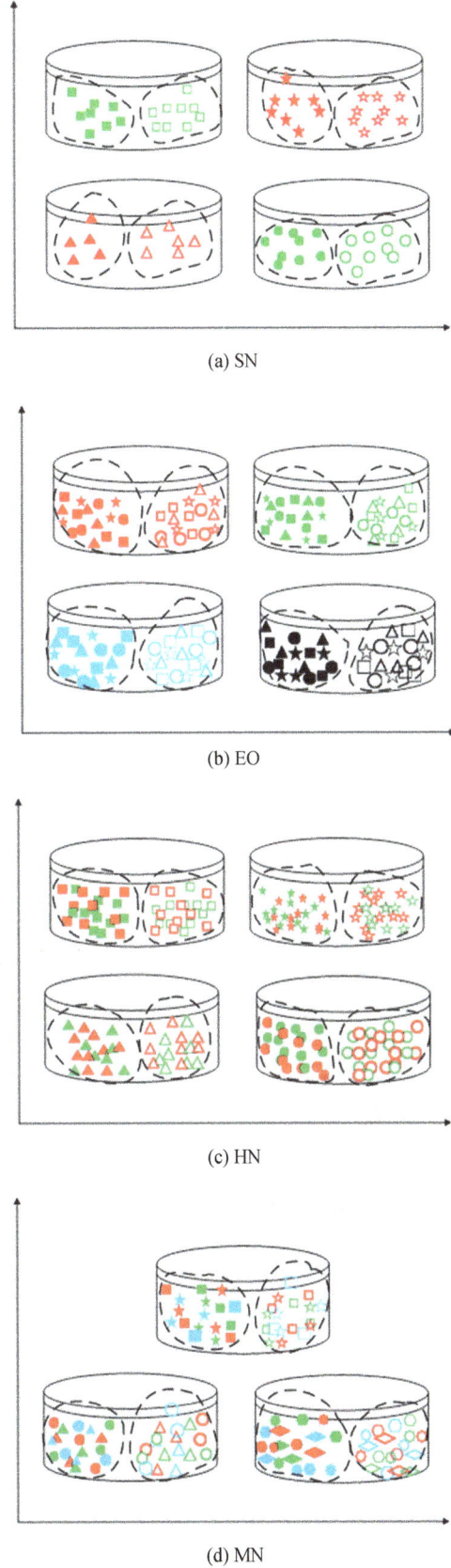

(a) SN

In this figure, f_i is the selected feature for the corresponding node and a_i is the threshold for f_i in the classification.

In the stage of classification, the category of each example is determined by voting of all the decision trees based on the following formula:

$$c_p = argmax_{r_i}\left(\frac{1}{N}\sum_{n=1}^{N}I\left(\frac{m_{h_n,r_i}}{m_{h_n}}\right)\right) \qquad (5)$$

where N is the number of the decision trees, $I(\cdot)$ is the indicator function, m_{h_n,r_i} is the result of category r_i predicted by the tree h_n, m_{h_n} is the number of leaf nodes in the tree h_n.

2.4.2. Recognition Strategy

To recognize the level of tone quality, we focus on solving a classification problem and predicting that which category the audio signal is classified into. However, the variance of tones is raised not only by the tone quality, but also by the variance of notes. Moreover, the latter may lead to larger difference between two tone examples than the former in many cases. The difference among notes mainly reflects in the octave and note name. To analyze how the notes effect the results and achieve the evaluation results independent of the notes, we present and compare the following 5 different strategies. 1) All-notes classification strategy (AN): To build a single classification model over the whole examples of different tone quality and different notes. 2) Single-note strategy (SN): To build multiple classification models, each corresponding to the examples of the same note. 3) Equal-octave strategy (EO): To build multiple classification models, each corresponding to the examples of 7 notes in the same octave. 4) Homonymic-note strategy (HN): To build multiple classification models, each corresponding to the examples of the notes with the same note name. For example, $\{c_1,c_2,c_3,c_4\}$ are the homonymic notes and there is an octave between two neighboring notes. 5) Mixture-note strategy (MN): To build multiple classification models, each corresponding to the examples of the combination of some notes, for example $\{c_i, c\#_i,\}$, $\{d\#_i, f_i\}$, $\{g_i, a_i, a\#_i\}$ are three sample groups to build the models. Fig. 5 shows the latter four strategies.

(b) EO

(c) HN

(d) MN

Figure 5. *The illustration of four classification strategies: (a) SN, (b) EO, (c) HN and (d) MN. Different shapes of points represent different note name and different colors represent different octave. Each cylinder represents a classifier, and each closed, dashed curve corresponds to a class, i.e., a level of tone quality.*

In the practical environment, we first split a continuous music signal into multiple fragments by note recognition and each fragment corresponds to a note. Then we use above recognition algorithm to handle each fragment. Finally, we fuse the recognition result of each fragment to achieve the final result. Suppose the continuous audio signal s_l can be split into k fragments denoted by $\{x_{l_1}, x_{l_2}, \cdots, x_{l_k}\}$. Based on the recognition algorithm, we can get the recognition matrix $P = [p_{ki}]_{K \times C}$, where p_{ki} denotes the probability of sample x_{l_k} belonging to class r_i:

$$\Pr(x_{l_k} \mid r_i) = \frac{1}{N} \sum_{n=1}^{N} I\left(\frac{m_{h_n, r_i}}{m_{h_n}} \right) \quad (6)$$

Like formula (5), we adopt the voting strategy to get the final evaluation result for the continuous music signal based on the following form:

$$c^* = argmax_{r_i} \left(\sum_{k=1}^{K} \Pr(x_{l_k} \mid r_i) \right) \quad (7)$$

Furthermore, the tone quality of an instrument generally does not just locate on the predefined discrete rank of quality. Therefore, we can evaluate the tone quality of an instrument with the continuous recognition result instead of discrete one shown in Eq. 7 by using the average value of a probability distribute overall the ranks. Given C ranks of tone quality named $r_1, r_2, \cdots, r_C$ in order of increasing tone quality, the continuous recognition result can be achieved by:

$$c^*_{co} = \frac{1}{K} \sum_{k=1}^{K} \sum_{i=1}^{C} i \cdot \Pr(x_{l_k} \mid r_i) \quad (8)$$

3. Experimental Result

3.1. Dataset and Experiment Parameter Selection

In this paper, we select 4 ranks of clarinets with different tone quality, and number them 1,2,3,4 in the order of increasing tone quality. 28 different notes in 4 neighboring octaves are performed, each note is repeated 30 times, and then 840 examples are obtained. According to the feature extraction method in 2.4, we extract features for each note sample. Then the dataset is split randomly into two sets, 70% is for training and the rest for test. In the experiment, we use 100 decision trees to build the random forest classifier, in each tree we choose 5 features and the depth of 4.

3.2. Experimental Result

To evaluate the classification results, we use the correct classification rate defined as follows:

$$R = \frac{CN}{TN} \quad (9)$$

where CN is the number of examples that are correctly classified and TN is the total number of the examples of one rank.

Based on the method proposed above, we take 5 classification strategy to recognize the tone quality of instruments. Table 1 shows the correct classification rate in each strategy. Rank1, Rank2, Rank3, Rank4 represent the correct classification rate in each corresponding level of tone quantity and AVGR represents the average rate of all level of tone quantity.

Table 1. The correct classification rate of five strategy.

	AN	SN	EO	HN	MN
Rank1	76.91%	88.65%	76.06%	84.11%	79.80%
Rank2	69.63%	80.56%	66.44%	78.62%	78.23%
Rank3	69.21%	85.11%	70.99%	76.00%	73.33%
Rank4	70.26%	82.67%	59.60%	72.11%	64.32%
AVGR	71.64%	84.24%	71.30%	77.71%	73.61%

From table 1 we can see that single-note strategy (SN) gets the best result. In this strategy, the difference among notes have been removed before the samples are used to build the model, and the classifier only needs to consider the difference of tone quality for one note. That is, each group of examples only include the examples with the same note, and then it has lower within-class divergence and higher between-class divergence. Therefore, the classifier can achieve better performance in the recognition of tone quality. Comparing equal-octave strategy (EO) to homonymic-note strategy (HN), we find that the latter strategy has higher accuracy rate. This result illustrates that the samples with same note name have lower within-class divergence. Mixture-note strategy (MN) combine the note name and the octave, so we can infer that the divergence of the samples is between EO and HN strategy. The table also shows that the result of mixture-note strategy is between that of equal-octave strategy and homonymic-note strategy.

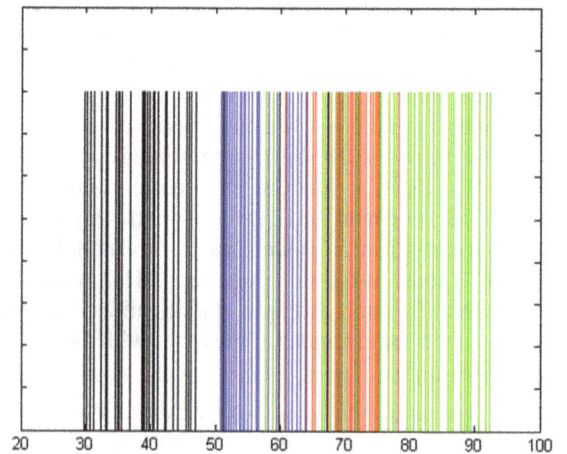

Figure 6. The evaluation result based on the continuous music. The horizontal axis represents the continuous result of the music signal. The black, blue, red and green part respectively denote four different ranks of clarinets in order of increasing tone quality. The center position of each rank is artificially located at 40,55,70,85 for an appropriate show.

We also evaluate the tone quality of four different rank of clarinets based on continuous music signal with Eq.8. Each music signal includes multiple different notes, and we fuse the evaluation for each notes together and achieve the final evaluation result of the instrument. The continuous evaluation result of each note is shown in Fig. 5. We notice that the all the classification results of different level of instruments distribute around the center of the corresponding category, and only a very low proportion of examples are misclassified into the other categories.

4. Conclusions

This paper presents an identifying and evaluating method of tone quality of wind instruments. In this method, we fuse 3 kinds of effective features and use the random forest as the basic classification. To remove the influence of different notes in the evaluation of tone quality, we introduce 5 recognition strategies and give the optimal classification strategy. The experimental result shows that our method can recognize the rank of wind instruments effectively. For future work, we will test the method on more wind instruments and try to explore more features to improve the accuracy.

Acknowledgements

Acknowledgements are usually put in the end of the paper, between the reference and the content. There should be no number in this part. This work is supported in part by the National Natural Science Foundation (6137540,61221063), Fundamental Research Funds for the Central Universities (xjj2013090) and 111 International Collaboration Program, of China.

References

[1] Yu-Hsiang H, Chao-Ton S. Multiclass MTS for saxophone timbre quality inspection using waveform-shape-based features [J]. IEEE Transactions on Systems Man & Cybernetics Part B Cybernetics A Publication of the IEEE Systems Man & Cybernetics Society, 2009, 39(3): 690-704.

[2] Paulus J, Klapuri A. Music Structure Analysis Using a Probabilistic Fitness Measure and an Integrated Musicological Model. [C] ISMIR 2008, 9th International Conference on Music Information Retrieval, Drexel University, Philadelphia, PA, USA, September 14-18, 2008. 2008: 369-374.

[3] Typke R, Veltkamp R C, Wiering F. Searching notated polyphonic music using transportation distances [C]. Acm Multimedia Conference. 2004: 128-135.

[4] Downie S, Nelson M. Evaluation of a Simple and Effective Music Information Retrieval Method [C] Research & Development in Information Retrieval. SIGIR '00 Proceedings of the 23rd annual international ACM SIGIR conference on Research and developm, 2000: 73-80.

[5] K. Roger B. Dannenberg, Ning Hu. Pattern Discovery Techniques for Music Audio [J]. Journal of New Music Research, 2003, 32(2): 63-70.

[6] Yu Y, Zimmermann R, Wang Y, et al. Recognition and Summarization of Chord Progressions and Their Application to Music Information Retrieval[C]. Multimedia (ISM), 2012 IEEE International Symposium on. 2012: 9-16.

[7] Guo J, Ding M, Guan X, et al. Timbre identification of instrumental music via energy distribution modeling[C]. Proceedings of the 7th International Conference on Internet Multimedia Computing and Service. ACM, 2015: 1-5.

[8] Cheveigné A D, Kawahara H. YIN, a fundamental frequency estimator for speech and music [J]. Journal of the Acoustical Society of America, 2002, 111(4): 1917-30.

[9] Valero X, Alias F. Gammatone Cepstral Coefficients: Biologically Inspired Features for Non-Speech Audio Classification [J]. IEEE Transactions on Multimedia, 2012, 14(6): 1684-1689.

[10] Logan B. Mel Frequency Cepstral Coefficients for Music Modeling [C]. In International Symposium on Music Information Retrieval. 2000.

[11] Sethares W A. Tuning, Timbre, Spectrum, Scale [M]. Springer London, 2005.

[12] Vassilakis P N. Perceptual and Physical Properties of Amplitude Fluctuation and their Musical Significance [J]. Acta ibérica radiológica-cancerológica, 2001, 28(4): 119-128.

[13] Biau G. Analysis of a Random Forests Model [J]. Journal of Machine Learning Research, 2010, 13(2): 1063-1095.

28

Results of Simulation Program for Pathological Index relating the Climate Factors

Javzmaa Tsend[1], Bat-Enkh Oyunbileg[2], Ajnai Luvsan[1], S. Battulga[1]

[1]Mongolian National University of Medical Science, Ulaanbaatar, Mongolia
[2]School of Information and Telecommunication Technology, the Mongolian University of Science and Technology, Ulaanbaatar, Mongolia

Email address:

javzmaa.ts@mnums.edu.mn (J. Tsend), o_bat_enkh@yahoo.com (Bat-Enkh O.)

Abstract: To protect and prevent human health, this study calculated pathological index using weather factors such as air average temperature, average wind speed, average relative humidity and air pressure at days between 2004 and 2014 in Ulaanbaatar city of Mongolia. Then, we developed software program that calculates above mentioned pathology index, statistic parameters, and correlation.

Keywords: Change of Climate Factor, Pathological Index, Program, Health, Prevent

1. Introduction

In medical Bio Climate sector that treats various diseases, and prevents it, programming discovery is very important to execute it [1-2].

2. Method and Material

2.1. Pathological total Index

Pathological total index is join of weather and helio-geo physics index [2].

$$I_{total} = I_{weather} + I_{helio} \qquad (1)$$

2.2. Weather Pathogenicity Index

When information of helio-geophysics is meagre, evaluate it to represent by below describing general index of weather factors. The formula (2) is shown below.

$$I_{weather} = I_t + I_r + I_v + I_{\Delta t} + I_{\Delta p} \qquad (2)$$

So, each index of weather factor is shown in formula from (3) and to (7) [2].

I_t – Pathological index of temperature

$$I_t = 0.02 * (18-t)^2 \qquad (3)$$

t - medium air temperature in a day

I_r – Pathological index of relative humidity

$$I_r = 10^{(r-70)/20} \qquad (4)$$

r- medium relative humidity in a day

I_v – Index of wind speed

$$I_v = 0.2 * v^2 \qquad (5)$$

v - medium wind speed in a day

$I_{\Delta p}$ – Index of pressure change

$$I_{\Delta p} = 0.06 * \Delta p^2 \qquad (6)$$

Δp- change of air pressure in a day

$I_{\Delta t}$ – Index of Temperature change

$$I_{\Delta t} = 0.3 * \Delta t^2 \qquad (7)$$

Δt - change of air temperature in a day

Depending on the value $I_{weather}$ the conditions are assessed as [2]:

I = 0 - 9 – optimal
I = 10 - 24 – irritant
I >24 – critical

So, in this study, we calculated general index of weather factors and developed simulation software program using C# programing language for ASP.NET web developer. Also had built weather factor database by the factors at days between 2004 and 2014 in Ulaanbaatar of Mongolia [3,5, 7-9].

Also estimated correlation between the index and infection disease at days between 2013 and 2014 and non infection disease such us diseases of the circulatory system (I00-I99), Diseases of the nervous system (G00-G99) at day between 2009 and 2013.

3. Result

Our simulation program calculated above mentioned pathology index, and automatically displayed (presented) below mentioned statistic specification, and estimated correlation between the weather index and these diseases.

3.1. Statistics

Some parameters such as mean index, temperature, humidity, wind speed of each month in last ten year are shown in below table. Indexing parameters are very important parameters to protect and prevent human health [4-5, 10]. These are results that automatically displayed by our simulation program.

Table 1. Mean index, temperature, humidity, wind speed of each month in last ten year.

Month	Mean	Range	t	r	v	Δp	Δt
1	38	52.7	-21.7	74.4	1.2	2.8	2.1
2	32.5	78.5	-17.9	69.5	0.7	2.9	2.4
3	20.3	98.1	-7.6	57.9	2.2	3.4	2.6
4	12.9	91.6	3.1	42.8	2.8	3.4	3.1
5	11.7	82.8	10.1	42.2	3.1	2.8	3.7
6	2.8	37.6	16.8	50.2	2.9	1.7	2.7
7	4.1	89.8	19.3	56.6	2.5	1.4	2
8	2.7	43.9	17	57.3	2.5	1.7	2.3
9	3.2	64	10.5	52.2	2.5	2.4	2.8
10	3.3	36.3	1.8	56.1	2.1	2.8	2.3
11	21.7	86.6	-9.5	65.9	1.7	2.9	2.4
12	34.4	69.9	-18.7	73.4	1.3	3	2.5

In "Fig. 1", when weather pathogenic mean index was placed by descending order, it is high and climate state is impressive at winter. Next, there are month of autumn and spring. Then there is summer. The climate state is optimal which give positive effect to human health. Horizontal axis is month.

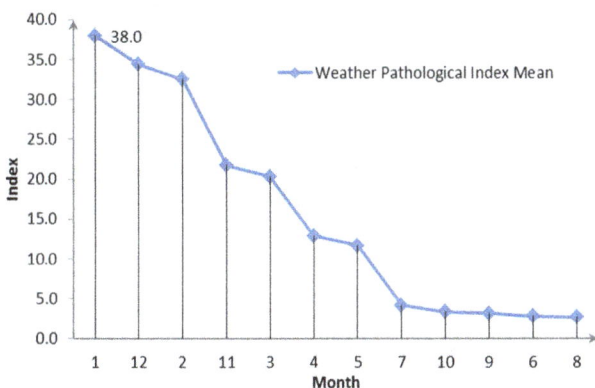

Figure 1. Weather Pathogenical mean index was placed by descending order.

Looking at the "Fig. 2", when difference of the maximum and minimum values of pathological index was placed by ascending order in horizontal axis, there is high fluctuation at autumn and spring. Next, in winter, it is more little fluctuation than spring and autumn. In summer, the fluctuation is lowest.

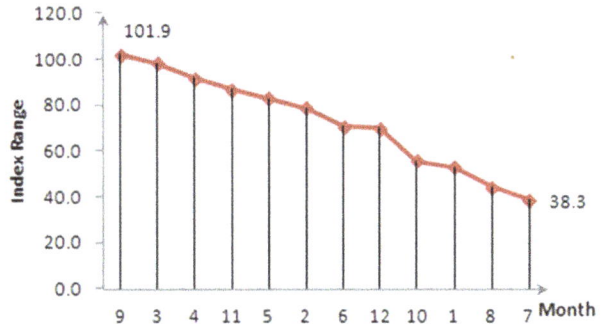

Figure 2. Weather Pathogenical index's range was placed by descending order.

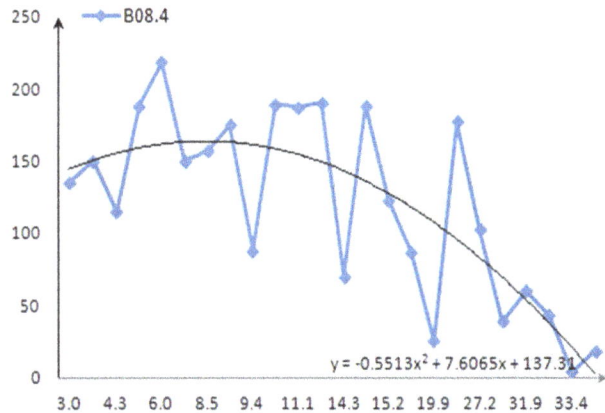

Figure 3. Weather Pathogenical index's skewnes coefficient was placed.

We had organized weather pathological index's maximum value from to 2011 and to 2013 by showed "Fig.3".

Looking at "Fig. 3" picture, 2011, 2012 and 2013, Pathological index maximum value is in May of 2011, next November of 2013. Then, in April of 2012, the value was the highest.

3.2. Correlation of Index and Climate Factor

In table 2, correlation between the index and climate factor are shown.

Table 2. Mean index, temperature, humidity, wind speed of each month in last ten year.

	T	R	v	Δp	Δt	Pato
T	1	-0.8	0.9	-0.7	0.2	-0.9
R	-0.8	1	-0.9	0.1	-0.7*	0.8
v	0.9	-0.9	1	-0.3	0.6	-0.8
Δp	-0.7*	0.1	-0.3	1	0.3	0.6*
Δt	0.2	-0.7*	0.6	0.3	1	-0.2
Pato	-0.9	0.8	-0.8	0.6	-0.2	1

Tempchange and Pathological Index

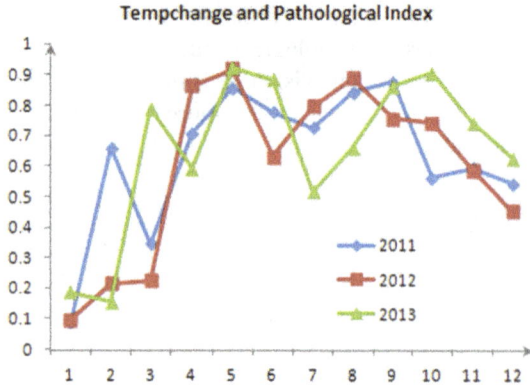

Figure 4. Weather Pathogenical index and Temp of change.

We have also organized correlation between temperature change and weather pathological index from to 2011 and to 2013 by showed "Fig.3".

From the figure, in January and February of winter, first month of spring, correlation between medium change of air temperature in a day and pathological index is weak and don't affect to disease. Sometime, for example, February of 2011, March of 2013, it is negative, strong. But it is affected to pathological index in most month of year.

3.4. Correlation Between the Index and the Factor and Infection Diseases

We have estimated correlation between the index and infection diseases at days between 2013 and 2014. In "Fig. 3", "Fig.4" it describes hand, foot and mouth disease and Shigellos's type and while sort ascending order to the index.

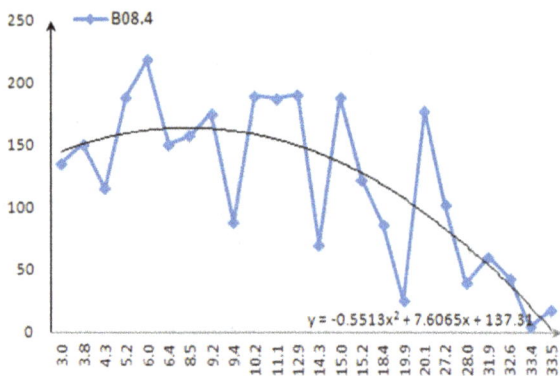

$$y = -0.5513x^2 + 7.6065x + 137.31$$

Figure 5. B08.4 Hand, foot and mouth disease.

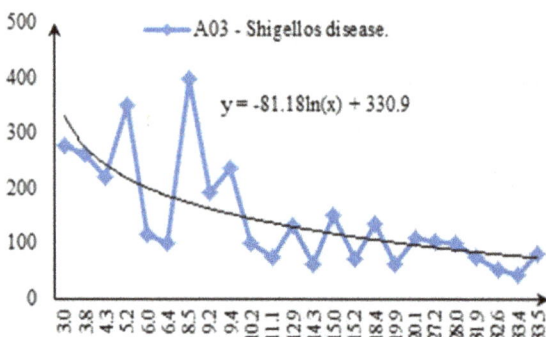

$$y = -81.18\ln(x) + 330.9$$

Figure 6. Shigellos disease.

Also we have estimated correlation between the index and disinfection disease at days between 2009 and 2013.

In "Fig. 7", it shows transient cerebral ischaemic attacks and related syndromes disease's type while sort decreasing order to the index.

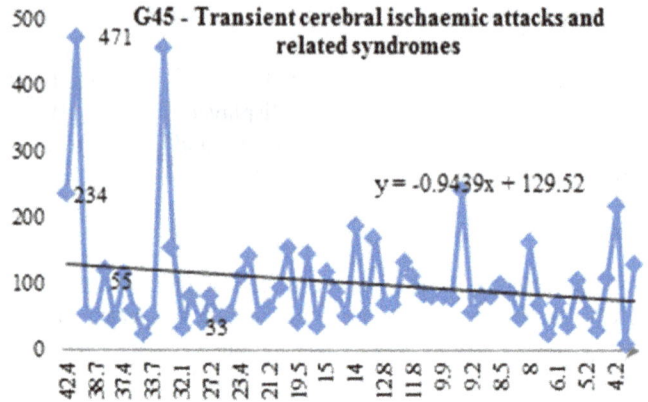

$$y = -0.9439x + 129.52$$

Figure 7. Transient cerebral ischaemic attacks and related syndromes.

In "Fig. 8", it also show hypertensive heart and renal disease's type while sort decreasing order to the index.

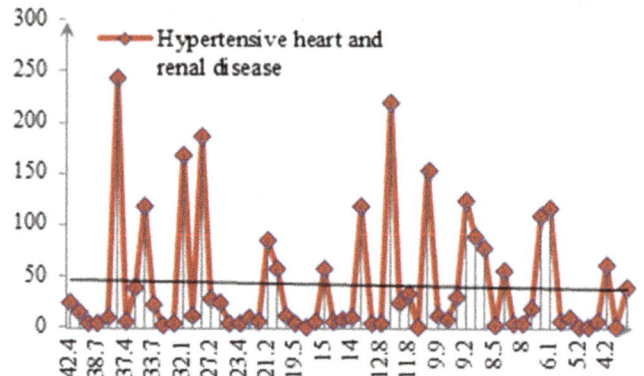

Figure 8. Hypertensive heart and renal disease.

4. Conclusion

We determined pathological index that connects climate factors. To show result of the research, Weather Pathological index was high in winter and then decreased in spring. It was the lowest level in summer and then again increased in autumn.

In correlation of climate factors, in most month of year, it is increased pathological index due to medium change of air temperature in a day. Specially in spring and autumn, the correlation is the highest. Seen here, in any season of year, when medium change of air temperature in a day grow, disease will increase.

Seen the pattern of fluctuation specification, range is the highest in spring, autumn and that means there are all climate state. In summer, it was more decreased. Also in winter, it was low than spring and autumn.

Generally, weather pathological index maximum value like as difference of minimum and maximum pathological index value.

Intestinal infectious diseases probably occur in thermal season. And we will more learn about diseases of the nervous system further.

In the further, we will estimate to add pathological index of geo magnetic field and then we will determine total pathological index. And also we will determine how much climatic state that gets pathology in which season in Mongolian country. So it is possible to protect and prevent human health.

References

[1]   Mongolia-Korea Conference on Biomedical Applied Science and Engineering, The Issue to organize Information Technology based Pathological Index relating the Climate Factors, 2014/07/09-2014/07/10

[2]   Бокша В.Г, Богуцкий Б.В. Медицинская климатология и климатотерапия,Киев Здоровя, 1980, 47-82

[3]   Нарантуяа Л., Купул Ж. Монголын хүн амын эрүүл мэнд, экологийн зарим хүчин зүйлийн харилцан хамаарал, эрүүл ахуйн үнэлгээ, УБ, 1999

[4]   Андреев, С.С. Экология человека / С.С. Андреев.- Ростов: н/д: Изд-во. Е.А. Турова, 2007.- 248с.

[5]   http://webcasting.mn/mn/news/, 2014.04.16

[6]   https://mn.wikipedia.org/wiki/%D3%A8%D0%B2%D1%87%D0%B8%D0%BD 2015.03.16

[7]   С. Баттулга. "Монгол улсад эмнэлгийн мэдээллийн системийг хөгжүүлэх зарим асуудал", 2009, диссертаци

[8]   Ч. Наранчимэг, "Эконометрикс" , 2003 он

[9]   Удирдлагын ерөнхий онол, Д. Нарангэрэл, Улаанбаатар,2005

[10] Study of Factors Influencing Mortaility from the Cerebral Stroke in Patients of Different Ages,Vazgen Martirosyan and Krupskaya ,British Journal of Medicine Research 3(4):1530-1557, 2013

29

Selection of efficient relay for energy-efficient cooperative ad hoc networks

Manish Bhardwaj

Department of Computer science and Engineering, SRM University, NCR Campus, Modinagar, Ghaziabad, India

Email address:

aapkaapna13@gmail.com (M. Bhardwaj)

Abstract: The Cooperative Communication (CC) is a technology that allows multiple nodes to simultaneously transmit the same data. It can save power and extend transmission coverage. However, prior research work on topology control considers CC only in the aspect of energy saving, not that of coverage extension. This paper identify the challenges in the development of a centralized topology control scheme, named Cooperative Bridges, which reduces transmission power of nodes as well as increases network connectivity. Previous research on topology control with CC only focuses on maintaining the network connectivity, minimizing the transmission power of each node, whereas ignores the energy efficiency of paths in constructed topologies. This may cause inefficient routes and hurt the overall network performance in cooperative ad hoc networks. With the help of studied topology control problem for energy-efficient topology with cooperative communication. This paper proposed optimum relay nodes selection for CC network to reduce overall power consumption of network.

Keywords: Cooperative Communication, Topology Control, Optimum Relay, Power Efficient, Greedy Algorithm

1. Introduction

Increasing demand for high-speed wireless networks has motivated the development of wireless ad-hoc networks. In order to fully exploit the technological development in radio hardware and integrated circuits, which allow for implementation of more complicated communication schemes, the fundamental performance limits of wireless networks should be reevaluated. In this context, the distinct characteristics of wireless networks compared to their wired counterpart lead to more sophisticated design of protocols and algorithms. Some of the most important inherent properties of the Physical Layer (PHY) that make the design more complicated include the attenuation of radio signals over long range communications called path loss, and the fading effect caused by multipath propagation. In order to mitigate these effects, the user has to increase its transmission power or use more sophisticated reception algorithms. Another important limitation of wireless performance caused mainly as a result of communication over a limited bandwidth is the interference from other users, communicating over the same frequency spectrum. Wireless ad hoc networks are multi-hop structures, which consist of communications among wireless nodes without infrastructure.

Therefore, they usually have unplanned network topologies. Wireless ad hoc networks have various civilian and military applications which have drawn considerable attentions in recent years. One of the major concerns in designing wireless ad hoc networks is to reduce the energy consumption as the wireless nodes are often powered by batteries only. Wireless nodes need to save their power as well as sustain links with other nodes, since they are battery powered. Topology control deals with determining the transmission power of each node so as to maintain network connectivity and consume the minimum transmission power. Using topology control, each node is able to maintain its connection with multiple nodes by one hop or multi-hop, even though it does not use its maximum transmission power. Consequently, topology control helps power saving and decreases interferences between wireless links by reducing the number of links. Topology control [1-4] is one of the key energy saving techniques which have been widely studied and applied in wireless ad hoc networks. Topology control lets each wireless node to select certain subset of neighbors or adjust its transmission power in order to conserve energy meanwhile maintain network connectivity. Topology control have been widely studied and applied in wireless ad hoc networks as one of the key energy saving techniques. In order to save energy and extend lifetime of networks topology control lets each wireless node to select certain subset of neighbors or adjust its transmission power meanwhile maintain network connectivity. Recently, a new class of communication techniques, cooperative communication (CC) [37], [38], has been introduced to allow single antenna devices to take the advantage of the multiple-input-multiple-output (MIMO) systems. This cooperative communication explores the broadcast nature of the wireless medium and allows nodes that have received the transmitted signal to cooperatively help relaying data for other nodes. Recent study has shown significant performance gain of cooperative communication in various wireless network applications: energy efficient routing [39]–[41] and connectivity improvement [42]. In this paper, we study the energy efficient topology control problem with CC model by taking the energy efficiency of routes into consideration. Taking advantage of physical layer design that allows combining partial signals containing the same information to obtain the complete data, we formally define cooperative energy spanner in which the least energy path between any two nodes is guaranteed to be energy efficient compared with the optimal one in the original cooperative communication graph. We then introduce the energy-efficient topology control problem with CC (ETCC), which aims to obtain a cooperative energy spanner with minimum total energy consumption,

The cooperative communication techniques can also be used in topology control. In [35], Cardei et al. first studied the topology control problem under cooperative model (denote by TCC) which aims to obtain a strongly-connected topology with minimum total energy consumption. They proposed two algorithms that start from a connected topology assumed to be the output of a traditional (without using CC) topology control algorithm and reduce the energy consumption using CC model. The first algorithm (DTCC) uses 2-hop neighborhood information of each node to reduce the overall energy consumption within its 2-hop neighborhood without hurting the connectivity under CC model. The second algorithm (ITCC) starts from a minimum transmission power, and iteratively increases its power until all nodes within its 1-hop neighborhood are connected under CC model. Observing that the CC technique can also extend the transmission range and thus link disconnected components. In [36], Yu et al. applied CC model in topology control to improve the network connectivity as well as reduce transmission power. Their algorithm first constructs all candidates of bidirectional links using CC model (called cooperative bridges) which can connect different disconnected components in the communication graph with maximum transmission power. Then they apply a 2-layer MST structure (one MST over the CC links to connect the components, the other is inside each component) to further reduce the energy consumption.

2. Related Work

Topology control has drawn a significant amount of research interests in wireless ad hoc networks [6-12]. Primary topology control algorithms aim to maintain network connectivity and conserve energy by selecting certain subset of neighbors and adjusting the transmission power of wireless nodes. Comprehensive surveys of topology control can be found in [1-4]. Cooperative communication (CC) exploits space diversity through allowing multiple nodes cooperatively relay signals to the receiver so that the combined signal at the receiver can be correctly decoded. Since CC can reduce the transmission power and extend the transmission coverage, it has been considered in topology control protocols. However, prior research on topology control with CC only focuses on maintaining the network connectivity, minimizing the transmission power of each node, whereas ignores the energy efficiency of paths in constructed topologies. This may cause inefficient routes and hurt the overall network performance in cooperative ad hoc networks. Paper [43] address this problem, author introduce a new topology control problem: energy-efficient topology control problem with cooperative communication, and propose two topology control algorithms to build cooperative energy spanners in which the energy efficiency of individual paths are guaranteed. Both proposed algorithms can be performed in distributed and localized fashion while maintaining the globally efficient paths. Cooperative communication (CC) allows multiple nodes to simultaneously transmit the same packet to the receiver so that the combined signal at the receiver can be correctly decoded. Since CC can reduce the transmission power and extend the transmission coverage, it has been considered in topology control protocol. However, prior research on topology control with CC only focuses on maintaining the network connectivity, minimizing the transmission power of each node, whereas ignores the energy-efficiency of paths in constructed topologies. This may cause inefficient routes and hurt the overall network performance. Paper [44] introduces a new topology control problem: energy-efficient topology control problem with cooperative communication, and propose two topology control algorithms to build cooperative energy spanners in which the energy efficiency of individual paths are guaranteed. Chen and Huang [5] first studied the strongly connected topology control problem, which aims to find a connected topology such that the total energy consumption is minimized. They proved such problem is NP-complete. Several following works [8-12] have focused on finding the minimum power assignment so that the induced communication graph has some "good" properties in terms of network tasks such as disjoint paths, connectivity or fault-tolerance. On the other hand, several localized geometrical structures [13-18] have been proposed to be used as underlying topologies for wireless ad hoc networks. These geometrical structures are usually kept as few links as possible from the original communication graph and can be easily constructed using location information. Recently, a new class of communication techniques, cooperative communication (CC) [19], [20], has been introduced to allow single antenna devices to take the advantage of the multiple-input-multiple-output (MIMO) systems. This cooperative communication explores the broadcast nature of the wireless medium and allows nodes that have received the transmitted signal to cooperatively help relaying data for other nodes. Recent study has shown significant performance gain of cooperative communication in various wireless network applications: energy efficient routing [21-24], broadcasting [25-27], multicasting [28], connectivity/coverage improvement [29], [30], and relay selection for throughput maximization or energy conservation [31-34].

3. Cooperative Communication

Wireless communication technique with a wireless network, of the cellular or ad hoc selection, where the wireless users, may increase their valuable quality of service via cooperation a cooperative communication system, each wireless user is assumed to transmit data as well as act as a cooperative agent for an additional user (Fig. 1). For example, in figure 1, node S is unable to communicate with node D, since D is out of its maximum transmission range of S. On the other hand, S can send a cooperation request message and data to adjacent connected nodes R as relay node and then the three nodes all together pass on the data to D. Therefore, D can receive it due to the extended transmission range of nodes S, R, and R.

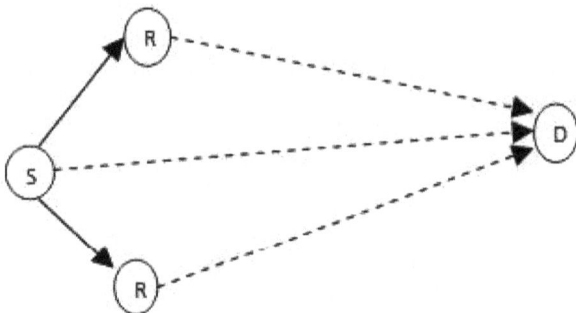

Figure 1. Coverage Extension using CC

Cooperative communication means in any system users share and cooperative their resources to enhance their performance jointly with help of each other. This method is very useful for enhance transmission range of a node in mobile adhoc network as diverse channel quality and limited energy and limited bandwidth limitations wireless environment. Due to cooperation, users that know-how a deep weaken in their connection towards the target can utilize quality channels provided by their partners to achieve the preferred quality of service (QoS). This is also identified like the spatial diversity gain, which is in the same way achieved in multiple-input-multiple-output (MIMO) wireless systems. Cooperation has an interesting trade-off between code rates and transmits power. In the case of power, extra power is

needed because to every user, when system is in cooperative mode, is transmitting for both users. But transmits power for both users will be reduced because of diversity. Due to this trade-off, one hopes for a net reduction of transmit power, given every-thing else being constant. In cooperative communication every user sends both his/her personal bits as well as a few data for his/her neighbor; one may believe this causes loss of rate in the system. However, the spectral efficiency of each user improves because; due to cooperation diversity the channel code rates are able to be improved. Hence one more trade-off is occurred. So whether cooperation is worth the incurred cost, has been studied positively by numerous research studies.

4. Cooperative Model

Here, we explain a cooperative communication model and a network representation for topology control system. In addition, we define two problems: Topology Control considering Extended Links caused by CC and Energy-Efficient Extended Link with CC.

4.1. Cooperative Communication Model

In Cooperative Communication Model PMAX represents every node's maximum transmission power limit. Pi is the transmission power of node i. α is the path loss exponent and τ is the minimum average SNR for decoding received data. dij is the distance between node i and node j. For a source node i to communicate with node j directly (figure 1), they must satisfy

$$Pi\ (dij) - H \geq \tau\ (Pi \leq PMAX).$$

H denotes the set of a source node and helper nodes. If nodes in H transmit simultaneously, i.e., use cooperative communication, the following formula must be satisfied for correct decoding at destination node j.

$$i{\in}H\ Pi(dij) - H \geq \tau\ (Pi \leq PMAX)$$

CC leads to extended transmission coverage. For example, in figure 1, node S cannot communicate with node D, since D is out of the maximum transmission range of S. Node S can send a cooperation request message and data to nodes R and R, and then the three nodes simultaneously transmit the data to D. Therefore, D can receive it due to the extended transmission range of nodes R, R, and S. The physical layer issues including synchronization for implementing the CC technique can be found in [8]. In figure 1, if node R applies CC with partner S in sort to communicate with D, which is already accessible to R by straight links, the network can decrease the sum of node transmission power. Cardei et al. [26] focus their problem formulation on saving power with CC, not extended CC links.

4.2. Network Model

The wireless network topology is form as a 2-dimensional graph is collection of vertices V and edges E, graph G = (V,

E). V = (v1, ..., vn) is a set of random nodes and E is a set of pairs of nodes as link between them (vi, vj), with vi, vj ∈ V. The notations V (G) and E (G) are used for the vertex- and edge-set of G. The weight of a directional link from u to v is denoted as w (u → v). Edge (u, v) has weight, w (u, v), which indicates the average power utilization for maintaining a bi-directional link (u, v). N (v) is the set of neighbor nodes within the maximum transmission range of node v. All elements in N (v) are the candidate nodes, which are eligible as helper nodes for v. Node v is capable to communicate directly with its neighbors within 1 hop. R(u) is the set of nodes which are accessible to node u by 1-hop or multi-hop, i.e., have a path to a node u.

4.3. Problem Formulation

Major difficulty in given a wireless multi-hop network G=(V,E) which is restricted under CC connection model, it that assign transmission power Pi for every node vi such that make topology G' from this power assignment is a cooperativeenergy t-spanner of G and the sum of transmission power of all nodes, ΣviεVPi , is minimized. Key point is that the spanner property also guarantees that the induced topology G' is strongly connected under CC model. Paper [43] presents an Energy-Efficient Topology Control in Cooperative Ad Hoc Networks, but if neighbor nodes are more for any node so they all will help to source node for transmitting data to destination whether only some nodes of them as capable to transmit data till destination so power of other nodes are unnecessarily used during this transmission as given in figure 2.

5. Proposed Work

This paper proposed efficient in two phase first phase is to Energy-efficient topology control with cooperative communication and then optimum relay node selection. First phase propose two topology control algorithms which build energy-efficient cooperative energy spanners. To keep the proposed algorithms simple and efficient, we only consider its one-hop neighbors as possible helper nodes for each node when CC is used [43]. Thus, the original cooperative communication graph G contains all direct links and CC links with one hop helpers, instead of all possible direct links and CC-links. In addition, for each pair of nodes vi and vj, we only maintain one link with least weight if there are multiple links connecting them. Here, all links are directional links. Both proposed algorithms are greedy algorithms. The major difference between them is the processing order of links. The first algorithm deletes links from the original graph G greedily, while the second algorithm adds links into G" greedily. Here, G00 is a basic connected sub graph of G. Both algorithms can guarantee the cooperative energy spanner property of the constructed graph G'.

5.1. Phase One

5.1.1. Greedy Method for Deleting Links from Network Graph

Step 1: Construction of G. Initially, G is an empty graph. First, add every direct links vivj into G, if node VI can reach node VJ when it operates with PMAX. Then, for every pair of nodes vi and vj, we select a set of helper nodes Hij for node vi from its one-hop neighbors N (vi), such that the link weigh w (vi,vj) of the constructed CC-link is minimized. Notice that this helper node decision problem is challenging even under our assumption that the transmission powers of VI and its helper node set to maintain CC-link are the same. If we try all combinations of the helper sets to find the optimal helper set which minimizes the total energy consumption of vi and its helpers, the computational complexity is exponential to the size of the one-hop neighborhood N (vi). It is impractical to do so in case of a large number of neighbors. Therefore, we directly use the greedy heuristic algorithm Greedy Helper Set Selection (vi, N (vi), vj), to select the helper set Hij. Then, we compare w (vivj) with p (PG(vi,vj)) which is the current shortest path from node vi to node vj in G. If w (vivj) ≤ p (PG (vi,vj)) and

$$\frac{\tau}{\sum_{v_k \in v_i \cup H_{ij}} (d_{kj})^{-\alpha}} \le P_{MAX},$$

Add this CC-linkg vivj into G. If there already exists a direct link vivj, delete it after the new CC-link g vivj is added (since it costs more energy than the CC-link). Notice that if

$$\frac{\tau}{\sum_{v_k \in v_i \cup H_{ii}} (d_{kj})^{-\alpha}} \le P_{MAX},$$

Node vi cannot communicate with node vj within one-hop even in CC model.

Step 2: Construction of G'. Copy all links in G to G', and sort them in the descending order of their weights. Start to process all links one by one and delete the link vi vj from G' if G-vivj is still a cooperative energy t-spanner of G. Hereafter, we use G- e or G+e to denote the graph generated by removing link e from G or adding link e into G, respectively. In addition, when a CC-link g vivj is kept in G', all its helper links must be kept in G' too.

Step 3: Power Assignment from G'. For each node vi, its transmission power is decided by the following equation:

Here $P_i^d(j) = \frac{\tau}{d_{ij}^{-\alpha}}$ and $P_i^{cc}(j) = \frac{\tau}{\sum_{v_k \in v_i \cup H_{ij}} (d_{kj})^{-\alpha}}$ are the energy consumption at vi for a direct link vivj and a CC-link vivj , respectively.

5.1.2. Greedy Method for Adding Links

The second topology control algorithm starts with a sparse topology G" which is strongly connected under CC model. We can use the output of the algorithm in [36] as the initial topology. Then, we gradually add the most energy-efficient link into G". Here, the energy-efficiency of a link is defined as the gain on reducing energy stretch factors by adding this link. Our algorithm will terminate until the constructed graph G' satisfies the energy stretch factor requirement. The detail steps are summarized as follows:

Step 1: Construction of G and G''. The step of constructing G is the same as the one in Algorithm 1. Then, we call the algorithm in [36] to generate G'', a connected sparse sub graph of G.

Step 2: Construction of G'. Initialize G'=G'', for every link vivjεG but not G', compute its stretch-factor-gain g G'G (vivj) as follows:

$$g_G^{G'}(v_i v_j) = \sum_{v_p, v_q \in V} \left(\rho_G^{G'}(v_p, v_q) - \rho_G^{G'+v_i v_j}(v_p, v_q) \right)$$

In other words, the total gain of a link vivj is the summation of the improvement of stretch factors of every pair of nodes in G' after adding this link In each step, we greedily add the link with the largest stretch-factor-gain into G'. If there is a tie, we use the link weight to break it by adding the link with the least weight. We repeat this procedure until G' meets the stretch factor requirement t.

Step 3: Power Assignment from G'. For every node VI, assign its power level Pi using equation for Pi.

5.2. Phase Two

5.2.1. Optimum Relay Nodes Selection

Once communication topology has been created optimum nodes can be selected from this topology for efficient transmission. As problem definition mention in example in figure 2(a) according to CC model if S sends packets to D which is not in transmission range of S because of power saving fixed transmission range but it can be increase its transmission range with help of its relay nodes and transmit packets. In this example node S uses its all 1-hop neighbors where as other hand only few nodes are enough for sending data till D. hence power of other nodes are useless for this communication if ΣviεVPi for selected neighbors of node S.

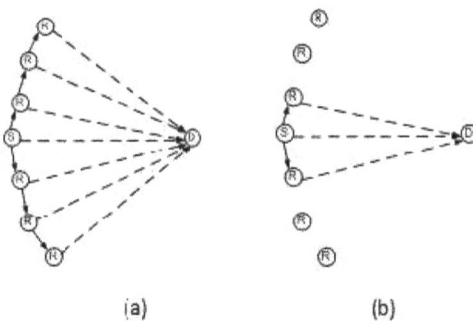

(a)   (b)

Figure 2. *Demonstration to reduce energy consumption in CC ad-hoc network*

We propose two topology control algorithms which build energy-efficient cooperative energy spanners. To keep the proposed algorithms simple and efficient, we only consider its one-hop neighbors as possible helper nodes for each node

whenCCis used. Thus, the original cooperative communication graph G contains all direct links and CClinks with one hop helpers, instead of all possible direct links and CC-links. In addition, for each pair of nodes vi and vj, we only maintain one link with least weight if there are multiple links connecting them.

6. Conclusion

In this paper, we deliberate a topology control problem in detailed, energy-efficient topology control problem with cooperative communication, which aims to keep the energy-efficient paths in the constructed topology. Also key point has been discussed as in wireless ad-hoc network for effective energy transmission. In this paper, we introduced a new topology control problem: optimum relay selection topology control problem with cooperative communication, which aims to keep the energy efficient paths in the constructed topology and reduce power consumption in network. In future this scheme is implemented and tested in real simulation for result gathering. This paper proposes novel algorithm for optimum relay selection rather selecting all nodes only those nodes will be selected which are capable for large enough to make transmission range within destination node to save power of other nodes hence overall network power consumption is minimize. Every node also store power level of every neighbor node in routing table with routing information. For transmit data packets relay selection is based on highest power level nodes. The nodes having maximum power level in direct neighbor selected for relay transmission. As given in figure 2(b) proposed algorithm can be given as follow. This will helpful for saving battery power for other nodes in to reduce overall network power consumption.

References

[1]   R. Rajaraman, "Topology Control and Routing in Ad Hoc Networks: A Survey," SIGACT News, vol. 33, pp. 60-73, 2002.

[2]   X.-Y. Li, "Topology Control in Wireless Ad Hoc Networks," Ad Hoc Networking, S. Basagni, M. Conti, S. Giordano, and I. Stojmenovic, eds., IEEE Press, 2003.

[3]   C.-C. Shen and Z. Huang, "Topology Control for Ad Hoc Networks: Present Solutions and Open Issues," Handbook of Theoretical and Algorithmic Aspects of Sensor, Ad Hoc Wireless and Peer-to-Peer Networks, J. Wu, ed., CRC Press, 2005.

[4]   A.E. Clementi, G. Huiban, P. Penna, G. Rossi, and Y.C. Verhoeven, "Some Recent Theoretical Advances and Open Questions on Energy Consumption in Ad-Hoc Wireless Networks," Proc. Workshop Approximation and Randomization Algorithms in Comm. Networks, 2002.

[5]   W.-T. Chen and N.-F. Huang, "The Strongly Connecting Problem on Multihop Packet Radio Networks," IEEE Trans. Comm., vol. 37, no. 3, pp. 293-295, Mar. 1989.

[6] L.M. Kirousis, E. Kranakis, D. Krizanc, and A. Pelc, "Power Consumption in Packet Radio Networks," Theoretical Computer Science, vol. 243, nos. 1/2, pp. 289-305, 2000.

[7] A.E.F. Clementi, P. Penna, and R. Silvestri, "On the Power Assignment Problem in Radio Networks," Proc. Electronic Colloquium on Computational Complexity (ECCC), 2000.

[8] D. Blough, M. Leoncini, G. Resta, and P. Santi, "On the Symmetric Range Assignment Problem in Wireless Ad Hoc Networks," Proc. Second IFIP Int'l Conf. Theoretical Computer Science, 2002.

[9] E. Althaus, G. Ca^linescu, I. Mandoiu, S. Prasad, N. Tchervenski, and A. Zelikovsly, "Power Efficient Range Assignment in Ad-Hoc Wireless Networks," Proc. IEEE Wireless Comm. and Networking (WCNC), 2003.

[10] R. Ramanathan and R. Hain, "Topology Control of Multihop Wireless Networks Using Transmit Power Adjustment," Proc. IEEE INFOCOM, 2000.

[11] M. Hajiaghayi, N. Immorlica, and V.S. Mirrokni, "Power Optimization in Fault-Tolerant Topology Control Algorithms for Wireless Multi-Hop Networks," Proc. ACM Mobicom, 2003.

[12] J. Cheriyan, S. Vempala, and A. Vetta, "Approximation Algorithms for Minimum-Cost K-Vertex Connected Subgraphs," Proc. Ann. ACM Symp. Theory of Computing (STOC), 2002.

[13] P. Bose, P. Morin, I. Stojmenovic, and J. Urrutia, "Routing with Guaranteed Delivery in Ad Hoc Wireless Networks," Proc. Int'l Workshop Discrete Algorithms and Methods for Mobile Computing and Comm., 1999.

[14] X.-Y. Li, Y. Wang, and W.Z. Song, "Applications of K-local MST for Topology Control and Broadcasting in Wireless Ad Hoc Networks," IEEE Trans. Parallel and Distributed Systems, vol. 15, no. 12, pp. 1057-1069, Dec. 2004.

[15] X.-Y. Li, P.-J. Wan, and Y. Wang, "Power Efficient and Sparse Spanner for Wireless Ad Hoc Networks," Proc. 10th Int'l Conf. Computer Comm. and Networks (ICCCN), 2001.

[16] R. Wattenhofer, L. Li, P. Bahl, and Y.-M. Wang, "Distributed Topology Control for Wireless Multihop Ad-Hoc Networks," Proc. IEEE INFOCOM, 2001.

[17] N. Li, J.C. Hou, and L. Sha, "Design and Analysis of a MST-Based Topology Control Algorithm," Proc. IEEE INFOCOM, 2003.

[18] Y. Wang and X.-Y. Li, "Localized Construction of Bounded Degree and Planar Spanner for Wireless Ad Hoc Networks," Mobile Networks and Applications, vol. 11, no. 2, pp. 161-175, 2006.

[19] N. Laneman, D. Tse, and G. Wornell, "Cooperative Diversity in Wireless Networks: Efficient Protocols and Outage Behavior," IEEE Trans. Information Theory, vol. 50, no. 12, pp. 3062-3080, Dec. 2004.

[20] Nosratinia, T.E. Hunter, and A. Hedayat, "Cooperative Communication in Wireless Networks," IEEE Comm. Magazine, vol. 42, no. 10, pp. 74-80, Oct. 2004.

[21] G. Jakllari, S.V. Krishnamurthy,M. Faloutsos, P.V. Krishnamurthy, and O. Ercetin, "A Framework for Distributed Spatio-Temporal Communications in Mobile Ad Hoc Networks," Proc. IEEE Infocom, 2006.

[22] Khandani, J. Abounadi, E. Modiano, and L. Zheng, "Cooperative Routing in Static Wireless Networks," IEEE Trans. Comm., vol. 55, no. 11, pp. 2185-2192, Nov. 2007.

[23] J. Zhang and Q. Zhang, "Cooperative Routing in Multi-Source Multi-Destination Multi-Hop Wireless Networks," Proc. IEEE INFOCOM, 2008.

[24] Ibrahim, Z. Han, and K. Liu, "Distributed Energy-efficient Cooperative Routing in Wireless Networks," IEEE Trans. Wireless Comm., vol. 7, no. 10, pp. 3930-3941, Oct. 2008.

[25] M. Agarwal, J. Cho, L. Gao, and J. Wu, "Energy Efficient Broadcast in Wireles Ad Hoc Networks with Hitch-hiking," Proc. IEEE INFOCOM, 2004.

[26] J. Wu, M. Cardei, F. Dai, and S. Yang, "Extended Dominating Set and Its Applications in Ad Hoc Networks Using Cooperative Communication," IEEE Trans. Parallel and Distributed Systems, vol. 17, no. 8, pp. 851-864, Aug. 2006.

[27] G. Jakllari, S. Krishnamurthy, M. Faloutsos, and P. Krishnamurthy, "On Broadcasting with Cooperative Diversity in Multi-Hop Wireless Networks," IEEE J. Selected Area in Comm., vol. 25, no. 2, pp. 484-496, Feb. 2007.

[28] F. Hou, L.X. Cai, P.H. Ho, X. Shen, and J. Zhang, "A Cooperative Multicast Scheduling Scheme for Multimedia Services in IEEE 802.16 Networks," IEEE Trans. Wireless Comm., vol. 8, no. 3, pp. 1508-1519, Mar. 2009.

[29] L. Wang, B. Liu, D. Goeckel, D. Towsley, and C. Westphal, "Connectivity in Cooperative Wireless Ad Hoc Networks," Proc. ACM Mobihoc, 2008.

[30] A.K. Sadek, Z. Han, and K.J.R. Liu, "Distributed Relay-Assignment Protocols for Coverage Expansion in Cooperative Wireless Networks," IEEE Trans. Mobile Computing, vol. 9, no. 4, pp. 505-515, Apr. 2010.

[31] Y. Shi, S. Sharma, and Y. Hou, "Optimal Relay Assignment for Cooperative Communications," Proc. ACM Mobihoc, 2008.

[32] Q. Zhang, J. Jia, and J. Zhang, "Cooperative Relay to Improve Diversity in Cognitive Radio Networks," IEEE Comm. Magazine, vol. 47, no. 2, pp. 111-117, Feb. 2009.

[33] Wang, Z. Han, and K.J.R. Liu, "Distributed Relay Selection and Power Control for Multiuser Cooperative Communication Networks Using Stackelberg Game," IEEE Trans. Mobile Computing, vol. 8, no. 7, pp. 975-990, July 2009.

[34] M. Veluppillai, L. Cai, J.W. Mark, and X. Shen, "Maximizing Cooperative Diversity Energy Gain for Wireless Networks," IEEE Trans. Wireless Comm., vol. 6, no. 7, pp. 2530-2539, July 2007.

[35] M. Cardei, J. Wu, and S. Yang, "Topology control in ad hoc wireless networks using cooperative communication," IEEE Trans. on Mobile Computing, 5(6):711-724, 2006.

[36] J. Yu, H. Roh, W. Lee, S. Pack, and D.-Z. Du, "Cooperative bridges: topology control in cooperative wireless ad hoc networks," in IEEE InfoCom, 2010.

[37] N. Laneman, D. Tse, and G. Wornell, "Cooperative diversity in wireless networks: efficient protocols and outage behavior," IEEE Trans.Information Theory, 50(12):3062-3080, 2004.

[38] Nosratinia, T.E. Hunter, and A. Hedayat, "Cooperative communication in wireless networks," IEEE Comm. Magazine, 42(10):74-80, 2004.

[39] Khandani J. Abounadi E. Modiano and L. Zheng, "Cooperative routing in static wireless networks," IEEE Trans. on Communications, 55(11):2185-2192, 2007.

[40] Ibrahim, Z. Han and K. Liu, "Distributed energy-efficient cooperative routing in wireless networks," IEEE Trans. on Wireless Communications, 7(10):3930-3941, 2008.

[41] M. Agarwal, J. Cho, L. Gao, and J. Wu, "Energy efficient broadcast in wireles ad hoc networks with hitch-hiking," in IEEE InfoCom, 2004.

[42] L. Wang, B. Liu, D. Goeckel, D. Towsley, and C. Westphal, "Connectivity in cooperative wireless ad hoc networks," in ACM Mobihoc, 2008.

[43] Ying Zhu, Minsu Huang, Siyuan Chen, and Yu Wang, "Energy-Efficient Topology Control in Cooperative Ad Hoc Networks", IEEE TRANSACTIONS ON PARALLEL AND DISTRIBUTED SYSTEMS, VOL. 23, NO. 8, page 1480-1491, IEEE, 2012

[44] Ying Zhu Minsu Huang Siyuan Chen Yu Wang, "Cooperative Energy Spanners: Energy-Efficient Topology Control in Cooperative Ad Hoc Networks", IEEE, 2010.

30

Combat-Sniff: A Comprehensive Countermeasure to Resist Data Plane Eavesdropping in Software-Defined Networks

Fan Jiang, Chen Song[*], Hao Xun, Zhen Xu

Institute of Information Engineering, Chinese Academy of Sciences, Beijng, China

Email address:

jiangfan@iie.ac.cn (Fan Jiang), songchen@iie.ac.cn (Chen Song), xunhao@iie.ac.cn (Hao Xun), xuzhen@iie.ac.cn (Zhen Xu)
[*]Corresponding author

Abstract: Software-defined networking (SDN), on account of its unprecedented capability of network traffic monitoring and data resource transferring, has been deployed into a wide range of application scenarios. However, typical cyber-attacks which prevail in traditional IP networks, have also mutated their implementation models adjusting to SDN environment. Eavesdropping is one of such attacks and causes severe information disclosure to different degree. In this paper, we focus on data plane eavesdropping in SDN and treat it on two levels according to the extent an adversarial sniffer can exploit a SDN switch. Then we introduce Combat-Sniff, a comprehensive countermeasure which includes two methods to deal with the two-level sniffing respectively. And later, we both theoretically and experimentally demonstrate their reliability and performance. Results represent that we can exert Combat-Sniff in SDN to satisfy different security requirements with an acceptable overhead.

Keywords: Eavesdropping, Software-Defined Networking (SDN), Flow Entries Integrity Verification, Moving Target Defense (MTD)

1. Introduction

SDN has offered traditional IP networks a brand new paradigm [1], with its separated control and forwarding planes, logically centralized controllers, and unified programmable interfaces. Since the original rigid closed network model suffers a hardship when keeping pace with the rapid expansion of network size and abrupt outburst of huge data volume, newly developed network practices, such as cloud service and big data analytics, turn to SDN for feasible solutions. Innovative as SDN infrastructure is, it fails to put an end to traditional typical cyber-attacks, which have exploited their distinctive realization methods against SDN-specific background. As the present mostly referenced implementation of SDN is OpenFlow protocol [2], we would focus our subsequent discussion on OpenFlow-based network.

Network eavesdropping [3] is a kind of packets interception attack in traditional IP network. So far, there are no recognized reliable detection methods to deal with it, and the accepted defense method is encryption [4]. However, situation changes when transiting to SDN, either for detection or defense. In this paper, we focus on data plane eavesdropping in SDN. According to what degree a malicious attacker can exploit a SDN switch, in figure 1, we classify the eavesdropping into two levels: flow entries compromised level and switch compromised level.

Flow entry compromised level. The decoupled control and data plane expose a forwarding rule inconsistency problem. Central controllers take charge of networking intelligence by means of installing flow entries on flow tables in switches to instruct traffic forwarding. Maliciously falsifying flow entries from switch side can cause the inconsistency between original flow entries delivered by controllers and the ones preserved by switches. For example, for the convenience of debugging networking, some current OpenFlow switches [5, 6] are left with a listening mode, through which network administrators can connect them from unauthenticated TCP port for manipulating, such as writing rules or reading information. Utilizing such interface, an intentional attacker can eavesdrop certain data streams by artificially adding a mirror port in a switch.

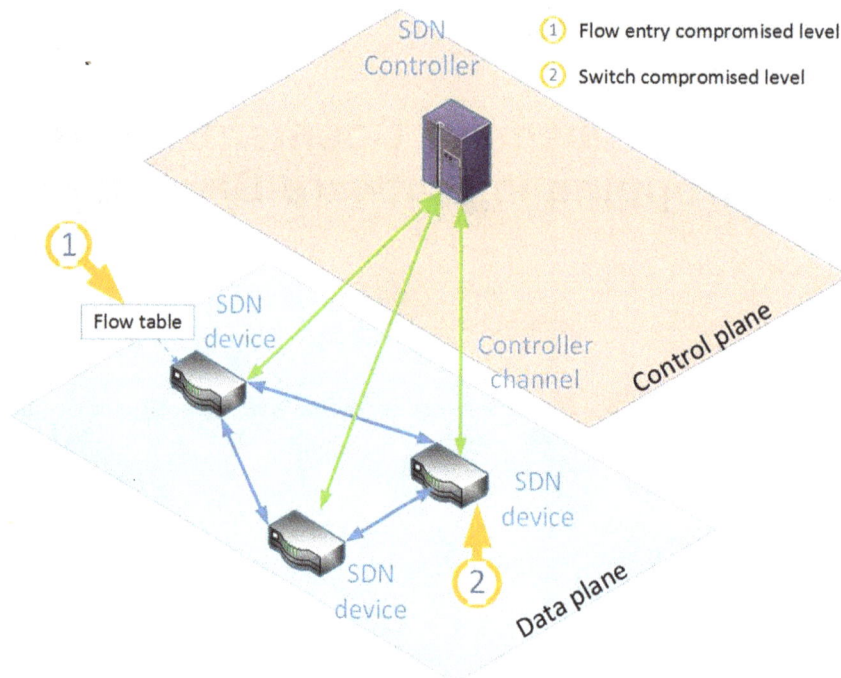

Figure 1. *Eavesdropping in SDN data plane through exploiting switch in two levels.*

Switch compromised level. Directly compromising a switch enables a hacker to monitor the whole traffic flowing through the device, even need not bother to modify the flow entries. Researchers [7] have exploited malware which could hunt out the switch within a targeted network and install a sneaky, second-stage piece of malware on the switch and push data to a command-and-control server. In addition, malicious entity can also pretend regular switches and thus have full visibility into all of the traffic running through the switch.

As for the flow entries compromised level, in traditional network, there are no efficient mechanisms to guarantee the validity of the forwarding rules in devices. However, in SDN, we can utilize its centralized management nature to inspect the integrity of flow entries preserved in switches. As to the switch compromised level, although encryption has demonstrated its reliability in traditional network, in SDN, the scope of protection through encryption is limited. Network encryption mostly is protocol dependent, eg. HTTPS. Since SDN is being designed to open network, it is applied into various areas where customized communication protocols are popular and such protocols haven't adopted corresponding encryption scheme. For example, Data center networking, which is one of the main application domains of SDN, is more frequently using Data Center Interconnect (DCI) protocols, such as Virtual Extensible LAN (VXLAN), Stateless Transport Tunneling (STT), which lack authentication and any form of encryption to secure the packet contents. Thus, we need a protocol oblivious solution.

Such two levels sniffing require different cost from attackers and also reward them distinct benefit. Thus in this paper, we introduce Combat-Sniff, a comprehensive countermeasure, which includes two methods to correspondingly cope with the above two levels situation.

Our contributions are as follows:

- We propose a flow entries integrity verification method to deal with flow entries compromised level eavesdropping.
- We propose an innovative protocol oblivious method to prevent data disclosure in switch compromised level eavesdropping.
- We implement the integrated countermeasure, Combat-Sniff, and demonstrate its reliability and performance in experiments.

The structure of the paper is as follows. Firstly, in section 2, we introduce related research works and knowledges. Secondly, we introduce the specifications of Combat-Sniff in Section 3. Later in Section 4, we testify the effectiveness of our methods using experiments. At last, we conclude our work and discuss the subsequent research direction in Section 5.

2. Background

2.1. Related Works

Eavesdropping in SDN can be exerted both within data plane and within the communication channel between controllers and switches. As for data plane eavesdropping, Kevin *et al.* [8] figure out the possibility of adversarial flow tables modifications in OpenFlow network through listening mode of switches. Markku *et al.* [9] analyze such threat in detail by showing us how attackers can utilize flow tables modification as the first-step attacking and then exert eavesdropping. But they didn't give the feasible solutions. Po-Wen *et al.* [10] design a detection mechanism to find compromised OpenFlow switches. Qi *et al.* [11] propose a proactive Random Route Mutation (RRM) technique to randomly change the routing of multiple flows to defend against eavesdropping. As for communication channel eavesdropping, Kreutz *et al.* [12] propose the vulnerability of the openflow channel, when the attacker captures the flow mod messages, it can modify the messages to add another mirror port. Daniel *et al.* [13] testify such vulnerability through experiments.

Except for eavesdropping, there are also many researches focusing on typical cyber-attacks, which cause serious damages to legacy networks and now developed transformed attacking schemes in SDN. Shin *et al.* [14] figure out how to consume control plane and data plane resource to exert Denial-of-service(Dos) [15] attack in OpenFlow. Later, Shin *et al.* also introduce AVANT-GUARD [16] to relieve such attack by expediting control plane's detection and response ability. Hong *et al.* [17] reveal how to exploit SDN's inherent topology discovery mechanism to poison network visibility and they also construct TopoGuard as an OpenFlow controller extension to secure network topology.

2.2. Background Knowledges

Since our countermeasure involves specific knowledges about OpenFlow protocol and Protocol Oblivious Forwarding [18] technology, we would introduce some related information below.

Flow entries in OpenFlow. In OpenFlow network, every switch preserves multiple flow tables, in which store flow entries installed by controller. Switches would forward data plane packets conforming to flow entries. Each flow entry contains match fields and instructions [19]. The match fields consist of packet headers (eg. TCP_SRC) and if a passing packet matches one of flow entries, i.e. the values of the packet headers equal the values of corresponding match fields of an entry, then the packet would be disposed according to the instructions in the flow entry. A packet not matching any of the flow entries would be sent to controller.

Protocol Oblivious Forwarding. Basing on OpenFlow v1.3, Song [18] proposes Protocol-Oblivious Forwarding. In such forwarding technology, the controller and forwarding elements communicate at a field-offset level rather than protocol semantics level. That is to say, the switch needs not to understand specific packet formats to extract certain search keys, and instead, the controller guides the switch to locate a certain key using a {offset, length} structure. This novel conception allows user to use their own network protocols without any need to go back to the device vendor. This protocol agnostic forwarding device, acts as a facilitating tool to realize our protocol oblivious defense method.

3. Assumptions and Design

3.1. Assumptions and Designing Objectives

Before introduction of our countermeasure, we need to state the following assumptions clearly:
- As for the first-level-based eavesdropping, we assume the sniffer who falsifies the flow entries can't hamper

the other OpenFlow functions of that switch.
- As for the second-level-based eavesdropping, we suppose the sniffer who encroaches the whole switch won't hinder the switch from forwarding data packets normally.

For the first assumption, since the first level sniffers' ability is limited to falsifying the flow entries through a feasible interface, it's reasonable for us to think they can't hamper switches' inherent mechanism. For the second assumption, because our focus is networking eavesdropping, we don't consider the other destructive attack from that sniffers. Besides, a switch which doesn't perform normally will easily be detected.

In the design of Combat-Sniff, we plan to achieve the following goals:
- For the first level eavesdropping, we aim to detect the falsifiers efficiently.
- For the second level eavesdropping, we aim to protect communication confidentiality.

3.2. The Sketch of Combat-Sniff

Our countermeasure, Combat-Sniff, includes an active detection method ,called flow entries integrity verification, to realize the sniffing detection goal, and a proactive defense method, called protocol fields randomization, to reach the information confidentiality protection goal.

Method I: Flow Entries Integrity Verification. For a scalable integrity verification solution, the amount of flow entries accesses and transmissions should be minimized, so we utilize a random sampling mechanism to verify part of the flow entries during each round. Besides, we also need to ease the additional burden loaded to switches so as not to influence traffic forwarding. Since the upper controllers can be a centralized cluster of nodes [20] or a physically distributed set of elements [21], we don't need to worry about the storage or the computing capability of the controllers. So we make controller undertake the role of verifier to store the original delivered flow entries and compute the related values.

Design of the method. We adopt a query-reply mechanism, within which the controller periodically sends query messages to switches. On receiving the query messages, switches send the appointed flow entries to controller, and controller verifies their integrity with message digest algorithm (MD5).

The process of verification is in Algorithm 1. Given the total number of flow entries, sampling ratio and total number of switches, in step 1, controller computes the sampling number for each switch according to their share of flow entries. Step 2-10 is the query process. Controller will firstly check the number of flow entries within that switch to inspect if there are maliciously added or deleted flow entries. If not, secondly, it randomly generate the flow entries IDs to be queried and send the message to switches. Step 11-18 is the verification process. Once controller detected inconsistency, it will find the specific flow entry and shut down the suspected ports.

Table 1. *The flow entries integrity verification algorithm.*

Algorithm 1 The flow entries sampling verification algorithm
input: total EntryNum, Sr, SwNum.
1: Computes the sampling number of flow entries for each switch: samp EntryNum;
2: for Sw in switches do
3: Query for the number of flow entries of the switch: entryNum
4:   if the value are correct then
5:     exert SelectEntry (0, entryNum, sampEntryNum);
6:     deliver query messages;
7:   else
8:     check flow entries item by item in that switch and shut down illegal port;
9:   end if
10: end for
11: if Receiving reply from switches then
12:   CompareMD5 (OriginalEntry, SampledEntry);
13:   if values are not equal then
14:     find the specific inconsistent flow entries and shut down illegal port;
15:   else
16:     pass;
17:   end if
18: end if

Method II: Protocol Fields Randomization.

To protect the confidentiality of the data packets which pass through a potential compromised switch, we need to make the switch partially blind. That is to say, the switch should know how to forward the packets, but not know what it is forwarding. As we have mentioned, though traditional communication encryption is a recognized reliable solution, it is protocol dependent.

Inspired by the idea of Address Space Layout Randomization (ASLR) [22], which randomizes the locations of executable segments of a running process to raise the bar for Return-oriented programming(ROP) attack. It occurs to us that we can artificially reorder the locations of protocol fields within a packet scope to enhance the difficulty for sniffers to parse the data packets. Our method has the following features:

- It's protocol oblivious.
It can be applied into any kind of transmission protocol.

- It's content oblivious.
Switch can hardly parse the packet content using regular protocol format knowledges.

Design of the method. We design a specific protocol fields randomization algorithm and a practical transmission scheme to forward the randomized packets.

Protocol fields randomization algorithm. Our randomization scheme includes a reordering and a XOR process. Algorithm 2 represents the specification. There are different levels of reordering according to different data granularities: protocol field level, byte level, bit level. The finer, the safer, but also more expensive for computing. In our defense method, we use byte level reordering.

Table 2. *The protocol fields randomization algorithm.*

Algorithm 2 Protocol fields randomization algorithm
input: A normal data packet with N bytes, within which the header part is M bytes.
output: A protocol fields randomized packet
1: Split the single packet into separated N bytes ignoring the header fields meaning.
2: Randomly reorder the M bytes header and distribute them into the whole N bytes scope.
3: Fill in the remaining N − M locations with the original payload of that packet, without changing their relative order.
4: Use a N byte randomly generated secure key to encode the whole packet with XOR

Figure 2 raises an example of the protocol fields reordering process. For simplicity, we only take one protocol field for illustration. In the original TCP packet, the field src IP is located at the offset of 26 bytes of the whole packet ,its length is 4 bytes and value is 10.0.0.2, i.e. 0x0a00 0002 in hexadecimal format. On exerting our protocol fields randomization, we firstly split the integrated field into 4 separate bytes and then reorder them within the whole packet scope. The result is 4 independently bytes, their offsets are k1, k2, k3, k4 respectively and values are 0x02, 0x00, 0x0a, 0x00.

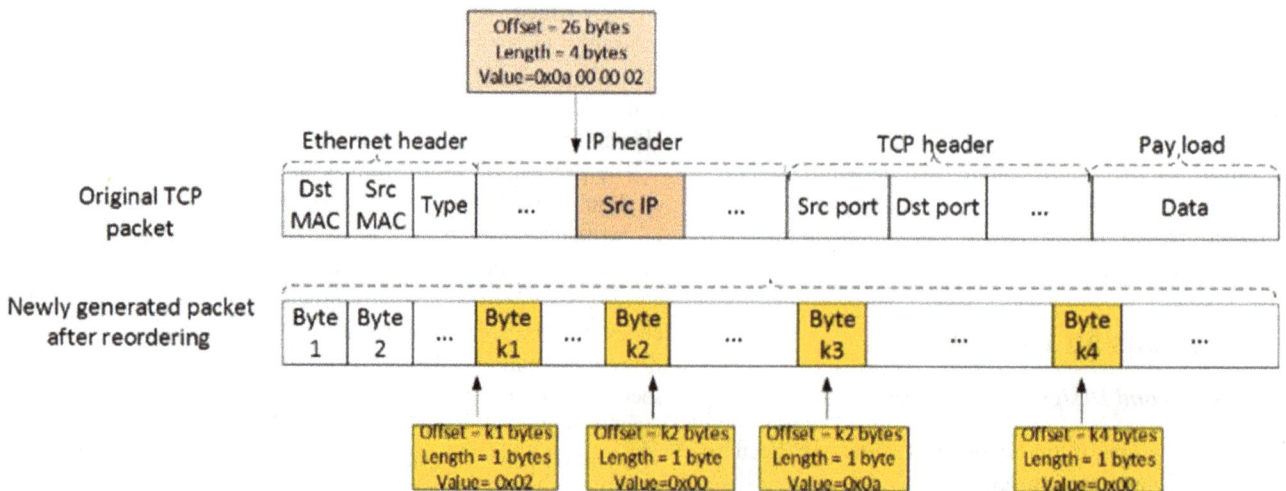

Figure 2. *The reordering mechanism of protocol fields randomization.*

Besides, we also display the actual effect through wireshark in figure 3. For clarity, we only encode the field of source address of IP. The left side is a transformed packet and the right side is an original one. The src IP is 0x0a00 0002. The original field of src IP is reordered and be encoded with XOR. The random key is 0xaaaa aaaa, and the value after XOR should be 0xa8, 0xaa, 0xa0 and 0xaa. We can see that the original field value is distributed into other locations of the packet. And the original location has been substituted by corresponding fields in the packet.

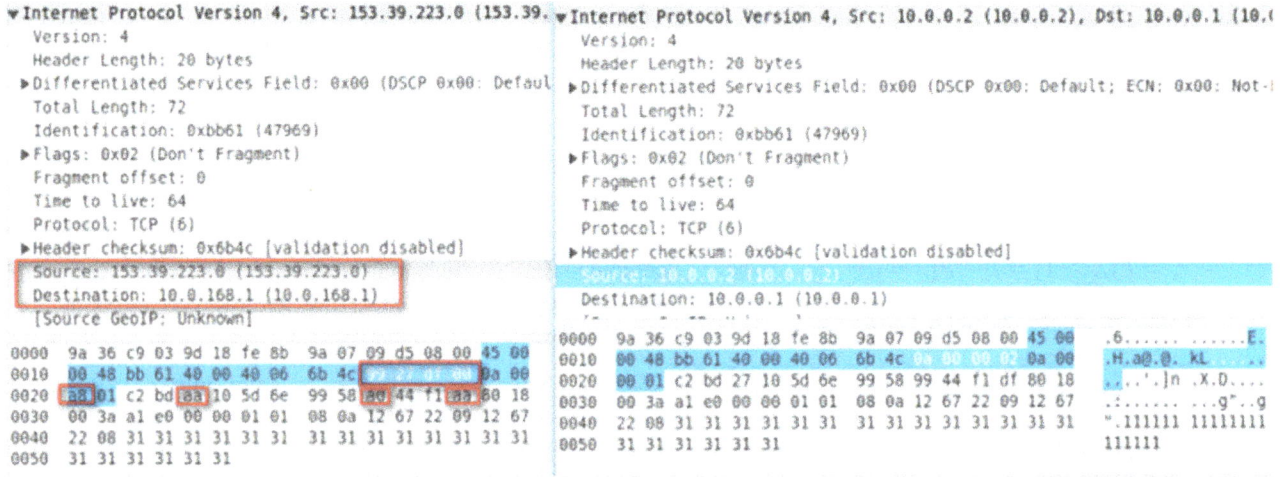

```
▼Internet Protocol Version 4, Src: 153.39.223.0 (153.39.    ▼Internet Protocol Version 4, Src: 10.0.0.2 (10.0.0.2), Dst: 10.0.0.1 (10.0
    Version: 4                                                  Version: 4
    Header Length: 20 bytes                                     Header Length: 20 bytes
  ▶Differentiated Services Field: 0x00 (DSCP 0x00: Defaul     ▶Differentiated Services Field: 0x00 (DSCP 0x00: Default; ECN: 0x00: Not-
    Total Length: 72                                            Total Length: 72
    Identification: 0xbb61 (47969)                              Identification: 0xbb61 (47969)
  ▶Flags: 0x02 (Don't Fragment)                               ▶Flags: 0x02 (Don't Fragment)
    Fragment offset: 0                                          Fragment offset: 0
    Time to live: 64                                            Time to live: 64
    Protocol: TCP (6)                                           Protocol: TCP (6)
  ▶Header checksum: 0x6b4c [validation disabled]              ▶Header checksum: 0x6b4c [validation disabled]
    Source: 153.39.223.0 (153.39.223.0)                         Source: 10.0.0.2 (10.0.0.2)
    Destination: 10.0.168.1 (10.0.168.1)                        Destination: 10.0.0.1 (10.0.0.1)
    [Source GeoIP: Unknown]

0000  9a 36 c9 03 9d 18 fe 8b  9a 07 09 d5 08 00 45 00    0000  9a 36 c9 03 9d 18 fe 8b  9a 07 09 d5 08 00 45 00   .6...... ......E.
0010  00 48 bb 61 40 00 40 06  6b 4c 99 27 01 00 0a 00    0010  00 48 bb 61 40 00 40 06  6b 4c 0a 00 00 02 0a 00   .H.a@.@. kL   ..
0020  a8 01 c2 bd a9 10 5d 6e  99 58 a0 44 f1 aa 80 18    0020  00 01 c2 bd 27 10 5d 6e  99 58 99 44 f1 df 80 18   ...'.]n .X.D....
0030  00 3a a1 e0 00 00 01 01  08 0a 12 67 22 09 12 67    0030  00 3a a1 e0 00 00 01 01  08 0a 12 67 22 09 12 67   .:...... ...g".g
0040  22 08 31 31 31 31 31 31  31 31 31 31 31 31 31 31    0040  22 08 31 31 31 31 31 31  31 31 31 31 31 31 31 31   ".111111 11111111
0050  31 31 31 31 31 31                                    0050  31 31 31 31 31 31                                  111111
```

Figure 3. The comparison of data packet before and after protocol fields randomization.

The practical transmission scheme. We represent how the protocol fields randomization can be used in practical data transmission in Figure 4.

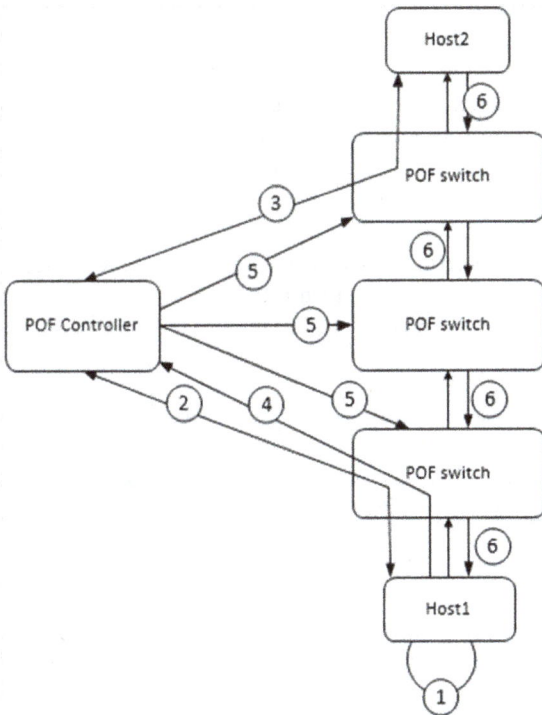

Figure 4. The practical process of data transmission for protocol fields randomized packets.

In step 1, host1 will generate its own randomization manners and also appoint some unused fields in a protocol field to mark the corresponding manners. In step 2, host1 shakes hands with controller to inform it the above information and also its communication object host2. Then in step 3, controller would exchange the same information with host2. When the process of handshake succeeds, host1 will communicate with host2 using randomized packets in step 4. And since switch doesn't understand the packet, it would send the packet to the controller. In step 5, controller delivers corresponding flow entries to switches to instruct them to forward the randomized packets between host1 and host2. In step 6, the two hosts can communicate with each other using our protocol fields randomization method.

4. Implementations and Evaluations

Due to our particular function requirement of protocol oblivious forwarding for switches, we use pof controller [23] and pof switch [24] to realize our two methods. And we use pof-mininet plugin to embed pof switch into Mininet to construct a topology. Although the working mechanism of pof is not identical with OpenFlow, but it's based on OpenFlow v1.3 [19] and their distinctions don't hinder our demonstration to our countermeasure.

In realization of the flow entries integrity verification method, we firstly construct a pair of new controller-to-switch OpenFlow messages, FLOW_QUERY and FLOW_REPLY. Secondly, we add a verification module in pof controller and a flow entry reply function in pof switch.

In realization of the protocol fields randomization method, at the host side, we use scapy [25] module of python to encapsulate and parse the randomized packet. At the controller side, we add a packet identification module, it will deliver related flow entries to switches to instruct the packet forwarding.

Our pof controller runs on a physical machine with Intel I5 3.1GHz CPU and 4GB memory. Pof switch-based Mininet runs on a physical machine with Intel E5-2600 v3 1.6GHZ CPU and 16GB memory.

4.1. Reliability Analysis

(1) Flow Entries Integrity Verification

Assume the attacker tampers E_t flow entries out of total E_n flow entries. And we sample E_s flow entries during every round query. We compute P_x, the probability that at least one of the flow entries selected by sampling matches one of the flow entries tampered by the attacker. Let X be a discrete random variable. It is the number of flow entries selected by sampling that match the flow entries tampered by attacker. So we have:

$$P_x = P\{X \geq 1\} = 1 - P\{x = 0\}$$

$$= 1 - \frac{E_n - E_t}{E_n} \cdot \frac{E_n - 1 - E_t}{E_n - 1} \cdot \frac{E_n - 2 - E_t}{E_n - 2} \cdots \frac{E_n - E_s + 1 - E_t}{E_n - E_s + 1} \quad (1)$$

Since $\frac{E_n - i - E_t}{E_n - i} \gg \frac{E_n - i - 1 - E_t}{E_n - i - 1}$, so $S_r \gg 1 - \left(\frac{E_n - E_t}{E_n}\right)^{E_s}$. We plot figure 5 to express the relationships among sampling ration S_r, detection probability P_x, and total number of flow entries E_n. When tampering ratio i.e. $\frac{E_t}{E_n}$, is a certain probability, say 1%, we can detect the flow entry tampering by sampling a constant amount of flow entries, independently of the total number of the flow entries. For example, if we require a detection probability of 99%, we need only to sample 460 flow entries.

Probabilistic analysis reveals that when the tampering ratio is a constant value, the detection ratio can be pretty high while maintaining a constant number of sampling quantity, independently of the total number of flow entries.

(2) Protocol Fields Randomization

Assume the total length of a data packet is N bytes, and the total number of protocol fields is M. Then the number of reordering schemes, Scheme_N, is computed in equation 2.

$$\text{Scheme}_N = A_{M+N}^M$$

$$= (M + N) \cdot (M + N - 1) \cdot (M + N - 2) \cdots (N + 1) \quad (2)$$

And after the reordering, we also use a secret key, whose length is 8*(N+M), to encode the packet with XOR. Then we compute P_{parse}, which represents the possibility that an attacker can parse the packet to get the correct value in equation 3:

$$P_{parse} = 1 / \{Scheme_N * 2^{8*(M+N)}\} \quad (3)$$

The possibility value demonstrates the difficulty for an attacker to parse the packet and obtain the private information. Of course, we are not specialized in encryption and we just give an instance of our protocol fields randomization method. Professional encryption can be combined to our protocol oblivious method.

4.2. Experiment Effect

(1) Flow Entries Integrity Verification

As the number of flow entries sampled is constant, analyzed in section 4.1, we measure the time needed for sampling different number of flow entries from different number of switches. Table 1 shows that when sampling the same number of flow entries, the more switches we sample, the time consumption is lower. So, we had better collect the sampled flow entries proportionately from as more switches as possible. Because the result data represents that the bottleneck of our method is in performance of a single switch.

Table 3. The time consumption of flow entries integrity verification in different situations.

Time(s) Sampling num Switch num	1	10	100
1	0.006	1.2	46.056
8	0.005	1.005	25.011
15	0.002	0.038	15.063

(2) Protocol Fields Randomization

We test the performance of our protocol fields randomization method with a file transmission experiment in a 3-switch linear topology. And the time delay comparison with the normal file transmission is in figure 6. When the file size varies from 1KB to 100KB, the transmission delay keeps at about 46%, compared to normal transmission time. The time consumption lies in randomization and de-randomization within communication hosts. And the split

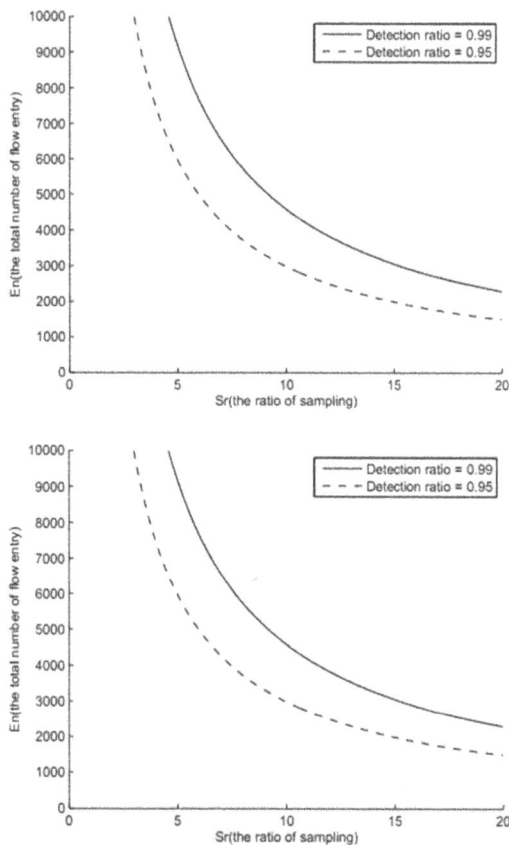

Figure 5. The probabilistic relationships among sampling ratio, detection probability and total number of flow entries.

protocol fields also increase the match fields for a switch to search. It is a tradeoff between randomization complexity and data transmission efficiency.

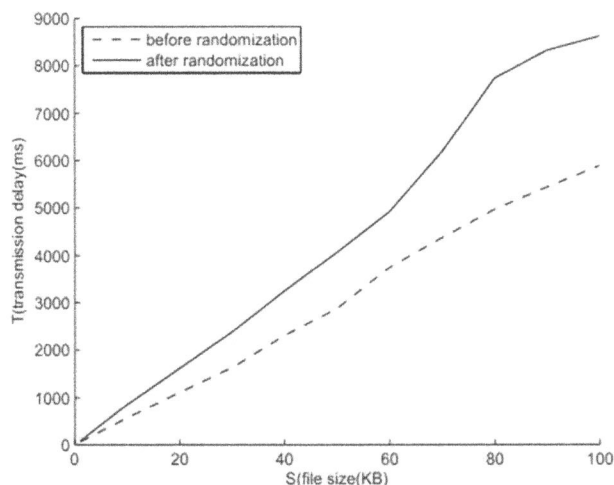

Figure 6. Comparing transmission delay of different size files.

5. Conclusion and Future Work

In this paper, we propose Combat-Sniff, a comprehensive countermeasure including two methods to resist data plane eavesdropping in SDN. The detection method can effectively detect the flow entries falsification with a constant number of sampled flow entries. And the defense method enhances a considerable difficulty for attackers to parse the sniffed packets.

In the future work, we plan to research the weight dependent sampling scheme according to the importance of a switch. Besides, we will also try to enhance the process of protocol fields randomization transmission to improve the performance.

Acknowledgements

This material is based upon work supported in part by the Institute of Information Engineering, Chinese Academy of Sciences under Y670021105 and Y5W0011105. Any opinions, findings, and conclusions or recommendations expressed in this material are those of the authors and do not necessarily reflect the views of Institute of Information Engineering, Chinese Academy of Sciences.

References

[1] Open Network Foundation. Software-defined networking: the new norm for networks [EB/OL]. [2012-04-13].

[2] MCKEOWN N, ANDERSON T, BALAKRISHNAN H, et al. OpenFlow: enabling innovation in campus networks [J]. ACM.

[3] https://www.owasp.org/index.php/Network_Eavesdropping.

[4] Schultz E E. Assessing and combating the sniffer threat [J]. Local Area Network Handbook, 1999: 85.

[5] Hp switch software - openflow supplement. http://h20000.www2.hp.com/bc/docs/support/SupportManual/c03170243/c03170243.pdf, Feb 2012.

[6] Open vSwitch, 2013. [Online]. Available: http://vswitch.org/

[7] http://www.pcworld.com/article/2957175/sdn-switches-arent-hard-to-compromise-researcher-says.html.

[8] Benton K, Camp L J, Small C. Openflow vulnerability assessment[C]//Proceedings of the second ACM SIGCOMM workshop on Hot topics in software defined networking. ACM, 2013: 151-152.

[9] M. Antikainen, T. Aura, and M. S̈arel̈a, "Spook in your network: Attacking an SDN with a compromised openflow switch," in Secure IT Systems - 19th Nordic Conference, NordSec 2014, Tromsø, Norway, October 15-17, 2014, Proceedings, 2014, pp. 229–244.

[10] Chi P W, Kuo C T, Guo J W, et al. How to detect a compromised SDN switch[C]//Network Softwarization (NetSoft), 2015 1st IEEE Conference on. IEEE, 2015: 1-6.

[11] Duan Q, Al-Shaer E, Jafarian H. Efficient random route mutation considering flow and network constraints[C]//Communications and Network Security (CNS), 2013 IEEE Conference on. IEEE, 2013: 260-268.

[12] D. Kreutz, F. M. Ramos, and P. Verissimo. Towards secure and dependable software-defined networks. in Proc.2nd ACM SIGCOMM Workshop Hot Topics Softw. Defined Netw., 2013, pp. 55–60

[13] Romão D, van Dijkhuizen N, Konstantaras S, et al. practical security analysis of OpenFlow [J]. 2013.

[14] Shin S, Gu G. Attacking software-defined networks: A first feasibility study[C]//Proceedings of the second ACM SIGCOMM workshop on Hot topics in software defined networking. ACM, 2013: 165-166.

[15] https://en.wikipedia.org/wiki/Denial-of-service_attack

[16] Shin S, Yegneswaran V, Porras P, et al. Avant-guard: Scalable and vigilant switch flow management in software-defined networks[C]//Proceedings of the 2013 ACM SIGSAC conference on Computer & communications security. ACM, 2013: 413-424.

[17] Hong S, Xu L, Wang H, et al. Poisoning Network Visibility in Software-Defined Networks: New Attacks and Countermeasures [C]. NDSS, 2015.

[18] Song H. Protocol-oblivious forwarding: Unleash the power of SDN through a future-proof forwarding plane[C]//Proceedings of the second ACM SIGCOMM workshop on Hot topics in software defined networking. ACM, 2013: 127-132.

[19] Specification, OpenFlow Switch. v1.3.0. (2012).

[20] S. Jain and al., "B4: Experience with a Globally-Deployed Software Defined WAN," in ACM SIGCOMM, 2013.

[21] Berde P, Gerola M, Hart J, et al. ONOS: towards an open, distributed SDN OS[C]//Proceedings of the third workshop on Hot topics in software defined networking. ACM, 2014: 1-6.

[22] M. Miller, T. Burrell, and M. Howard. Mitigating software vulnerabilities, July 2011. http://www.microsoft.com/download/en/details.aspx?displaying=en&id=26788.

[23] http://www.poforwarding.org/pofcontroller-1-1-7-released/

[24] http://www.poforwarding.org/pofswitch-1-3-4-released/

[25] http://www.secdev.org/projects/scapy/doc/usage.html

31

Design of secure Ad Hoc network using three dimensional discrete wavelet transformation based on performance enhancement

Laith Ali Abdul-Rahaim, Ammar Abdulrasool Muneer

Department of Electrical Engineering, University of Babylon, Babil, Iraq

Email address:

drlaithanzy@yahoo.com (L. A. Abdul-Rahaim), ammar.abdulrasool@gmail.com (A. A. Muneer)

Abstract: This work shows new and efficient algorithm of cryptographic purpose based symmetric and conventional techniques that considers the representation of the cipher text by using the three dimensional Discrete Wavelet Transform to find the wavelet decomposition vector containing the approximation and the detail coefficients then build the three dimensional data structure approach. The decryption is done by extracting the encrypted data from the wavelet decomposition vector using the algorithm of inverse Discrete Wavelet Transformation. The encrypted message consists the wavelet decomposition vector. The key is used for authorization purpose to access the network. Results shows great data security and BER over wireless channels based Ad Hoc network.

Keywords: 3D-Discrete Wavelet Transform, Cryptography, Signal Processing, Symmetric Cryptography, Wavelet Decomposition

1. Introduction

Cryptography is one of the most important tools that provide data and information confidentiality by hiding it. It is usually done through mathematical manipulation of the data with an incomprehensible format for unauthorized users.

In this work, a cryptographic techniques that based on Transformation of Three Dimensional-Discrete Wavelet is presented. Section 2 explains theory of wavelet transformation. Section 3 the process of encryption and decryption of the transmitting of real time information between two clients. Section 4 shows the results of the algorithm using different topologies. Section 5 is an analytic discussion on the technique with the conclusion and future scopes of this work.

2. The Basics of Wavelet

Wavelet is mathematical tool that analysis data into different frequency component values, and then check each component with a resolution matched to its scale value. They have benefits over customary Fourier methods in analyzing physical states where the signal has discontinuities and sharp

points [1]. Wavelets were advanced individually in the fields of quantum, physics mathematics and electrical engineering. Swaps between these fields during the last thirty years have controlled to many new wavelet uses such as human vision, turbulence, image compression, earthquake prediction, and radar [2, 3].

The necessary idea behind using wavelets is to analyze according to predefined scale value. Definitely, researchers in the wavelet application field sense that, using wavelets, one is implementing a whole new perspective in processing data.

The DWT variables (scale and timing window) are defined as discrete in time and scale, means that the DWT coefficients could have real values (floating-point), but the scale and time values used to guide these coefficients are integers [4 , 5].

An information is analyzed by Discrete wavelet transformation into different resolution of one or more levels (called octaves), as presented in Fig. 1, where a 1-dimensional signal is analyzed into three octaves. Figure 2 expressed a one-dimensional, one- octave discrete wavelet transform. It contains the decomposing on the left side and the synthesis on the right side. The low-pass filter generates the average signal, while the high-pass filter produces detail signal. In multi-resolution analysis, the average signal at one level is sent to another set of filters (Fig. 1), which produces the average and detail signals at the next octave [6, 12].

The detail signals are retained, anyway, the higher octave averages could be discarded, because they could be re-calculated through the inverse transform process. Every output of channel has only amount of half data input (plus a few coefficients due to the filter process). However, the wavelet illustration is approximately the same size as the original. The discrete wavelet transform can be 1-dimensional, 2-D, 3-D, etc. dependent on dimensions of the signal [7, 8].

The two dimensional transformation is merely an use of the one dimensional discrete wavelet transformation in the horizontal and vertical dimensions [8]. The illustration in Figure 3 shows the two dimensional transform (separable) for one octave (level). The non-separable two dimensions transform is different from the one shown, since it calculates the transformation based on a two dimensional signal of the input convolved with a matrix, but the outputs are equal. The separable method could be extended to the three dimensional discrete wavelet transform, as illustrated in Fig.4.

The low-pass filter related to scaling function of the signal, while the high-pass filter related to the wavelet function. The scaling function lets approximation of any given information with a variable value of precision [9, 12].

Putting on the below difference equations with the coefficients of scaling function, h, gives a calculation of the signal. This is also known as the low-pass output, where W are the coefficients of scaling function, while j represents the octave, except in the case of W(0, n), which is the original signal:

$$W(j,n)=\sum_{m=0}^{2n} W(j-1,m)h(2n-m) \qquad (1)$$

The Convolution with the wavelet function's coefficients, g, produces the detail signal, called high-pass output W_h

$$W_h(j,n) = \sum_{m=0}^{2n} W(j-1,m)g(2n-m) \qquad (2)$$

The DWT of a 1-D signal can be computed recursively using a filter pair with the fast pyramid algorithm, by Mallat and Meyer [10], Fig.1. It has a complexity of O(N), with an input of N sample. Other transforms normally require O(N2) calculations. Even the Fast Fourier Transform proceeds O(N log N) computations.

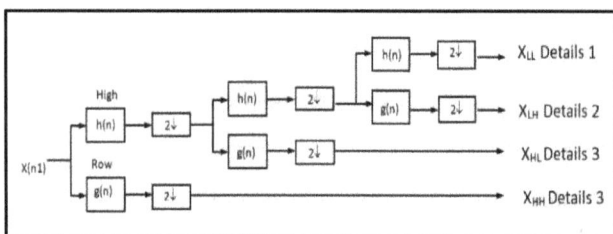

Figure 2. *A 1-Dimensional, 1-octave DWT and Inverse DWT.*

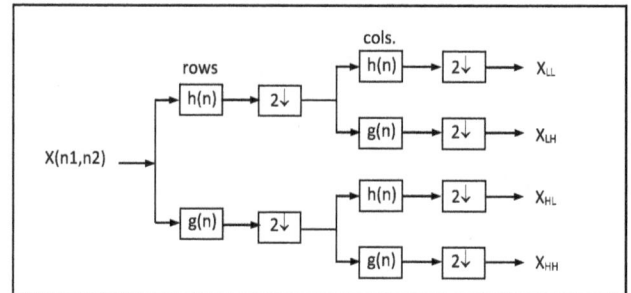

Figure 3. *A 2-Dimensional, 1-octave DWT.*

The fast pyramid algorithm gets its efficiency by $2^J=$ splitting the output data of each channel, otherwise known as down sampling. Then every octave (levels) uses half the number of data as the previous octave, the maximum number of octaves (levels), J , can be found by setting 2 equal to the input length,i.e. $2^J= N$, and the discrete wavelet transformation generates approximately N/2j outputs for each octave j. However, practical using limit the number of octaves (levels) depending on real time processing and other criteria [10 ,11].

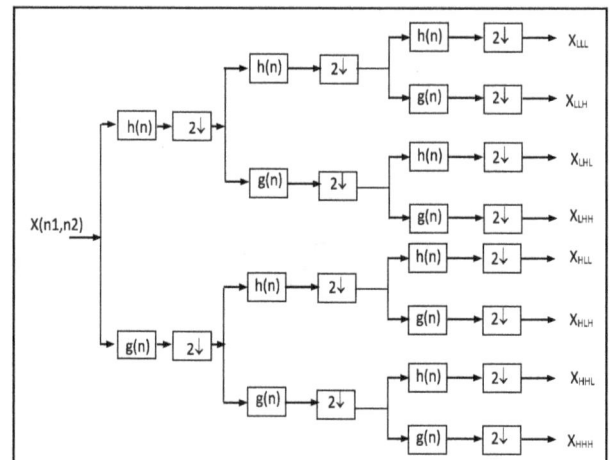

Figure 4. *A 3-Dimensional, 1-octave DWT.*

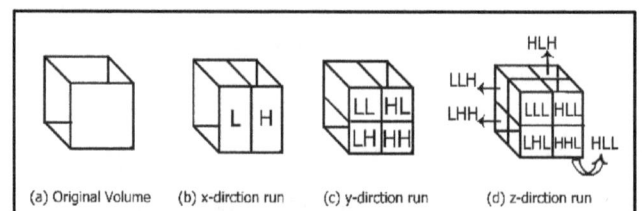

Figure 1. *Three Octave of decomposition of a 1-D signal.*

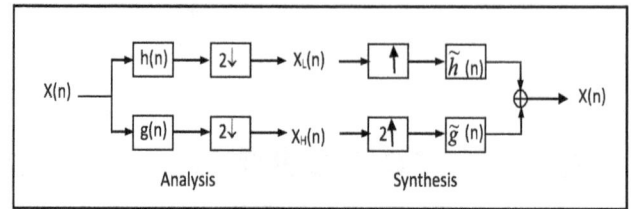

Figure 5. *Frequency sub bands produced by single level of wavelet decomposition of a 3-D image.*

2.1. Prosperities of Wavelets

Not each waveform provides a wavelet function. Nevertheless, the function should have few characteristics to be a wavelet. Two of the best essential possessions of wavelets are the admissibility and the regularity conditions. It can be shown that square integrable functions Ψ (t) sustaining the admissibility condition which could be used to first analyses and then reconstruct a signal without loss of information content [12 , 13].

$$\int \frac{|\psi(\omega)|^2}{|\omega|} d\omega \prec +\infty \qquad (3)$$

The above inequality $\Psi(w)$ views the Fourier transform of $\Psi(t)$. The admissibility condition means that the square of Fourier transform of Ψ (t) vanishes at the zero frequency, i.e.

$$|\psi(\omega)|^2 \Big|_{\omega=0} = 0 \qquad (4)$$

This means that wavelets must have a band-pass like spectrum. This is a very important observation, which is used to construct efficient wavelet transforms [14]. Furthermore, at the zero frequency also means that the average value of the wavelet in the time domain should be equal to zero and therefore it must be oscillatory (oscillating wave). In other words, Ψ (t) must be a wave in Continuous time domain. Similarly, Ψ (n) will be a wave in discrete time domain. Mathematically,

$$\int \psi(t)dt = 0, and, \sum \psi(n) = 0 \qquad (5)$$

In addition, the regularity of wavelet corresponds to the number of vanishing moments [15]. Therefore, if a wavelet has N vanishing moments, then the approximation order of the wavelet transform is also N as compared with N vanishing moment. A small value is often good instead of exactly zero of vanishing moment. The suggestions from experimental research show that the number of vanishing moments required are application dependent. The first moment to be vanished is corresponding to admissibility condition [16,17,18]

3. Encryption and Decryption for Wireless

The planned DSP based security of communication system and Transmitter and Receiver using UDP protocol, the system is shown as in Figure (6).

At the receiving end, all the steps of proposed algorithm applied but in reversely order to retrieve the original data. However frames are divided into 4096 samples to offer(16! x 16!) possible permutation for each page. This changeability of the scrambled signal is increased to be (16! x16! x16!) when interring the total frame permutation (the third dimension permutation) is also performed. Not all permutations gives good security quality so the effective permutation used [19]. Fortunately, results offer high security

level to the system. The block diagrams of the proposed scrambler and descrambler based on this new scheme are shown in Fig.(7).

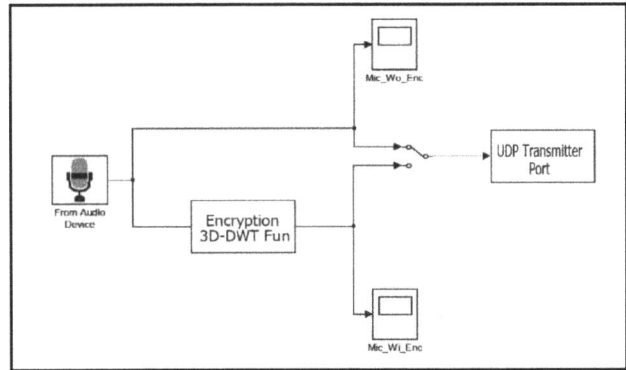

Figure 6-a. *Purposed encryption System based Wireless Link (802.11n) over UDP port.*

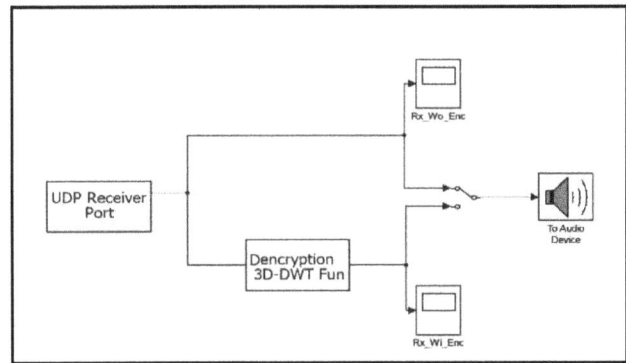

Figure 6-b. *Purposed decryption System based Wireless Link (802.11n) over UDP port.*

Although using the subject scheme, in order to have insight of time and frequency domain analysis, the recorded wave files of the various speech segments are implemented and evaluated using Simulink MATLAB® (R2013a). To reinforce the quality of the obtained results, the experiments are conducted not only in English but in Arabic languages as well.

Pauses between talk burst cannot be sensed and there is no residual intelligibility.

In term of security system, the concept of the residual of intelligibility while the quality of the recovered data are subjective quantity, thus the scramble and descramble process techniques are estimated on the results of expert listeners during the test [20, 21, 22].

Table 1. *Purposed System Specifications.*

Parameter	Range
Input speech	300 Hz to 3400 Hz
Sampling frequency	44100 Hz
Type of transformation	3D-Discrete Wavelet Transformation
Frame length	4096 sample
Frame duration	92.879 m sec
Total Changeable coefficients	16x16x16 = 4096

The tests were made tighter by adopting the following steps:

(1) Separating digits that are pronounced via male and

female as well. (2) Examinations are applied for digits also sentences are included too. (3) Those segments also are proved by male and female too. (4) However, to have more practical results, exchanged segments are carried out via two languages for male and female individually. The results are calculated in terms of correctly identified words Q, which is equal to Q:

$$Q=\frac{(R-W)}{T}\times100\% \qquad (6)$$

Where R, right words, W, wrong words and T is total words [23].

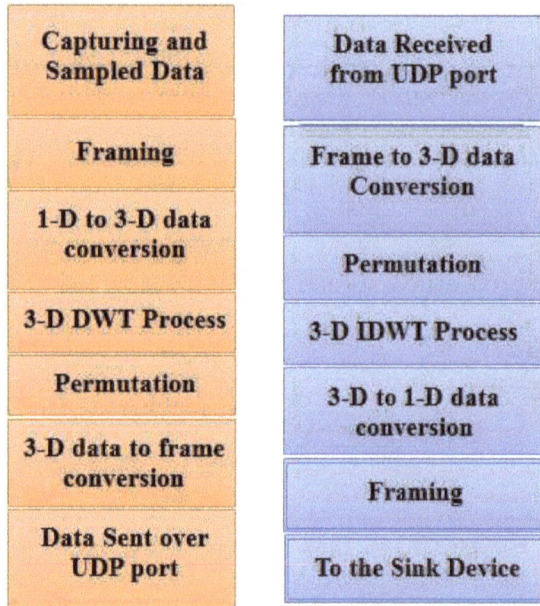

Capturing and Sampled Data	Data Received from UDP port
Framing	Frame to 3-D data Conversion
1-D to 3-D data conversion	Permutation
3-D DWT Process	3-D IDWT Process
Permutation	3-D to 1-D data conversion
3-D data to frame conversion	Framing
Data Sent over UDP port	To the Sink Device

Figure 7. *Encryption algorithm at Transmitter (Red) and Receiver (Blue).*

4. Real Time Wireless Communication

The below graphs shows the results of real time communication based on Ad-hoc network and star topology using time and spectrum expressions.

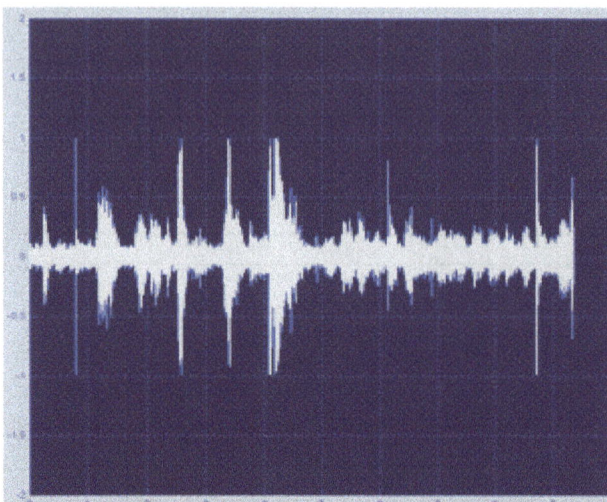

Figure 8. *Transmitted data over Ad Hoc network without encryption.*

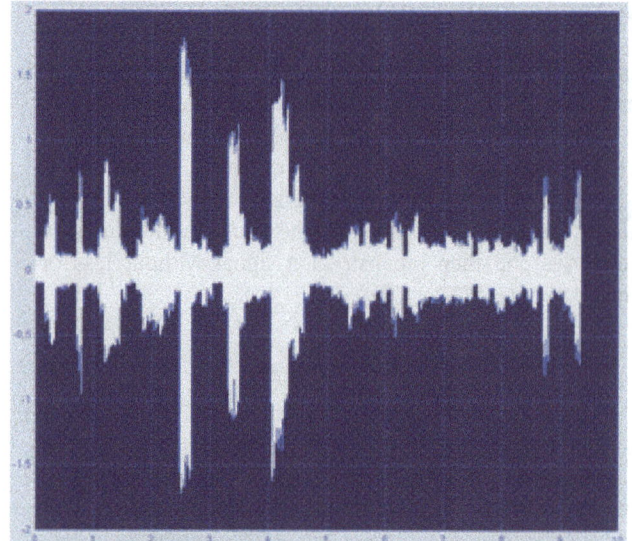

Figure 9. *Transmitted data over Ad Hoc network with encryption.*

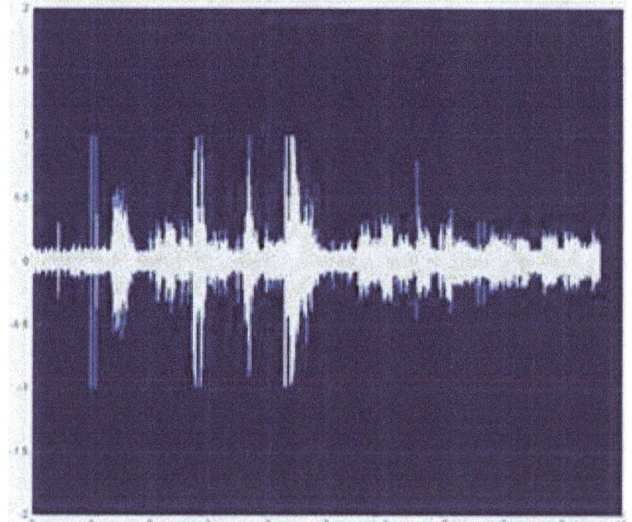

Figure 10. *Received data over Ad Hoc network without decryption.*

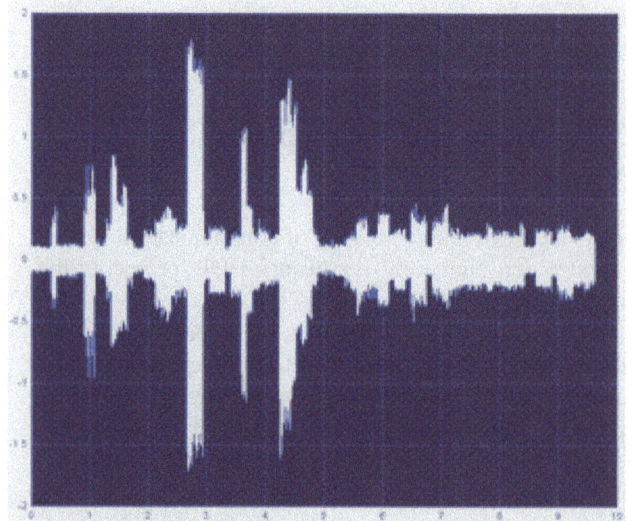

Figure 11. *Received data over Ad Hoc network with decryption.*

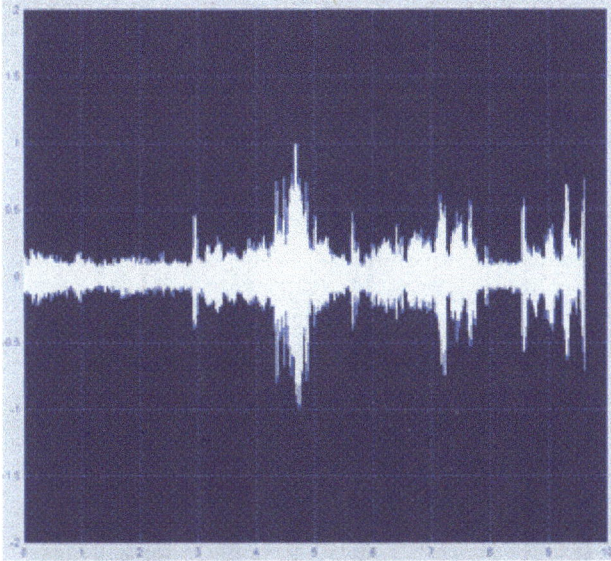

Figure 12. *Transmitted data over Star network without encryption.*

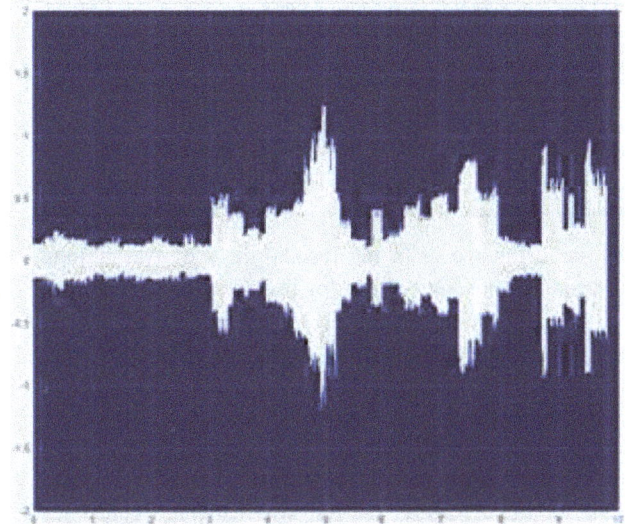

Figure 13. *Transmitted data over Star network with encryption.*

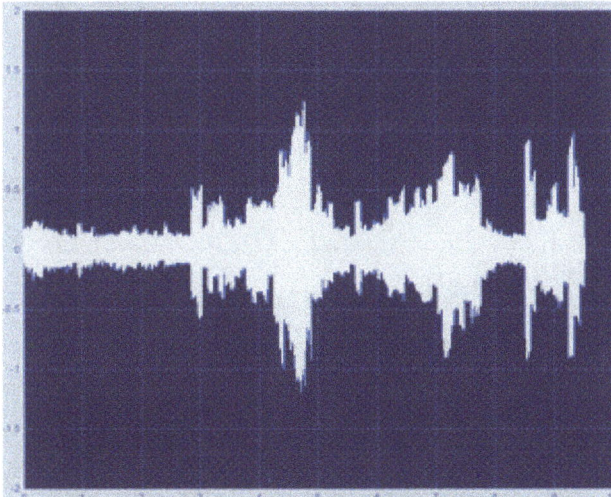

Figure 14. *Transmitted data over Star network with encryption.*

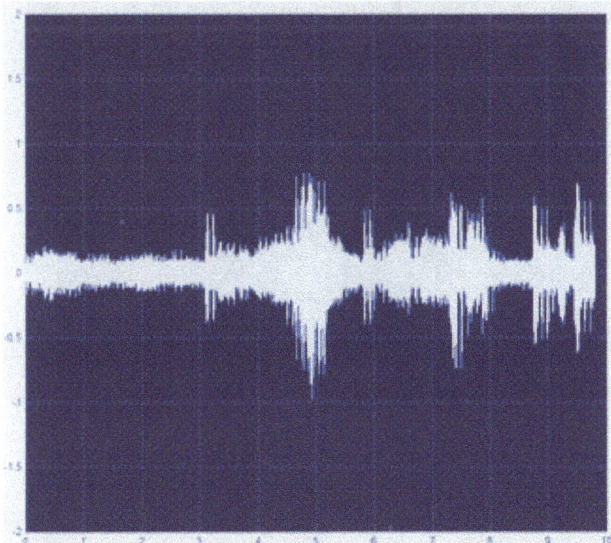

Figure 15. *Received data over Star network without decryption.*

5. The Performance Enhancement in Simulation Results

In this section, the contents show the simulation results of OFDM with proposed encryption based 3D-DWT method.

However, for time-domain it is clearly represented as discrete-time signals. In frequency domain the division of energy is not as original as before the encryption process application. The spectrum is reversed altogether which inverts the distribution of energy level with respect to function of frequency. However, the data is mixed in frequency domain which is similar to convolution in time domain. While transmitted signal is represented in time-domain this leads that any unauthorized access that tries to de-ciphering the data without knowledge about the used scheme, would have to convolve in time-domain which, without doubt would be time consuming process based real-time systems. Furthermore, non- knowledge the permutation order of the system that's why he would have to apply on each frame could be recognized to take infinite time [20, 23] as shown in fig.(16).

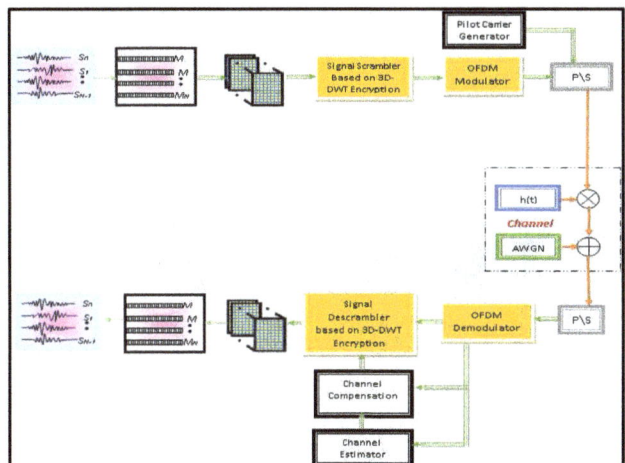

Figure 16. *Block diagram of real time encryption system.*

These parameters are shown in table (2)

Table 2. Simulation Parameters.

25 MHz	Bandwidth
AWGN	
Flat fading+AWGN	Channel model
Frequency selective fading+AWGN	
0.1μsec	Delay spreading *(T_d)*
64	FFT Points
26	Symbol Number

After an extensive tests the results showed over a long period to create those tests clear for listeners, the speech files of that contain wave signals are played and listened by listeners. By following mechanism steps, thirty listeners who are all listened to 50 encrypted wave segments. Segment consists of the digits 0 to 9 is spoken in cluster of four digits. Additional, tests are not restricted to spoken digits only but also to sentence segments. So as to make test stringent and result has oriented feature, the test was hard and consumes time, and the tests are implemented in English as well in Arabic language. However, duplication of spoken digits of the same position is avoided. The tests were inflexible by using:

i. Separating the digits that were spoken by male and female as well.

ii. Test is done for not limited to digits only but also for sentences. The test segments also recorded.

iii. The recorded segments are tested via two languages by male and female.

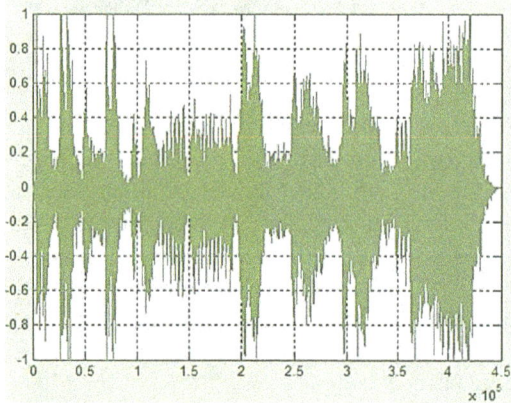

Figure 17. Time Domain of Original Audio File to be encrypted.

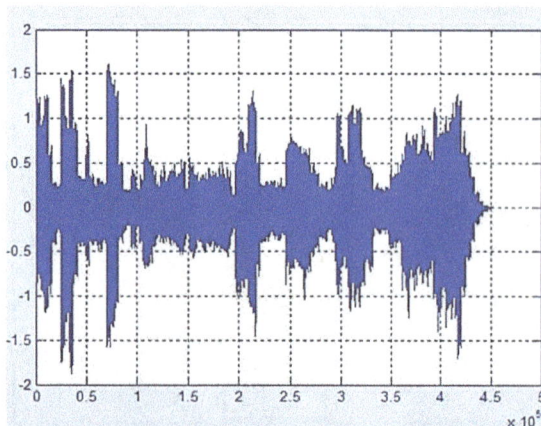

Figure 18. Time Domain of Original Audio File to be encrypted.

The recorded data file, contains spoken digits "Zero, One, Two" vocal by a male is showed in. Figs. (17) – (18) which represent time-domain representation of original ciphered data and deciphered files respectively. On the rest, Figs (20) – (22) reveal distribution of power as a function of frequency of original encrypted and retrieved speech, respectively.

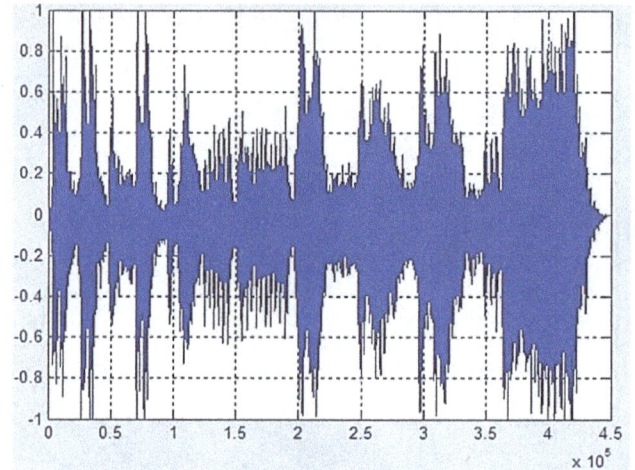

Figure 19. Time Domain of Recovered Audio wave.

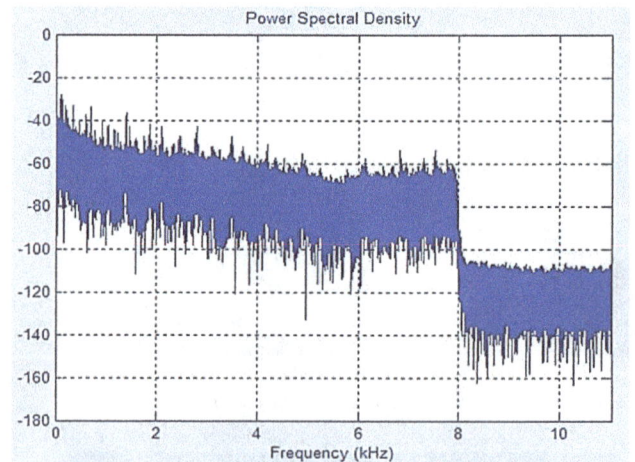

Figure 20. Original Power Spectral Density of Audio wave.

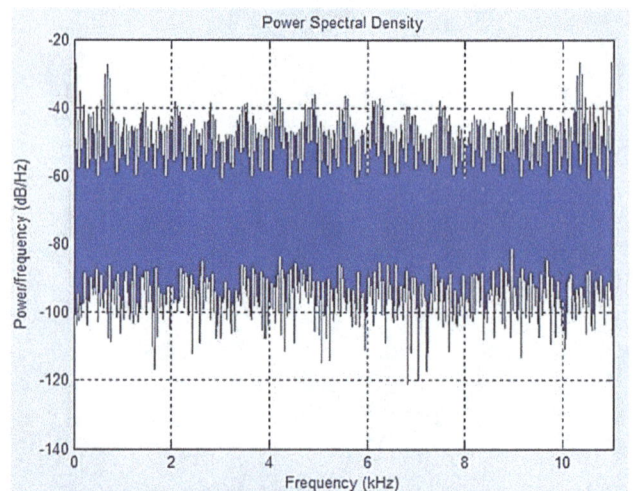

Figure 21. Encrypted Power Spectral Density of Audio wave.

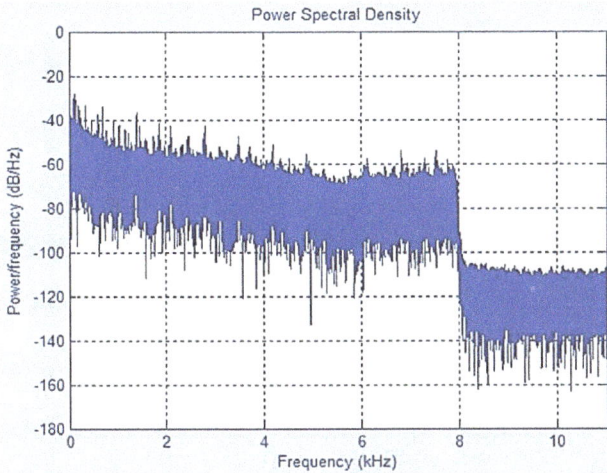

Figure 22. Recovered Power Spectral Density of Audio wave.

A. The Encryption-OFDM In AWGN Channel

The MATLAB V8.1 is used to simulate the Encryption - OFDM transceiver proposed system as shown in Fig.(16). Most MATLAB functions are written to simulate the encryption system as shown in Fig.(16). The functions include frame resizing, Encryption-description, the using of pilot carriers, etc. the output of the simulated proposed system is estimated and represented in Fig.(23), and gives the performance of BER for the Encryption-OFDM using discrete wavelet transformation and OFDM system in AWGN channel. It is represented clearly that the Encryption-OFDM system using 3D-DWT Encryption gives much better results than OFDM transceiver and the Encryption of 2D-DWT OFDM.

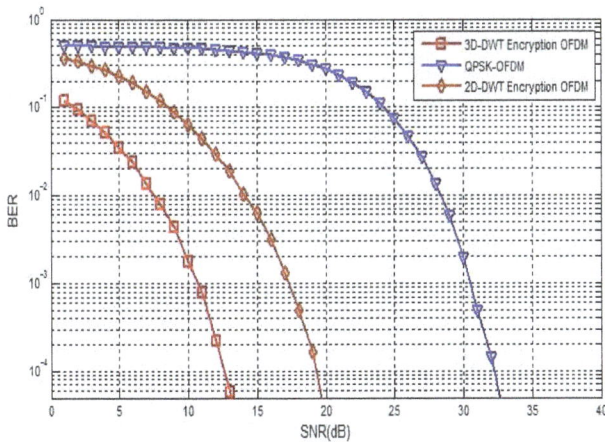

Figure 23. BER performance of Encryption-OFDM using 3D-DWT Encryption in AWGN channel model.

B. The Flat Fading Encryption Channel based 3D-DWT

MATLAB V8.1 simulated the results as in Fig. (16) is used here to mimic the results in flat fading channel additional to AWGN excluding a flat fading channel is added to the channel model. For AWGN and flat fading types of channel, the signal is influenced by the fading effect add to AWGN. However, all the frequency assembled of the signal will be influenced with an attenuation and linear distortion for assumed channel and this leads to a Rayleigh's distribution.

The assumption of 10 Hz is used for Doppler frequency which leads to BER of 10-4 and the SNR required for Ciphering –OFDM using 3D-DWT is about 17 dB could be seen from Fig(24), while 2D-DWT OFDM scrambling of transceiver is about 23 dB and the SNR in OFDM transceiver is about 36dB.

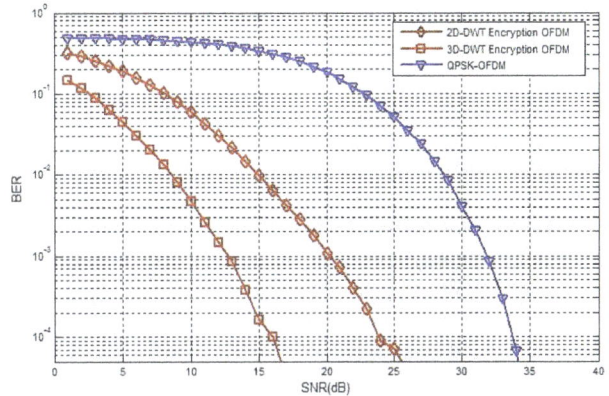

Figure 24. performance of 3D-DWT Encryption for Flat Fading Channel with Doppler Shift =10 Hz.

The same thing are shown in from fig. (25) and fig. (26), therefore from fig. (24) fig. (25) and fig.(26) a gain of 19dB and 6dB for the Encryption-OFDM using 3D-DWT Encryption against OFDM 2D-DWT Encryption transceivers are obtained respectively.

Therefore the Encryption-OFDM using 3D-DWT Encryption outdone dramatically for this model channel.

C. The Frequency Selective Fading Channel

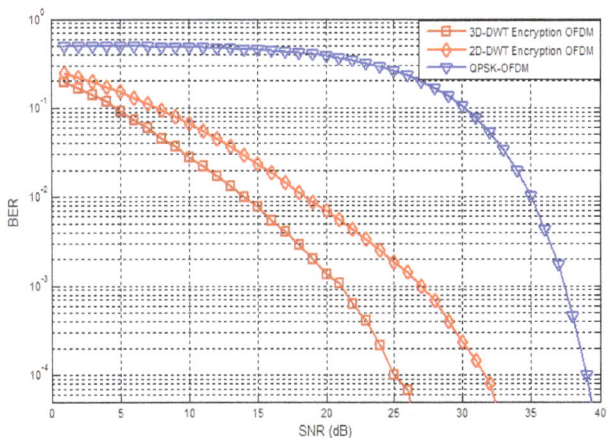

Figure 25. performance of 3D-DWT Encryption for Flat Fading Channel with Doppler Shift =100 Hz.

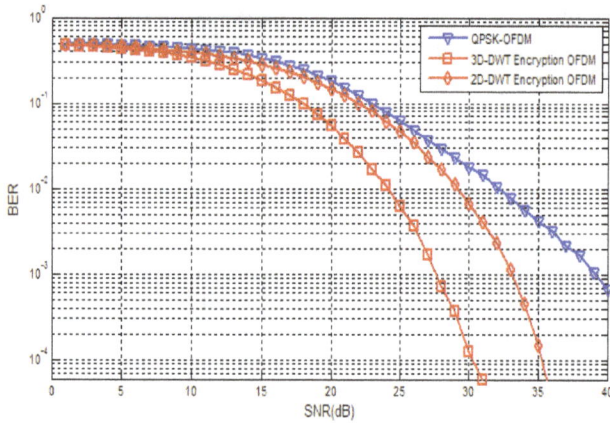

Figure 26. *performance of 3D-DWT Encryption for Flat Fading Channel with Doppler Shift =500 Hz*

D. Encryption based 3D-DWT

According to BER performances of Encryption-OFDM using 3D-DWT are mimicked for AWGN with multi-path frequency selective Rayleigh distributed channels. Assuming two ray channel with gain of -8dB for the second path, the second path would have maximum delay of τmax=0.1μsec for range of values of signal to the noise ratio. Fig. (27) represents mimic results of fDmax =10Hzas maximum Doppler shift. It could be seen clearly from Fig.(27) the BER=10e-4 wouldrequire SNR for Encryption-OFDM using 3D-DWT about 19dB, however Encryption-OFDM utilizing 2D-DWT and OFDM transceivers,the SNR areabout 31dB and 37dB respectively.Therefore from figs.(26) the gain of 18dB of the Encryption-OFDM using 3D-DWT against OFDM transceiver which obtained. In Figs(27-29) the same thing can noted that Encryption-OFDM using 3D-DWT Encryption system outperforms significantly for this channel model. In this sections the results are briefed in table (3), also those results are computed later by testing the system via transferring approximately 1M symbols. Table (3) presents SNR values corresponding to BER.

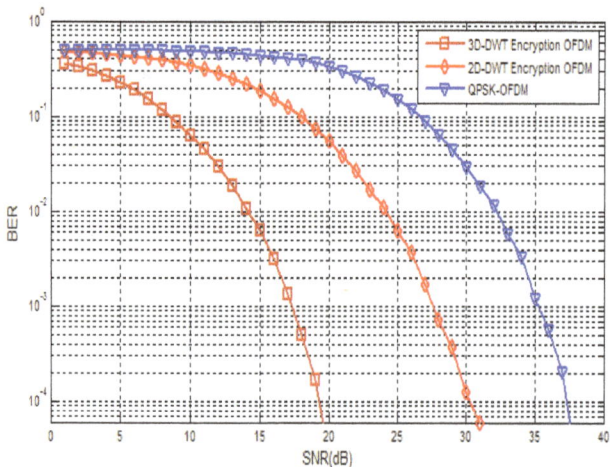

Figure 27. *Performance of 3D-DWT Encryption- for Selective Fading Channel with Max. Doppler Shift=10Hz.*

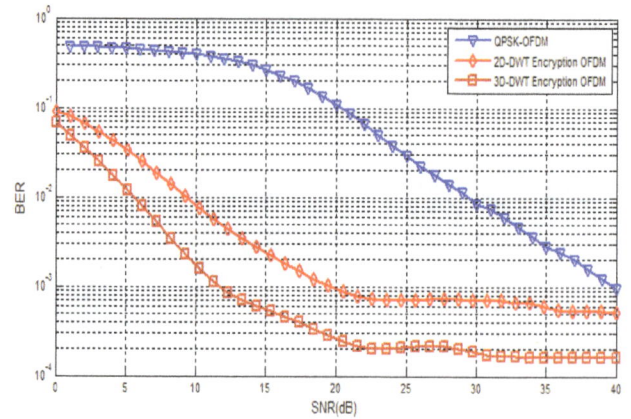

Figure 28. *Performance of 3D-DWT Encryption- for Selective Fading Channel with Max. Doppler Shift=100Hz.*

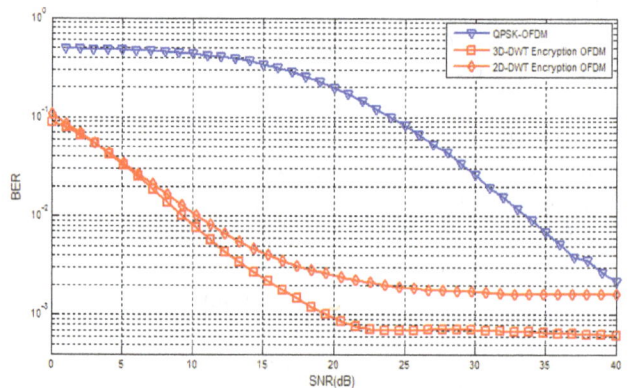

Fig. 29. *Performance of 3D-DWT Encryption- for Selective Fading Channel with Max. Doppler Shift=500Hz.*

Since the essential goal of communication security is the hiding of the fact that a secret message is transmitted, then it is very important to make the recovered process at receiver. we present a list of interpretations.

1. This work is novel in wireless security based DSP techniques, where most algorithms that were used based on inserting dummy packets or based permutation of Fourier Transform, while in this work, Three Dimensional Transformation based Discrete Wavelet Transformation is adapted and has decrease BER over OFDM modulation in different modulated channels as descripted above.

2. It obvious that the proposed system that based on 3D-DWT with permutations is secure against brute force attack, when the required process to retrieve data is governed by permutation preprocess it gives system robust against those kind of attacks.

3. The key length based on the length of the message which is close to the best sec urity algorithm (One Time Pad) according to what Shannon showed. The length of the key = 16x16x16 = 4096 =212.

4. The BER shows great interesting results, the system seems robust and has the immunity for wide range of SNR and this gives two advantages: security and noise immunity.

5. The delay time in Ad hoc network is less 50 % than Star

network, this results with two nodes. For more than two nodes, this results is reversed, the Star network topology would be faster than Ad hoc, the ration depends on the distance, number of hops between the two nodes, indoor or outdoor, finally, interference existence. The diversity in test shows great results in both topologies with delay distinction in real time environment.

6. The proposed system fulfills most of the Kerchoff's principles which state that the security of the system has to be based on the assumption that the enemy has full knowledge of the design and implementation details of the security system, i.e. the algorithm could be published in public, only the length of key, permutation algorithm and cipher mode type are kept secret, the resultant message is in format which is suitable for transmission, the system is practicallyunbreakable, system implementation is easy and it requires a short time.

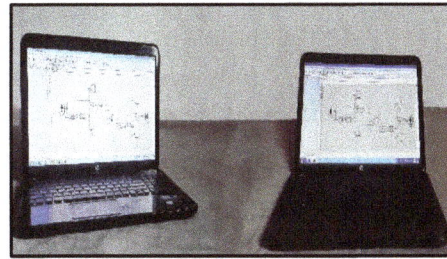

Figure 30. The connection of Two Stations in Ad Hoc topology.

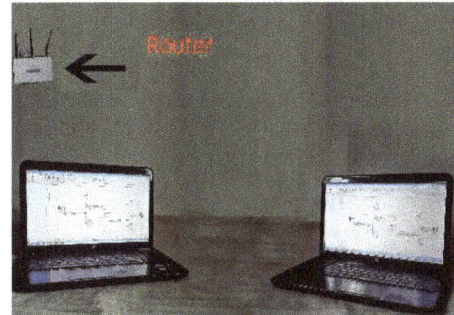

Figure 31. The connection of Two Stations in Star topology.

Table 3. The results for all systems.

| System name | AWGN | Flat Fading | | | Selective Fading | | |
| | | Max. Doppler Shift | | | Max. Doppler Shift | | |
		10 Hz	100 Hz	500 Hz	10 Hz	100 Hz	500 Hz
OFDM- transceiver	32	33	39	non	37	Non	non
2D-DWT ENCRYPTION- transceiver	19	24	32	36	31	Non	non
3D-DWT ENCRYPTION- transceiver	13	15	25	31	19	Non	non

6. Conclusion

The scrambling possibilities based on 3D-DWT matrix show the following features are:

a) Bandwidth is preserved.

b) There is no noise expansion, and quality of the recovered is preserved.

c) There exist fast algorithms, chapter four contains description to the forward algorithm.

d) Inverse transform is found easily, and has the same fast algorithm. Chapter four contains description to the reversed algorithm.

e) The encrypted data is meaningless and thus the residual intelligibility is considerably very low.

f) Permutation are better than inserted dummy components.

g) cryptanalytic efforts are considerably increased due to altered data components in such away will take infinite time to retrieve data if they do not know the structure of the system or one of its limits;

h) Implementation of the new scrambling concept into all existing data scramblers is straightforward, i.e., this concept is fully compatible with conventional systems.

References

[1] Amara Graps, "An Introduction to Wavelets," IEEE Computational Science and Engineering, vol. 2, num. 2, published by the IEEE Computer Society, Summer 1995.

[2] S. Mallat. A Theory for Multiresolution Signal Decomposition: the Wavelet Representation. IEEE Transaction on Pattern Analysis and Machine Intelligence, 11, pp. 674-693, 1989 H. Simpson, Dumb Robots, 3rd ed., Springfield: UOS Press, 2004, pp.6-9.

[3] M. Vishwanath and C. Chakrabarti, "A VLSI Architecture for Real-Time Hierarchical Encoding/Decoding of Video using the Wavelet Transform," IEEE International Conference on Acoustics, Speech and Signal Processing (ICASSP '94), Adelaide, Australia, vol. 2, April 19–22, 1994, pp. 401–404.

[4] M. Weeks, et al, "Discrete Wavelet Transform: Architectures, Design and Performance Issues," Journal of VLSI Signal Processing 35, 155–178, 2003.

[5] Md. ShoaiburRahmanl, Md. AynalHaque, "Introduction to a Novel Wavelet," IEEE/OSA/IAPR International Conference on Informatics, Electronics & Vision, 2012.

[6] V.Senk, V. D. Deli´c,V. S. Milo˘sevi´c, "A New Speech Scrambling Concept Based on Hadamard Matrices," IEEE SIGNAL PROCESSING LETTERS, VOL. 4, NO. 6, JUNE 1997.

[7] H. Ojanen. "Orthonormal compactly supported wavelets with optimal sobolev regularity. Applied and Computational Harmonic Analysis, 10, pp. 93-98,2001.

[8] A.R. Calderbank, I. Daubechies, W. Sweldens, and B.L. Yeo. "Wavelet transforms that map integers to integers,". Appl. Comp. Harm. Anal., 5 (3), pp. 332-369,1998.

[9] Shannon, C. E., "Communication Theory of Secrecy Systems, "Bell System Technical Jo urnal, Vol. 28, 1949, pp. 656–715.

[10] Dr. Jameel Ahmed, "Transform-Domain and DSP Based Secure Speech Communication". Ph.D. dissertation, Hamdard Institute of Information Technology, 2007.

[11] D.J.H. Garling, D. Gorenstein, T. Tom Dieck, P. Walters, "WAVELETS AND OPERATORS,", Cambridge University Press 1992

[12] Tuan Van Pham, "Wavelet Analysis for Robust Speech Processing and Applications," VDM Verlag, Germany, 2008.

[13] StephaneMallat, "A Wavelet Tour of Signal Processing," Elsevier, 1999.

[14] Richard E. Blahut, "Fast Algorithms for Signal Processing," CAMBRIDGE UNIVERSITY PRESS, 2010.

[15] John J. Benedetto, "Applied and Numerical Harmonic Analysis: Frames and Bases," Birkha¨user Boston, 2008.

[16] Mladen Victor W. ,"Adapted Wavelet Analysis from Theory to Software," A K Peters,Ltd 1994.

[17] Michel Misiti, Yves Misiti, Georges Oppenheim, Jean-Michel Poggi "Wavelet Toolbox For Use with MATLAB ®," The MathWorks, Inc, 2002.

[18] Alfred Mertins, "Signal Analysis: Wavelets, Filter Banks, Time-Frequency Transforms and Applications,",Mertins, Signaltheorie, 1996.

[19] Ali N. Akansu, Michael J. Medley, "WAVELET, SUBBAND AND BLOCK TRANSFORMS IN COMMUNICATIONS AND MULTIMEDIA," Kluwer Academic / Plenum Publishers, New York, 2002.

[20] A.Jensen, A.la Cour-Harbo, "Ripples in Mathematics The Discrete Wavelet Transform,", Springer.Verlag Berlin Heidelberg 2001.

[21] CHARLESK.CHUI, "An Introduction to Wavelets,", Academic Press 1992.

[22] C. Sidney Burrus, Ramesh A. Gopinath, and .HaitaoGuo, "Introduction to Wavelets and Wavelet Transforms," Prentice-Hall, Inc., 1998.

[23] "Wavelets and Multiscale Analysis Theory and Applications,"SpringerScience+Business Media, LLC, 2011.

32

Research of Information Recovery Methods Corresponding to Up-shift Part in NB-PLC

Chao Li[1, *], Yongjian Jing[1, 2]

[1]Inner Mongolia Xinyuan Information Technology Co., LTD, Huhhot, China

[2]Center of Information and Network Technology, Inner Mongolia Agricultural University, Huhhot, China

Email address:

Digital9898@sina.com (Chao Li), wshbupt@163.com (Yongjian Jing)

*Corresponding author

Abstract: Power line communication (PLC) is applied in various areas, such as Internet access, smart home, automatic meter reading system, security monitoring system, etc. The introduction of Orthogonal Frequency Division Multiplexing (OFDM) improves the reliability and efficiency of PLC substantially, and the standards of wide-band PLC and narrow-band PLC are becoming more complete. This paper is based on the receiver designing work according to the transmitter flow described in ITU-T G.9955. Regarding to the up-shift part of transmitter, two different methods to recover the useful information are provided including Hilbert transform and sub-carrier calculation. Both the theoretic analysis and simulation result shows that the two methods can recover the transmitted data exactly, while the second method can decrease the complexity of receiver and improve the efficiency of PLC system greatly.

Keywords: Narrow-Band Power Line Communication, G.9955, Up-shift, Hilbert Transform, Sub-carrier Calculation

1. Introduction

Power Line Communication (PLC) technology is widely distributed, coaxial cable and telephone lines are used as a transmission medium for data transmission and voice communications, also widely used in the last kilometer broadband access in remote areas. According to the different band, PLC is usually divided into broadband PLC (1.8 ~ 250MHz) and narrowband PLC (3 ~ 500KHz). Broadband power line communications used for broadband data and multimedia signal transmission, and narrowband power line communication provides a relatively low transmission rate, used for transmission of control information in smart grid and industrial control.

The first generation of narrowband power line communication using single-carrier FSK or PSK modulation, the transmission rate is low, and with low anti-interference ability. With the introduction of OFDM (Orthogonal Frequency Division Multiplexing) technology, the second-generation narrowband PLC is formed. The main standards include G3-PLC standard and PRIME standards. In

early 2010, to make the product worldwide, P1901.2 and G.9955 (G.hnem physical layer specification) are promoted by IEEE and ITU-T separately, this paper is based on research of G.9955.

According to standard G.9955 [1], NB-PLC frequency band is divided into two categories, one is CENELEC (European Committee for Electro technical Standardization) band (3KHz-148.5KHz), the other is FCC band (9KHz -490KHz).

2. The PLC Transmit System

Narrow band PLC system use OFDM modulation to transmit information. It's physical frame including Preamble, PFH, CES and payload sections, as shown in figure 1. Preamble and CES does not carry any information, the function is synchronization and channel estimation, and transmit data is in payload section.

Figure 1. The frame structure.

With Different number of symbols in PFH, the modulation and coding scheme is different. Payload length may vary in different frame, the number and bandwidth of subcarriers in each OFDM symbol is shown in table 1.

Table 1. OFDM parameter.

Variable	parameter	value
N	Number of subcarriers	2k, k=7,8
FSC	Bandwidth of subcarriers [KHz]	15.625/n, n=5,10
FUS	Shift frequency [KHz]	(N/2)* FUS

In different band of CENELEC and FCC, the number of subcarriers in each OFDM symbol is different.

Table 2. OFDM subcarriers in different band.

Band	Start frequency	End frequency	Number of subcarriers	Empty subcarriers
CENELEC-A	35.9375KHz	90.625KHz	23~58	0-22,59-127
CENELEC-B	98.4375KHz	123.4375KHz	63~79	0-62,80-127
CENELEC-CD	125KHz	143.75KHz	80~92	0-79,93-127
FCC	34.375KHz	478.125KHz	11~153	0-10,154-255
FCC-1	34.375KHz	137.5KHz	11~43	0-10, 45-255
FCC-2	150 KHz	478.125KHz	48~153	0-47,154-255

The transmit process of PLC system is shown in figure 2.

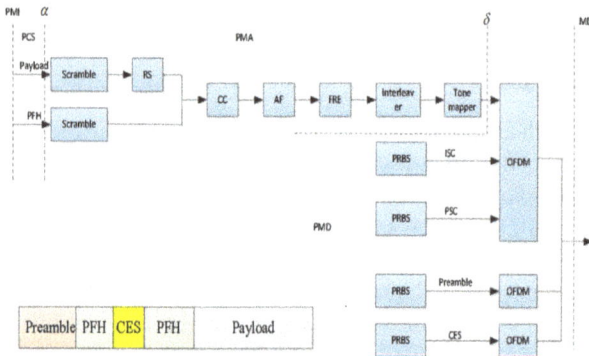

Figure 2. The transmit process.

As shown in figure 2, the payload data is the research focus. At the transmitting end, the data need to be transmitted first perform scrambling, using pseudo-random sequence generated by shift register (LFSR) [2-5]. then forward error correction coding (FEC), including RS coding and convolution coding (CC), then the packet aggregation (AF), block repetition coding, interleaving and channel ranks to improve data anti-interference performance.

In a typical modulation technique for link-level system, the real part of the OFDM symbol is transmitted directly, and receive end can use oversampling and Hilbert transform to recover data [6-7]. In G.9955 standard, the transmitter-side first oversampling OFDM symbol, then frequency shift, and then take the real part into the channel for transmission, as shown in figure 3.

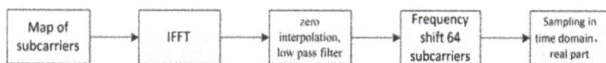

Figure 3. The transmit end.

In OFDM modulation first turn the frequency domain discrete points N (128 or 256) into time-domain signal by IFFT transform. For example, just take a point in OFDM symbol. Set the points in the frequency domain as

$$X(k,m) = a_m + jb_m (0 \leq m \leq N-1) \quad (1)$$

Get N discrete time-domain signal sample points by IFFT.

$$u_k(n) = IFFT(X(k)) = \frac{1}{N}\sum_{m=0}^{N-1}(a(k,m)+jb(k,m))e^{j\frac{2\pi mn}{N}}$$

$$\frac{1}{N}\sum_{m=0}^{N-1}\{a(k,m)\cos(2\pi\frac{mn}{N})-b(k,m)\sin(2\pi\frac{mn}{N})\}+ \quad (2)$$

$$j\frac{1}{N}\sum_{m=0}^{N-1}\{a(k,m)\sin(2\pi\frac{mn}{N})+b(k,m)\cos(2\pi\frac{mn}{N})\}, (n=0,1,...N-1)$$

Perform $p(p = 2^\alpha)$ oversampling to time domain sampling points, get discrete data.

$$u_k(n) =$$
$$\frac{1}{N}\sum_{m=0}^{N-1}\{a(k,m)\cos(2\pi\frac{mn}{pN})-b(k,m)\sin(2\pi\frac{mn}{pN})\}+ \quad (3)$$
$$j\frac{1}{N}\sum_{m=0}^{N-1}\{a(k,m)\sin(2\pi\frac{mn}{pN})+b(k,m)\cos(2\pi\frac{mn}{pN})\}, (n=0,1,2...pN-1)$$

Performs frequency shift operation $F_{US} = m \times F_{SC}$, where m is an integer no less than N/2, G.9955 specified m = N / 2, after the frequency shift, time domain sampling point becomes:

$$s_n = u_k(n)e^{j\frac{2\pi mn}{Np}} = \text{Re}(s_n) + j\text{Im}(s_n), (0 \leq n \leq 2N-1) \quad (4)$$

Transmit signal is the real part of s_n:

$$s_{out} = \text{Re}(s_n) = \text{Re}(u_k(n))\cos(\frac{2\pi mn}{Np}) - \text{Im}(u_k(n))\sin(\frac{2\pi mn}{Np}) \quad (5)$$

3. Two Recover Solutions

Two ways are proposed to recover the original data respectively, one is the Hilbert transform detection algorithms; the other is subcarriers calculation solution according to the frequency domain.

3.1. Hilbert Transform Solution

At the receiving end, first test the received data, then recover the frequency-domain signal $X(k,m)$.

Take $p(p=2^{\alpha})$ and m=N/2 into receive signal

$$r_n = s_{out} = \text{Re}(u_k(n))\cos(\frac{n\pi}{2^{\alpha}}) - \text{Im}(u_k(n))\sin(\frac{n\pi}{2^{\alpha}}), (0 \le n \le pN-1) \quad (6)$$

When $n = 2^{\alpha}i(i=0,1,2..N-1)$

$$r_{2i} = \text{Re}(u_k(2^{\alpha}i))\cos(i\pi) = (-1)^i \text{Re}(u_k(2^{\alpha}i)) \quad (7)$$

When $n = (i+\frac{1}{2})\cdot 2^{\alpha}(i=0,1,2..N-1)$

$$r_{2i+1} = -\text{Im}(u_k(2^{\alpha}(i+\frac{1}{2})))\sin(\frac{(2i+1)\pi}{2})$$
$$= (-1)^{i+1}\text{Im}(u_k(2^{\alpha}(i+\frac{1}{2}))) \quad (8)$$

The Hilbert transform [3]is:

$$\tilde{f}(t) = H[f(t)] = \frac{1}{\pi}\int_{-\infty}^{\infty}\frac{f(\tau)}{t-\tau} = f(t) \otimes \frac{1}{\pi t} \quad (9)$$

According to the define of resolve signal,

$$f_a(t) = f(t) + j\tilde{f}(t) \quad (10)$$

By equation (3), $u_k(n)$ is resolve signal, the imaginary part is the Hilbert transform of the real part, get all real part of sampling points of (N) (as in formula (7)), and imaginary part (such as the formula (8)) of N over-sampling points. With inverse Hilbert transform of imaginary part, real part of 2N-point can obtained.

$$r_{2i+1}' = H^{-1}(r_{2i+1}\cdot(-1)^{i+1}) \quad (11)$$

Connect these real part point, get $r_i = (r_0, r_1', r_2, r_3'...)(i=0,1,2...2N-1)$, perform 2N-point FFT transform, derived the original frequency domain signal.

$$y_l = \sum_{n=0}^{2N-1} r_n e^{-j\frac{2\pi nl}{2N}} = \sum_{n=0}^{2N-1}\{\frac{1}{N}\sum_{m=0}^{N-1}[a(k,m)\cos(2\pi\frac{mn}{2N}) - b(k,m)\sin(2\pi\frac{mn}{2N})]\}e^{-j\frac{2\pi nl}{2N}}$$
$$==\begin{cases} 2a(k,0) & l=0 \\ a(k,l)+jb(k,l) & 1 \le l \le N-1 \\ \text{other} & N \le l \le 2N-1 \end{cases} \quad (12)$$

Thus, in addition to the first data restoring the real part, the other N-1 point complex data can be recovered. As no data was put in the first sub-carrier, there is no impact on the demodulation of received data.

3.2. Subcarriers Calculation Solution

In this section, after the analysis of frequency domain change, can get information from received signal by FFT transform, at different locations of subcarriers.

After modulation, transmit signal was mapped to the corresponding sub-carriers, for example, CENELEC A band use No.24-59 subcarriers, CENELEC B use No.64-80 subcarriers, CENELEC CD use No.81-93 subcarriers. Then perform oversampling and interpolation. When the interpolation factor is p, insert (p-1) zeros between every two time domain sampling point, corresponding to changes in the frequency domain of the original signal, which is called mirroring, the process is shown in figure 4.

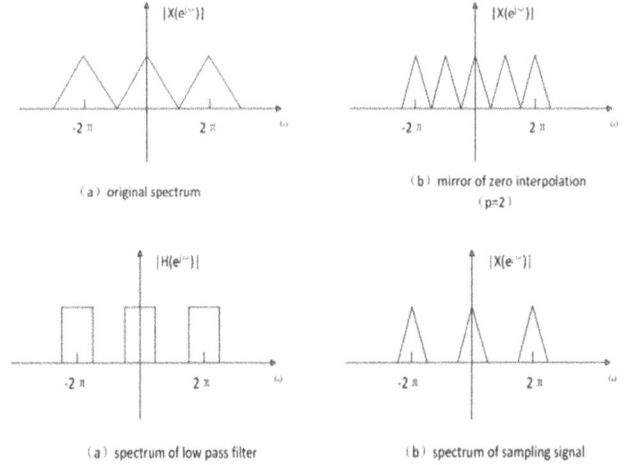

(a) original spectrum

(b) mirror of zero interpolation (p=2)

(a) spectrum of low pass filter

(b) spectrum of sampling signal

Figure 4. The mirroring process.

After over-sampling, move $m \ge N/2$ subcarriers, then take the real part, replicate the spectrum on the complex frequency domain again, and conjugate.

Take CENELEC A band as an example; put data in 24-59 subcarriers, as shown in figure5.

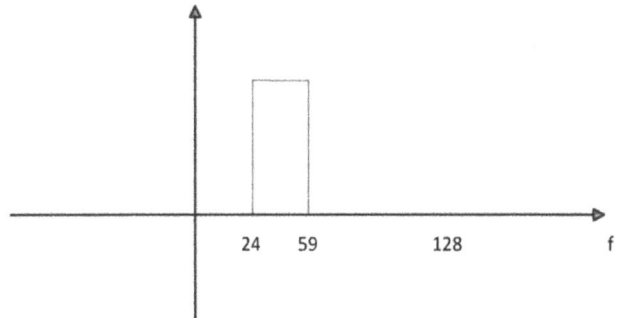

Figure 5. Data of CENELEC A.

Then zero interpolation, low pass filter.

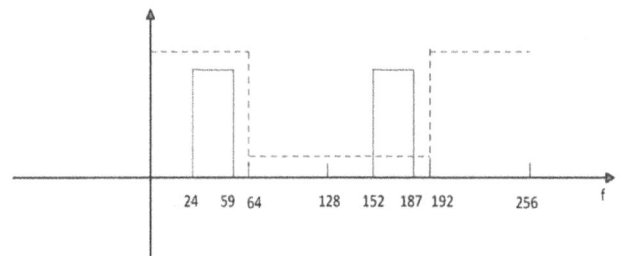

Figure 6. Spectrum of oversampling.

As shown in figure 6, the dotted line is shape of filter, data in 24-59 sub-carriers can pass, and the data in subcarrier 155-187 will be filtered out.

Then is the frequency shift, move 64 subcarriers:

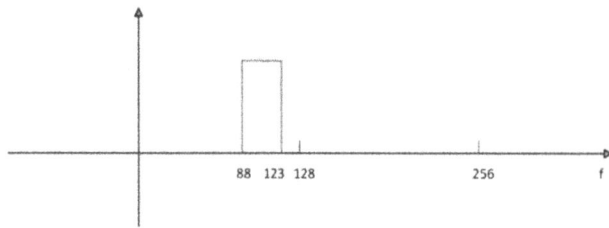

Figure 7. Spectrum of shift.

And the real part is

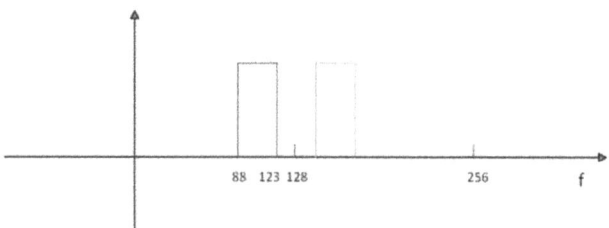

Figure 8. Spectrum of real part.

The green part is the conjugate of the original data, that is the negative frequency and positive frequency spectrum is symmetric.

In summary, in the receiving end, get 88-123 point out, which is the data required.

The change process of CENELEC B is as follows:

First, data is placed in 64-80 subcarriers, shown in figure9.

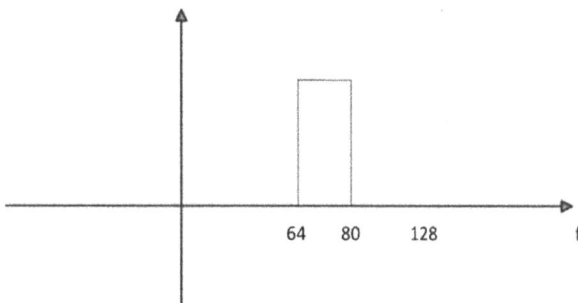

Figure 9. Data placed in 64-80 subcarriers.

Then, zero interpolation and low-pass filter, through the low-pass filter, the data in 192-208 is reserved; data in 64-80 is filtered, as shown in figure 10.

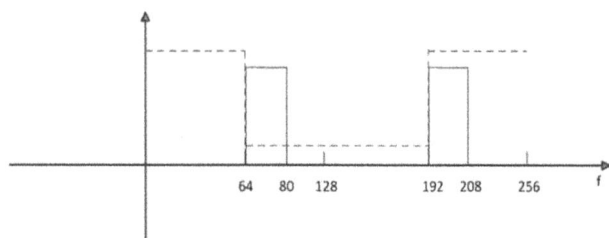

Figure 10. Low pass filter.

Then moved 64 sub-carrier frequency, shown in figure 11.

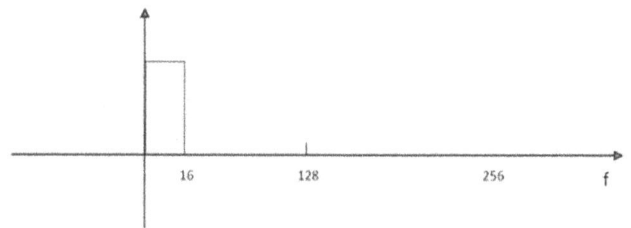

Figure 11. 64 frequency move.

Take real part, get figure 12.

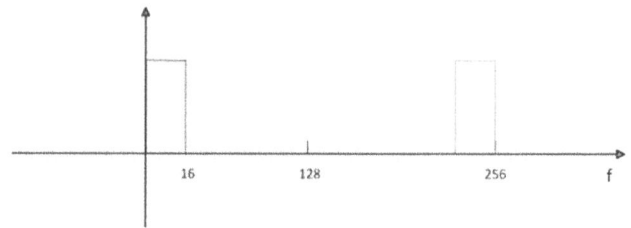

Figure 12. Real part.

CENELEC CD similar to CENELEC B, after a low-pass filter, the data in 128-256 is retained, while others is filtered, and then frequency shift, the data bits back to the first 128 sample points. Transmitted data is located on the positive frequency spectrum regardless of band.

Therefore, the receiving end can calculate subcarriers occupied by the data, take the data from corresponding sub-carriers to operate, without Hilbert transform.

4. Simulation Results and Conclusion

From the above analysis, the two solutions didn't process the noise and interference in the subcarriers, they should have same performance. Simulation platform was setup to verify the solutions, the receiver process is shown in figure13.

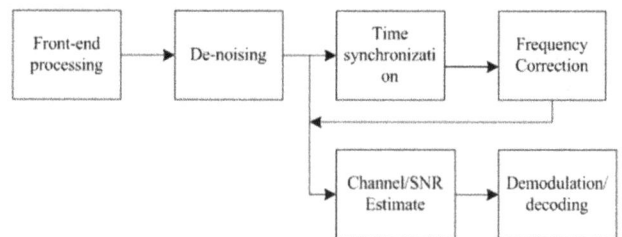

Figure 13. The receiver process.

As shown in figure 13, the receiving end is the reverse process of the transmitting end, the receiving end perform demodulation and decoding. The de-noising module includes interference filter in narrowband, and against burst; time synchronization includes coarse synchronization and fine synchronization; decoding includes convolution decoding and RS decoding.

There are five kind of noise in PLC channels, including colored background noise, narrowband interference, periodic impulse noise, asynchronous random impulse noise and periodic random impulse noise [8-10]. Gaussian white noise is set in the channel, use the CENELEC A band to transmit data, each OFDM symbol has 128 sub-carriers, No.23-58 sub-carriers are used to transmit data. The performance comparison of the two solutions is shown in figure 14.

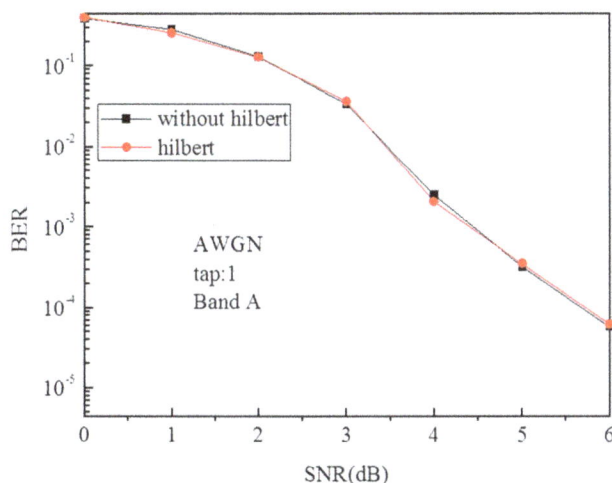

Figure 14. *Comparison of Hilbert transform and subcarrier computing.*

In the above figure, the red line is the Hilbert transform performance curve, the black line is subcarrier computing performance curve. The two solutions, whether theoretical analysis or link-level simulation, can recovery original data, and no difference in performance. But the sub-carriers calculation solution can filter data from frequency domain directly, and quickly, which can greatly improve the system efficiency.

References

[1] ITU-T G.9955,Narrowband OFDM Power Line Communication Transceivers-Physical Layer Specification, 2011. 12.

[2] Xiaoyu Wen, Oversampling and Hilbert transform in DAMB system. Journal of Nan yang Normal university. 2008.3.

[3] Jiongpan Zhou, The principle of communication, pp 27-31.

[4] Xiuqing yuan, Study of channel estimation in PLC based on OFDM. Harbin University of Science and Technology, 2008.

[5] Stefania Sesia, Issam Toufik, Matthew Baker. LTE-The UMTS Long Term Evolution From Theory to Practice. Pp 91.

[6] J. Maxwell, A Book on Electricity, 9rd ed., Oxford: Clarendon, 1892, pp. 88.

[7] J. Young, A Technical Write, 5rd ed., vol. 9. Oxford: Clarendon, 2002, pp. 73.

[8] Feng Ye, Study of PLC simulation, Xinan Jiaotong University, 2008.

[9] Jianhua Zhao, Zhi Li, Simulation of OFDM in PLC system, Computer and Digital Engineer, 2011, Vol. 39, No. 9.

[10] V. Degardin, M. Lienard and P. Degauque et al, "Impulsive noise on indoor power lines: characterization and mitigation of its effect on PLC systems", Electromagnetic Compatibility, IEEE, May 2003, vol. 1, pp. 166-169.

33

Optimization Algorithm of Resource Allocation IEEE802.16m for Mobile WiMAX

Bat-Enkh Oyunbileg[1], Otgonbayar Bataa[1], Tuyatsetseg Badarch[2], Baatarkhuu Tsagaan[1]

[1]Department of Information Technology, School of Information and Telecommunication, MUST, Ulaanbaatar, Mongolia
[2]Department of Information Technology, Mongolian National University, Ulaanbaatar, Mongolia

Email address:

o_bat_enkh@yahoo.com (O. Bat-Enkh), b_otgonbayar2002@yahoo.com (B. Otgonbayar)

Abstract: Multi user resource allocation is one of the key features towards high speed wireless network based on Orthogonal Frequency Division Multiplexing Access (OFDMA). According to IEEE802.16m (Mobile WiMAX) standard resource allocation problem has to be performed on a frequency and time two-dimensional space with the Physical and logical resource units (PRU and LRU) including Distributed logical resource unit (DRU and CRU) and Contiguous logical resource unit In this paper we analysed the WiMAX frame structure in IEEE802.16m based on the Mobile WiMAX Standard. We apply novel resource allocation algorithm are used for managing two dimensional resources (time and frequency) to maximizing system capacity depending on mobile user's data rate. This paper proposes a resource allocation algorithm for radio resource allocation in the downlink/uplink cellular networks using Orthogonal Frequency Division Multiple Access of Mobile WiMAX. Radio Resource Management (RRM) is fundamental in cellular networks. Many optimization problems involving allocation strategies of various types of resource appear in RRM. In this paper we determine the Fractional Frequency Reuse (FFR) and analyze the resources in Mobile WiMAX.

Keywords: IEEE802.16m, Mobile WiMAX, Resource Allocation, PRU, DRU, CRU and LRU, Physical Resource Units, LTE

1. Introduction

The most suitable physical layer for the future deployment of Broadband Wireless Access networks is OFDMA. OFDMA has a few benefits large data rates, robustness against multi-path, flexible bandwidth allocation, and the possibility of exploiting multi-user diversity through an intelligent scheduling and resource allocation [1,3,4,9]. One of the most important advantages of OFDMA systems is included flexible resource allocation in two dimensional spaces.

The purpose of this paper is to provide resource allocation procedure and to give comparative analysis that need to be considered in developing distributed resource allocation for IEEE 802.16m. In OFDMA, the available spectrum (frequency domain) is divided into orthogonal sub-carriers, which are combined into groups, usually referred as sub-channels. Furthermore, the time domain structure is segmented into consecutive frames, each one containing multiple OFDM symbols. A slot delimited by one sub-channel and by a given number of OFDM symbols is the smallest resource unit that a Base Station (BS) can allocate to the users. OFDMA systems can provide multi-user access assigning different sub-channels and OFDM symbols to different users. Moreover, the system can exploit multi-user diversity, allocating in each frequency and time region (two dimensions) the user with the best channel conditions. This feature makes OFDMA systems, such as Long Term Evolution (LTE) or WiMAX, more flexible than 3G solutions as High Speed Packet Access (HSPA) networks. However, these degrees of freedom present new challenges for the resource assignment and they require more sophisticated algorithms, capable of handling efficiently two-dimensional allocations.

In IEEE 802.16m for Mobile WiMAX [2,3,4,9], the case study in this paper, two constraints affect the allocation strategies. For OFDM, each subchannel can be modulated differently, but it only in the time domain. Because in

OFDMA is considered both time and frequency domains. The OFDMA scheduler is the one of most complex problems because each MS can receive some portions of the allocation for the combination of time and frequency so that the channel capacity is efficiently utilized.

2. Downlink Physical Structure

Each downlink AAI sub frame is divided into 4 or fewer frequency partitions; each partition consists of a set of physical resource units across the total number of OFDMA symbols available in the AAI sub frame. Each frequency partition can include contiguous (localized) and/or non-contiguous (distributed) physical resource units. Each

frequency partition can be used for different purposes such as fractional frequency reuse (FFR). "Fig. 1" illustrates the downlink physical structure in the example of two frequency partitions with frequency partition 2 including both contiguous and distributed resource allocations, where Sc stands for subcarrier.

The DL/UL subcarrier to resource unit mapping process, which is used to our research, is defined as following and illustrated in the Fig. 2.

1. Outer permutation is applied to the PRUs in the units of N1 and N2 PRUs. Direct mapping of outer permutation can be supported only for CRU.

2. Distributing the reordered PRUs into frequency partitions.

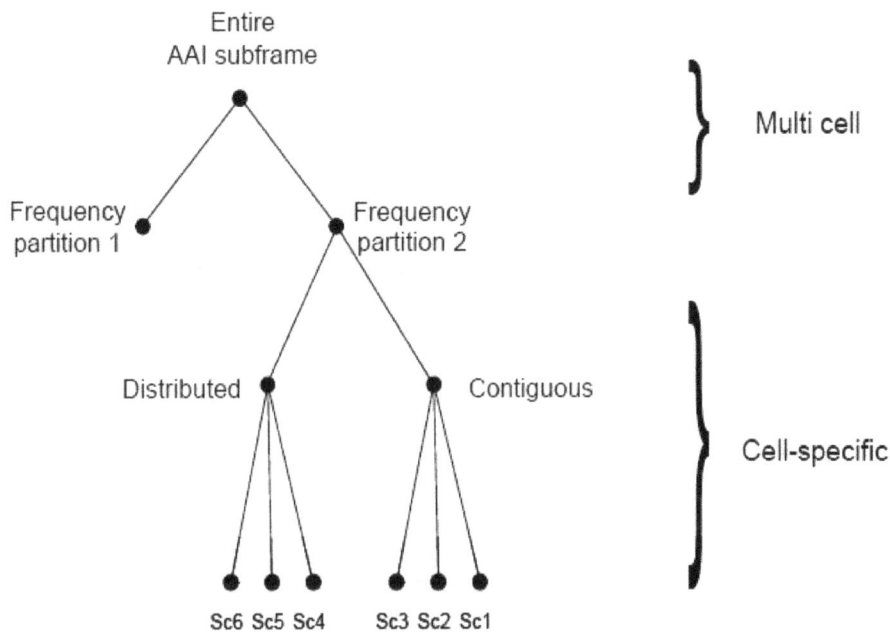

Figure 1. Downlink physical structure.

3. The frequency partition is divided into localized and/or distributed groups. Sector specific permutation can be supported and direct mapping of the resources can be supported for localized resources. The sizes of the distributed/localized groups are flexibly configured per sector.

4. The localized and distributed groups are further mapped into LRUs (by direct mapping of CRU and by "Subcarrier permutation" for DRUs) as shown in the following figure [2].

Physical and logical resource unit: A physical resource unit (PRU) is the basic physical unit for resource allocation that comprises Psc consecutive subcarriers by Nsym consecutive OFDMA symbols. Psc is 18 subcarriers and Nsym is 6 OFDMA symbols for type-1 sub-frame, respectively. A logical resource unit (LRU) is the basic logical unit for distributed and localized resource al-locations.

Distributed logical resource unit: The downlink distributed logical resource unit (DLRU) contains a group of subcarriers that are spread across the distributed resource allocations within a frequency partition. The downlink DLRUs are obtained by subcarrier permuting on the data subcarriers of the

distributed resource units (DRUs). The size of the DRU equals the size of PRU, i.e., P_{sc} subcarriers by N_{sym} OFDMA symbols.

Contiguous logical resource unit: The localized logical resource unit, also known as contiguous logical resource unit (CLRU) contains a group of subcarriers that are contiguous across the localized resource allocations. CLRU consists of the data subcarriers only in the contiguous resource unit (CRU) of which size equals the size of the PRU, i.e., P_{sc} subcarriers by N_{sym} OFDMA symbols. The CLRUs are obtained from direct mapping of CRUs. Two types of CLRUs, subband LRU (SLRU) and miniband LRU (NLRU), are supported according to the two types of CRUs, sub band and miniband based CRUs, respectively.

The Advanced Air Interface supports TDD and FDD duplex modes, including H-FDD AMS operation. Unless otherwise specified, the frame structure attributes and baseband processing are common for all duplex modes. The Advanced Air Interface uses OFDMA as the multiple access schemes in the downlink and uplink.

Both the uplink and downlink for WiMAX TDD are divided

into radio frames, each 10 ms in length. "Fig. 1" shows the frame structure for Mobile WiMAX TDD [1]. The frame consists of two "half-frames" of equal length, with each half-frame consisting of either 10 slots or 8 slots plus the three special fields downlink pilot time slot, guard period and uplink pilot time slot in a special sub frame. Each slot is 0.5 ms in

length and two consecutive slots form exactly one sub frame, just like with FDD. The lengths of the individual special fields depend on the uplink/downlink configuration selected by the network, but the total length of the three fields remains constant at 1 ms.

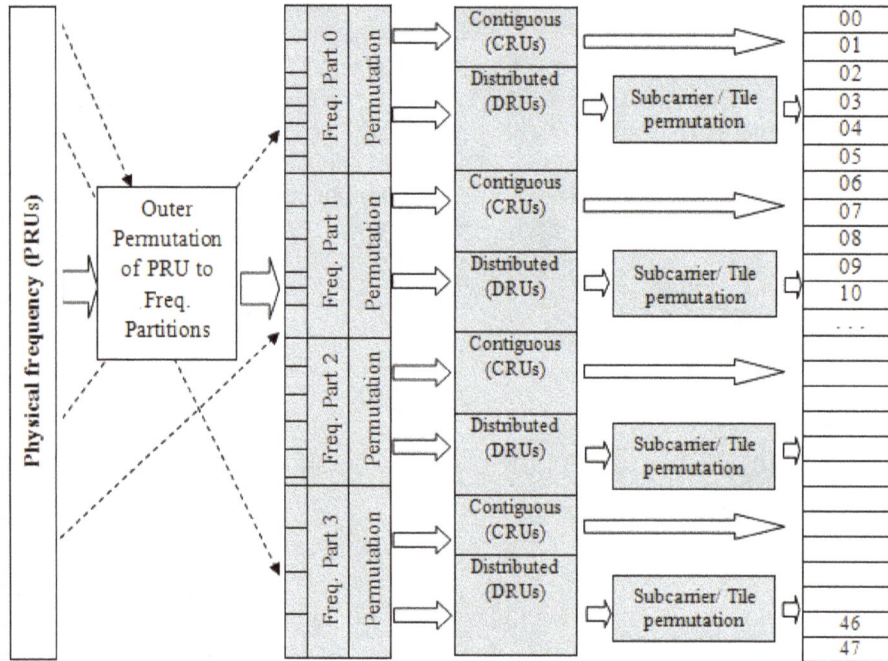

Figure 2. *DL and UL subcarrier to resource unit mapping.*

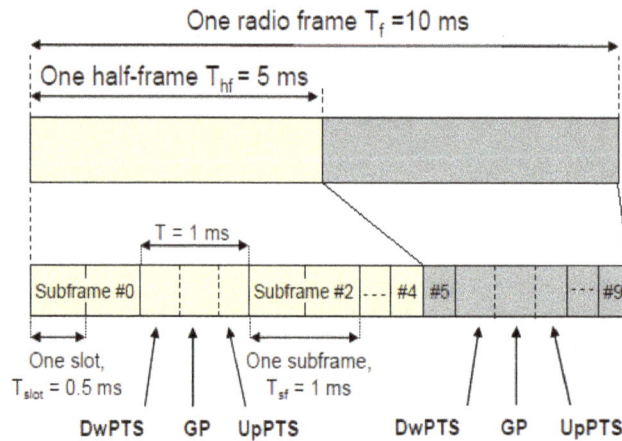

Figure 3. *TDD frame structure.*

The Physical Resource Units (PRU) are first subdivided into sub bands and minibands. The number of sub bands is denoted by KSB. The number of minibands is denoted by KMB. Since the maximum valid value of KSB is 12 or NPRU/N1, we choose 10 MHz bandwidth with KBS equal to 7. In case of KBS =7, values of parameters used multi –cell resource mapping are specified Table 2. PRUs are partitioned and reordered into two groups: sub band PRUs and miniband PRUs, denoted by PRUSB and PRUMB.

Equation (1) defines the mapping of PRUs to $PRU_{SB}s$. Equation (2) defines the mapping of PRUs to $PRU_{MB}s$.

$$PRU_{SB}[j] = PRU[i]; \quad j=0,1, ..., L_{SB}-1. \tag{1}$$

$$PRU_{MB}[k] = PRU[i]; k=0,1,...,L_{MB}-1. \tag{2}$$

The miniband permutation maps the PRUMBs to Permuted $PRU_{MB}s$ ($PPRU_{MB}s$) to ensure frequency diverse PRUs are allocated to each frequency partition. Equation (3) describes the mapping from PRU_{MB} to $PPRU_{MB}s$:

$$PPRU_{MB}[j] = PRU_{MB}[i]; \quad j=0,1, ..., L_{MB}-1. \tag{3}$$

The number of sub bands in i^{th} frequency partition is denoted by $K_{SB,FPi}$. The number of minibands is denoted by

$K_{MB,FPi}$, which is determined by the FPS_i and $DFPSC$ fields. The number of subband PRUs in each frequency partition is denoted by $L_{SB,FPi}$, which is given by $L_{SB,FPi} = N_1 \cdot K_{SB,FPi}$.

The number of miniband PRUs in each frequency partition is denoted by $L_{MB,FPi}$, which is given by $L_{MB,FPi} = N_2 \cdot K_{MB,FPi}$. The mapping of subband PRUs and miniband PRUs to the frequency partition is given by Equation (4).

$$PRU_{FPi} (j) = PRU_{SB} (k_1) \text{ for } 0 \leq j < L_{SB,FPi} \qquad (4)$$

$$PPRUMB (k_2) \text{ for } L_{SB,FPi} \leq j < L_{SB,FPi} + L_{MB,FPi}$$

It in fact produces the harmony or timbre. The method to find the structure of the harmony of the sounds is disintegration of the frequency.

Table 1. Parameters for sub band and miniband partition.

Parameters	Values
N_1	4
N_2	1
K_{SB}	7
L_{SB}	28
L_{MB}	20
DSAC	7
N_{PRU}	48
N_{PRU}	48

3. Parameters of for Resource Allocation

First, we need to calculate summary of the parameters. Table 2 presents a summary of the parameters used to configure the downlink physical structure. In other words, these parameters are used to physical and logical unit for resource allocation.

Primitive parameters

The following four primitive parameters characterize the OFDMA symbol:

—BW: The nominal channel bandwidth.

—$Nused$: Number of used subcarriers (which include the DC subcarrier).

—n: Sampling factor. This parameter, in conjunction with BW and $Nused$ determines the subcarrier spacing and the useful symbol time.

—G: This is the ratio of CP time to "useful" time. The following values shall be supported: 1/8, 1/16, and 1/4.

Derived parameters

The following parameters are defined in terms of the primitive parameters of 16.3.2.3:

—N_{FFT}: Smallest power of two greater than N_{used}

—Sampling frequency: $F_s = floor(n \cdot BW/8000) \times 8000$

—Subcarrier spacing: $\Delta f = F_s/N_{FFT}$

—Useful symbol time: $Tb = 1/\Delta f$

—CP time: $G = Tg \cdot Tb$

—OFDMA symbol time: $T_s = T_b + T_g$

—Sampling time: Tb/N_{FFT}

Table 2. Some parameters.

Operation Procedure	Related Signaling Field (BW 10 MHz)	Parameters Calculated from Signaled Fields	Definition	Units
Subband partitioning	DSAC (5/4/3bits)	$L_{SB} = N_1*K_{SB}$	Number of PRUs assigned to subbands	PRUs
Mini-band partitioning	DSAC (5/4/3bits)	$L_{MB} = N_2*K_{MB}$	Number of PRUs assigned to minibands	PRUs
Frequency partitioning	DFPC (4/3/3bit)	$FPCT$ FPS_i	Number of frequency partitions Number of PRUs in FP_i	Frequency partitions
Frequency partitioning	DFPSC (3/2/1bit)	$K_{SB, FPi}$ $K_{MB, FPi}$ $L_{SB\,FPi} = N_1*K_{SB, FPi}$ $L_{MB, FPi} = N_2*K_{MB,FP}$	Number of subbands assigned to FPi Number of minibands assigned to FP_i Number of PRUs assigned to be subbands in FP_i Number of PRUs assigned to be minibands in FP_i	Sub bands Minibands PRUs PRUs
CRU/DRU allocation	DCASMB,0 (5/4/3bit) DCASi (3/2/1bit) DCASSB,0 (5/4/3 bit)	$L_{SB-CRU, FPi}$ $L_{MB-CRU, FPi}$ $L_{CRU,FPi} = L_{SB-CRU,FPi}$ $+L_{MB-CRU, FPi}$ $L_{DRU,FPi} = FPS_i - L_{CRU,FPi}$	Number of subband-based CRUs in FPi Number of miniband-based CRUs in FPi Number of miniband-based CRUs in FPi Number of DRUs in FPi	CRUs CRUs CRUs DRUs

4. Modeling Optimization Algorithm of Resource Allocation

A downlink resource mapping algorithms support both localized and distributed subcarriers, subcarriers permutation,

the mapping between LRU and PRU. In addition, we can reduce the information bit of S-SFH 101 by using new algorithm.

The SFH is divided into two parts: Primary Super frame Header (P-SFH) and Secondary Super frame Header (S-SFH) 101. The information transmitted in S-SFH is divided into three sub-packets (S-SFH SP1 102, S-SFH SP2 103, S-SFH SP3). The sub-packets of S-SFH are transmitted periodically where each sub-packet has a different transmission periodicity (Fig. 4).

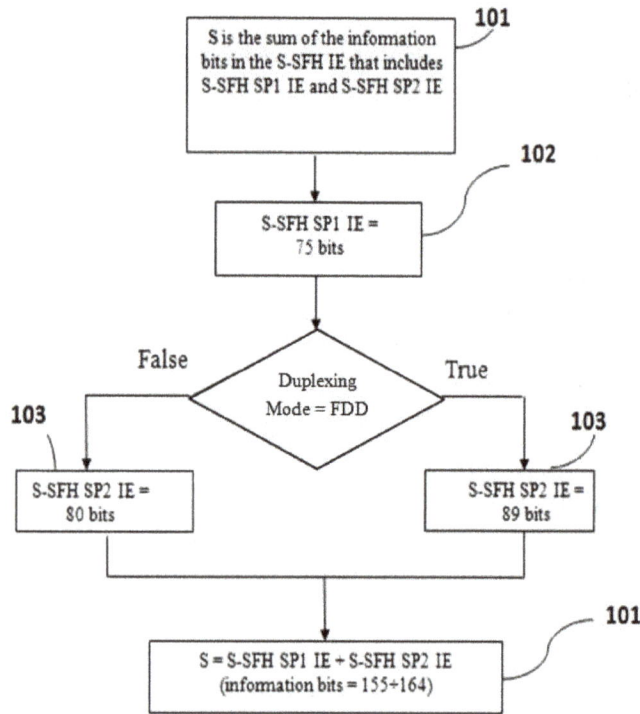

Figure 4. Optimization Algorithm.

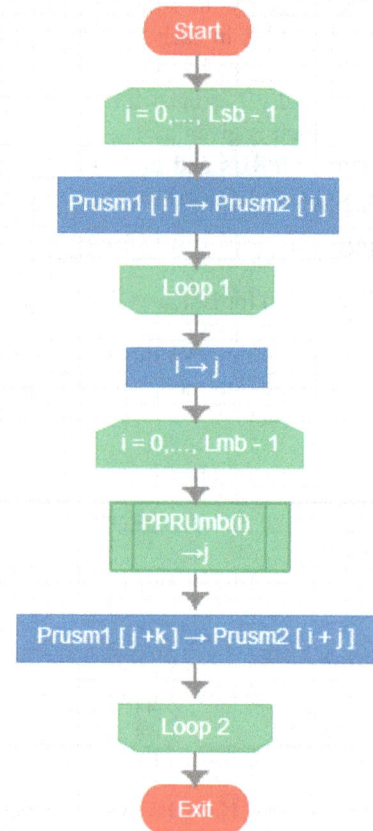

Figure 5. Sub band and Miniband partition's flowchart.

The SFH occupies the first N_{SFH} distributed LRUs in the first AAI sub frame of a super frame where N_{SFH} is no more than 24. The remaining distributed LRUs in the first AAI sub frame of a super frame are used for other control and data transmission. The SFH is divided into two parts: Primary Super frame Header (P-SFH) and Secondary Super frame Header (S-SFH) (101).

Table 2 includes the parameters and values for resource allocation of the SFH.

Some source code for new algorithm of resource allocation:

```
for(i=0; i<FPCT(); i++)
{   for(j=0;j<Lcrufp[i];j++)
{   Prusm4[ind+j]=Prusm3[ind+j];   }
ind=ind+Lcrufp[i];
SetPremSeq(FPS[i]-Lcrufp[i]);
for(j=0;j<Ldrufp[i];j++)
    {   jj = GetPremSeq ( j+ ( Lcrufp[i] - Lsbcrufp[i] ) );
        jj = jj % ( FPS [i]-Lsbcrufp [i]);
        Prusm4[ind+j] = Prusm3[ind+jj];
    }   ind=ind+Ldrufp[i]; }
```

"Fig. 5" shows the flowchart of the miniband permutation, which is used to our research.

5. Conclusions

The novel resource allocation algorithm in uplink and downlink is proposed to optimize resource allocation procedure for use in 802.16m system. A downlink resource mapping algorithms supports both localized and distributed sub carriers, subcarrier permutation, the mapping between LRU and PRU, supports different fractional frequency reuse (FFR) group allocations. In a future work to developing that algorithm we can possible to reduce redundancy of transmitted control bit field in super frame header.

By using this algorithm, we may be reduce information bit by 10 bits.

References

[1]   3GPP TS 36.211; LTE; Evolved Universal Terrestrial Radio Access (E-UTRA); Physical channels and modulation (version 11.1.0 Release 11)

[2]   Harri Holma, Antti Toskala, "LTE for UMTS-OFDMA and SC-FDMA Based Radio Access", pp. 71, John Wiley & Sons Ltd, 2009

[3]   Young- il Kim, Otgonbayar Bataa, Bat-Enkh Oyunbileg, Khishigjargal Gonchigsumlaa, "The analysis of two dimensional resource allocation procedures for IEEE802.16m of Mobile WiMAX systems", 2011 International Conference on Networking, VLSI and Signal Processing (ICNVSP 2011), Bangkok, Thailand.

[4]   Otgonbayar Bataa, Bat-Enkh Oyunbileg, Baatarkhuu Tsagaan, "A Resource Allocation for Mobile IPTV", ICKI 2011, Proceedings of International Conference on knowledge Based Industry, Ulaanbaatar, Mongolia

[5]   IEEE P802.16m/D7 July 2010, DRAFT Amendment to IEEE Standard for Local and metropolitan area networks, Part 16: Air Interface for Broadband Wireless Access Systems, July 29, (2010)

[6]   IEEE 802.16m System Description Document, IEEE 802.16m-08/003r9a, (2009)

[7]   IEEE 802.16m Evaluation Methodology Document (EMD), January 15, (2009)

[8]   Bat-Enkh Oyunbileg, Purevtseren Bayarsaikhan, Baatarkhuu Tsagaan, Optimization algorithm of resource allocation in LTE (Long Term Evolution) , IFOST-2013, Ulaanbaatar, Mongolia.

[9]   Young-il Kim, Otgonbayar Bataa, Bat-Enkh Oyunbileg, Khishigjargal Gonchigsumlaa,. The Analysis of Two Dimensional Resource Allocation Procedures for IEEE802.16m of Mobile WiMAX Systems. *Software Engineering and Knowledge Engineering: Theory and Practice Advances in Intelligent and Soft Computing,*. Vol. 115, 2012, pp. 343-350. doi: 10.1007/978-3-642-25349-2_46.

34

A Malware Analysis Using Static and Dynamic Techniques

Bymbadorj Dondogmegd[1, 2, *], Usukhbayr B.[2], Nyamjav J.[2]

[1]Department of Electrical Engineering, Ulaanbaatar State University, Ulaanbaatar, Mongolia
[2]Department of Electronic and Communication Engineering, National University of Mongolia, Ulaanbaatar, Mongolia

Email address:

pheelectro@gmail.com (B. Dondogmegd), usukhbayar@num.edu.mn (Usukhbayr B.), nyamjav@num.edu.mn (Nyamjav J.)

Abstract: In this survey work we analyze "win 32 malware gen" it's genre, procedure, harm using static and dynamic techniques. Static and dynamic methods were used to analyze a software program for any threats to the system. Static analysis involves testing its own source code and analyzing the threat itself, dynamic analysis involves specific secure, keeping threats within a system and analyzing the working progress of threats within the system.

Keywords: Malware, Threat - Data Fail Safe

1. Introduction

Malware spreads over LAN, online networks intended to steal information or spy on computer users for an extended period without their knowledge. Malware, short for malicious software, is any software used to disrupt computer operation, gather sensitive information, or gain access to private computer systems. 'Malware' is an umbrella term used to refer to a variety of forms of hostile or intrusive software, including computer viruses, worms, trojan horses, ransomware, spyware, adware, scareware and other malicious programs. Malware could be arranged into 14 main types [2]. It can take the form of executable code, scripts, active content, and other software. To test whether the software may contain potential threats we used static source code and dynamic code analysis [2, 3]. Win32.Malware-Gen refers to a range of malware applications that infect computers running a 32-bit version of the Windows operating system. Depending on which version of Win32, Malware-Gen of your computer is infected with, the virus may download and install other viruses, monitor computer activities, log keystrokes or corrupt your system registry and files. Programs that contain malicious code use this technique for analyzing the static and dynamic core of your computer.

Dynamic analysis involves the testing and evaluation of a program by executing data analysis in real-time. The objective is to find errors in a program while it is running, rather than by repeatedly examining the program's code offline. Snapshots of the uninfected original source code are used to compare and analyze the infected system. Dynamic analysis compares the system's processes, system registry, and downloaded networks [4-6]. The testing and analysis process cannot be completed by only doing one attempt, successful analysis often requires multiple runs.

Static code analysis records the HTML, GUI, Scripts, passwords, control string, and other commands [4]. By accumulating and analyzing these codes, static analysis creates certain "signatures" (algorithms, codes), describes which files are malware related, and identifies which commands the malware is using.

2. Related Works

The power of such test programs to isolate and destroy harmful malware is needed to effect control of the spread of viruses and create a transparent protective environment. Create a first state to collect information using the following program, it is controlled following the operating system, isolated environment can be carried out by a malicious program.

- PEview program's viewing of structure and content of 32 bit. Portable Executable.(PE) file. Show date and time stamp of the malware complied and created.
- PEiD program's malware authors often pack the malware
- Detect packers used (if any)
- Detect the language being used to write the malware
- DependsWalker program's scan any 32 bit or 64 bit windows module

- Build a hierarchical tree diagram of all dependent modules.
- Used to identify any suspicious API(s) and DLL(s) imported
- Process Explorer program's The unique capabilities of *Process Explorer* make it useful for tracking down DLL-version problems or handle leaks, and provide insight into the way Windows and applications work [11].
- Process Monitor is an advanced monitoring tool for Windows that shows real-time file system, registry and process/thread activity [9].
- Wireshark is a free and open-source packet analyzer. It is used for network troubleshooting, analysis, software and communications protocol development, and education. Originally named Ethereal, the project was renamed Wireshark in May 2006 due to trademark issues. [10].
- RegShot program's windows registry comparison tool that allows you to take and compare two registry snapshots.[12]

3. Experimental Result

In this survey work we tested static code analyses and dynamic one too. Rig used for test hardware has 2nd Generation i7 CPU, 4 GB ram and Windows 7, Virtual XP are for software. In Virtual XP we used Reshot software for collecting registry data. Then we used Review, Pied, and Depends Walker software's for comprehensive study. That malware is complete date: 2013.11.04.

Has an ability to work on windows XP and NT software. Written by C++. DLL and functions for threat are described above in shows Figure 1.

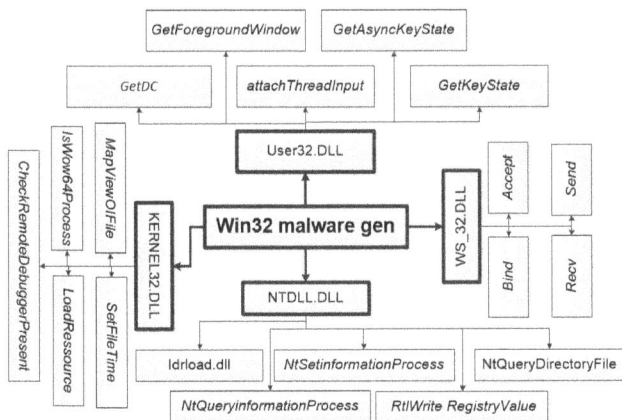

Figure 1. *Destructive load function of DLL files and system software.*

Figure 1 Win32 malware gen program uses for those DLL files:

Kernel32.dll is most often used file that contains:
- IsWoW64Process: Used by a 32-bit process to determine if it is running on a 64-bit operating system [12].
- MapViewOfFile: Maps a file into memory and makes the contents of the file accessible via memory addresses. Launchers, loaders, and injectors use this function to

read and modify PE files. By using MapViewOfFile, the malware can avoid using Write File to modify the contents of a file [12].
- Load Resource: Loads a resource from a PE file into memory. Malware sometimes uses resources to store strings, configuration information, or other malicious files [12].
- SetFileTime: Modifies the creation, access, or last modified time of a file. Malware often uses this function to conceal malicious activity [12].

User32.dll contains user interface codes.
- AttachThreadInput: Attaches the input processing for one thread to another so that the second thread receives input events such as keyboard and mouse events. Key loggers and other spyware use this function [12].
- GetAsyncKeyState: Used to determine whether a particular key is being pressed. Malware sometimes uses this function to implement a key logger [12].
- GetDC: Returns a handle to a device context for a window or the whole screen. Spyware that takes screen captures often uses this function [12].
- GetForegroundWindow:Returns a handle to the window currently in the foreground of the desktop. Key loggers commonly use this function to determine in which window the user is entering his keystrokes [12].
- GetKeyState: Used by key loggers to obtain the status of a particular key on the keyboard [12].

Ntdll.dll The DLL ntdll.dll is primarily concerned with system tasks. It includes a number of kernel-mode functions which implements much of the functionality of the Windows Application Programming Interface (API). As usual Ntdll.dll doesn't work alone, works through kernel.dll. If it is used via any software alone that software aims hiding malicious acts. Hiding progress, controlling process etc.
- RtlWriteRegistryValue: Used to write a value to the registry from kernel-mode code [12].
- NtQueryDirectoryFile: Returns information about files in a directory. Rootkits commonly hook this function in order to hide files [12].

Ws2_32.dll File that contains the Windows Sockets API used by most Internet and network applications to handle network connections. This is a module that contains many different internet functions, like all DLL's, many of them are used to share functions for various applications. Such as FTP, HTTP, NTP
- Connect: Used to connect to a remote socket. Malware often uses low-level functionality to connect to a command-and-control server [12].
- Recv: Receives data from a remote machine. Malware often uses this function to receive data from a remote command-and-control server [12].
- Send: Sends data to a remote machine. Malware often uses this function to send data to a remote command-and-control server [12].
- Bind: Used to associate a local address to a socket in order to listen for incoming connections [12].

We analyzed Win32 Malware gen has functions to damage

OS, hiding itself, using network ports. As for dynamic analysis we used Process monitor, Process explorer and We Shack on virtual OS. We found which area Win32 Malware Gen affects. Image 2 displays areas affected in Table 1.

Table 1. Changes to the system contains the parts.

File system	Malware gen
C:\WINDOWS\WinSxS\x86_Microsoft.Windows.Comon-Controls_6595b64144ccf1df_6.0.2600.6028_x-ww_61e65202	X
C:\Documents and Settings\Administrator\Local Settings\Temp\	X
C:\Documents and Settings\Administrator\Application Data	X
C:\Windows\Prefetch\	X
Windows registry	X

As shown is image 2 Win32 Malware Gen copies ScreenSaver.scr into C:\Documents and Settings\Administrator\ApplicationData. OS registry affected by Win32 Malware gen displayed above. Windows Registry is a hierarchical database that stores configuration settings and options on Microsoft Windows operating systems Windows operating system has six type of registry special Om their own [7]. Table 2 shown, Win32 malware gen related registries are shown above.

Table 2. Registry changed section.

File system	Malware gen
HKLM\Software\Microsoft	X
HKLM\System\ControlSet001\Control\	X
HKLM\Hardware\	
HKLM\System\CurrentControlset\Services\	X
HKLM\Software\Microsoft\Cryptography\	X
HKLM\Software\Microsoft\Windows NT\CurrentVersion\	X
HKU\S-1-5-21-602162358-492894223-2995022267-500\Software\Microsoft\Windows\CurrentVersion\Run	X
HKLM\SOFTWARE\Microsoft\DirectDraw\MostRecentApplication	X
HKLM\SOFTWARE\Microsoft\Windows\CurrentVersion\Extensions\	X
HKLM\SYSTEM\ControlSet001\Enum\Root\	X

The Windows Registry contains a root key titled HKEY_LOCAL_MACHINE, or HKLM. The HKLM root key contains settings that relate to the local computer software is and drivers [8]. HKEY_USERS contains user-specific configuration information for all currently active users on the computer [8].

By editing HKLM and HKU malware called

screensaver.scr and changed its code. Network activities collected by Wireshark software are shown in Table 3.

Table 3. Host a weak port is used to traffic.

Port	Protocol	Process
1140	TCP	Windows\syswow64\vmnate.exe
1140	TCP	Windows\syswow64\vmnate.exe

Win32 malware gen infects active networks, multiplies itself, and calls certain websites to exacerbate OS.

4. Conclusion

Analyzed Win32 Malware gen uses static and dynamic methods. Based on static analyze, we knew win32 malware gen was a threat. By the dynamic method, we studied how malware affecting OS. As a result of these methods, the main result proves malware copies screensaverpro.scr to system, changes HKLM. HKEY_USERS registry hides itself, works through network.

References

[1]   www.symantec.com/connect/articles/malware-analysis-administrators.

[2]   Chris Gates, "Hacker Defender Rootkit for the Masses", 2007.

[3]   "Malware challenge" , jerome.segura@gmail.com.

[4]   Dean De Beer, "Malware Analysis Challenge III", 2007.

[5]   Bill Arnold, David Chess, John Moral, "An Environment for Controlled Worm Replication and Analysis", 2008.

[6]   Alla Segal, "Reverse-Engineering Malware", 2006.

[7]   http://en.wikipedia.org/wiki/Windows_Registry

[8]   http://kb.chemtable.com/ru/windows-registry-main-keys.htm#hkcu

[9]   https://technet.microsoft.com/en-us/library/bb896645.aspx

[10]   "Wireshark FAQ", Retrieved 31 December 2011.

[11]   https://technet.microsoft.com/en-us/sysinternals/bb896653.aspx

[12]   "Practical Malware Analysis" The Hands-On Guide to Dissecting Malicious Software.

Permissions

All chapters in this book were first published in AJNC, by Science Publishing Group; hereby published with permission under the Creative Commons Attribution License or equivalent. Every chapter published in this book has been scrutinized by our experts. Their significance has been extensively debated. The topics covered herein carry significant findings which will fuel the growth of the discipline. They may even be implemented as practical applications or may be referred to as a beginning point for another development.

The contributors of this book come from diverse backgrounds, making this book a truly international effort. This book will bring forth new frontiers with its revolutionizing research information and detailed analysis of the nascent developments around the world.

We would like to thank all the contributing authors for lending their expertise to make the book truly unique. They have played a crucial role in the development of this book. Without their invaluable contributions this book wouldn't have been possible. They have made vital efforts to compile up to date information on the varied aspects of this subject to make this book a valuable addition to the collection of many professionals and students.

This book was conceptualized with the vision of imparting up-to-date information and advanced data in this field. To ensure the same, a matchless editorial board was set up. Every individual on the board went through rigorous rounds of assessment to prove their worth. After which they invested a large part of their time researching and compiling the most relevant data for our readers.

The editorial board has been involved in producing this book since its inception. They have spent rigorous hours researching and exploring the diverse topics which have resulted in the successful publishing of this book. They have passed on their knowledge of decades through this book. To expedite this challenging task, the publisher supported the team at every step. A small team of assistant editors was also appointed to further simplify the editing procedure and attain best results for the readers.

Apart from the editorial board, the designing team has also invested a significant amount of their time in understanding the subject and creating the most relevant covers. They scrutinized every image to scout for the most suitable representation of the subject and create an appropriate cover for the book.

The publishing team has been an ardent support to the editorial, designing and production team. Their endless efforts to recruit the best for this project, has resulted in the accomplishment of this book. They are a veteran in the field of academics and their pool of knowledge is as vast as their experience in printing. Their expertise and guidance has proved useful at every step. Their uncompromising quality standards have made this book an exceptional effort. Their encouragement from time to time has been an inspiration for everyone.

The publisher and the editorial board hope that this book will prove to be a valuable piece of knowledge for researchers, students, practitioners and scholars across the globe.

List of Contributors

Huthaifa Ahmad Al_Issa
Department of Electrical and Electronics Engineering, Faculty of Engineering, Al-Balqa Applied University, Irbid Al-Huson, Jordan

Vahid Haji Hashemi
Computer Engineering, Faculty of Engineering, Kharazmi University of Tehran,Tehran, Iran

Abdorreza Alavi Gharahbagh
Department of Electrical and Computer Engineering, Islamic Azad University, Shahrood, Iran

Anuj Lohani, Aditi Lohani, Jitendra Singh and Manish Bhardwaj
Dept. of Computer Science and Engineering, SRM University, NCR Campus, Modinagar, India

Ali M. Alsahlany and Hayder S. Rashid
Department of Communication Engineering, Engineering Technical College / Najaf, Al-Furat Al-Awsat Technical University, Najaf, Iraq

Nosiri Onyebuchi Chikezie, Ezeh Gloria Nwabugo, Agubor Cosmos Kemdirim and Nkwachukwu Chukwuchekwa
Department of Electrical & Electronic Engineering, Federal University of Technology, Owerri, Nigeria

Pourush, Naresh Sharma and Manish Bhardwaj
Department of Computer Science and Engineering, SRM University, NCR Campus, Modinagar, India

Chinyere Onyemaechi Agabi and Nwachukwu Prince Ololube
Department of Educational Foundations and Management, Faculty of Education, Ignatius Ajuru University of Education, Port Harcourt, Rivers State, Nigeria

Comfort Nkogho Agbor
Department of Environmental Education, Faculty of Education, University of Calabar, Calabar, Cross River State, Nigeria

Ho Khanh Lam
Faculty of Information Technology, Hung Yen University of Technology and Education, Hung Yen, Vietnam

Khawlah Hussein Alhamzah
Computer Science Department, Basrah University, Basrah, Iraq

Tianjiang Wang
School of Computer Science, HUST University, Wuhan, China

Edgar Manuel Cano Cruz and Juan Gabriel Ruiz Ruiz
Department of Computer Science, University of the Istmo, Ixtepec, Mexico

Laith Ali Abdul-Rahaim
Electrical Engineering Department, Babylon University, Babil, Iraq

Bourdilllon Odianonsen Omijeh and Philip Ogah
Department of Electronic & Computer Engineering, University of Port Harcourt, Port Harcourt, Nigeria

Liton Chandra Paul
Department of Electronic and Telecommunication Engineering, Pabna University of Science & Technology, Pabna, Bangladesh

Md. Sarwar Hosain and Sohag Sarker
Department of Information and Communication Engineering, Pabna University of Science & Technology, Pabna, Bangladesh

Makhluk Hossain Prio and Ajay Krishno Sarkar
Department of Electrical and Electronic Engineering, Rajshahi University of Engineering & Technology, Rajshahi, Bangladesh

Monir Morshed
Department of Information and Communication Technology, Mawlana Bhashani Science and Technology University,Tangail, Bangladesh

Purvi Garg, Shivangi Varshney and Manish Bhardwaj
Department of Computer Science and Engineering, SRM University, Modinagar, Utter Pradesh, India

Binyam Shiferaw Heyi
Department of Computer Engineering, Addis Ababa Science and Technology University (AASTU), Addis Ababa, Ethiopia

Ho Khanh Lam
Faculty of Information Technology, Hung Yen University of Technology and Education, Hung Yen, Vietnam

Nguyen Xuan Truong
Training Department, Hung Yen University of Technology and Education, Hung Yen, Vietnam

Yasuji Murakami
Telecommunications and Computer Networks, Osaka Electro-Communication University, Osaka, Japan

Shivani Rohilla, Megha Sharma, A. Kulothungan and Manish Bhardwaj
Department of Computer science and Engineering, SRM University, NCR Campus, Modinagar, Ghaziabad, India

IS Amiri and H. Ahmad
Photonics Research Centre, University of Malaya (UM), 50603 Kuala Lumpur, Malaysia

Hamza M. R. Al-Khafaji
Wireless Communication Centre, Faculty of Electrical Engineering, Universiti Teknologi Malaysia (UTM), 81310 UTM Skudai, Johor, Malaysia

Geetanjali tyagi, kumar kaushik, Arnika Jain and Manish Bhardwaj
Department of Computer science and Engineering, SRM University, NCR Campus, Modinagar, Ghaziabad, India

Sanjana Lakkadi, Amit Mishra and Manish Bhardwaj
Department of Computer science and Engineering, SRM University, NCR Campus, Modinagar, Ghaziabad, India

Yanhua Zhang and Yupu Hu
State Key Laboratory of Integrated Service Networks, Xidian University, Xi'an, China

Tuyatsetseg Badarch
Department of Information Technology, Mongolian National University, Ulaanbaatar, Mongolia

Otgonbayar Bataa and Bat-Enkh Oyunbileg
Department of Information Technology, Mongolian University of Science and Technology,Ulaanbaatar, Mongolia

Megha Sharma, Shivani Rohilla and Manish Bhardwaj
Department of Computer science and Engineering, SRM University, NCR Campus, Modinagar, Ghaziabad, India

Bat-Enkh Oyunbileg and Baatarkhuu Tsagaan
Department of Information Technology, School of Information and Telecommunication, MUST, Ulaanbaatar, Mongolia

Battugs Oyunbileg and Chuluuntsetseg Jamyaan
Mongolian University of Art and Culture, Ulaanbaatar, Mongolia

Tuyatsetseg Badarch
Mongolian National University, Department of Information Technology, Ulaanbaatar, Mongolia

I. S. Amiri and H. Ahmad
Photonics Research Centre, University of Malaya (UM), 50603 Kuala Lumpur, Malaysia

Hamza M. R. Al-Khafaji
Wireless Communication Centre, Faculty of Electrical Engineering, Universiti Teknologi Malaysia (UTM), 81310 UTM Skudai, Johor, Malaysia

Zhe Lei, Mengying Ding, Xiaohong Guan and Youtian Du
The School of Electronic and Information Engineering, Xi'an Jiaotong University, Xi'an, China

Jicheng Feng, Qinping Gao and Zheng Liu
Xi'an Conservatory of Music, Xi'an, China

Javzmaa Tsend, Ajnai Luvsan and S. Battulga
Mongolian National University of Medical Science, Ulaanbaatar, Mongolia

Bat-Enkh Oyunbileg
School of Information and Telecommunication Technology, the Mongolian University of Science and Technology, Ulaanbaatar, Mongolia

Manish Bhardwaj
Department of Computer science and Engineering, SRM University, NCR Campus, Modinagar, Ghaziabad, India

Fan Jiang, Chen Song and Hao Xun, Zhen Xu
Institute of Information Engineering, Chinese Academy of Sciences, Beijng, China

Laith Ali Abdul-Rahaim and Ammar Abdulrasool Muneer
Department of Electrical Engineering, University of Babylon, Babil, Iraq

Chao Li
Inner Mongolia Xinyuan Information Technology Co., LTD, Huhhot, China

Yongjian Jing
Inner Mongolia Xinyuan Information Technology Co.,
LTD, Huhhot, China
Center of Information and Network Technology, Inner
Mongolia Agricultural University, Huhhot, China

**Bat-Enkh Oyunbileg, Otgonbayar Bataa and
Baatarkhuu Tsagaan**
Department of Information Technology, School
of Information and Telecommunication, MUST,
Ulaanbaatar, Mongolia

Tuyatsetseg Badarch
Department of Information Technology, Mongolian
National University, Ulaanbaatar, Mongolia

Bymbadorj Dondogmegd
Department of Electrical Engineering, Ulaanbaatar
State University, Ulaanbaatar, Mongolia

Department of Electronic and Communication
Engineering, National University of Mongolia,
Ulaanbaatar, Mongolia

Usukhbayr B. and Nyamjav J.
Department of Electronic and Communication
Engineering, National University of Mongolia,
Ulaanbaatar, Mongolia

Index